Plant Molecular Biology

Edited by **Agatha Wilson**

SYRAWOOD
PUBLISHING HOUSE
New York

Published by Syrawood Publishing House,
750 Third Avenue, 9th Floor,
New York, NY 10017, USA
www.syrawoodpublishinghouse.com

Plant Molecular Biology
Edited by Agatha Wilson

International Standard Book Number: 978-1-68286-353-4 (Hardback)

Printed in China.

Contents

Preface

This book aims to highlight the current researches and provides a platform to further the scope of innovations in this area. This book is a product of the combined efforts of many researchers and scientists from different parts of the world. The objective of this book is to provide the readers with the latest information in the field.

Researches and studies of plants at molecular levels of structure and functioning have facilitated the development of significant techniques for plant breeding, crop protection, etc. It has also influenced the discipline of biotechnology to a great extent. Replete with significant details and examples, this book covers the structure, function, mechanisms and regulatory processes which occur within plants at molecular levels. Also included in this book are detailed explanations of the various concepts and applications of plant genomics which would enhance the knowledge of the readers. This book strives to aid the students and researchers of botany and associated fields of biology.

I would like to express my sincere thanks to the authors for their dedicated efforts in the completion of this book. I acknowledge the efforts of the publisher for providing constant support. Lastly, I would like to thank my family for their support in all academic endeavors.

Editor

1

Tomato (*Solanum lycopersicum* L.) SlIPT3 and SlIPT4 isopentenyltransferases mediate salt stress response in tomato

Eva Žižková[1*], Petre I Dobrev[1], Yordan Muhovski[2], Petr Hošek[1,3], Klára Hoyerová[1], Daniel Haisel[1], Dagmar Procházková[1], Stanley Lutts[4], Václav Motyka[1] and Imène Hichri[4,5*]

Abstract

Background: Cytokinins (CKs) are involved in response to various environmental cues, including salinity. It has been previously reported that enhancing CK contents improved salt stress tolerance in tomato. However, the underlying mechanisms of CK metabolism and signaling under salt stress conditions remain to be deciphered.

Results: Two tomato isopentenyltransferases, SlIPT3 and SlIPT4, were characterized in tomato and Arabidopsis. Both proteins displayed isopentenyltransferase (IPT) activity *in vitro*, while their encoding genes exhibited different spatio-temporal expression patterns during tomato plant development. *SlIPT3* and *SlIPT4* were affected by the endogenous CK status, tightly connected with CKs feedback regulation, as revealed by hormonal treatements. In response to salt stress, *SlIPT3* and *SlIPT4* were strongly repressed in tomato roots, and differently affected in young and old leaves. *SlIPT3* overexpression in tomato resulted in high accumulation of different CK metabolites, following modifications of CK biosynthesis-, signaling- and degradation-gene expression. In addition, *35S::SlIPT3* tomato plants displayed improved tolerance to salinity consecutive to photosynthetic pigments and K^+/Na^+ ratio retention. Involvement of *SlIPT3* and *SlIPT4* in salt stress response was also observed in Arabidopsis *ipt3* knock-out complemented plants, through maintenance of CK homeostasis.

Conclusions: SlIPT3 and SlIPT4 are functional IPTs encoded by differently expressed genes, distinctively taking part in the salinity response. The substantial participation of *SlIPT3* in CK metabolism during salt stress has been determined in *35S::SlIPT3* tomato transformants, where enhancement of CKs accumulation significantly improved plant tolerance to salinity, underlining the importance of this phytohormone in stress response.

Keywords: Cytokinin, Isopentenyltransferase, Salt stress, SlIPT3, SlIPT4, Tomato

Background

Tomato (*Solanum lycopersicum* L.) is one of the most popular vegetable species grown world-wide because of its edible fruit and agronomic importance. Nevertheless, its production is frequently threatened by environmental stresses such as drought and salinity. High salt concentration reduces tomato germination, leaf number and area, slows down shoot and root growth and increases root/shoot ratio, induces leaf senescence, and ultimately impairs crop production [1]. Salinity imposes both an osmotic stress due to water shortage, followed by an ionic stress resulting from disproportionate nutrients accumulation and an augmentation of toxic ion concentrations, such as sodium [2,3]. Modifications of plant physiological processes following salinization are partly attributed to alterations in phytohormone metabolism. This may be true especially for cytokinins (CKs). Indeed, under long-term salt stress the level of bioactive CKs in tomato plants was reduced down to 50% both in roots and leaves [2,4,5].

The ability to control cell division and differentiation, root growth, leaf senescence, apical dominance, branching,

* Correspondence: zizkovae@ueb.cas.cz; imene.hichri@sophia.inra.fr
[1]Institute of Experimental Botany, Academy of Sciences of the Czech Republic, Prague 165 02, Czech Republic
[4]Groupe de Recherche en Physiologie Végétale (GRPV), Earth and Life Institute - Agronomy (ELI-A), Université catholique de Louvain (UCL), Louvain-la-Neuve 1348, Belgium
Full list of author information is available at the end of the article

flower and seed development as well as germination and nutrients uptake into sink organs defines CKs as essential regulatory substances in plants [6]. Chemically, CKs are derivatives of adenine substituted at the N^6 position either with isoprenoid or an aromatic side chain. Both isoprenoid and aromatic CKs are present in plants in bioactive forms, as free bases and corresponding nucleosides and nucleotides, and in non-active or storage forms as conjugates with glucose (CK-*O*- and CK-*N*-glucosides) for instance. The typical representatives of isoprenoid CKs are derivatives of *trans*-zeatin (*t*Z), *cis*-zeatin (*c*Z), dihydrozeatin (DHZ) and N^6-(Δ^2-isopentenyl)adenine (iP), occuring throughout the plant kingdom [7,8]. The first step of CK biosynthesis is performed by isopentenyltransferases (IPTs). Plant *IPTs* belong to multigenic families, chiefly described in Arabidopsis (*AtIPT1-AtIPT9*) [9,10] and in maize, for instance [11]. Recently, *IPT* genes have also been identified in tomato (*SlIPT1-SlIPT6*), and their expression patterns specifically investigated during fruit set and development [12]. IPTs catalyze the transfer of the isoprenoid moiety from precursors dimethylallyl diphosphate (DMAPP) or (*E*)-4-hydroxy-3-methyl-but-2-enyl diphosphate (HMBDP) to adenine nucleotide forms (AMP, ADP, ATP) [9,10]. Modulation of CK contents through modification of *IPT* expression strongly impacts plants development. Indeed, while overexpression of *IPT* in plants resulted in faster shoot formation, shorter internodes, loss of apical dominance, delay of leaf senescence, higher photosynthetic rates and accumulation of *t*Z and its riboside [13-16], Arabidopsis *ipt* deficient plants showed strong inhibition of shoot growth, elongation of primary and lateral roots, and reduction of *t*Z and iP contents [17].

Recent progress in genetic engineering of CKs enabled to control plant CK contents, affected plant traits, increased yield production and improved plant adaptation to enviromental stresses such as salinity [18-20]. Indeed, responsiveness of Arabidopsis *IPT* genes to salt stress has been demonstrated by transcriptome analyses, where upregulation of *AtIPT1*, *AtIPT2* and *AtIPT8* and downregulation of *AtIPT3*, *AtIPT5*, *AtIPT7* and *AtIPT9* were reported [21]. Regulation of *IPT* expression through inducible promoter in tomato roots improved tolerance to salt stress and led to higher yield compared to the wild-type plants [3,4]. Similarly, plant salinity tolerance was enhanced in transgenic cotton expressing *IPT* under control of a *cysteine proteinase* (*Ghcysp*) promoter, delaying the salt-induced senescence of leaves [22]. Positive correlation between *IPT* expression and salt resistance has been demonstrated in transgenic tobacco by introducing *Agrobacterium tumefaciens IPT* under control of the stress-inducible promoter *rd29A* as well [23].

In addition to CK biosynthesis, CK signaling steps are also critical for plant response to salinity. CKs are perceived at the plasma membrane by specific receptors, and the signal is transduced *via* type-B Arabidopsis Response Regulators (ARRs) controlling transcription of type-A ARRs, which act as negative feedback regulators of CK signaling [24]. ARR transcription factors respond in different ways to salt stress and were reported to regulate sodium accumulation in Arabidopsis [25]. Moreover, a transcriptome analysis of Arabidopsis CK deficient *ipt1,3,5,7* mutants exposed to salinity revealed the importance of CK regulation on stress-responsive signaling pathways, even under normal conditions of growth [21].

Although CKs assume crucial functions in tomato salt tolerance, data concerning the underlying molecular cues involved in this process remain unclear. To better understand salt impact on CK metabolism in tomato, a thorough characterization of genes coding for IPTs and their response to salinity is required. The present study reports the functional characterization of two tomato IPT encoding genes, *SlIPT3* and *SlIPT4*, both *in vitro* and *in planta*. Spatio-temporal expression profiles during plant development, enzymatic activity of both proteins and their involvement in CK biosynthesis were determined. Likewise, SlIPT3 and SlIPT4 participation in response to salt stress was investigated in Arabidopsis and tomato.

Results

Identification of two putative non-redundant tomato isopentenyltransferases

To identify IPT encoding genes in tomato, Arabidopsis *IPT1-9* coding sequences were used to screen the tomato genome database [26] (www.solgenomics.net). Among the selected expressed sequence tags (ESTs), two were apparently encoding full length proteins and could be amplified by PCR. For the first gene (GenBank accession JF423320), a 1122 bp coding sequence was cloned, including a 990 bp open reading frame (ORF) defining a 329 amino acid (aa) polypeptide of 37.5 kDa (Figure 1A). This gene was recently annotated as *SlIPT3* [12]. For the second gene (GenBank accession JF433930), a 1073 bp coding sequence was amplified, exhibiting a 972 bp ORF predicted to encode a 323 a protein of 37.1 kDa. This gene was classified as *SlIPT4* [12].

SlIPT3 is located on tomato chromosome 1, while *SlIPT4* is situated on chromosome 9. In addition, *SlIPT3* includes a 293 bp intron localized in the 5′ UTR region. SlIPT3 and SlIPT4 display 52% identity and 72% similarity. Alignment of SlIPT3 and SlIPT4 proteins with Arabidopsis IPT3 and IPT5 indicates that the IPT catalytic domain represents almost the entirety of the proteins, with exception of the N- and C-termini (Figure 1A). A phylogenetic analysis of SlIPT3 and SlIPT4 and nineteen additional orthologs belonging to different species was constructed by means of the neighbour-joining method using full-length amino acid sequences (Figure 1B). It

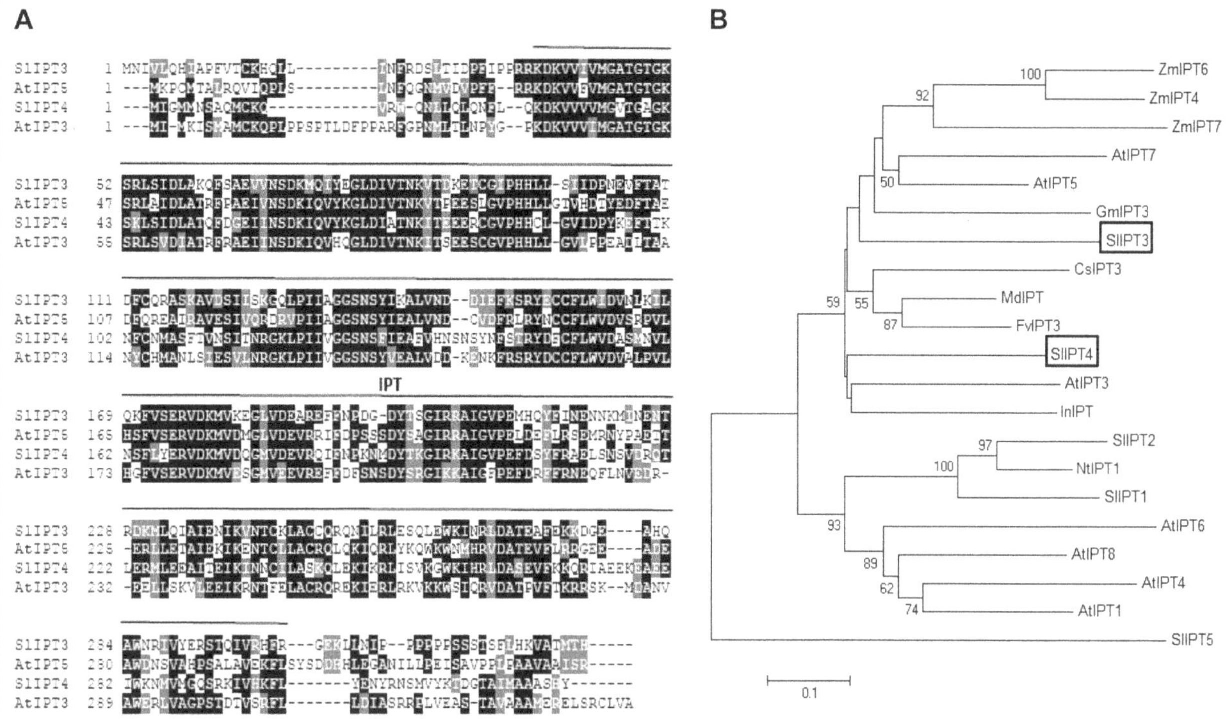

Figure 1 Identification of SlIPT3 and SlIPT4. (A) Full-length protein sequence alignment of SlIPT3, SlIPT4 and Arabidopsis AtIPT3 and AtIPT5. Identical residues are colored in black and conserved residues in dark gray. The isopentenyltransferase (IPT) domain is labeled. **(B)** Phylogenetic relationships between SlIPT3 [ADZ28498], SlIPT4 [AEE39459] and related proteins of *Cucumis sativus* CsIPT3 [XP_004136062], *Ipomoea nil* InIPT [BAG55006], *Malus domestica* MdIPT [ADY80558], *Fragaria vesca* FvIPT3 [XP_004290672], *Arabidopsis thaliana* AtIPT3 [Q93WC9], AtIPT4 [Q9SB60], AtIPT5 [Q94ID2], AtIPT6 [Q9C6L1], AtIPT7 [Q94ID1] and AtIPT8 [Q9LJL4], *Glycine max* GmIPT3 [XP_003528670], *Zea mays* ZmIPT4 [ABY78883], ZmIPT6 [ABY78885] and ZmIPT7 [ABY78886], *Solanum lycopersicum* SlIPT1 [BAM08995], SlIPT2 [BAM08996], SlIPT5 [BAM08998] and SlIPT6 [XP_004251888], *Nicotiana tabacum* NtIPT1 [AFV15392]. The phylogenetic tree was constructed according to the neighbor-joining method, using MEGA5 [27]. The percentage of reliability of each branch point of the rooted tree, as assessed by the analysis of 1000 trees (bootstrap replicates), is shown on the branch stem.

indicated that SlIPT3 mainly clusters with soybean GmIPT3 and Arabidopsis AtIPT5, while SlIPT4 clusters with Arabidopsis AtIPT3 and Japanese Morning Glory InIPT. Except from Arabidopsis and maize, full characterization of IPT enzymes in other plant species is missing.

Tomato *SlIPT3* and *SlIPT4* show isopentenyltransferase activity *in vitro*

To examine the ability of SlIPT3 and SlIPT4 to catalyze CK biosynthesis, a functional analysis of both proteins was carried out. As SlIPT3 and SlIPT4 are highly insoluble when produced in bacteria, both proteins were produced *in vitro*. The enzymatic activity of SlIPT3 and SlIPT4 was determined using *in vitro* assays based on conversion of radiolabeled adenylated substrates ([^{3}H] AMP, [^{3}H]ADP or [^{3}H]ATP) in the presence of DMAPP to isopentenylated products. Both enzymes were able to convert tritium-labeled AMP/ADP/ATP to the corresponding iPRMP/iPRDP/iPRTP in time-dependent manner (Figure 2A and B). The prevailing metabolites formed by both SlIPT3 and SlIPT4 were iPRMP and iPRDP, with almost 10-fold higher catalytic activity shown by SlIPT4 compared to SlIPT3. These results clearly demonstrate that SlIPT3 and SlIPT4 display CK biosynthetic activity in *in vitro* conditions.

Subcellular localization of SlIPT3 and SlIPT4

An analysis of the deduced SlIPT3 and SlIPT4 sequences revealed the presence of chloroplast transit peptides [28] (pSORT, http://psort.ims.u-tokyo.ac.jp/). To investigate protein subcellular localization, *SlIPT3* and *SlIPT4* cDNA were fused at their C-terminal end to green fluorescent protein (GFP) coding sequence, placed under cauliflower mosaic virus *35S* promoter and used to agro-infiltrate tomato leaves. As observed in Figure 3 (a, a'), despite a strong auto-fluorescence, SlIPT4 showed localization predominantly in chloroplasts and in the cytosolic area surrounding the chloroplast. Indeed, GFP fluorescence colocalized with chlorophyll fluorescence in this cellular compartment. In contrast, free GFP was restricted to the cytoplasm (Figure 3 (b, b')), while no fluorescence at 520 nm (GFP emission wavelength) was detected in untransformed tomato leaves (Figure 3 (c, c')). Despite several attempts, we could not

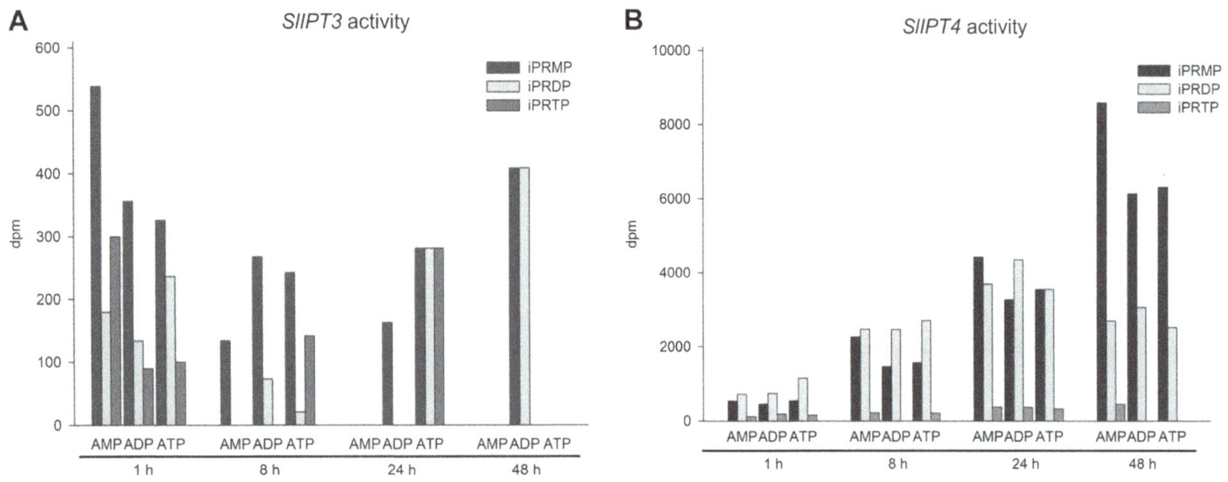

Figure 2 ***In vitro*** **determination of DMAPP:AMP, ADP and ATP isopentenyltransferase activity of SlIPT3 (A) and SlIPT4 (B) proteins.** ADP: adenosine diphosphate, ATP: adenosine triphosphate, iPMP: isopentenyladenosine-5'monophosphate; iPDP: isopentenyladenosine-5'-diphosphate, iPTP: isopentenyladenosine-5'-triphosphate. Note that vertical scales are not the same for SlIPT3 and SlIPT4.

establish subcellular localization for SlIPT3, and no GFP signal of the recombinant protein could be detected in transformed tomato cells. These data show that at least SlIPT4 is localized in the chloroplast and in the cytosolic area surrounding chloroplasts.

Characterization of SlIPT3 and SlIPT4 in Arabidopsis *ipt3* mutants

In parallel, tomato *SlIPT3* and *SlIPT4* were functionally characterized *in planta via* complementation of Arabidopsis *ipt3* knock-out (*ipt3* KO) mutants. Five days after sowing (DAS), the Arabidopsis wild-type (WT) plants, *ipt3* KO mutants and Arabidopsis *ipt3* KO complemented plants showed a similar phenotype on half-strength MS medium (Additional file 1). We thus suggested to determine the function of SlIPT3 and SlIPT4 in Arabidopsis complemented *ipt3* KO mutants grown on salt medium and to follow their response to salinity. Based on the highest phenotype discrimination, 100 mM NaCl medium was used for further testing of

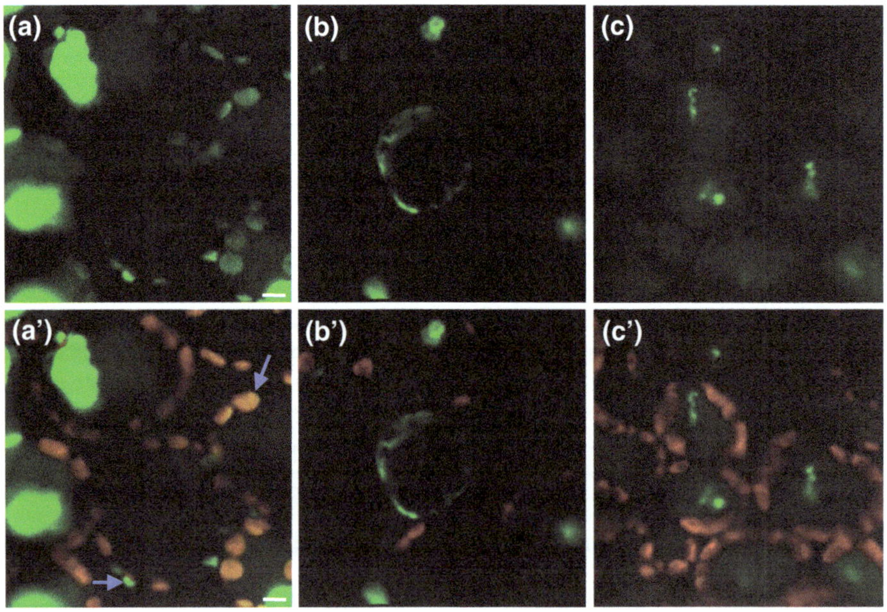

Figure 3 Subcellular localization of SlIPT4-GFP by tomato leaves agroinfiltration. (a, a') SlIPT4-GFP fluorescence and chlorophyll/GFP fluorescence, respectively. **(b, b')** Control GFP fluorescence and chlorophyll/GFP fluorescence, respectively. **(c, c')** Control untransformed tomato leaves and chlorophyll autofluorescence, respectively. *Bar* = 4 μm.

seed germination and plant survival percentages, primary root elongation, seedlings CKs content and CK-related genes expression profiles.

On the control medium, germination and survival frequencies were not significantly different in between the variants reaching 100%. On the saline medium, seed germination percentage was reduced to 67% for both lines of Arabidopsis *SlIPT3* complemented plants (9AT and 10AT lines), significantly less than that of *ipt3* KO and WT plants (Figure 4A). However, in Arabidopsis *SlIPT4* complemented plants (3AT and 5AT lines), seed germination reached 89% and 90%, respectively, while WT seeds displayed 78% and *ipt3* KO 93% of germination rate, respectively. The same trend was detected for seedlings' survival percentage on 100 mM NaCl (Figure 4B) (Fisher's exact test with $p \leq 0.05$). Similarly, primary root elongation was estimated by screening 10 DAS Arabidopsis *SlIPT3* and *SlIPT4* complemented plants (Figure 4C). On control medium, the primary root length among the 9AT and 10AT lines significantly differed compared to WT while on the salt-containing medium, no significant changes between the *SlIPT3* complemented and WT plants were observed. Arabidopsis plants complemented with *SlIPT4* showed significantly longer primary root (3.77 ± 0.41 and 3.85 ± 0.8 cm for 3AT and 5AT, respectively) than WT (3.35 ± 0.58 cm) on control medium, while antagonistic effects were detected on salt-containing medium (Figure 4C, Mann–Whitney U Test with $p \leq 0.05$). These results demonstrate that both tomato IPT enzymes are functional in heterologous system. However, in reaction to salt stress the *SlIPT4* complemented plants displayed the same characteristics as WT in germination and survival assays, while both of the *SlIPT3* complemented lines differed in their response to salinity.

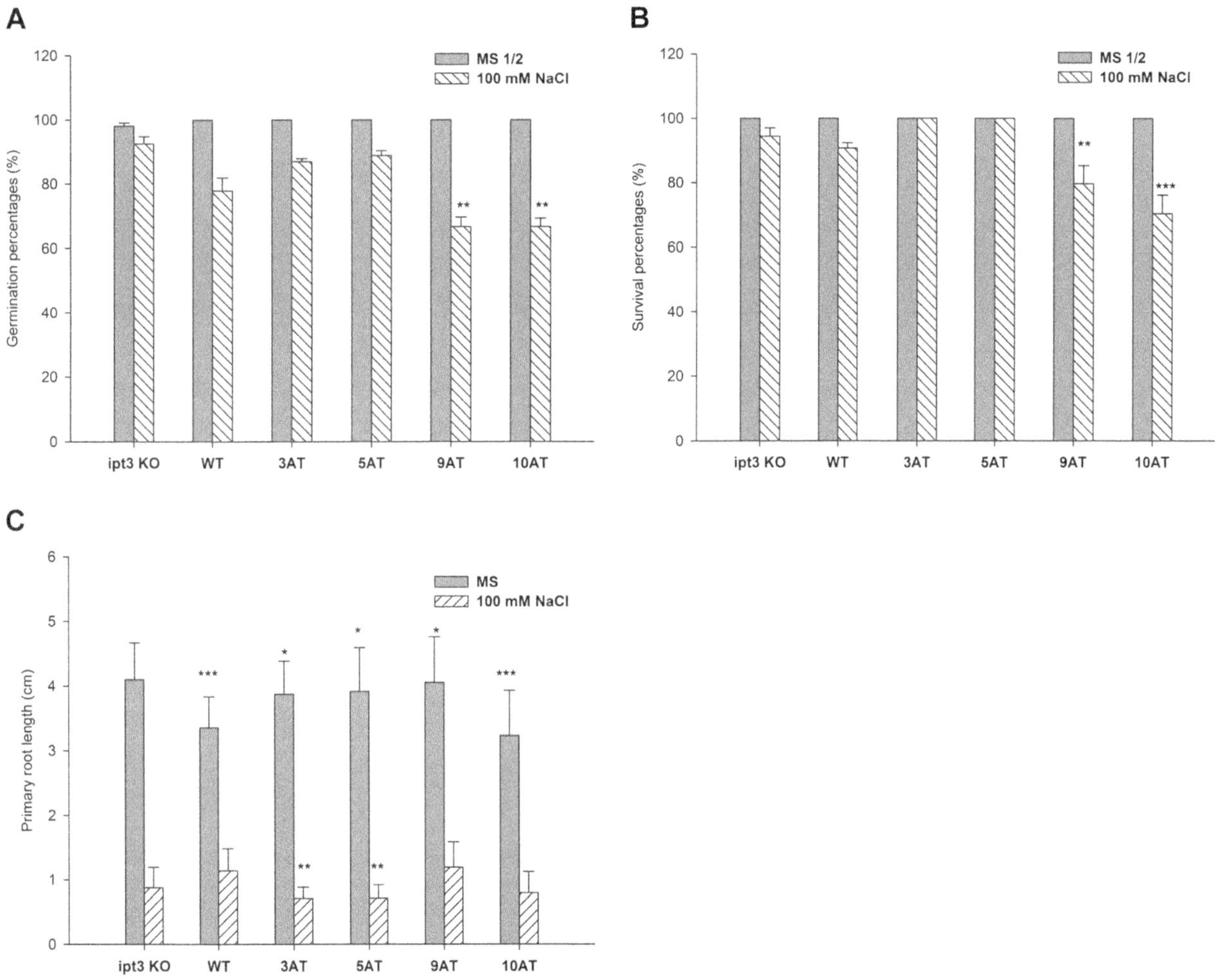

Figure 4 Salinity response of SlIPT3 and SlIPT4 in Arabidopsis. Germination percentage **(A)** and survival percentage **(B)** of *SlIPT4* (lines 3AT and 5AT) and *SlIPT3* (lines 9AT and 10AT) Arabidopsis *ipt3* complemented plants. **significantly lower germination and survival percentages compared to *ipt3* KO. ***significantly lower survival percentages compared to WT and *ipt3* KO. **(C)** Primary root length of Arabidopsis *SlIPT3* and *SlIPT4* complemented plants. *significantly longer primary root compared to WT. **significantly shorter primary root compared to WT. ***significantly shorter primary root compared to *ipt3* KO plants (Mann–Whitney U Test with $p \leq 0.05$).

To assess whether salt treatment affected CK metabolism in Arabidopsis *SlIPT3* and *SlIPT4* complemented plants, CK contents and gene expression profiles of CK biosynthetic genes were investigated on control and salt medium (100 mM NaCl). Generally, in both types of media, the total content of CKs of Arabidopsis *SlIPT3* and *SlIPT4* complemented plants exceeded that of *ipt3* KO plants. However, in comparison to WT plants, Arabidopsis *ipt3* complemented plants showed lower CK concentrations in transgenic lines with exception of the 9AT line on salt medium (Figure 5A). Proportionally, CK-*N*-glucosides accumulated strongly in all tested variants (149.01 – 254.81 pmol/g FW), with iP7G as the predominant metabolite regardless of growth medium (Additional file 2 and Additional file 3). Salt treatment substantially increased the levels of CK free bases and ribosides, while *tZ*-type CKs were mainly enhanced among the CK groups (Figure 5B).

The expression profiles of CK biosynthetic genes were investigated in Arabidopsis seedlings. No expression in vegetative phase of development was detected for *AtIPT4, AtIPT6*, and *AtIPT8*, as had been previously observed [17]. On the control medium, complemented plants showed differential expression of the remaining *IPT* genes compared to WT or *ipt3* KO plants. *AtIPT2* and *AtIPT5* were repressed in Arabidopsis *SlIPT3* complemented

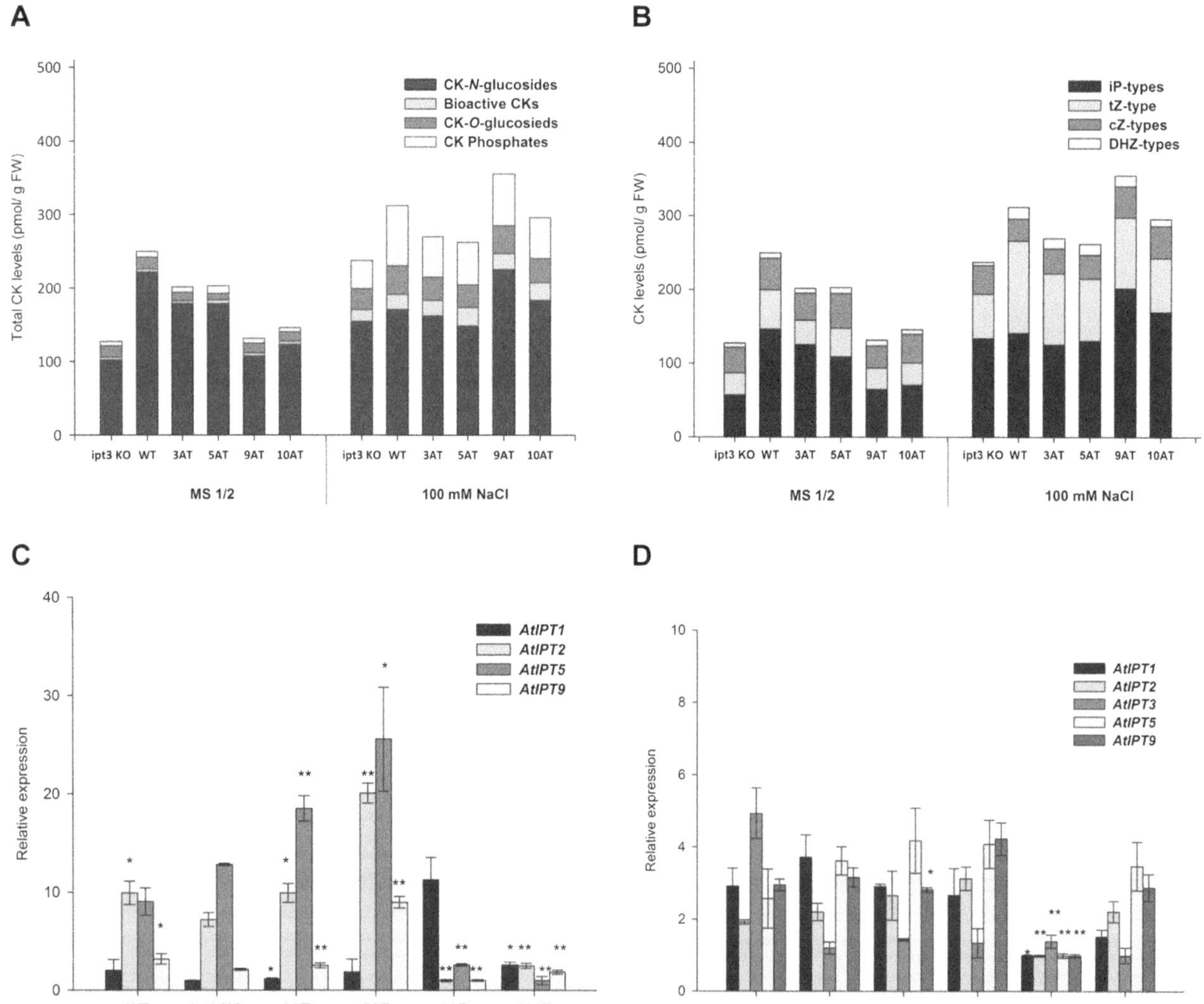

Figure 5 CKs accumulation in Arabidopsis *ipt3* complemented plants. Total endogenous CK levels of *SlIPT4* (lines 3AT and 5AT) and *SlIPT3* (lines 9AT and 10AT) *ipt3* KO complemented plants divided according to the chemical structure **(A)** and contents of bioactive CKs **(B)**. Relative expression of CK biosynthetic genes in Arabidopsis *ipt3* KO mutants complemented plants cultivated on the control medium **(C)** and on the salt-containing (100 mM NaCl) medium **(D)**. *Actin* and *EF1* were used as internal controls for normalization of *AtIPT* transcript levels. Data represent means and SD of two replicates. *statistically significant difference from *ipt3* KO (unpaired two-tailed Student's *t*-test, $p \leq 0.05$). **statistically significant difference from *ipt3* KO after Šidák correction for testing of multiple lines (multiple unpaired two-tailed Student's *t*-test, overall $\alpha = 0.05$, individual $p \leq 0.01$).

plants, while their expression increased in Arabidopsis *SlIPT4* complemented plants, in comparison with WT or *ipt3* KO plants (Figure 5C). Salinity strongly downregulated *AtIPT2* and *AtIPT5* expression, while it increased expression of *AtIPT3* in WT, *ipt3* KO and *SlIPT4* complemented plants (Figure 5D). In *SlIPT3* complemented plants, genes were differentially regulated depending on the transgenic line considered (9AT or 10AT). These results demonstrate that salinity differently affects *IPT* genes expression and CK status in Arabidopsis, and that SlIPT3 and SlIPT4 show distinct regulatory mechanisms in heterologous system. Therefore we propose that Arabidopsis plants respond to salt stress by the elevation of CK levels, with predominant accumulation of *N*-glucosides suggesting overabundance of CKs that is followed by downregulation of some IPT encoding genes to maintain CK homeostasis.

SlIPT3 and *SlIPT4* are differently expressed in tomato organs and participate in cytokinin homeostasis

The expression patterns of *SlIPT3* and *SlIPT4* in various tomato vegetative and reproductive organs from plants cultivated in the greenhouse under normal growth conditions, including three stages of fruit development and ripening, were investigated by qRT-PCR (Figure 6A and B). In general, *SlIPT3* transcripts were much more abundant than *SlIPT4* ones in all the tested organs. Both genes were expressed in young and old leaves, roots and stems of tomato plants. In vegetative tissues, *SlIPT3* transcripts were weakly detected in young leaves but highly accumulated in

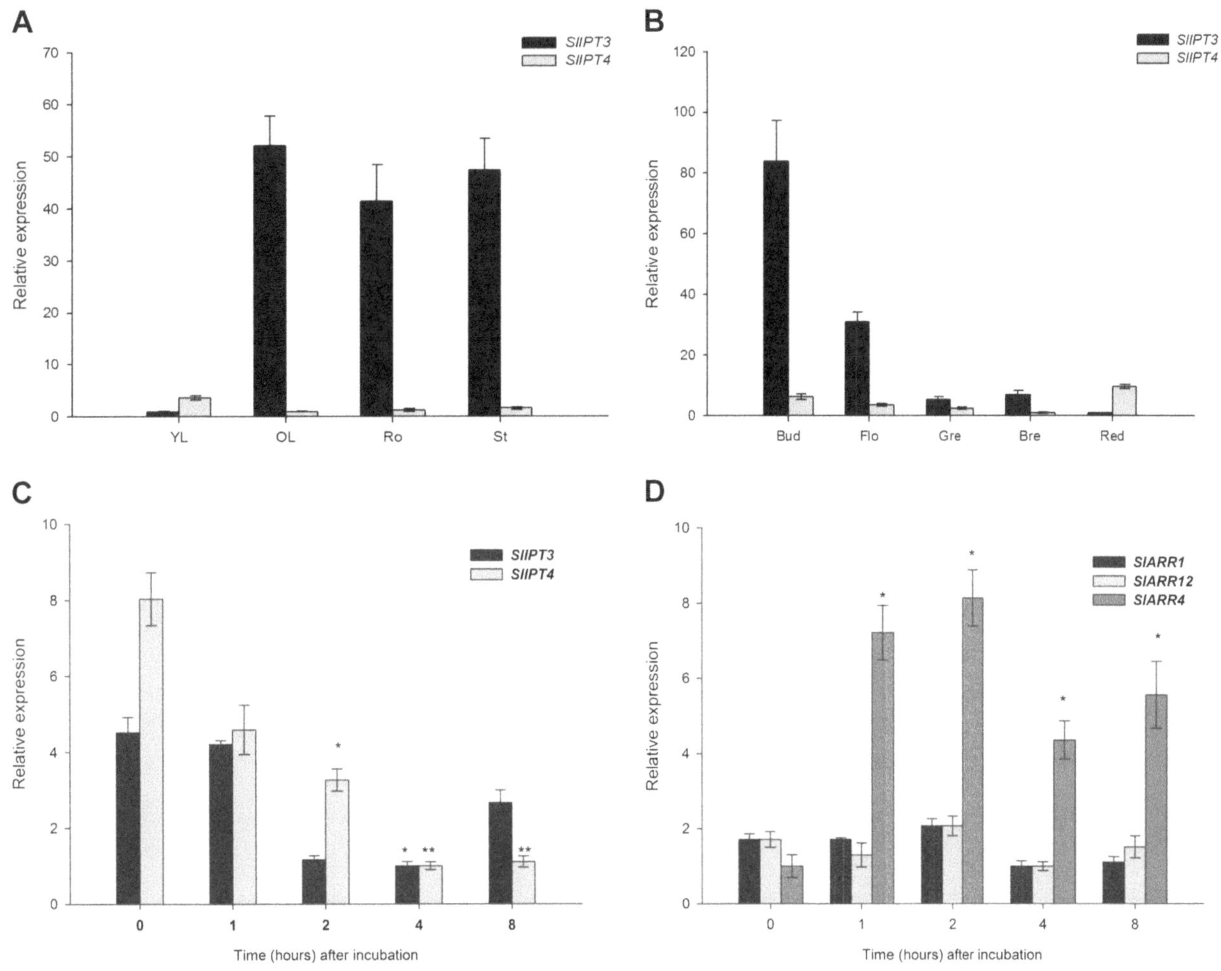

Figure 6 QRT-PCR analysis of *SlIPT3* and *SlIPT4* expression. Spatio-temporal expression profile during tomato plant development in vegetative organs **(A)** and reproductive organs **(B)**. Regulation of *SlIPT3* and *SlIPT4* expression by CK treatment: Effects of *tZ* (10 μM) treatment on *SlIPT3* and *SlIPT4* expression **(C)** and the expression levels of CK response regulators *SlARR1*, *SlARR4* and *SlARR12* **(D)**. *Actin* and *GAPDH* were used as internal controls for normalization of *SlIPT3* and *SlIPT4* transcript levels. Data represent means and SD of three replicates. *statistically significant difference from time 0 h (unpaired two-tailed Student's *t*-test, $p \leq 0.05$). **statistically significant difference from time 0 h after Šidák correction for testing of multiple points (multiple unpaired two-tailed Student's *t*-test, overall $\alpha = 0.05$, individual $p \leq 0.0127$). YL, young leaves; OL, old leaves; Ro, roots; St, stem; Bud, bud; Flo, flower; Gre, green stage; Bre, breaker stage; Red, red stage of tomato development.

stems, roots and old leaves. In contrast, *SlIPT4* was preferentially expressed in young leaves (Figure 6A). In reproductive tissues, *SlIPT3* transcripts were abundant in buds and flowers, then rapidly decreased during tomato fruit maturation with higher levels of transcripts in the breaker stage. *SlIPT4* was preferentially expressed in buds, then continuously declined until the breaker stage and reached its peak expression at the red stage of tomato fruit (Figure 6B). Taken together, these data indicate that *SlIPT3* and *SlIPT4* display distinct expression profiles during tomato development.

In order to clarify a putative feedback regulation by CKs on *SlIPT3* and *SlIPT4* expression, WT tomato roots from hydroponic plants were treated with *tZ*. CK treatment gradually repressed *SlIPT3* (up to 4-fold) and *SlIPT4* (up to 8-fold) expression, the inhibitory effect being most noticeable 4 h after treatment onset. However, *SlIPT3* transcripts then started to accumulate again, which was not the case for *SlIPT4* (Figure 6C). Likewise, the exogenous application of *tZ* strongly induced expression of the CK negative-feedback ARR type-A *SlARR4*, in contrast to its more discrete effect on ARR type-B *SlARR1* and *SlARR12* (Figure 6D). Altogether, these data suggest a regulation of *SlIPT3* and *SlIPT4* transcripts accumulation through plant endogenous CK status, in addition to the feedback regulation *via* the CK signaling pathway.

Overexpression of *SlIPT3* strongly impacts tomato phenotype and cytokinin status

To understand the impact of *SlIPT3* and *SlIPT4* overexpression on CK metabolism and verify the enzymes activity *in planta*, we attempted to generate transgenic lines constitutively expressing *SlIPT3* or *SlIPT4*. Several *35S::SlIPT3* transgenic lines were able to develop and survive repotting into soil, while no *35S::SlIPT4* tomato plants could be regenerated from transgenic calli *in vitro*. For further characterization, three *35S::SlIPT3* transgenic lines (L6, L8, L9) were selected, with transgene expression in leaves increasing up to 14-fold (L9) compared to WT (Figure 7A).

The phenotype of *35S::SlIPT3* tomatoes substantially differed from that of the WT plants. *SlIPT3* overexpression induced plant dwarfism, release from apical dominance with a branchy phenotype, thick and ligneous stems, thicker leaves with punctual shape modification and shorter internodes (Figure 8A (a-d)). A significant yellowing of leaves, as well as accumulation of anthocyanin, was also observed. Additionally, despite inflorescence formation, no flowers were visible in most of *SlIPT3* transgenic lines (Figure 8A (f)) compared to WT plants (Figure 8A (g)). Only a single transgenic line (L1) showing the weakest phenotype (Additional file 4) was able to develop normal fruits containing seeds. Generally, during several months of cultivation, fruit formation was either significantly reduced or completely aborted. When fruits did develop, ripening was incomplete in transgenic plants. Taken together, these observations indicated that constitutive overexpression of *SlIPT3* strongly affects tomato plant development, morphogenesis and reproduction.

The strong effects of *SlIPT3* overexpression on tomato phenotype were simultaneously reflected in CK metabolism. The total CK contents extracted from young and fully expanded tomato leaves increased up to 12-fold in *35S::SlIPT3* lines (Figure 8B, Additional file 5) compared to WT plants. Proportionally, CK-*N*-glucosides represented the main group of CKs, especially in *35S::SlIPT3* lines, with iP7-glucoside (iP7G) being the predominant metabolite (Figure 8B, Additional file 5). In addition, the highest concentration of *tZ* was observed in WT plants (Figure 8C, Additional file 5). Interestingly, the CK storage forms (CK-*O*-glucosides) only slightly increased (up to 2.6-fold) in all transformants (Figure 8B). These results show that *SlIPT3* overexpression leads to a dramatic increase in CKs, particularly to the enhanced formation of iP7G throughout *N7*-glucosylation pathway, which seems to be the major metabolic pathway for bioactive CKs inactivation. Indeed, no significant differences in CK oxidase/dehydrogenase (CKX) activity were found between the control and transgenic tomatoes (Additional file 6).

SlIPT3 overexpression modifies cytokinin-related genes expression and plant tolerance to salinity

To elucidate the role of CK biosynthesis (*IPT*), signaling (*ARRs*) and catabolism (*CKX*) related genes in the response to *SlIPT3* overexpression in tomatoes, the expression patterns of corresponding genes were determined by qRT-PCR in young leaves of transgenic tomatoes and WT plants. In *35S::SlIPT3* lines, the transcript abundance of *SlIPT6* (which encodes a tRNA dimethylallyltransferase) was higher than in WT, whereas lower expression of *SlIPT1* was observed in all transgenic lines (Figure 7A). Likewise, upregulation of *SlARR4*, functioning in feedback regulatory loops of CKs (Figure 7B), and downregulation of CK deactivating genes *SlCKX4* and *SlCKX5* (Figure 7C) were observed in all transformants, relative to WT plants. Thus, CK-related genes expression analysis revealed upregulation of some genes involved in CK biosynthesis (*SlIPT6*) and signaling (*SlARR4*), but downregulation of genes involved in CK degradation pathways (*SlCKX4* and *SlCKX5*).

We hypothesized that the upregulation of some CK metabolic genes and enhancement of CKs content could be elementary prerequisites of tomato salt stress tolerance. As only L1 could successfully reproduce, T2 homozygous transformants from this transgenic line were further analyzed for their tolerance to salinity. For this purpose, L1 and WT lines were cultivated *in vitro* for 4 weeks on the

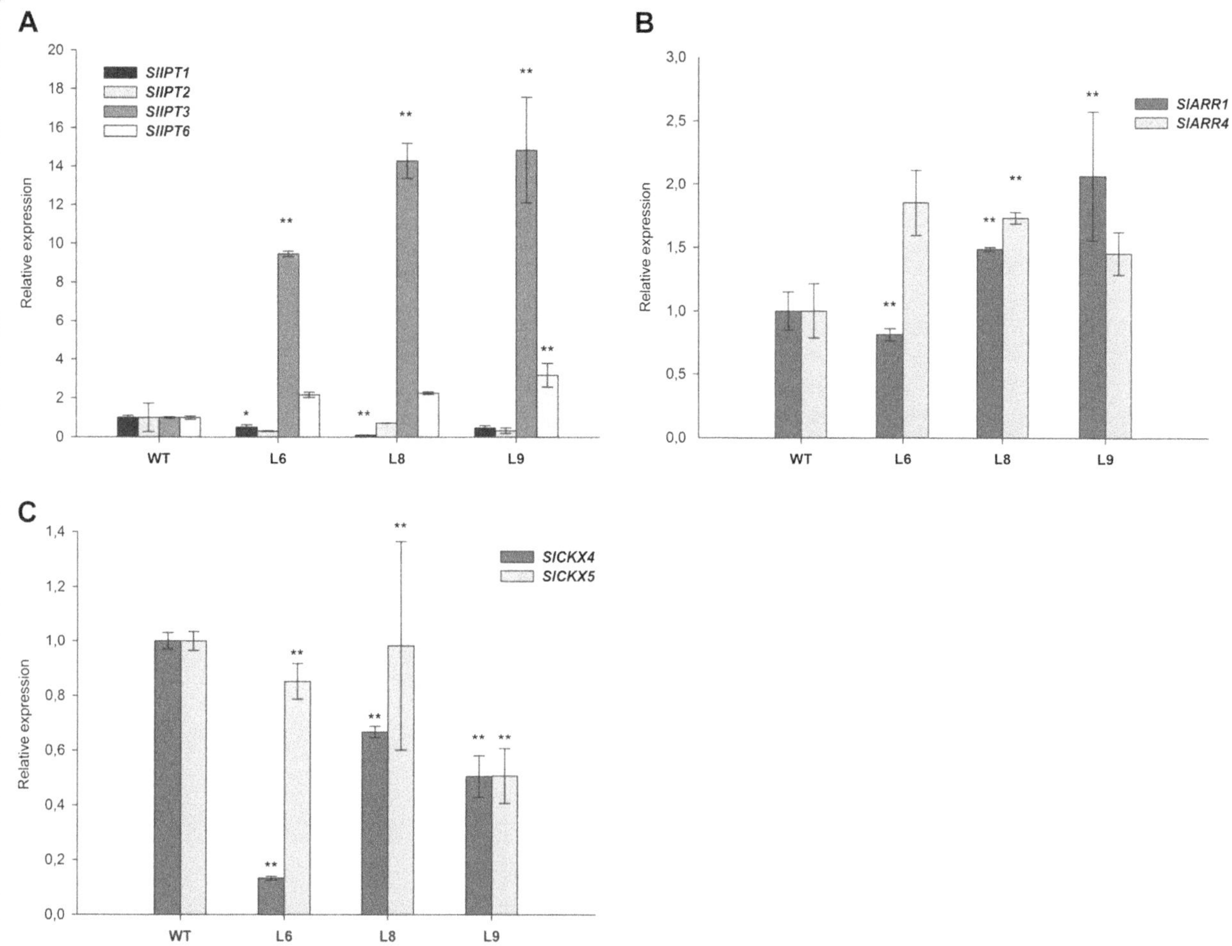

Figure 7 Relative expression of CK metabolic and response genes in tomato young leaves of *35S::SlIPT3* and WT plants. Transcript abundance of CK biosynthetic genes **(A)**, genes involved in CK signaling **(B)** and degradation **(C)** pathways. *Actin* and *GAPDH* were used as internal controls for normalization of candidate genes expression. Data represent means and SD of two replicates. *statistically significant difference from WT (unpaired two-tailed Student's *t*-test, $p \leq 0.05$). **statistically significant difference from WT after Šidák correction for testing of multiple lines (multiple unpaired two-tailed Student's t-test, overall $\alpha = 0.05$, individual $p \leq 0.017$).

100 mM NaCl medium. Compared to WT plants, *35S::SlIPT3* L1 plants exhibited a larger shoot with extending leaves, whereas WT plants showed more compact shoot and smaller leaves as a result of limited growth in presence of salt (Figure 9A). Subsequently, the effects of salt stress induced either by 100 mM and 150 mM NaCl concentration on photosynthesis and nutrients balance were evaluated in WT and L1 by measurements of photosynthetic pigments and macronutrients contents.

In WT plants, chlorophyll *a* (Chl *a*) and chlorophyll *b* (Chl *b*) significantly decreased with the addition of NaCl into the growth media. In L1 plants, the content of both types of chlorophyll was significantly lower on control medium relative to WT, but remained nearly constant on NaCl containing media, and a significant decrease in Chl *a* content was only apparent on 150 mM NaCl (Figure 9B). Likewise, the levels of carotenoids (Car), especially lutein and β-carotene displaying photoprotective roles in photosynthesis, exhibited almost identical contents in *35S::SlIPT3* L1 plants regardless of growth media. In contrast, WT plants showed a significant decrease of β-carotene on both types of NaCl media, and of lutein on 150 mM NaCl. Increasing contents of Car involved in the xanthophyll cycle, violaxanthin (V), antheraxanthin (A) and zeaxanthin (Z) were as well observed in presence of salt in both WT and transgenic tomatoes, significant for WT at 150 mM NaCl and *35S::SlIPT3* grown on both concentrations of salt (Figure 9C and Additional file 7).

Shoot sodium (Na^+), potassium (K^+), magnesium (Mg^{2+}) and calcium (Ca^{2+}) concentrations were assessed in plants exposed to 0 (control), 100 mM and 150 mM NaCl (Figure 9D and E). As observed in Figure 9D, Na^+ and K^+ concentrations were similar in WT and L1 under control conditions of growth. Salinity induced Na^+ accumulation and substantially reduced K^+ contents in both types of tomato plants. However, Na^+ accumulated to a lesser

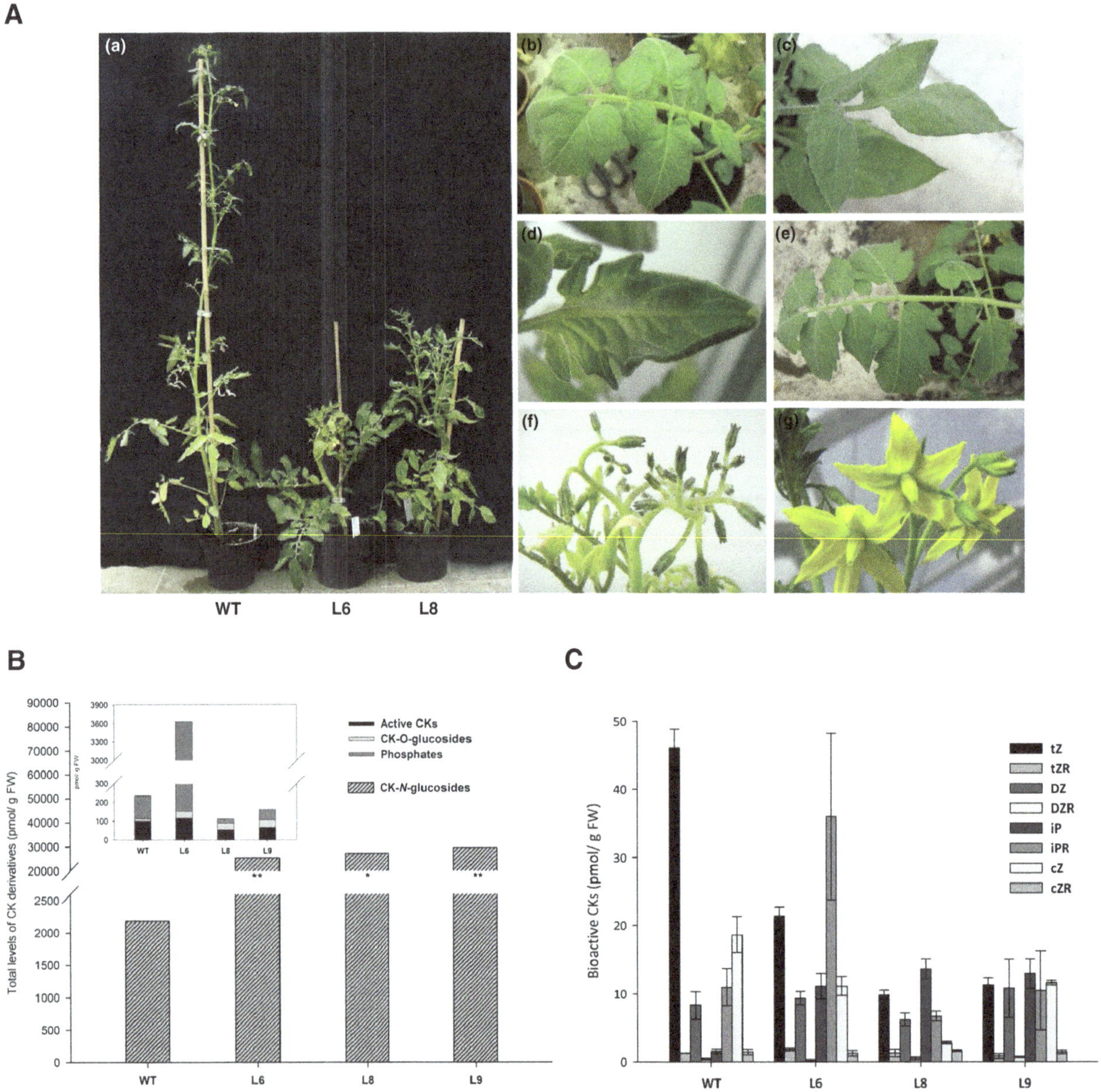

Figure 8 Phenotype of T0-generation *35S::SlIPT3* tomato transformants. Transformants showed dwarfed aerial part with a branching phenotype (a), modified leaf shape (b, c, d) and flower inflorescence (f) compared to WT plants (e and g) **(A)**. Endogenous CK levels in tomatoes divided according to the chemical structure **(B)** and contents of bioactive CKs **(C)**. WT: wild type; L6, L8 and L9: independently regenerated *35S::SlIPT3* transformants. Data represent means and SD of two replicates. *statistically significant difference datasets from WT (unpaired two-tailed Student's *t*-test, $p \leq 0.05$). **statistically significant difference from WT after Šidák correction for testing of multiple lines (multiple unpaired two-tailed Student's *t*-test, overall $\alpha = 0.05$, individual $p \leq 0.017$).

extent in L1 (0.33 and 0.37 mmol/g DW) than in WT (0.38 and 0.48 mmol/g DW) on the 100 and 150 mM NaCl medium, respectively. Similarly, K^+ concentration was significantly higher in L1 (0.27 and 0.22 mmol/g DW) than in WT (0.216 and 0.154 mmol/g DW) in presence of 100 and 150 mM NaCl, respectively, and these results are also visible by the K^+/Na^+ ratio, which remained significantly higher in L1 than in WT in salt stress conditions (Figure 9E). Finally, no difference of Ca^{2+}contents between WT and transgenic plants was noticed, while the Mg^{2+} concentration was significantly higher in transgenic tomatoes on control and 100 mM NaCl conditions, compared to WT (Figure 9D).

Overall, plant morphological changes in presence of salt, roughly constant levels of photosynthetic pigments, especially Chl *a*, Chl *b*, lutein and β-carotene in *35S::SlIPT3* tomatoes regardless on the growth medium and higher K^+ contents in presence of NaCl indicate that *SlIPT3* overexpression could raise plant tolerance to salt stress.

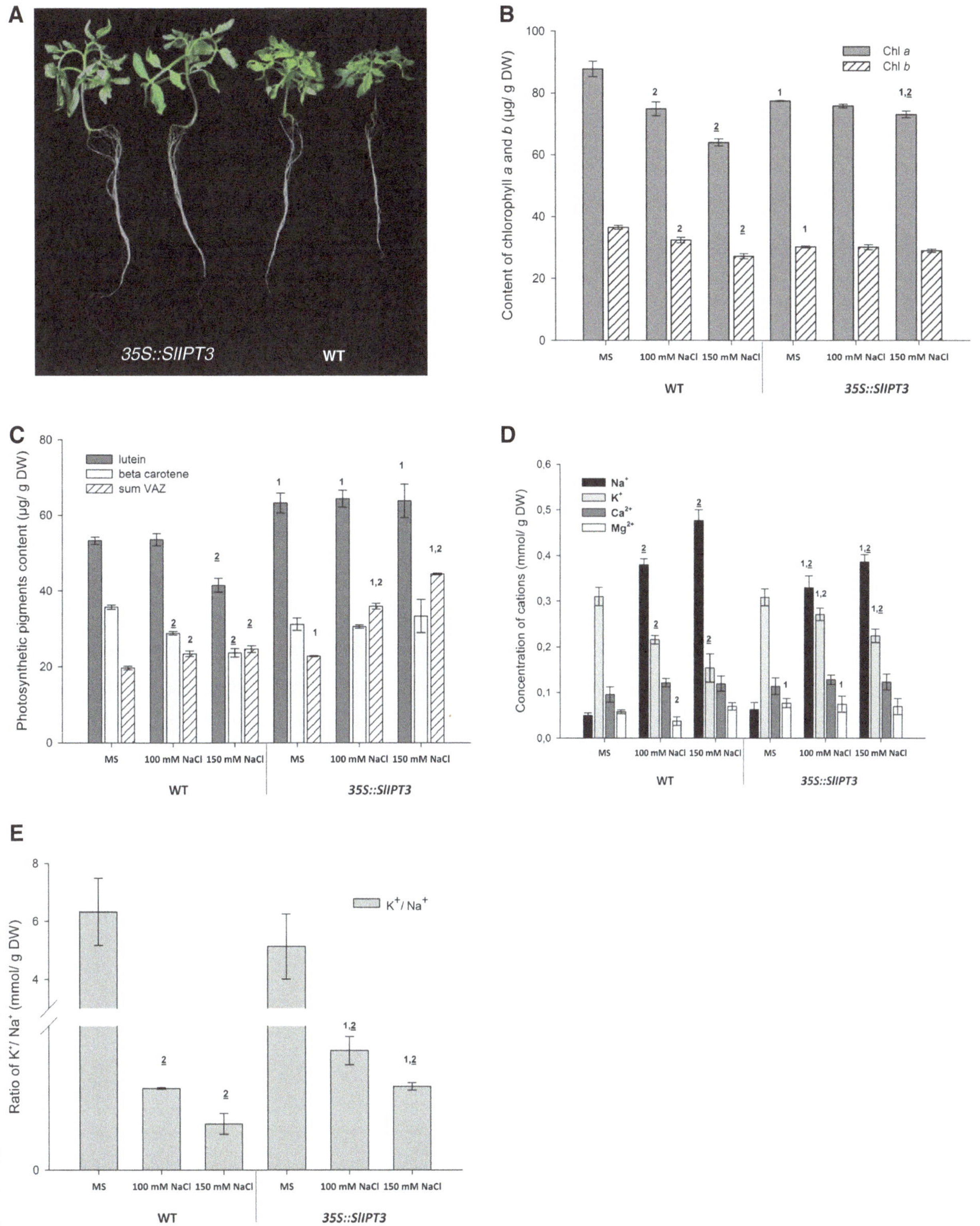

Figure 9 (See legend on next page.)

(See figure on previous page.)
Figure 9 Effect of salt stress on WT and *35S::SlIPT3* (L1) tomato plants cultivated on control and salt-containing growth media. (A) Phenotype of WT and *35S::SlIPT3* (L1) transgenic plants grown 4 weeks on the 100 mM NaCl medium. Chlorophyll content **(B)**, carotenoid content **(C)**, cation concentration **(D)** and K^+/Na^+ ratio **(E)** in 4 weeks-old tomato developing leaves of WT and trangenic plants. Sum VAZ, sum violaxanthin, antheraxanthin and zeaxanthin. Data represent means and SD of two replicates for photosynthetic pigments while three replicates represent data from cation concentration. [1]statistically significant difference from the same treatment of WT (unpaired two-tailed Student's *t*-test, $p \leq 0.05$). [2]statistically significant difference from MS medium within the same line (unpaired two-tailed Student's *t*-test, $p \leq 0.05$), [2]statistically significant difference from MS medium within the same line after Šidák correction for testing of multiple NaCl concentrations (multiple unpaired two-tailed Student's *t*-test, overall $\alpha = 0.05$, individual $p \leq 0.0253$).

Both *SlIPT3* and *SlIPT4* are involved in salt stress response

To further understand the participation of *SlIPT3* and *SlIPT4* in salt stress response, their expression pattern was examined by qRT-PCR in roots, young leaves and old leaves of tomato WT plants exposed to salinity (150 mM NaCl) for 12 h. In roots (Figure 10A), which are the first organ perceiving salinity, *SlIPT3* and *SlIPT4* shared very similar expression profiles. A rapid decrease in transcript levels of both genes within 2 h after salt exposure was followed by a very small peak at 8 h for

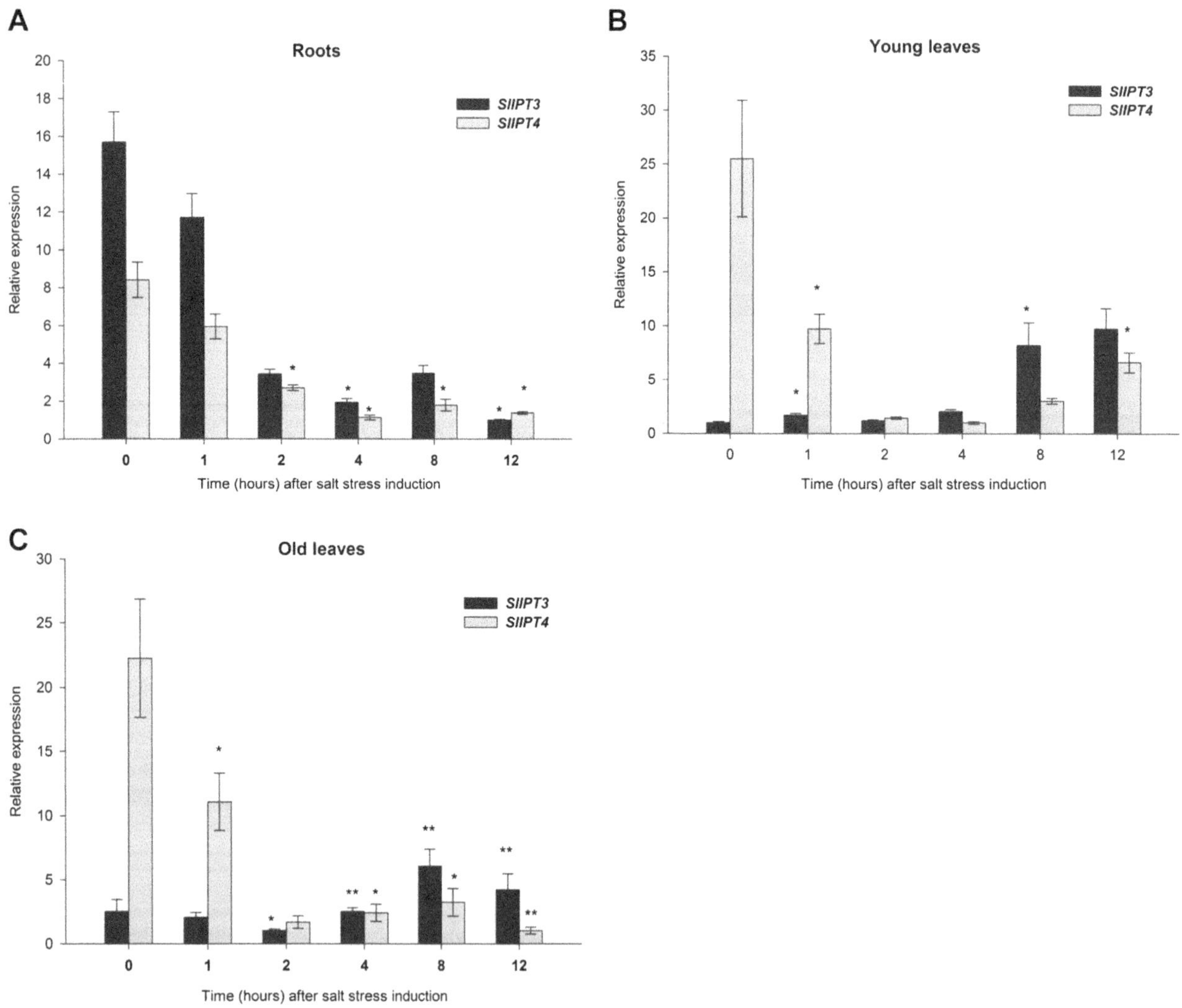

Figure 10 Salinity response of *SlIPT3* and *SlIPT4* in tomato WT plants. Analysis of *SlIPT3* and *SlIPT4* expression levels in tomato WT roots **(A)**, young leaves **(B)** and old leaves **(C)** in response to salt stress (150 mM NaCl). *Actin* and *GAPDH* were used as internal controls for normalization of *SlIPT3* and *SlIPT4* transcript levels. Data represent means and SD of two replicates. *statistically significant difference from time 0 h (unpaired two-tailed Student's *t*-test, $p \leq 0.05$). **statistically significant difference from time 0 h after Šidák correction for testing of multiple time points (multiple unpaired two-tailed Student's *t*-test, overall $\alpha = 0.05$, individual $p \leq 0.01$).

both genes, before a new decline. In young leaves, *SlIPT3* and *SlIPT4* showed partially divergent transcript accumulation profiles (Figure 10B). Indeed, *SlIPT3* transcripts gradually and regularly increased with incubation time. In contrast, *SlIPT4* was repressed 2.5-fold as rapidly as 1 h after stress initiation. This repression seemed transient, however, as a slight increase in *SlIPT4* transcripts was visible 8 h after stress onset (Figure 10B). Similar expression profiles were observed in old leaves (Figure 10C). These results indicate that *SlIPT3* and *SlIPT4* directly participate in salt stress response in tomato *via* a differential modulation of their expression, depending on the organ and duration of the stress.

Discussion

CK homeostasis during plant development and in response to changing environmental conditions is, among other mechanisms, properly maintained by the activity of CK biosynthetic enzymes, IPTs. It was previously reported that root-localized *IPT* expression essentially improved salt stress tolerance in tomato [3]. However, the genes involved and the mechanisms of CKs regulation under salt stress conditions have not yet been elucidated. Therefore, based on functional characterization of *SlIPT3* and *SlIPT4* we provide insights into both genes involvement in the response to salt stress in tomato plants.

Tomato SlIPT3 and SlIPT4 are non-redundant functional isopentenyltransferases

In the present study, we have demonstrated that tomato *SlIPT3* and *SlIPT4* encode two functional isopentenyltransferases (Figure 2A). Multiple alignment of proteins sequences and phylogenetic analysis revealed that SlIPT3 and SlIPT4 are closely related to Arabidopsis ortholog AtIPT3 and AtIPT5 (Figure 1B), as previously reported [12]. Interestingly, we observed identical chloroplast localization of GFP-tagged SlIPT4 protein in tomato leaf cells (Figure 3), as described in Arabidopsis where AtIPT3 was localized to plastids in roots and leaf cells [29]. These data indicate that SlIPT4 and AtIPT3 partly share similar function and protein subcellular localization. We therefore speculate that SlIPT4, and probably also SlIPT3, due to the presence of a chloroplastic transit peptide as well, may have other comparable regulatory mechanisms for protein chloroplast targeting such as farnesylation for instance, as was demonstrated for AtIPT3 [30].

Distinct spatio-temporal expression profiles of *SlIPT3* and *SlIPT4* have been observed in tomato vegetative organs and during fruit onset, development and ripening, indicating that these two genes have non-redundant functions (Figure 6A and B). In general, *SlIPT4* expression was lower relative to that of *SlIPT3* and showed its highest levels in young leaves and red fruits whereas *SlIPT3* substantially predominated in the remaining tested organs. Our data are in agreement with a previous report describing very low transcript abundance of *SlIPT4* in all tomato tested organs with flowers exhibiting the highest *SlIPT4* expression, and a greater accumulation of *SlIPT3* transcripts, especially in the very early stages of flower and fruit development [12]. Interestingly, although *SlIPT4* is less expressed than *SlIPT3* in tomato, SlIPT4 showed ten times greater activity *in vitro* than SlIPT3, which may explain why no transgenic plants could be regenerated from calli. Indeed, SlIPT4 is probably at the origin of a greater hormonal imbalance than SlIPT3, which may have impaired proper shoot and root systems regeneration.

Differences between *SlIPT3* and *SlIPT4* in organ expression specificity point out their distinct physiological roles in tomato developmental processes. However, it seems rather difficult to assign a specific role to each IPT during tomato plant development, due to their overlapping functions. Indeed, although *AtIPT3* is one of the three most expressed *IPT* genes in *Arabidopsis* during the vegetative phase, *ipt3* knock-out mutant exhibited no distinguishable phenotype from WT plants, indicating functional redundancy between the Arabidopsis IPT proteins [17]. Therefore, none of the visible phenotypes or characteristics in *SlIPT3* and *SlIPT4* complemented Arabidopsis *ipt3* knock-out mutants were apparent in comparison to WT plants under control conditions (Additional file 1, Figure 4A and B).

Ectopic expression of *SlIPT3* alters tomato phenotype and CK metabolism

Overexpression of *SlIPT3* under *35S CaMV* promoter and consequent CK over-accumulation had extensive negative effects on tomato growth under control conditions (Figure 8), resulting in dwarf phenotype and delayed senescence, substantially limiting formation of tomato generative organs. Only one transgenic line (L1) showing the weakest phenotypic differences from WT was able to produce seeds (Additional file 4). A similar effect of CK overabundance in flowering delay was reported in stably transformed tobacco plants with *35S::MdIPT3a* [31]. Likewise, phenotype alterations such as a strong delay in growth, shorter rosettes, serrated leaves and decreased apical dominance were observed in Arabidopsis plants overexpressing maize *ZmIPT2* [32]. In contrast, no changes in flowering time were evident in Arabidopsis and tobacco transgenic plants expressing *IPT* under *HSP70* [33], underlining the crucial role of CKs accumulation and homeostasis during flowering.

We found out that constitutive expression of *SlIPT3* increased CK production in tomatoes by up to 12-fold relative to WT plants, primarily iP7G (Figure 8B, Additional file 5). Similarly, high concentrations of iP7G in Arabidopsis complemented plants cultivated on both control and salt media suggest that the formation of

CK-*N7*-glucosides may be a sufficient pathway for downregulation of bioactive CKs because no significant enhancement of CKX activity was detected in *35S::SlIPT3* tomatoes when compared to WT (Additional file 6). In accordance with our results, *IPT* overexpression resulted in increased levels of bioactive iP, *tZ* and their ribosides in tomato [3] and tobacco [31,32]. However, detailed data regarding all CK derivatives were not reported in these previous studies. Based on the observation that iPRMP/iPRDP/iPRTP are the first products in CK biosynthesis [9,10], that iP and iPR are the best substrates for CKX activity [34] and are in massive amounts metabolized to CK-*N*-glucoside forms (Additional file 5), we conclude that iP-type CKs have a prominent role in fast regulation and balancing of the CK metabolism.

SlIPT3 and SlIPT4 participate in salinity response in tomato and Arabidopsis

Expression analyses of *SlIPT3* and *SlIPT4* in tomato vegetative organs confirmed the participation of both genes in response to salt stress. After stress onset, salinity evoked repression of *SlIPT3* and *SlIPT4* in roots, and differently affected *SlIPT3* and *SlIPT4* in young and old leaves. After 6 to 8 h of exposure to salinity, increasing levels of both transcripts were detected in young leaves, suggesting preferential CK biosynthesis in photosynthetic organs and thus underlining CKs stress-prevention and anti-senescence properties (Figure 10B). Similarly, an augmentation of *IPT* expression was apparent both in maize leaves and roots following longer exposure to salinity whereas shorter salt treatment (30 min) induced a repression of *IPT* encoding genes (3 out of 4) in maize leaves [11]. Likewise, two weeks-old Arabidopsis plants showed downregulation of *AtIPT1* and *AtIPT3* as early as 1 h after salt treatment, whereas the expression of *AtIPT5* and *AtIPT7* was elevated. Conversely, with longer salt stress, the majority of *AtIPT* transcripts were repressed [35]. Together, these data indicate a dynamic regulation of CK biosynthesis/metabolism over the stress period.

Moreover, multiple putative stress-responsive *cis*-regulatory elements were identified in the promoter regions of *BrIPTs* (and *BrCKXs*) promoters, pointing to the response of these genes to stress stimuli and a possible response to unfavourable conditions *via* changing CK status [36]. Similarly in this study, an *in silico* analysis of *SlIPT3* and *SlIPT4* promoters [37] (www.dna.affrc.go.jp/PLACE/index.html) revealed the presence of putative stress response elements and many ARR binding *cis*-elements, indicating at least for *SlIPT3* the existence of salt stress tolerance mechanisms similar to those in *BrIPTs*. Our findings in tomato plants illustrated that the regulation of *SlIPT3* and *SlIPT4* in response to salt stress is organ specific, depends on the duration of stress exposure and potentially also on the presence of *cis*-regulatory elements in the promoters of CK-related genes. In addition, diverse roles of both tomato *SlIPTs* in salt stress responses have been characterized in Arabidopsis *ipt3* KO complemented plants under longer period of salt treatment (Figures 4 and 5).

Transgenic plants ectopically expressing *SlIPT3* under *35S* promoter showed superior above-ground growth after several weeks on salt medium, relative to WT plants (Figure 9A). Beneficial effects of CKs overproduction in response to salt stress were reported in tomato [3]. CKs elevation enhanced shoot growth and prevented senescence under salinity conditions due to the specific induction of *IPT* in roots by a *heat-shock* promoter (*HSP70::IPT*). Likewise, higher CK concentration was associated with salt tolerance in tomatoes overexpressing a cysteine-2/histidine-2-type zinc finger transcription factor [38]. Similarly, expression of *Agrobacterium tumefaciens IPT* under the stress inducible promoter *rd29A* resulted in higher salinity tolerance in transgenic tobacco [23]. Using specific promoters for *IPT* overexpression also successfully improved tolerance to drought in rice [39], peanut [40], cassava [41], tobacco [42] and to the cold stress in tobacco [43]. On the other hand, tobacco plants with reduced CK contents showed smaller osmotic potential and higher drought resistance than WT plants [44]. Likewise, Arabidopsis knock-out mutants for two CK histidine kinase receptors (*ahk2* and *ahk3*) showed tolerance to drought and salinity [45]. Aspects of plant phenotype, such as root architecture modifications and growth inhibition in transgenic plants, influenced by imbalance between CKs accumulation and distribution can partly explain tolerance to various environmental stresses. Thus, root-specific *CKX* overexpression enhanced root system development and consequently increased drought tolerance, nutrient uptake and leaf nutrient enrichment in *Arabidopsis* [43,46]. Consequently, it is still difficult to estimate the precise nature of CK effects due to the complexity of phytohormones metabolism and other cross-talk networks.

Salinity tolerance of *35S::SlIPT3* plants is correlated, among others, with improved nutrients balance and retention of the photosynthesis capacity, in addition to the senescence-protective effects of the photosynthetic pigments (Figure 9). Indeed, salinity strongly impacts tomato plants ionic status, as Na^+ concentration considerably increases in leaves, inducing a premature senescence, while K^+ content differently declines depending on the leaf age [4]. Transgenic tomato plants *HSP70::IPT* showing improved growth and final yield under long-term salinity treatment consequently accumulate less Na^+ in leaves and roots than WT, but higher K^+ contents, underlining the importance of the limitation of toxic ions accumulation for plant development [3]. In addition, salinity strongly affects plant photosynthesis, and salt-induced reduction in photosynthetic pigments such as Chl and Car have been reported in several studies [47]. Variations in Chl and Car contents under different salt concentrations and time exposure were observed in three varieties of tomato cultivars, suggesting that sensitivity to salt stress depends on plant genotype [48]. Our results indicate that the contents of Chl *a* and Ch *b* as well as lutein and β-carotene are maintained in *35S::SlIPT3* plants at the same levels regardless of the growth media in contrast to WT, in which salt treatments decreased both Chl and Car contents (Figure 9B and C). Similarly, transgenic tomatoes overexpressing *IPT* within the root zone maintained their photosynthesis ability under long-term salinity treatement [3], while transgenic tobacco expressing stress-inducible promoter *rd29A-IPT* showed only slight decrease in Chl contents in comparison to WT plants [23]. Significant accumulation of pigments involved in the xanthophyll cycle in *35S::SlIPT3* plants exposed to salt may reflect higher intensity in photoprotective function of Car in transgenic plants than in WT plants.

In summary, our results indicate that constitutive expression of *SlIPT3* induces several changes in plant growth, nutrients accumulation, hormonal metabolism and photosynthesis maintenance, leading to enhanced tolerance to salt stress.

Mechanism of SlIPT3 and SlIPT4 regulation under salt and CK treatments

The comprehensive mechanism of CK regulatory feedback loop is supported by our findings that exogenous application of *tZ* inhibited the expression of *SlIPT3* and *SlIPT4* within 4 h, enhanced the expression of negative-feedback regulator type-A *SlARR4* and slightly moderated the expression levels of CK primary response genes type-B *SlARR1* and *SlARR12* (Figure 6C and D). Comparable data reporting downregulation of the *IPT* genes after CK treatment have been reported for Chinese cabbage [36] and maize [11]. The connection between CK homeostasis maintenance *via* type-B response regulators and salinity response was reported in Arabidopsis where ARR1 and ARR2 were shown to control sodium accumulation in shoots by regulating expression of genes encoding high-affinity K^+ transporter 1;1 responsible for Na^+ exclusion from the root xylem [25]. These data suggest that CK homeostasis may be effectively regulated, not only by CKs molecules, but also environmental stimuli.

Taken together, we have shown the mechanism of *SlIPT3* and *SlIPT4* response to early salt stress and their feedback regulation by exogenous CK treatment in tomato plants. Based on the data obtained in this study, we propose a model illustrating the complexity of CK networks in tomato plants in response to salinity (Additional file 8). In this model, we only included relations supported by our experimental data. We outline an organ specific downregulation of *SlIPT3* and *SlIPT4* transcripts immediately after stress onset, followed by transcripts increase with the duration of the stress treatment. The complexity of CK networks is demonstrated by the repression of *SlIPT3* and *SlIPT4* after exogenous application of *tZ*, while an overabundance of CKs (especially iP7G) in *35S::SlIPT3* may ensure a fitter phenotype under salt stress conditions. Similarly, the response to salinity by *SlIPT3* and *SlIPT4* was paralleled in Arabidopsis *ipt3* KO complemented plants where salinity resulted in elevation of CKs (mainly *N*-glucosides) with successive downregulation of *AtIPTs* to maintain CK homeostasis (Figure 5, Additional file 8).

Conclusions

In summary, we have characterized *SlIPT3* and *SlIPT4* response to early salt stress in tomato, and identified *SlIPT3* as a key player in the CK metabolism of tomato plants under salinity conditions. Our results contribute to the understanding of CK regulation at the molecular level and provide a potentially useful tool to obtain and improve high-quality stress-tolerant crops in agriculture.

Methods

Plant materials and growth conditions

Tomato (*Solanum lycopersicum* L. cv Ailsa Craig) seeds were pre-germinated on Whatman 3MM paper soaked with sterile water in a sterile petri dish until the radicle was a few mm long. Seeds were sown in trays filled with a perlite-vermiculite mix (1/3, v/v), as previously described [4]. Four weeks later, tomato plantlets cultivated in a growth chamber with a 16 h light (24°C)/8 h dark (22°C) photoperiod were transferred into the 52 L tanks containing aereted half-strength modified Hoagland nutrient solution, and kept in hydroponic system for an additional ten days before applying the salt stress, as previously described [38].

Salt stress treatment of wild-type tomatoes was evoked by direct addition of 150 mM NaCl into the culture tank. For hormonal assays, plants were grown in smaller tanks with a capacity of 4.5 L. Zeatin (mixed *cis-* and *trans-*isomers including approximately 80% of *trans*-zeatin, Sigma Z0164) was used at a final concentration of 10 μM. In order to investigate the spatio-temporal expression profiles of *SlIPT3* and *SlIPT4*, plants were grown in soil in the greenhouse during 8 weeks. Three plants were collected for each time point of the experiment (0 h, 1 h, 2 h, 4 h, 8 h and 12 h). After harvesting, samples were immediately frozen in liquid nitrogen and stored at −80°C until further analyses. For tomato *35S::SlIPT3* (L1) *in vitro* culture, seeds were surface sterilized with a mixture of 2.5% (v/v) potassium hypochlorite and 0.02% (v/v) Triton X-100 for 15 min and rinsed 5 times with sterile water. Seeds were sown on full MS media and the medium for transgenic line selection was supplemented with kanamycin (100 mg/L). Two weeks later, plants were transferred into magenta boxes on the control medium and the medium containing NaCl (100 mM) and cultivated under the same conditions as before (16 h light (24°C)/8 h dark (22°C) photoperiod in the growth chamber). Phenotypic characteristics were evaluated after four weeks.

Arabidopsis thaliana ecotype Columbia was used as the wild-type for all Arabidopsis experiments. The *ipt3* T-DNA insertion line N553810 was obtained from the Nottingham Arabidopsis Stock Center. Seeds were surface sterilized as described for tomato and kept at 4°C for 48 h before being sown on Petri dishes containing half-strength MS with 0.9% (w/v) agar. Additionally, the medium for *ipt3* KO mutants selection contained 35 mg/L kanamycin and *SlIPT3* and *SlIPT4* Arabidopsis complemented plants were selected on the medium supplemented with 15 mg/L hygromycin. Plates were then transferred to the growth chamber with a 16 h light (20°C)/8 h dark (18°C) photoperiod. For the analysis of salt stress tolerance, half-strength MS medium containing 100 mM NaCl was used. The root length (from hypocotyl base to the root tip) of at least twenty vertically growing ten-days-old plants was measured. Differences in root length were tested for significance using a Mann–Whitney U test and data representing two biological replicates. Germination percentage was determined for eighteen seedlings after 24 h using a magnifying glass, when the radicle occurs. Survival percentage of eighteen seedlings was evaluated in thirteen-day-old plants. Total numbers of germinated and non-germinated seeds (54) were cross tabulated for each pair of groups and Fisher's exact test was then applied to the resulting 2x2 contingency tables. For both assays, the data represent three biological replicates.

Plasmid constructions and plant transformation

SlIPT3, SlIPT4, SlARR1 (XP_004239797), *SlARR4* (XP_004238726) and *SlARR12* (XP_004251765) cDNAs were amplified with the *Pfu* DNA polymerase (Promega) in a final reaction volume of 50 μL following the manufacturer's instructions, using 300 ng of cDNAs synthesized from total RNA extracted from tomato grown *in vitro* as template. Specific primers used to amplify the coding sequence of *SlIPT3* and *SlIPT4* contain anchors: forward (5′-GGGGACAAGTTTGTACAAAAAAGCAGGCT-3′) and reverse (5′-GGGGACCACTTTGTACAAGAAAGC TGGGT-3′) allowing BP cloning (Invitrogen) in the pDONR221 vector. Specific primers for every coding sequence were as follows: IPT3F (5′-TATGAATATTGT GTTACAACATATTG-3′), IPT3R (5′-CTAGTGCGTCA TAGTAGCAAC-3′), IPT4F (5′-ATGATTGGCATGATG AACTCT-3′), IPT4R (5′-TTAATAGTGAGATGCTGCT GCC-3′). The PCR product was then transferred to the pDONR221 entry vector (Invitrogen) by a BP recombination reaction prior to DNA sequencing (Macrogen). *SlIPT3* and *SlIPT4* were subsequently transferred either to the pK7WG2D or to the pH7WG2D binary vectors [49]. *SlARR* coding sequences were amplified with the following primers: ARR1F (5′-AAAAAATTCAAAATTT TGAAGAAAGT-3′) and ARR1R (5′-AACTACGCTCT TTTCGCATC-3′), ARR4F (5′-ATTTGGTGAAATATT TGTGGGTT-3′) and ARR4R (5′-GCTTAAACGACCC CGGAAGTA-3′), ARR12F (5′-ATGACTGTGGAGGAA ATTAGA-3′) and ARR12R (5′-TCATAAACCTGAACC AAGTGAA-3′), prior to cloning into the pGEM-T-easy vector (Promega).

SlIPT3 and *SlIPT4* constructs were introduced into the *Agrobacterium tumefaciens* strain GV3101 for Arabidopsis transformation and in the LBA44A4 for tomato transformation. Competent cells were prepared as previously described [50]. For subcellular localization, *SlIPT3* and *SlIPT4* cDNA sequences lacking a stop codon were cloned in frame with the *GFP* sequence in the pK7FWG2 binary vector [49].

Tomato stable transformation was adapted from a published method [51] and tomato genomic DNA extraction was performed as described before [52].

Arabidopsis plants were transformed by the floral dip method [53]. For Arabidopsis *ipt3* complementation, a combination of 3 primers LBb1.3 (5′-ATTTTGCCGAT TTCGGAAC-3′), IPT3LP (5′-TGGAATGGTTGAGGA AGTCAG-3′), IPT3RP (5′-CATTGGCTTAGAAATTT GTGTCC-3′) was used to identify homozygous plants in the segregating F3 population.

RNA extraction and real-time PCR analysis

Tomato total RNAs were extracted from all tissues, treated with the RNase-free DNase I (Promega) and purified according to a previous reference [54]. An aliquot of 2 μg was subsequently used as a template for reverse-transcription (RevertAid™ H Minus First Strand cDNA Synthesis Kit, Fermentas) using oligo $d(T)_{18}$ according to the instructions in the manual. Transcript levels of the different genes were measured by qRT-PCR using SYBR Green on a LightCycler 480 II (Roche). PCR reactions were performed in triplicate using 0.2 μM of each primer, 5 μL SYBR Green mix (Promega), and 300 ng of DNAse-treated cDNA in a final volume of 10 μL. Negative controls were included in each run. PCR conditions were: initial denaturation at 95°C for 120 s followed by 45 cycles of 95°C for 10 s, and 58°C for 15 s. Amplification was followed by melting curve analysis to check the specificity of each reaction. The primers IPT3PCRQF (5′-CCTTCT TGCACAAAGTTGCT-3′) and IPT3PCRQR (5′-TGAGG TTATTGATATTAGCAAATA-3′) were used to amplify a

107 bp sequence, while IPT4PCRQF (5′-GGACAGAGCA GAAAGAT-3′) and IPT4PCRQR (5′-TAATAGTGAGAT GCTGCTGCCA-3′) allowed amplification a 108 bp PCR product.

Data obtained from tomato were normalized according to *SlGAPDH* and *SlActin* expression levels. The primers used were GAPDHF (5′-GGTGCCAAGAAGGTTGTG AT-3′) and GAPDHR (5′-TTTTCTGGGTGGCAGTCA T-3′) that generated a 217 bp PCR product, ActinF (5′-ATGGTGGGTATGGGTCAAAA-3′) and ActinR (5′-GAGGACAGGATGCTCCTCAG-3′) that allowed the formation of a 183 bp PCR product. Normalized expression of *SlIPT3* and *SlIPT4* was calculated using the Gene Expression Analysis for iCycle iQ_ Real Time PCR Detection System software from Bio-Rad with a method derived from the algorithms previously outlined [55].

A similar procedure was adopted for the qRT-PCR analysis of Arabidopsis samples. The list of primers used for both tomato and Arabidopsis qRT-PCR analysis is attached as Supplementary data (Additional files 9 and 10). Statistical differences in target gene transcripts were evaluated using unpaired Student's *t*-test $p \leq 0.05$ to compare ΔCt values.

In vitro isopentenyltransferase activity assay

SlIPT3 and *SlIPT4* coding sequence were respectively amplified by PCR with the following primers: SlIPT3F (5′- GGATCCTAATACGACTCACTATAGGGAACAG CCACCATGAATATTGTGTTACAACATATT-3′) and SlIPT3R (5′-TTTTTTTTTTTTTTTTTTTTTTTTTTT TTTTTTAGTGCGTCATAGTAGCAA-3′) SlIPT4F (5′-GGATCCTAATACGACTCACTATAGGGAACAGCCAC CATGATTGGCATGATGAACTCT -3′), SlIPT4R (5′-TTTTTTTTTTTTTTTTTTTTTTTTTTTTTTTTTAAT AGTGAGATGCTGCTG-3′). The forward primer included a sequence of the T7 RNA polymerase promoter, while the reverse primer contained a stop codon and a polyadenylated extremity. Generated amplicons were subcloned into pGEMT-T Easy vector (Promega) before sequencing. 800 ng of purified PCR product were used for *in vitro* protein translation in rabbit reticulocyte cells (TnT® T7 Quick Coupled Transcription/Translation System, Promega), following the manufacturer's instructions, in a final reaction volume of 50 μL.

SlIPT3 and SlIPT4 were examined for DMAPP:[^{3}H] AMP, DMAPP: [^{3}H]ADP and DMAPP: [^{3}H]ATP isopentenyltransferase activity using an adapted method [9,32]. The reaction mixture, consisting of 50 μL protein produced *in vitro* as described above, 100 μL of salts and buffer containing 37.50 mM KCl, 5 mM $MgCl_2$ and 12.5 mM Tris–HCl (pH 7.5) and 100 μL of substrates DMAPP (34 nmol) plus [^{3}H]AMP, [^{3}H]ADP or [^{3}H]ATP (each 1 nmol, 20 Ci/mmol) was incubated at 30°C for 1 h, 2 h, 8 h, 24 h and 48 h. Negative controls included either a crude extract of non-transformed rabbit reticulocyte cells or no cells at all. To stop the reaction and to precipitate the SlIPT3 or SlIPT4 proteins, cold ethanol (150 μL) was added into an aliquot of 50 μL of reaction mixture at the indicated times. The samples were vortexed and stored at −20°C for 2 h. The mixture was then centrifuged for 25 min, 20 000 g at 4°C. The supernatant was recovered in clean tubes and evaporated in a vacuum concentrator (Alpha RVC, Christ) to dryness. The pellet was resuspended in 50 μL of 5% MeOH. Each sample was analyzed by HPLC (Perkin Elmer) coupled to a radioactivity flow detector (Ramona 2000, Raytest). The radioactive metabolites were identified on the basis of comparison of their retention times with authentication standards. IPT activity was determined in four independent experiments, which showed the same tendencies although with different absolute values (Additional files 11). Therefore the results of one representative experiment are presented.

Cytokinins extraction and quantification

Endogenous CKs were extracted from homogenized young fully expanded tomato leaves (15 mg of dry weight) and 17 DAS Arabidopsis plants (150 mg of fresh weight) using an extraction buffer consisting of methanol/formic acid/water (15/1/4, v/v/v) according to the published method [56]. Following the addition of stable isotope labeled internal standards (10 pmols), samples were extracted for 1 h at −20°C. The solids were separated by centrifugation (20 000 g, 20 min, 4°C) with a subsequent collection of supernatants. The pellets were re-extracted with an additional 0.5 mL of extraction buffer (30 min at −20°C). The supernatants were collected after re-centrifugation, and incubated 30 min at −80°C. Samples were evaporated in a vacuum concentrator (Alpha RVC, Christ), re-dissolved in 0.5 mL of 1 M formic acid and applied to a mixed mode reversed phase-cation exchange SPE column (Oasis-MCX, Waters).

The CK fraction was sequentially eluted with 0.35 M NH_4OH in 60% methanol. This fraction was evaporated to dryness in a vacuum concentrator and dissolved in 5% MeOH. An aliquot (0.01 mL) from each sample was separately analyzed on a high-performance liquid chromatography (HPLC) (Ultimate 3000, Dionex) coupled to a hybrid triple quadrupole/linear ion trap mass spectrometer (3200 Q TRAP, Applied Biosystems) using a multilevel calibration graph with [^{2}H]-labeled internal standards as described before [57,58]. Data are presented as mean ± standard error.

CKX activity assay *in vitro*

CKX enzymes were extracted and partially purified from the same tomato leaves samples used for hormones extraction (WT and *35S::SlIPT3*) according to a published reference [59]. CKX activity was determined by *in vitro* assays based on the conversion of [2-^{3}H]iP (prepared by the Isotope Laboratory, IEB ASCR, Prague, Czech Republic) to [^{3}H]adenine and expressed as pmol adenine mg protein^{-1} h^{-1}. The CKX activity was determined in two biological replicates for each variant.

Photosynthetic pigments and ion extraction, quantification and analysis

The content of photosynthetic pigments (Chl *a* and Chl *b*, β-carotene, lutein, neoxanthin, violaxanthin, zeaxanthin and antheraxanthin) was determined in acetone extracts made from the lyophilized developing leaves of 4 weeks-old plants analyzed by a high-performance liquid chromatography (ECOM, Czech Republic). The analysis was made using a reversed phase column (Watrex Nucleosil 120 5 C18, 5 μm particle size, 125 × 4 mm, ECOM, Czech Republic). The solvent system comprised acetonitrile/methanol/water (80/12/10, v/v/v) followed by methanol/ethylacetate (95/5, v/v), the total analysis time was 25 min, and the linear gradient was run from 2 to 6 min (the flow rate was 1 cm^{3} min^{-1}, the detection wavelength was 445 nm). Data were captured and calculated by PC-software Clarity (DataApex, Czech Republic). The photosynthetic pigments were determined in two biological replicates of each variant measured independently by two times.

Shoot mineral quantification was conducted on the same samples as for pigments analysis using an atomic absorption spectrometer (Thermo Scientific ICE3300) as described previously [38].

Availability of supporting data

Phylogenetic data are available in Treebase (www.treebase.org) database under submission identity 16841.

Additional files

Additional file 1: Phenotype of 5 DAS Arabidopsis WT, *ipt3* KO, and *SlIPT3* or *SlIPT4* complemented plants cultivated on control medium.

Additional file 2: Endogenous CKs content (pmol/g FW) of 17 DAS Arabidopsis *ipt3* plants complemented with *SlIPT3* or *SlIPT4* and grown on control medium. The system of abbreviations was adopted and modified according to published reference [60].

Additional file 3: Endogenous CKs content (pmol/g FW) of 17 DAS Arabidopsis *ipt3* plants complemented with *SlIPT3* or *SlIPT4* and grown on salt medium (100 mM NaCl). The system of abbreviations was adopted and modified according to published reference [60].

Additional file 4: Phenotype of T1-generation tomatoes *35::SlIPT3* Line 1 (L1) transformants.

Additional file 5: Endogenous CKs content (pmol/g FW) in young leaves of T0 *35S::SlIPT3* tomatoes. The system of abbreviations was adopted and modified according to published reference [60].

Additional file 6: Cytokinin oxidase/dehydrogenase (CKX) activity in *35S::SlIPT3* tomato young leaves (L6 and L7).

Additional file 7: The content of photosynthetic pigments (μg/g DW) in 4 weeks-old tomato developing leaves of WT and T2-generation of *35S::SlIPT3* (L1) tomato plants cultivated on the control and salt growth media. DEPS (De-epoxidation state) is calculated according to formula: DEPS = (0.5*A + Z)/(V + A + Z); DEPS (De-epoxidation state per chlorophyll) is calculated according to formula: DEPSC = (0.5*A + Z)/(Chl *a* + *b*). ^ expressed in relative unit. *Data represent means and SD of two replicates. *significantly different datasets from WT (unpaired Student's *t*-test, P ≤ 0.05).

Additional file 8: Hypothetical scheme of *SlIPT3* and *SlIPT4* response to early salt stress and the feedback regulation by exogenous CK treatment in tomato plants. After salt stress (150 mM NaCl) treatment, immediate down-regulation of *SlIPT3* and *SlIPT4* transcripts with a subsequent up-regulation in tomato vegetative organs was determined. Repression of both genes after exogenous application of *t*Z demonstrated the complexity of CK networks, while the overabundance of CKs (specially iP7G) in *35S::SlIPT3* may ensure a stronger phenotype under salt stress (100 mM NaCl) conditions.

Additional file 9: Sequences of tomato primers used for qRT-PCR analysis.

Additional file 10: Sequences of Arabidopsis primers used for qRT-PCR analysis.

Additional file 11: *In vitro* enzymatic activity of SlIPT3 and SlIPT4. Conversion of non-labeled AMP added into the water and cells containing SlIPT3 (A) or cells containing SlIPT4 (B) measured by IP-HPLC with UV detector. Conversion of radioactive labeled precursors AMP, ADP and ATP added into the water and cells containing SlIPT3 or SlIPT4 were measured by LCMS (C).

Competing interests
The authors declare that they have no competing interests.

Authors' contributions
EZ and IH designed the experiments and wrote the manuscript. KH drafted the manuscript and helped with figure preparation. PID and YM performed or assisted in the execution of specific experiments. PH helped with the statistical analyses. DH and DP provided help with photosynthetic pigments analyses. VM and SL guided the study and revised the manuscript.
All authors read and approved the manuscript.

Acknowledgements
The authors thank Dr Miroslav Kamínek for critical reading of manuscript, Dr Said Hafidh and Dr James Stones for language editing and Marie Korecká for excellent technical support. We also thank Abdelmounaim Errachid for his help with confocal microscope, and Dr Etienne G. Danchin for his help with phylogenetic analyses.
This work was supported by the Czech Science Foundation [P506/11/0774]; the Fonds National de la Recherche [FNRS; convention n° 1.5117.11]; and the Wallonie-Bruxelles International program [Rhéa 2011/35047 and WBI/14-3].

Author details
[1]Institute of Experimental Botany, Academy of Sciences of the Czech Republic, Prague 165 02, Czech Republic. [2]Département Sciences du vivant, Centre wallon de Recherches Agronomiques, Gembloux B-5030, Belgium. [3]Department of Biomedical Informatics, Faculty of Biomedical Engineering, Czech Technical University in Prague, Kladno 272 01, Czech Republic. [4]Groupe de Recherche en Physiologie Végétale (GRPV), Earth and Life Institute - Agronomy (ELI-A), Université catholique de Louvain (UCL), Louvain-la-Neuve 1348, Belgium. [5]Institut National de la Recherche Agronomique, Institut Sophia Agrobiotech (ISA), UMR INRA 1355, CNRS 7254, Université de Nice-Sophia Antipolis, 400 route des Chappes, BP167, F-06903 Sophia-Antipolis Cedex, France.

References

1. Cuartero J, Fernández-Muñoz R. Tomato and salinity. Sci Horticult. 1999;78:83–125.
2. Albacete A, Ghanem ME, Martínez-Andújar C, Acosta M, Sánchez-Bravo J, Martínez V, et al. Hormonal changes in relation to biomass partitioning and shoot growth impairment in salinized tomato (*Solanum lycopersicum* L.) plants. J Exp Bot. 2008;59:4119–31.
3. Ghanem ME, Albacete A, Smigocki AC, Frébort I, Pospísilová H, Martínez-Andújar C, et al. Root-synthesized cytokinins improve shoot growth and fruit yield in salinized tomato (*Solanum lycopersicum* L.) plants. J Exp Bot. 2011;62:125–40.
4. Ghanem ME, Albacete A, Martínez-Andújar C, Acosta M, Romero-Aranda R, Dodd IC, et al. Hormonal changes during salinity-induced leaf senescence in tomato (*Solanum lycopersicum* L.). J Exp Bot. 2008;59:3039–50.
5. Cueno ME, Imai K, Ochiai K, Okamoto T. Cytokinin dehydrogenase differentially regulates cytokinin and indirectly affects hydrogen peroxide accumulation in tomato leaf. J Plant Physiol. 2012;169:834–8.
6. Spíchal L. Cytokinins – recent news and views of evolutionally old molecules. Funct Plant Biol. 2012;39:267–84.
7. Mok DW, Mok MC. Cytokinin metabolism and action. Annu Rev Plant Physiol Plant Mol Biol. 2001;52:89–118.
8. Gajdošová S, Spíchal L, Kamínek M, Hoyerová K, Novák O, Dobrev PI, et al. Distribution, biological activities, metabolism, and the possible function of *cis*-zeatin-type cytokinins in plants. J Exp Bot. 2011;62:2827–40.
9. Kakimoto T. Identification of plant cytokinin biosynthetic enzymes as dimethylallyl diphosphate:ATP/ADP isopentenyltransferases. Plant Cell Physiol. 2001;42:677–85.
10. Takei K, Sakakibara H, Sugiyama T. Identification of genes encoding adenylate isopentenyltransferase, a cytokinin biosynthesis enzyme, in *Arabidopsis thaliana*. J Biol Chem. 2001;276:26405–10.
11. Vyroubalová S, Václavíková K, Turecková V, Novák O, Smehilová M, Hluska T, et al. Characterization of new maize genes putatively involved in cytokinin metabolism and their expression during osmotic stress in relation to cytokinin levels. Plant Physiol. 2009;151:433–47.
12. Matsuo S, Kikuchi K, Fukuda M, Honda I, Imanishi S. Roles and regulation of cytokinins in tomato fruit development. J Exp Bot. 2012;63:5569–79.
13. Smigocki AC, Owens LD. Cytokinin gene fused with a strong promoter enhances shoot organogenesis and zeatin levels in transformed plant cells. Proc Natl Acad Sci USA. 1988;85:5131–5.
14. Smart CM, Scofield SR, Bevan MV, Dyer TA. Delayed leaf senescence in tobacco plants transformed with *tmr*, a gene for cytokinin production in *Agrobacterium*. Plant Cell. 1991;3:647–56.
15. Ebinuma H, Sugita K, Matsunaga E, Yamakado M. Selection of marker-free transgenic plants using the isopentenyl transferase gene. Proc Natl Acad Sci USA. 1997;94:2117–21.
16. Sýkorová B, Kurešová G, Daskalova S, Trčková M, Hoyerová K, Raimanová I, et al. Senescence-induced ectopic expression of the *A. tumefaciens ipt* gene in wheat delays leaf senescence, increases cytokinin content, nitrate influx, and nitrate reductase activity, but does not affect grain yield. J Exp Bot. 2008;59:377–87.
17. Miyawaki K, Tarkowski P, Matsumoto-Kitano M, Kato T, Sato S, Tarkowska D, et al. Roles of Arabidopsis ATP/ADP isopentenyltransferases and tRNA isopentenyltransferases in cytokinin biosynthesis. Proc Nat Acad Sci USA. 2006;103:16598–603.
18. Ma QH. Genetic engineering of cytokinins and their application to agriculture. Crit Rev Biotechnol. 2008;28:213–32.
19. Peleg Z, Apse MP, Blumwald E. Engineering salinity and water-stress tolerance in crop plants: getting closer to the field. Adv Bot Res. 2011;57:405–28.
20. Zalabák D, Pospíšilová H, Šmehilová M, Mrízová K, Galuszka P. Genetic engineering of cytokinin metabolism: prospective way to improve agricultural traits of crop plants. Biotechnol Adv. 2013;31:97–117.
21. Nishiyama R, Le DT, Watanabe Y, Matsui A, Tanaka M, Seki M, et al. Transcriptome analyses of a salt-tolerantcytokinin-deficient mutant reveal differential regulation of salt stress response by cytokinin deficiency. PLoS One. 2012;7:e32124.
22. Liu YD, Yin ZJ, Yu JW, Li J, Wei HL, Han XL, et al. Improved salt tolerance and delayed leaf senescence in transgenic cotton expressing the Agrobacterium IPT gene. Biol Plant. 2012;56:237–46.
23. Qiu W, Liu M, Qiao G, Jiang J, Xie L, Zhuo R. An isopentenyl transferase gene driven by the stress-inducible rd29A promoter improves salinity stress tolerance in transgenic tobacco. Plant Mol Biol Rep. 2012;30:519–28.
24. Hwang I, Sheen J, Müller B. Cytokinin signaling networks. Annu Rev Plant Biol. 2012;63:353–80.
25. Mason MG, Jha D, Salt DE, Tester M, Hill K, Kieber JJ, et al. Type-B response regulators ARR1 and ARR12 regulate expression of AtHKT1;1 and accumulation of sodium in Arabidopsis shoots. Plant J. 2010;64:753–63.
26. The Tomato Genome Consortium. The tomato genome sequence provides insights into fleshy fruit evolution. Nature. 2012;485:635–41.
27. Tamura K, Peterson D, Peterson N, Stecher G, Nei M, Kumar S. MEGA5: Molecular evolutionary genetics analysis using maximum likelihood, evolutionary distance, and maximum parsimony methods. Mol Biol Evol. 2011;28:2731–9.
28. Nakai K, Horton P. PSORT: a program for detecting sorting signals in proteins and predicting their subcellular localization. Trends Biochem Sci. 1999;24:34–6.
29. Kasahara H, Takei K, Ueda N, Hishiyama S, Yamaya T, Kamiya Y, et al. Distinc isoprenoid origins of cis- and trans-zeatin biosyntheses in Arabidopsis. J Biol Chem. 2004;279:14049–54.
30. Galichet A, Hoyerová K, Kamínek M, Gruissem W. Farnesylation directs AtIPT3 subcellular localization and modulates cytokinin biosynthesis in Arabidopsis. Plant Physiol. 2008;146:1155–64.
31. Zhu YD, Jin YS, Wei S, Li H, Zhang W. Functional analysis of the isopentenyltransferase gene *MdIPT3a* from apple (*Malus pumila* Mill.). J Horticultural Sci. Biotech. 2012;87:478–84.
32. Brugière N, Humbert S, Rizzo N, Bohn J, Habben JE. A member of the maize isopentenyl transferase gene family, Zea mays isopentenyl transferase 2 (ZmIPT2), encodes a cytokinin biosynthetic enzyme expressed during kernel development. Cytokinin biosynthesis in maize. Plant Mol Biol. 2008;67:215–29.
33. Medford JI, Horgan R, El-Sawi Z, Klee HJ. Alterations of endogenous cytokinins in transgenic plants using a chimeric isopentenyl transferase gene. Plant Cell. 1989;1:403–13.
34. Armstrong DJ. Cytokinin oxidase and the regulation of cytokinin degradation. In: Mok DWS, Mok MC, editors. Cytokinins: Chemistry, Activity and the Function. Boca Raton, FL: CRC Press; 1994. p. 139–54.
35. Nishiyama R, Watanabe Y, Fujita Y, Le DT, Kojima M, Werner T, et al. Analysis of cytokinin mutants and regulation of cytokinin metabolic genes reveals important regulatory roles of cytokinins in drought, salt and abscisic acid responses, and abscisic acid biosynthesis. Plant Cell. 2011;23:2169–83.
36. Liu Z, Lv Y, Zhang M, Liu Y, Kong L, Zou M, et al. Identification, expression, and comparative genomic analysis of the IPT and CKX gene families in Chinese cabbage (Brassica rapa ssp. pekinensis). BMC Genomics. 2013;14:594.
37. Higo K, Ugawa Y, Iwamoto M, Korenaga T. Plant *cis*-acting regulatory DNA elements (PLACE) database:1999. Nucleic Acids Res. 1999;27:297–300.
38. Hichri I, Muhovski Y, Žižkova E, Dobrev PI, Franco-Zorilla JM, Solano R, et al. The *Solanum lycopersicum* Zinc Finger2 cysteine-2/histidine-2 repressor-like transcription factor regulates development and tolerance to salinity in tomato and Arabidopsis. Plant Physiol. 2014;164:1967–90.
39. Peleg Z, Reguera M, Tumimbang E, Walia H, Blumwald E. Cytokinin mediated source/sink modifications improve drought tolerance and increase grain yield in rice under water-stress. Plant Biotech J. 2011;9:747–58.
40. Qin H, Gu Q, Zhang J, Sun L, Kuppu S, Zhang Y, et al. Regulated expression of an isopentenyltransferase gene (*IPT*) in peanut significantly improves drought tolerance and increases yield under field conditions. Plant Cell Physiol. 2011;52:1904–14.
41. Zhang P, Wang WQ, Zhang GL, Kaminek M, Dobrev P, Xu J, et al. Senescence-inducible expression of isopentenyl transferase extends leaf life, increases drought stress resistance and alters cytokinin metabolism in cassava. J Integr Plant Biol. 2010;52:653–69.
42. Rivero RM, Kojima M, Gepstein A, Sakakibara H, Mittler R, Gepstein S, et al. Delayed leaf senescence induces extreme drought tolerance in a flowering plant. Proc Natl Acad Sci USA. 2007;104:19631–6.
43. Guo JC, Duan RJ, Hu XW, Li KM, Fu SP. Isopentenyl transferase gene (*ipt*) downstream transcriptionally fused with gene expression improves the growth of transgenic plants. Transgenic Res. 2010;19:197–209.
44. Macková H, Hronková M, Dobrá J, Turečková V, Novák O, Lubovská Z, et al. Enhanced drought and heat stress tolerance of tobacco plants with ectopically enhanced cytokinin oxidase/dehydrogenase gene expession. J Exp Bot. 2013;64:2805–15.

45. Tran LS, Urao T, Qin F, Maruyama K, Kakimoto T, Shinozaki K, et al. Functional analysis of AHK1/ATHK1 and cytokinin receptor histidine kinases in response to abscisic acid, drought, and salt stress in Arabidopsis. Proc Natl Acad Sci USA. 2007;104:20623–8.
46. Werner T, Nehnevajova E, Köllmer I, Novák O, Strnad M, Krämer U, et al. Root-specific reduction of cytokinin causes enhanced root growth, drought tolerance, and leaf mineral enrichment in Arabidopsis and tobacco. Plant Cell. 2010;22:3905–20.
47. Ashraf M, Harris PJC. Photosynthesis under stressful environments: an overview. Photosynthetica. 2013;51:163–90.
48. Doganlar ZB, Demir K, Basak H, Gul I. Effects of salt stress on pigment and total soluble protein contents of three different tomato cultivars. Afr J Agric Res. 2010;5:2056–65.
49. Karimi M, Inzé D, Depicker A. GATEWAY vectors for Agrobacterium-mediated plant transformation. Trends Plant Sci. 2002;7:193–5.
50. Jyothishwaran G, Kotresha D, Selvaraj T, Srideshikan SM, Rajvanshi PK, Jayabaskaran C. A modified freeze-thaw method for efficient transformation of *Agrobacterium tumefaciens*. Current Sci. 2007;93:770–2.
51. Ellul P, Garcia-Sogo B, Pineda B, Ríos G, Roig LA, Moreno V. The ploidy level of transgenic plants in Agrobacterium-mediated transformation of tomato cotyledons (*Lycopersicon esculentum* Mill.) is genotype and procedure dependent. Theor Appl Genet. 2003;106:231–8.
52. Fulton TM, Chunwongse J, Tanksley SD. Microprep protocol for extraction of DNA from tomato and other herbaceous plants. Plant Mol Biol Rep. 1995;13:207–9.
53. Clough SJ, Bent AF. Floral dip: a simplified method for agrobacterium-mediated transformation of *Arabidopsis thaliana*. Plant J. 1998;16:735–43.
54. Hichri I, Heppel SC, Pillet J, Léon C, Czemmel S, Delrot S, et al. The basic helix-loop-helix transcription factor MYC1 is involved in the regulation of the flavonoid biosynthesis pathway in grapevine. Mol Plant. 2010;3:509–23.
55. Vandesompele J, De Preter K, Pattyn F, Poppe B, Van Roy N, De Paepe A, et al. Accurate normalization of real-time quantitative RT–PCR data by geometric averaging of multiple internal control genes. Genome Biol. 2002;3:Research0034.
56. Dobrev PI, Kamínek M. Fast and efficient separation of cytokinins from auxin and abscisic acid and their purification using mixed-mode solid-phase extraction. J Chromatogr A. 2002;950:21–9.
57. Dobrev PI, Vankova R. Quantification of abscisic acid, cytokinin, and auxin content in salt-stressed plant tissues. In: Shabala S, Cuin TA, editors. Plant salt tolerance: methods and protocols. Methods in molecular biology, vol. 913. New York: Springer Science + Business Media; 2012. p. 251–61.
58. Djilianov DL, Dobrev PI, Moyankova DP, Vankova R, Georgieva DT, Gajdošová S, et al. Dynamics of endogenous phytohormones during desiccation and recovery of the resurrection plant species *Haberlea rhodopensis*. J Plant Growth Regul. 2013;32:564–74.
59. Motyka V, Vaňková R, Čapková V, Petrášek J, Kamínek M, Schmülling T. Cytokinin-induced upregulation of cytokinin oxidase activity in tobacco includes changes in enzyme glycosylation and secretion. Physiol Plant. 2003;117:11–21.
60. Kamínek M, Březinová A, Gaudinová A, Motyka V, Vaňková R, Zažímalová E. Purine cytokinins: a proposal of abbreviations. Plant Growth Regul. 2000;32:253–6.

2

Transcriptomic changes due to water deficit define a general soybean response and accession-specific pathways for drought avoidance

Jin Hee Shin†, Justin N Vaughn†, Hussein Abdel-Haleem, Carolina Chavarro, Brian Abernathy, Kyung Do Kim, Scott A Jackson and Zenglu Li*

Abstract

Background: Among abiotic stresses, drought is the most common reducer of crop yields. The slow-wilting soybean genotype PI 416937 is somewhat robust to water deficit and has been used previously to map the trait in a bi-parental population. Since drought stress response is a complex biological process, whole genome transcriptome analysis was performed to obtain a deeper understanding of the drought response in soybean.

Results: Contrasting data from PI 416937 and the cultivar 'Benning', we developed a classification system to identify genes that were either responding to water-deficit in both genotypes or that had a genotype x environment (*GxE*) response. In spite of very different wilting phenotypes, 90% of classifiable genes had either constant expression in both genotypes (33%) or very similar response profiles (*E* genes, 57%). By further classifying *E* genes based on expression profiles, we were able to discern the functional specificity of transcriptional responses at particular stages of water-deficit, noting both the well-known reduction in photosynthesis genes as well as the less understood up-regulation of the protein transport pathway. Two percent of classifiable genes had a well-defined *GxE* response, many of which are located within slow-wilting QTLs. We consider these strong candidates for possible causal genes underlying PI 416937's unique drought avoidance strategy.

Conclusions: There is a general and functionally significant transcriptional response to water deficit that involves not only known pathways, such as down-regulation of photosynthesis, but also up-regulation of protein transport and chromatin remodeling. Genes that show a genotypic difference are more likely to show an environmental response than genes that are constant between genotypes. In this study, at least five genes that clearly exhibited a genotype *x* environment response fell within known QTL and are very good candidates for further research into slow-wilting.

Keywords: Drought stress, Canopy-wilting, Glycine max, RNA-Sequencing, Quantitative trait loci (QTL), Genotype x environment

Background

Soybean is a primary contributor to worldwide food production. Water deficit dramatically limits growth and yield in crop plants, particularly for soybean, and the problem will likely be exacerbated by climate change. Irrigation is costly and often not a viable option for many soybean farmers. According to the USDA Economic Research Service report, only 8% of the U.S. soybean acreage is irrigated (http://www.ers.usda.gov/). Therefore, the development of drought-tolerant cultivars is critical in order to reduce the impact of drought stress on soybean production.

From a soybean breeding perspective, cultivar development is limited by the narrow diversity of elite germplasm, particularly with regard to drought tolerance [1]. Fortunately, a small number of land-races exhibit drought tolerance. One Japanese lace-race, PI 416937, retains yields in spite of drought [2] and was initially identified due to its slow-wilting phenotype. Further physiological characterization showed that PI 416937

* Correspondence: zli@uga.edu
†Equal contributors
Center for Applied Genetic Technologies & Department of Crop and Soil Science, University of Georgia, Athens, GA 30602, USA

has lower stomatal conductance [3], constant transpiration rate under vapor pressure deficit (VPD) above 2.0 kPa [4], and lower radiation use efficiency [5].

VPD is the difference between the water-vapor pressure in the air and the vapor pressure at which water-vapor condenses. At low VPD, dew forms, and, as VPD rises, plants transpire due to evaporation from the stomata. Interestingly, PI 46937 initially exhibits a conventional, linear increase in transpiration rate in response to VPD; yet, as the VPD continues to rise, the transpiration rate of PI 46937 stabilizes - a response that differentiates it from elite cultivars [4]. Transpiration rate is reduced within 40 minutes after exposure to cycloheximide, a bacterially-derived compound which inhibits protein translation [6]. This result indicates that symplastic/transcellular water pathway is maintained by continuous protein turnover. One explanation for PI 416937's unique response to increased VPD is that the transcription of proteins mediating transpiration rate is being modulated relative to elite cultivars. To examine this possibility, we used deep sequencing of mRNAs (RNA-seq) to assay the transcriptomic response to water deficit in both PI 416937 and Benning, a common drought-sensitive cultivar.

Plant breeders are interested in identifying genes that confer drought-tolerance that can then be used for marker assisted selection. Since drought-tolerance is a highly complex trait, a whole-genome perspective is required. Still, previous attempts to understand drought tolerance using whole-genome transcript profiles often relied on the relative difference in pre- versus post-drought conditions for a single genotype [7]. Observing the final product of an elaborate chain of transcriptional events does not easily translate to either a better understanding of the plant's responses or to improved plant varieties. One way to focus the search for useful drought tolerance genes is to compare differential expression of genes between genotypes that exhibit varying levels of drought tolerance. Indeed, this has been done previously in soybean for a relatively uncharacterized soybean variety [8]. While this study hinted genetic mechanisms that may confer drought resistance, the resistant variety used had not been extensively characterized in terms of its physiological response to water deficit, thus limiting the ability to connect genetic and physiological pathways. The study also illustrated the analytical difficulties of emphasizing only pairwise differences for samples that range across genotypes and environmental conditions. Here we apply a classification system to categorize genes based on the combination of genotypic and environmental response data. This approach allowed us to differentiate gene expression patterns that characterize a general soybean response from patterns that may be conferring PI 416937's distinct transpiration rate profile. An additional benefit of comparing PI 416937 and Benning transcriptional profiles is that they are the parents for a mapping population previously used to identify slow-wilting QTL [9]; thus, genotypic differences in expression could be correlated to genetic polymorphisms segregating between the two lines.

Results

PI 416937 exhibits a slow-wilting phenotype

As described in Methods, to create rapid water deficit, each genotype was gently removed from soil, washed, and exposed to constant ambient air for the remainder of the experiment. After 6 and 12 h of drying treatment, both genotypes did not show differences in wilting phenotype (Figure 1). However, the slow wilting genotype PI 416937 still maintained its shape whereas the fast wilting Benning was wrinkled and wilted after 24 hr of drying, clearly representing different levels of drought avoidance between two genotypes. After 36 hr, genotype PI 416937 also showed a wilting phenotype and Benning showed severe leaf curling.

Transcriptome data for sensitive and tolerant soybean genotypes is highly reproducible

A total of 24 samples comprised of two soybean genotypes without drying treatment (controls, 0 h) and imposed drought stress (6, 12, and 24 hr) were used for transcriptome sequencing using Illumina HiSeq2000 system (Table 1). One library of PI 416937 6 hr replicate 3 was lost during library preparation procedure, thus PI 416937 replicate 2 was sequenced twice. Hiseq 2000 sequencing resulted in from 9.5 million (M) to 26.4 M reads per sample. The reads for each biological replicate were mapped independently to the reference genome. There were no genes with significant differences at the transcriptional level between PI 416937 6 hr replicate 2 analyzed in two different lanes, showing that the sequencing reaction and subsequent analysis introduced very little error (Additional file 1). Moreover, across biological replicates, the number of gene models with no significant difference ranged from 99.10% and 99.98% (Additional file 1), indicating high reproducibility.

PI 416937 and Benning have similar transcriptional response to water deficit but exhibit numerous genotypic differences

We attempted to combine data across genotypes and time-points in order to classify these expression profiles of expression into biologically relevant categories. Our categories were based on varying degrees of genotypic versus environmental responses (Table 2 and [10]). Generally, the classification system took into account the coefficient of variation across time-points as well as the statistical significance as assessed by *cuffdiff* (see Methods).

Figure 1 Phenotypic response of Benning (sensitive) and PI 416937 (tolerant) soybeans after 0, 6, 24, and 36 hours of drying treatment. Genotypes are shown as rows and time-points as columns. For 0 hr, leaflets at their widest point measured ~5 cm and ~7.5 cm for Benning and PI 416937, respectively.

Figure 2 illustrates gene expression profiles and their classification. *G-only* genes differed by genotype, but were relatively constant with regard to environmental change. *E-only* genes showed similar levels for both genotypes at individual time-points, but varied between time-points. *G + E* genes had both a genotypic difference and an environmental response. For *GxE* genes, the genetic background conditioned the environmental response. *GxE* genes had highly variable differences between the two genotypes at different time-points; for example, a *GxE* gene might have a $\log^2$ ratio $FPKM_{Benning}$ to $FPKM_{PI\text{-}416937}$ of 1.2 at 6 hr, but a difference of 3.3 at 12 hr. These *GxE* were particularly interesting because they suggest the genes that might be mediating phenotypic differences in wilting response. Because of the highly stringent criteria used to define the above categories, there were many cases where ambiguous gene expression profiles clearly exhibited a response, but were undefined. We further categorized these genes depending on whether they exhibited an extreme environmental or genotypic response for at least one time point. These are defined with the *ambiguous* suffix in Table 2 and Figure 2 (see Methods for more details).

Table 1 Total read counts for treatments, genotypes, and replicates

Cultivar	Treatment	Bio Rep 1 reads	Bio Rep 2 reads	Bio Rep 3 reads	Total
Benning	0	24.8 M	19.5 M	22.8 M	67.1 M
	6	25.4 M	18.1 M	18.7 M	62.2 M
	12	17.8 M	20.5 M	20.9 M	59.2 M
	24	21.0 M	26.5 M	15.4 M	62.9 M
PI 416937	0	22.6 M	20.5 M	26.4 M	69.5 M
	6	19.2 M	18.6 M	18.1 M[a]	54.1 M
	12	20.9 M	25.6 M	16.4 M	62.9 M
	24	15.4 M	21.6 M	12.7 M[b]	37.0 M

[a]PI 416937 was sequenced twice.
[b]PI 416937 Bio Rep 3 was an outlier relative to Rep1 and Rep2, thus excluded.

A large fraction of gene models were not tested due to lack of transcript from sampled tissues and conditions (Table 2). Thirty-three percent of classifiable genes (all genes except *Untested, Low-expression,* and *Ambiguous*) were expressed at constant levels regardless of drought stress or genotype. Even with the large number of genes showing constant expression, very few exhibited a *G-only* response: 1%, [100 * (*G-only*/(*G-only* + *Constant*)]. Indeed, 96% of classifiable genes that were differentially expressed between genotypes – *G-only, GxE, G + E,* and *G + E-ambiguous* - exhibited an environmental response. Therefore, assuming the ratio of *GxE* to *G + E* genes holds for the *G + E-ambiguous* category, genotype generally appears to interact with the environment in a nonlinear way. All genes are listed along with their categories and expression profiles (Additional file 2).

E gene profiles define a general soybean response

Because we used two diverse soybean genotypes in this study, we could postulate a generic transcription response of soybean to water deficit. In order to elucidate this response, we further characterized the expression profiles of genes that showed a shared environmental response but little (*E-ambigous*) or no (*E-only*) genotypic difference (Figure 2), which we refer to as *E* genes. We formalized eight models to represent the average expression profile of these genes (Figure 3C): Up-early, in which genes were expressed to their maximum level within the first 6 hrs; Up-linear, in which genes continually increased over the time-course; and Up-late, in which genes stayed constant till the 24 hr time-point. We similarly defined a Down-early, Down-linear, and Down-late. Peak and Trough expression patterns were either up-then-down or down-then-up, respectively, across the time-course. Note that the shape of the

Table 2 Expression types for all genes in the study

Type	Interpretation	Count
Untested	Few transcripts present in any sample	19,391
Low-expression	Expression was too low to classify, but clearly present	8,488
E-only	Environmental response; gene expression levels change over the time-course, but there were no genotypic differences	9,208
E-ambiguous	Expression levels change over the time-course; genotypic differences may be present but minor.	3,619
G-only	No environmental response, but a constant genotypic difference across time-points	75
GxE	A substantial genotypic difference between two time-points; genotype is conditioning environmental response	542
G + E	An environmental response and a constant genotypic difference across time-points	84
G + E- ambiguous	Response is either a *GxE* or *G + E*, but difficult to specify which.	1,437
Constant	Expression was constant between genotypes and across time-points	7,290
Ambiguous	Expression was too erratic across replicates to classify.	3,511
Total		53,645

expression profile, not its absolute level, dictates its classification.

The fraction of up and down-regulated genes was similar (Figure 3A). Roughly half of the up-regulated genes exhibited a linear increase in expression. In contrast, the down-regulated genes were more evenly divided between early and linear responses. We additionally assessed the maximum magnitude relative to the control (0 hr) of all *E*-genes. Most genes had a range of between 1 and 3 $\log_2$ units (2 to 8-fold greater or less than 0 hr), but some exhibited very high changes in expression, on the order of 6 to 8 $\log_2$ units (Figure 3B). While there was the expected correlation between set size and range, both linearly and late down-regulated genes appear to

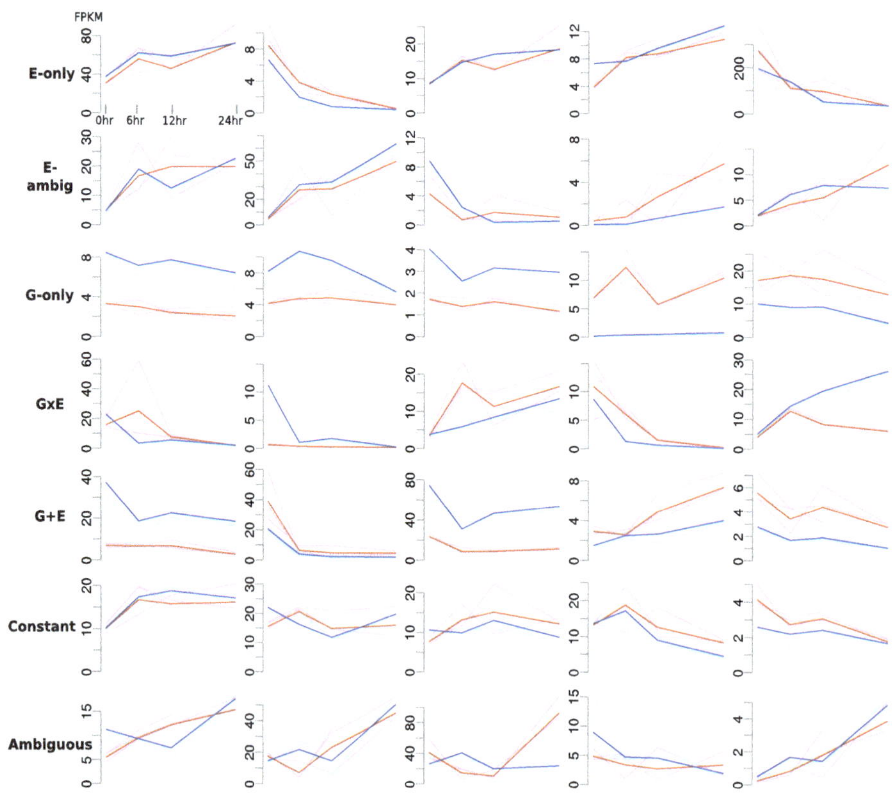

Figure 2 Classification system for gene expression profiles. Exhibited genes were randomly chosen from all genes within a category. Each row represents a single category. Blue and red colors indicate Benning and PI 416937, respectively. Light coloration indicates an individual replicate. Dark coloration indicates the mean profile across all replicates. Axis are labeled in the top right panel. Note, the scale of the y-axis differs for every plot.

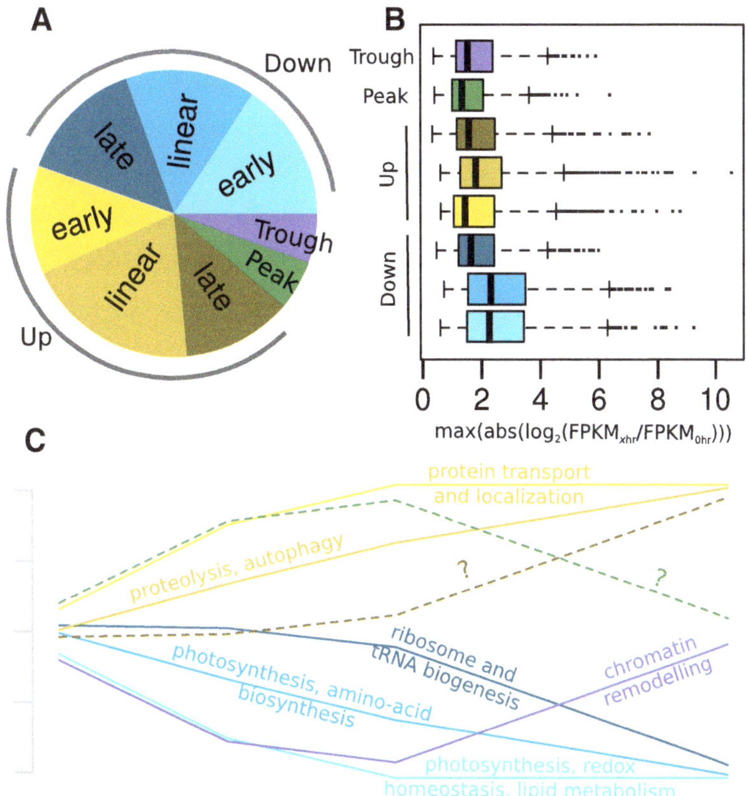

Figure 3 General soybean transcriptional response to water deficit. Color coding is consistent throughout the figure and defined in the pie chart. **(A)**, The distribution of E-type (E-only and E-ambiguous) genes are indicated as the proportion of the circle; n = 12,827. **(B)**, The maximum difference relative to 0 hr control of each gene is plotted with regard to its expression profile type. For each profile type, the mean, variance, and skewness of a distribution is estimated. Boxes indicate the middle quartile range of this distribution; lines indicate the highest and lowest quartile range. Dots indicate expression levels that extend beyond the estimated distribution. **(C)**, Expression profile models are illustrated, with functional enrichment categories labeling each profile.

change more extensively than up-regulated genes with similar profiles. Thus, on balance, the number of transcripts in the leaf should decline with time under drought.

We assessed each profile set separately for possible enrichment in functionally related genes. Using AgriGO, we found distinct and highly significant patterns of functional bias (Table 3). Indeed, the fact that these categories are quite distinct indicates that our choice to group *E* genes by expression profiles was generally valid. Genes associated with photosynthesis and lipid metabolism were rapidly reduced and remain low (Figure 3C). A distinct set of photosynthesis genes were also continually reduced across the time-course. Towards 24 hr, genes involved in translation were down-regulated, resulting in a general decline in cellular metabolism. On the other hand, protein transport genes were up-regulated rapidly and stayed at relatively high levels. As cell metabolism declined, proteolysis and autophagy genes were increasingly transcribed. No significant categories were associated with Up-late genes. This observation stands to reason as most cellular processes appeared to decline in activity as water deficit continued. Somewhat surprisingly, Peak genes also show little or no functional enrichment. Interestingly, a clear drop in the transcription of chromatin remodeling genes was observed at 6 to 12 hr; transcription returned to 0 hr levels at 24 hr time point. Both Peak and Trough genes may represent genes that are oscillating in circadian cycles, and have little to do with drought response. Chromatin remodeling genes generally appear to be constant regardless of time of day [11], suggesting that this response is a reaction to initial water deficit and downstream physiological symptoms.

We additionally assessed the GO enrichment of genes with very high-dynamic range in a category-wise fashion. These results generally bore out the functional enrichment analysis performed above, but were often less definitive (data not shown).

Genotypic differences in transcription

Given the utility of characterizing *E* gene profiles, we extended this analysis to *GxE* genes. In this case we

Table 3 GO categories significantly associated with particular *E*-type expression profiles

Type	GO accession	Description	% in group	% BG[b]	FDR
Down-early (1400[a])	GO:0015979	photosynthesis	3.29	0.63	1.8E-14
	GO:0055114	oxidation reduction	13.86	8.71	2.3E-07
	GO:0006629	lipid metabolic process	6.57	3.45	6.5E-06
	GO:0018130	heterocycle biosynthetic process	1.57	0.47	5.0E-04
	GO:0034641	cellular nitrogen compound metabolic process	4.36	2.30	5.9E-04
	GO:0045454	cell redox homeostasis	2.14	0.83	8.5E-04
Down-linear (1054)	GO:0015979	photosynthesis	3.51	0.63	3.0E-12
	GO:0006520	cellular amino acid metabolic process	4.08	1.60	2.5E-05
	GO:0043039	tRNA aminoacylation	1.52	0.35	3.0E-04
Down-late (1213)	GO:0006412	translation	20.86	3.21	4.5E-111
	GO:0034660	ncRNA metabolic process	3.30	0.85	9.0E-10
	GO:0044106	cellular amine metabolic process	4.04	1.66	3.9E-06
	GO:0015684	ferrous iron transport	0.74	0.12	2.2E-03
Up-early (952)	GO:0006886	intracellular protein transport	3.15	0.96	3.2E-05
Up-linear (1494)	GO:0044257	cellular protein catabolic process	2.61	0.61	7.3E-10
	GO:0016236	macroautophagy	0.60	0.05	4.7E-05
	GO:0015991	ATP hydrolysis coupled proton transport	1.07	0.21	5.4E-05
	GO:0007264	small GTPase mediated signal transduction	2.21	1.02	4.2E-03
Up-late (827)	No significant enrichment				
Peak (448)	No significant enrichment				
Trough (432)	GO:0006333	chromatin assembly or disassembly	2.23	0.37	1.6E-03

[a]Total number of genes within a category that have a GO annotation.
[b]Background model (BG) comprises all 29,641 soybean genes with a GO annotation.

included a ninth model, Constant, in addition to those described above. No *E* genes should be constant, so the Constant model was not applied to that group, whereas one of the two genotypes of a *GxE* gene might show constant expression across time points.

We initially characterized the relative frequency for each possible combination of environmental responses specific to each *GxE* gene (Figure 4A). The pattern observed deviates strongly from random expectation (p-value < 10^{-69}, Chi-squared test). As shown, most combinations fall along the linear axis, indicating that, even for *GxE* genes, the basic environmental response is the same, differing only by magnitude at a particular point. Indeed, there are very few examples of up-regulation in one genotype and down-regulation in another. In terms of combinations that are enriched but do not fall on the linear axis, most of these are not far from the axis, indicating that, even when expression profiles are distinct, they are not dramatically different. The most aberrant combination involves genes that are down-regulated late in Benning and show up-regulation and then down-regulation, or 'peak' profiles, in PI 416937. In examining the profiles of these genes, we found that PI 416937 genes most commonly peaked at a much higher levels than the relatively constant Benning genes. Note, this did not have to be true, as a gene could start higher in Benning than PI 416937 and then decline late as in Glyma07g01940 (Figure 4B); the absolute value of a profile is normalized by the maximum expression value, thus only the shape of the profiles are considered. Though the number was too small for robust enrichment statistics, of the seven genes that did show a sharp peak in early expression in PI 41937, such as Glyma17g05520 or Glyma07g17361, most are annotated as being transcription factors or as having some regulatory function at the protein level.

One hypothesis to explain PI 416937's slow-wilting response is that genes associated with water transport in PI 416937 have reduced expression during water deficit, thus reducing transpiration and facilitating water retention (see Introduction). Only a very small fraction of *GxE* genes that were down regulated in PI 416937 had strikingly different expression profiles in Benning. It is possible that the functionally significant changes in gene expression are not qualitative, such as differences in profile, but quantitative, as suggested by the sharp diagonal in Figure 4A. Thus, given that most *GxE* genes exhibited similar profiles, we looked for time points that were commonly differentiating the two genotypes.

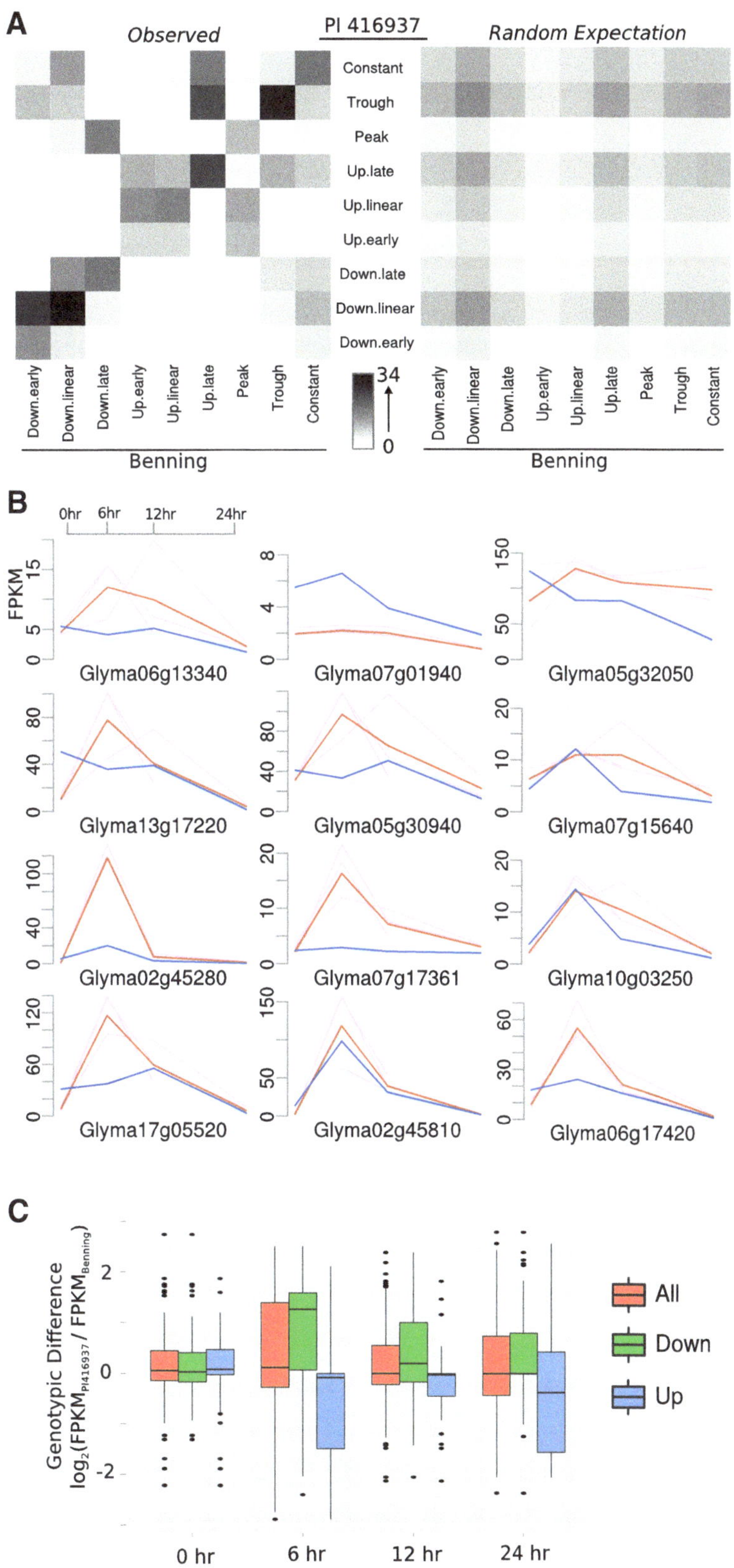

Figure 4 (See legend on next page.)

(See figure on previous page.)
Figure 4 Characteristics of response profiles of *GxE* genes. (A), Left panel shows a heat map reflecting the distribution of response profiles for all *GxE* genes in terms of their response in the two genotypes. The right panel shows the random expectation based on marginal frequencies of different profiles in the two genotypes. **(B)**, Twelve randomly sampled FPKM profiles for combinations of Peak and Down.late *GxE* profiles. Blue and red colors indicate Benning and PI 416937, respectively, as in Figure 2, where darker curves represent the mean of biological replicates shown in a lighter shade. **(C)**, Boxplots (as in Figure 3B) showing the genotypic difference at different timpoints for *GxE* genes that have the same response profiles, such as Up.late in Benning and Up.late in PI 416937. 'All' indicates both up and down-regulated genes while 'Up' and 'Down' indicate combined sets of up and down-regulated groups. The units of the y-axis are $\log_2(FPKM_{PI\ 416937}/FPKM_{Benning})$; positive values indicate that PI 416937 genes had higher expression than Benning at a given time-point.

Looking only at genes that had the same profile – that fell along the diagonal in Figure 4A – we analyzed the genotypic differences for each gene at each time-point (Figure 4C). For example, because the units of the y-axis are $\log_2(FPKM_{PI\ 416937}/FPKM_{Benning})$, positive values indicate that PI 416937 genes had higher expression than Benning genes at a given time-point. We observed that no particular time-point had a biased genotypic difference when considering all profiles regardless of profile type ('All' in Figure 4C). When we grouped genes based on up or down-regulation, we observed a small bias at the 6 hr time-point; in other words, for genes that were similarly down-regulated in both genotypes, PI 416937 genes were not down-regulated as substantially as Benning, particularly at 6 hr. It is possible that these genes represent, in effect, a delayed response to water deficit. Whether this response is causally related to resistance to wilting, or is merely a byproduct of undergoing less water deficit is unknown. The lack of any visual phenotype at this stage would suggest the former (Figure 1). Still, this observation is the opposite of what would be expected under a model in which PI 416937 differentially down-regulates expression of a subset of genes in order to reduce transpiration levels.

Genomic bias of *GxE* genes and known QTLs for slow canopy wilting

In our previous QTL study using 150 recombinant inbred lines (RILs) derived from a Benning and PI 416937 cross, seven QTL responsible for canopy wilting were identified. Of those, two and five favorable QTL alleles were found from Benning and PI 416937, respectively [9].

We compared the distributions of genes across the genome with those genes found within QTL intervals. There was no significant deviation from the expectation predicted by the genome-wide distribution (Figure 5). This finding is not surprising given that the QTL intervals are large and the majority of genes within a given interval are not expected to deviate sharply from the genome-wide distributions. Still, several genes within the known QTL have a clear *GxE* signal (Additional file 3 and Additional file 4) and are promising candidates for further investigation.

The distribution and/or expression levels of aquaporins are thought to be important in mediating PI 416937's unique response to drought [4,6]. We additionally compared the categorical distribution of aquaporins to the genome-wide expectation (Figure 5). Though the sample is small, the distribution is significantly different than background (p-val < 0.05, Chi-squared test), indicating that aquaporins are more likely to respond transcriptionally to water deficit and also that they are more likely to have genotypic differences in their response. No aquaporin genes classified as being *GxE*-type genes fell within the known QTL interval.

Discussion

Large-scale transcriptional reprogramming has long been interpreted as a mechanism of minimizing the effect of drought stress in plants [12,13]. The aim of this study was to identify a general response to drought stress in soybean and to compare differences at the transcriptional level between two accessions differing in canopy wilting phenotype. Although the drying treatment in

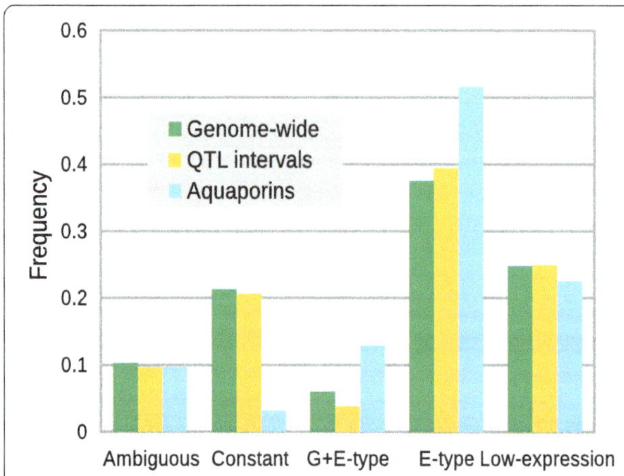

Figure 5 Categorical distribution of genes across the genome (n = 34,178), within QTLs intervals previously identified (n = 755), and among aquaporins (n = 31). Untested genes are not included in frequency calculation. Because the number of Aquaporins is small, all categories that showed a genotypic and environmental response – *GxE*, *G + E*, and *G + E-ambiguous* - were combined (*G + E*-type), as were categories that had an environmental response but no or small genotypic effects (*E*-only + *E-ambiguous* = *E*-type).

this study is far from the actual drought stress under field conditions, it allowed us to measure transcriptional responses to water deficit, a major component of drought stress.

The majority of genes that we could confidently characterize a drought response were classified as *E* genes, indicating that they had roughly identical expression patterns for both genotypes (Figure 2 and Table 2). Prior to any noticeable phenotypic effect (Figure 1), dramatic transcriptional changes were occurring in both genotypes (Figure 2). While genes that are up or down-regulated late may be due to the physiological repercussions of canopy wilting, both early and linearly responsive genes are abundant (Figure 3) and likely responding to immediate water-deficit.

The most obvious response shared by sensitive and tolerant genotypes was down-regulation of photosynthesis related genes (Figure 3). There have been contradictory observations with regard to photosynthesis under drought stress, and this discrepancy is thought to be caused by differences in the severity of stress imposed on plants [14]. When plants encountered mild or moderate drought stress, photosynthetic acclimation was observed [12,15-17]. In contrast, photosynthesis has been reported as one of the primary process to be adversely affected under severe drought [16-19]. Thus, our treatment appears to be simulating severe drought.

Another response shared by sensitive and tolerant genotypes was up-regulation of genes associated with autophagy and nutrient starvation. Autophagy is an essential protein degradation process induced by abiotic stresses such as starvation, drought, salt, pathogen, and oxidative stress [20,21]. Photosynthetic constraint is one cause of carbon starvation, and carbon starvation induces autophagy [22]. The breakdown of oxidized proteins during oxidative stress and aggregated proteins in nutrient-starved cells can ensure cellular survival by maintaining cellular energy levels [23].

Prior to autophagy-related gene up-regulation, there was a rapid increase in genes involved in protein localization (Figure 3), primarily within the vesicular trafficking pathway. To our knowledge, this has not been observed in soybean, but has some precedent in *Arabidopsis* where up-regulation of related genes promoted osmotic stress tolerance [24]. Interestingly, other reports in *Arabidopsis* have implicated the downregulation of vesicle-trafficking-related SNARE protein in salt tolerence [25]; suppression of the gene in roots suppressed the production of reactive oxygen species by preventing vesicle fusion with the tonoplast. The connection between salt and water stress is complex [18], but the above findings in conjuction with those presented here, indicate that the shoot and root are exhibiting very distinct vesicle-trafficking profiles.

Chromatin remodeling genes have an unusual Trough expression pattern in both genotypes (Figure 3 and Table 3). Chromatin regulation responses to drought, cold, and salinity stress have been described in Arabidopsis [26,27]. It was reported that the histone H3 modification correlates with gene activation of the drought stress-inducible genes, such as responsive to dehydration (RD) 29A, RD29B, and related to AP2.4 (RAP2.4) [28]. Moreover some chromatin remodeling and modifying enzymes such as histone modification enzymes, linker histone H1, and components of chromatin remodeling complex have been shown to function in plant abiotic stress responses [27]. The initial down-regulation of these genes may reflect the expansion of euchromatin associated with the major transcriptional reprogramming that is occurring even at early stages of water-deficit, while the late up-regulation counters this trend, returning much of the genome to heterochromatin, under extreme physiological stress [29].

We had strong evidence for the differential expression between genotypes for 2,138 transcripts for at least one time-point (Table 2). For 25% of these, we could say with confidence that the genotype was conditioning the environmental response (*GxE* genes in Table 2). Less than 4% of these genotypically different genes had a constant expression in both genotypes (*G-only* genes in Table 2). Note, this result is not predicted by the ratio of *Constant* to *E-only* genes (Table 2), suggesting that genes that differ between genotypes are generally disposed to be stress responsive. This stands to reason in that stress-response regimes are likely to be selected under unique local environmental conditions [14].

The three major categories enriched in *GxE* genes were photosynthesis, innate immune response, and apoptosis genes, with a FDR of $5.2E^{-06}$, $2.3E^{-07}$, and $4.9E^{-06}$, respectively. Photosynthesis genes were substantially down-regulated under drought stress in both soybeans, however, photosynthesis genes of tolerant soybean were less affected at an early stage (6 hr) of water-deficit (Additional file 5). This is supported by prior studies that showed lower decrease of net photosynthesis rate or chlorophyll content in tolerant versus sensitive genotype under salt or drought stress [26,30].

Perhaps more interesting are the innate immune response and apoptosis genes, which show dramatic differences between genotypes and across conditions. Immune response genes are also a major target of local adaptation and have been previously identified as eQTLs for differential drought response [31]. Contrary to the expectation based on *E-only* profiles, apoptotic *GxE* genes are primarily down-regulated and vary most commonly in their initial expression levels (Additional file 6), indicating that physiological responses to wilting are not manifesting these differences. Still, the biochemical connection between water-deficit and apoptotic/immune response is tenuous, and functional enrichment in *GxE* categories may reflect overlapping local adaptations to stress in general, and not drought specifically. We anticipate that further fine-mapping studies will help resolve these questions.

To that end, one motivation for this study was the prior development of a genetic mapping population generated from a cross of the two lines assayed herein [9]. We did not identify a significant relationship between genes within previously identified QTL regions and *GxE* genes (Figure 5). Additionally, the region containing the strongest QTL, qSW-Gm12, with an R^2 of 0.27 [9], did not have a significant enrichment in *GxE* or *G + E* genes (not shown). This result is not unexpected given that the QTL are not particularly well resolved and they could be mediating differences in slow-canopy wilting through any number of mechanisms [32,33]. Still, each of the QTL regions did contain *GxE* genes, and we propose these genes to be prime targets for fine-mapping, particularly those that have strikingly distinct expression profiles and act early in water deficit (Additional file 3 and Additional file 4).

The large majority of *GxE* genes exhibited quantitative differences in expression levels at particular points rather than qualitatively different profiles (Figure 4A). The exception to this trend was a small group of regulatory proteins that peaked in PI 416937 and remained relatively low and constant in Benning until 24 hr (Figure 4B). Though none of these genes fell directly within the range specified by the QTL mapping discussed above, chromosomes 5 and 17 contain QTLs nearby two of the most striking *GxE* profiles, Glyma02g45280 and Glyma17g05520. These genes are another set of promising leads in identifying solutions to problems posed by drought.

Conclusions

Drought reduces yield in all crops, particularly soybeans. The response to drought is biochemically complex and entails major changes in gene expression. To that end, genome-wide expression data can be useful in improving plants to be robust to drought. However, it is difficult for plant researchers and breeders to employ genome-wide data because the results, in isolation, are often impressionistic and the experimental design does not focus on refining genomic loci that are causally underlying phenotypic variation. Here we used two relevant breeding lines, Benning and PI 416937 that have been used previously by our group as parents in a mapping population. These two lines exhibit strikingly different wilting responses, as shown here and in previous work, and their progeny were used to identify QTL underlying the slow-canopy wilting trait. We could therefore compare genes that have strikingly different profiles between genotypes with these QTL in order to resolve those QTL further and to understand their functions. To facilitate this comparison, we also developed a computational pipeline that allowed us to characterize the transcriptional response of each gene based on observations across the entire time-course and between the two genotypes. This approach allowed us to differentiate between genes that form a shared response and those that distinguish genotypes.

Taken together, we feel this study offers the following insights: 1) There is a general and functionally significant transcriptional response to water deficit that involves not only known pathways, such as down-regulation of photosynthesis, but also up-regulation of protein transport and chromatin remodeling; 2) Genes that show a genotypic difference are more likely to show an environmental response than genes that are constant between genotypes; 3) At least five genes that clearly exhibited a *GxE* response fell within the known QTL and are very good candidates for further research into slow-canopy wilting.

Methods

Plant materials and drought stress treatment

Both Benning (drought sensitive, elite US soybean cultivar) and PI 416937 (drought tolerant, Japanese landrace) were planted in the greenhouse on June 18, 2012 with 12/12 hours light/dark regime. At the R2 stage of flowering (September 7, 2012), plants were removed from pots, roots were washed, and the whole plants exposed to air. After 0, 6, 12, and 24 hr intervals, leaves were collected from both Benning and PI 416937 with three biological replicates, frozen in liquid nitrogen and stored at –80°.

Total RNA extraction and library preparation

Tissues were ground under liquid nitrogen. The total RNA from leaf tissues was extracted using Trizol reagent (Invitrogen) and RNA-Seq libraries were prepared using TruSeq RNA Sample Prep Kits (Illumina) according to the manufacture's recommendations. RNA-Seq libraries were constructed from two genotypes, four treatment time (0, 6, 12, and 24 hr), and three biological replicates. All libraries were barcoded using 24 index adapters, quantified using Bioanalyzer DNA 1000 Chip (Agilent Technology 2100 Bioanalyzer) and normalized to 10 nM.

RNA sequencing and sequence analysis

All libraries were sequenced using the HiSeq2000 at the Genomics and Microarray Core at the University of Colorado Denver. Three lanes of HiSeq were used and each biological replicates was sequenced in different lanes according to proper blocking and randomization procedures [34]. Libraries were pooled equimolarly. Using

TopHat2 (version 2.0.8b), sequencing reads (Table 1) were aligned to the Glyma 1.1 transcriptome (54,175 genes; 73,320 transcripts) determined by the Soybean Genome Project, US Department of Energy, Joint Genome Institute [35] (http://www.phytozome.net/soybean.php). Technical replicates (PI 416937 6 h replicate 2 and 3) were merged prior to testing the transcriptome variances. Mapped reads were provided as an input to Cufflinks (version 2.1.1) for transcript assembly, and then, differential expression of genes between genotypes and between treatments were analyzed by CuffDiff [36]. After consolidation of identical genes, 53,645 genes were included in the Cuffdiff analysis. Gene abundance is given as fragments per kilobase of exon per million fragments mapped (FPKM). Differential expression statistics were based on the $\log_2$ FPKM ratios [36].

Classifier system for genotypic response profiles

All genes are classified via the following set of rules. For brevity, the following acronyms are used:

GD, Genotypic difference at a single time-point calculated as $\log_2$ FPKM ratio of Benning genotype to PI 416937. In other words, a GD > 2 indicates a more than 4-fold difference in mean genotypic expression levels at a single time-point.
ED, Environmental difference calculated as $\log_2$ FPKM ratio of 0 hr to 6, 12, or 24 hr for a given genotype. All genes have a 0 hr ED of 0.
SGD, Significant GD with q-value <0.05.
SED, Significant ED with q-value <0.05)
CV$^+$, 7 out of 8 time-points (4 time-points × 2 genotypes) had a coefficient of variation of <0.4 across the 3 replicates.
SS$^+$, Same sign for GDs; not a mixture of positive and negative GDs.

The classification key is described below as pseudocode using underlined control words 'IF', 'THEN', and 'ELSE IF':

IF No time-point in either genotype had an average FPKM of >1 **THEN** call *Untested*
ELSE IF the highest FPKM of any time-point was <4 & the average FPKM across all time-points was <2 **THEN** call *Low-expression*
ELSE IF CV$^+$ & at least one SGD was >2.8-fold the GD at another time-point THEN call *GxE*
ELSE IF CV$^+$ & ≥3 SGDs & both genotypes have ≥1 SEDs & SS$^+$ **THEN** call *G + E*
ELSE IF CV$^+$ & ≤1 SGDs & both genotypes have ≥1 SEDs **THEN** call *E-only*
ELSE IF CV$^+$ & 0 SEDs for at least one genotype & ≥3 SGDs & SS$^+$ **THEN** call *G-only*
ELSE IF CV$^+$ & 0 SEDs for at least one genotype & ≤1 SGDs & no average ED >2 **THEN** call *Constant*
ELSE IF At least one average ED >2 & ≥1 SGD had a q-value <0.001 **THEN** call *G + E-ambiguous*
ELSE IF At least one average ED >2 **THEN** call *E-ambiguous*
ELSE call *Ambiguous*

Of note, log scores are used in expression data because fold changes in expression are more meaningful than absolute changes. Yet, as commonly occurs in expression data, expression of one gene may be, for example, 0 at 0 hr and 50 at 6 hr. Though this is an important result with regard to our biological questions, it presents conceptual and, certainly mathematical, problems. For our purposes we used the following heuristic: if expression (FPKM) was <1 for one entry and >5 for the compared entry, then we gave the difference a score of 5, which represents an upper limit for the expression range across the time-course (Figure 3B). Alternatively, if the entry was <1 and the compared entry was <5, then we gave it a score of 0. An identical approach was taken for comparisons that would result in a negative score. For range calculations (Figures 3B and 4C), such values were ignored.

Defining water-deficit response curves

The shape of expression profiles were characterized further. ED values in a time-course were normalized by the maximum ED. These profiles were then compared to nine explicit models, described graphically in Figure 3C. Explicit models are given as Additional file 7. The profile was classified as the model with the lowest sum of squared differences between model and observation for the given time-points.

GO analysis

We use AgriGO web service to perform GO enrichment analysis on specific expression groups using Singular Enrichment Analysis [37]. Each analysis used 'Glycine max v1.1' as the species and 'Soybean genome locus (phytozome v1.1)' as the reference.

Availability of supporting data

The raw data sets supporting the results of this article are available in the Short Read Archive under BioProject accession: PRJNA259941 (http://www.ncbi.nlm.nih.gov/bioproject/?term=PRJNA259941).

Additional files

Additional file 1: Gene models with no significant difference at the transcriptional level between biological replicates.

Additional file 2: An Excel spreadsheet containing all genes, their expression categories, and log_2 ratios for genotypic difference and environmental responses. 'Untested' genes are not included.

Additional file 3: Profiles of *GxE* and *G+E* genes that overlap previously identified QTL. See Figure 2 for details.

Additional file 4: Genes that overlap previously identified QTL and exhibit *GxE* or *G+E* expression profiles.

Additional file 5: Profiles of photosynthesis genes exhibiting *GxE* patterns. See Figure 2 for details.

Additional file 6: Profiles of apoptosis genes exhibiting *GxE* patterns. See Figure 2 for details.

Additional file 7: Explicit models used to characterize the expression profile of a particular gene. A value of 1 represents the maximum expression level relative to control.

Abbreviations

VPD: Vapor pressure deficit; QTL: Quantitative trait locus; RIL: Recombinant inbred line; FPKM: Fragments per kilobase of transcript per million mapped reads.

Competing interests

The authors declare that they have no competing interests.

Authors' contributions

JHS performed sequence analysis, interpreted the results, and wrote the manuscript. JNV performed expression analysis, interpreted the results, and wrote the manuscript. HAH performed drought treatment and prepared RNA samples, and CC prepared library for RNA-seq. BA and KDK provided assistance in the data analysis, management, and data interpretation. ZL and SAJ designed the experiment and provided guidance of the study as well as the manuscript. All authors have read and approved the final manuscript.

Acknowledgments

Funding for this research was provided by the United Soybean Board. We thank Dr. Roger Boerma for his advice on the manuscript.

References

1. Sneller C, Dombek D. Use of irrigation in selection for soybean yield potential under drought. Crop Sci. 1997;37(4):1141–7.
2. Sloane RJ, Patterson RP, Carter TE. Field drought tolerance of a soybean plant introduction. Crop Sci. 1990;30(1):118–23.
3. Tanaka Y, Fujii K, Shiraiwa T. Variability of Leaf Morphology and Stomatal Conductance in Soybean [Glycine max (L.) Merr.] Cultivars All rights reserved. Crop Sci. 2010;50(6):2525–32.
4. Fletcher ALR, Jr ALH ST. Transpiration responses to vapor pressure deficit in well watered'slow-wilting' and commercial soybean. Environ Exp Bot. 2007;61:145–51.
5. Ries LL, Purcell LC, Carter TE, Edwards JT, King CA. Physiological traits contributing to differential canopy wilting in soybean under drought. Crop Sci. 2012;52(1):272.
6. Sadok W, Sinclair TR. Transpiration response of 'slow-wilting' and commercial soybean (Glycine max (L.) Merr.) genotypes to three aquaporin inhibitors. J Exp Bot. 2010;61(3):821–9.
7. Le DT, Nishiyama R, Watanabe Y, Tanaka M, Seki M, Ham LH, et al. Differential gene expression in soybean leaf tissues at late developmental stages under drought stress revealed by genome-wide transcriptome analysis. PLoS One. 2012;7(11):e49522.
8. Chen LM, Zhou XA, Li WB, Chang W, Zhou R, Wang C, et al. Genome-wide transcriptional analysis of two soybean genotypes under dehydration and rehydration conditions. BMC Genomics. 2013;14:687.
9. Abdel-Haleem H, Carter Jr TE, Purcell LC, King CA, Ries LL, Chen P, et al. Mapping of quantitative trait loci for canopy-wilting trait in soybean (Glycine max L. Merr). Theor Appl Genet. 2012;125(5):837–46.
10. Grishkevich V, Yanai I. The genomic determinants of genotype x environment interactions in gene expression. Trends Genet. 2013;29(8):479–87.
11. Covington MF, Maloof JN, Straume M, Kay SA, Harmer SL. Global transcriptome analysis reveals circadian regulation of key pathways in plant growth and development. Genome Biol. 2008;9(8):R130.
12. Thimm O, Bläsing O, Gibon Y, Nagel A, Meyer S, Krüger P, et al. Mapman: a user-driven tool to display genomics data sets onto diagrams of metabolic pathways and other biological processes. Plant J. 2004;37(6):914–39.
13. Yamaguchi-Shinozaki K, Shinozaki K. Organization of cis-acting regulatory elements in osmotic- and cold-stress-responsive promoters. Trends Plant Sci. 2005;10(2):88–94.
14. Des Marais DL, McKay JK, Richards JH, Sen S, Wayne T, Juenger TE. Physiological genomics of response to soil drying in diverse Arabidopsis accessions. Plant Cell. 2012;24(3):893–914.
15. Yamaguchi-Shinozaki K, Shinozaki K. A novel cis-acting element in an Arabidopsis gene is involved in responsiveness to drought, low-temperature, or high-salt stress. Plant Cell. 1994;6(2):251–64.
16. Watkinson JI, Sioson AA, Vasquez-Robinet C, Shukla M, Kumar D, Ellis M, et al. Photosynthetic acclimation is reflected in specific patterns of gene expression in drought-stressed loblolly pine. Plant Physiol. 2003;133(4):1702–16.
17. Chaves MM, Flexas J, Pinheiro C. Photosynthesis under drought and salt stress: regulation mechanisms from whole plant to cell. Ann Bot. 2009;103 (4):551–60.
18. Zhou J, Wang X, Jiao Y, Qin Y, Liu X, He K, et al. Global genome expression analysis of rice in response to drought and high-salinity stresses in shoot, flag leaf, and panicle. Plant Mol Biol. 2007;63(5):591–608.
19. Degenkolbe T, Do P, Zuther E, Repsilber D, Walther D, Hincha D, et al. Expression profiling of rice cultivars differing in their tolerance to long-term drought stress. Plant Mol Biol. 2009;69(1–2):133–53.
20. Liu Y, Xiong Y, Bassham DC. Autophagy is required for tolerance of drought and salt stress in plants. Autophagy. 2009;5(7):954–63.
21. Vásquez-Robinet C, Watkinson JI, Sioson AA, Ramakrishnan N, Heath LS, Grene R. Differential expression of heat shock protein genes in preconditioning for photosynthetic acclimation in water-stressed loblolly pine. Plant Physiol and Biochem. 2010;48(4):256–64.
22. Han S, Yu B, Wang Y, Liu Y. Role of plant autophagy in stress response. Protein Cell. 2011;2(10):784–91.
23. McDowell NG. Mechanisms linking drought, hydraulics, carbon metabolism, and vegetation mortality. Plant Physiol. 2011;155(3):1051–9.
24. Mazel A, Leshem Y, Tiwari BS, Levine A. Induction of salt and osmotic stress tolerance by overexpression of an intracellular vesicle trafficking protein AtRab7 (AtRabG3e). Plant Physiol. 2004;134(1):118–28.
25. Leshem Y, Melamed-Book N, Cagnac O, Ronen G, Nishri Y, Solomon M, et al. Suppression of Arabidopsis vesicle-SNARE expression inhibited fusion of H2O2-containing vesicles with tonoplast and increased salt tolerance. Proc Natl Acad Sci U S A. 2006;103(47):18008–13.
26. Colom MR, Vazzana C. Photosynthesis and PSII functionality of drought-resistant and drought-sensitive weeping lovegrass plants. Environ Exp Bot. 2003;49(2):135–44.
27. Han S-K, Wagner D. Role of chromatin in water stress responses in plants. J Exp Bot. 2014;65(10):2785–99.
28. Kim J-M, To TK, Nishioka T, Seki M. Chromatin regulation functions in plant abiotic stress responses. Plant Cell Environ. 2010;33(4):604–11.
29. Kim JM, To TK, Ishida J, Matsui A, Kimura H, Seki M. Transition of chromatin status during the process of recovery from drought stress in Arabidopsis thaliana. Plant Cell Physiol. 2012;53(5):847–56.
30. Lu KX, Cao BH, Feng XP, He Y, Jiang DA. Photosynthetic response of salt-tolerant and sensitive soybean varieties. Photosynthetica. 2009;47 (3):381–7.
31. Lowry DB, Logan TL, Santuari L, Hardtke CS, Richards JH, DeRose-Wilson LJ, et al. Expression quantitative trait locus mapping across water availability environments reveals contrasting associations with genomic features in Arabidopsis. Plant Cell. 2013;25(9):3266–79.
32. Abdel-Haleem H, Lee GJ, Boerma RH. Identification of QTL for increased fibrous roots in soybean. Theor Appl Genet. 2011;122(5):935–46.
33. King CA, Purcell LC, Brye KR. Differential wilting among soybean genotypes in response to water deficit. Crop Sci. 2009;49(1):290–8.
34. Auer PL, Doerge RW. Statistical design and analysis of RNA sequencing data. Genetics. 2010;185(2):405–16.
35. Goodstein DM, Shu S, Howson R, Neupane R, Hayes RD, Fazo J, et al. Phytozome: a comparative platform for green plant genomics. Nucleic Acids Res. 2012;40(D1):D1178–86.

36. Trapnell C, Roberts A, Goff L, Pertea G, Kim D, Kelley DR, et al. Differential gene and transcript expression analysis of RNA-seq experiments with TopHat and Cufflinks. Nat Protoc. 2012;7(3):562–78.
37. Du Z, Zhou X, Ling Y, Zhang Z, Su Z. agriGO: a GO analysis toolkit for the agricultural community. Nucleic Acids Res. 2010;38(Web Server issue): W64–70.

3

Expression and characterization of protein disulfide isomerase family proteins in bread wheat

Shizuka Kimura, Yuki Higashino, Yuki Kitao, Taro Masuda and Reiko Urade*

Abstract

Background: The major wheat seed proteins are storage proteins that are synthesized in the rough endoplasmic reticulum (ER) of starchy endosperm cells. Many of these proteins have intra- and intermolecular disulfide bonds. In eukaryotes, the formation of most intramolecular disulfide bonds in the ER is thought to be catalyzed by protein disulfide isomerase (PDI) family proteins. The cDNAs that encode eight groups of bread wheat (*Triticum aestivum L.*) PDI family proteins have been cloned, and their expression levels in developing wheat grains have been determined. The purpose of the present study was to characterize the enzymatic properties of the wheat PDI family proteins and clarify their expression patterns in wheat caryopses.

Results: PDI family cDNAs, which are categorized into group I (*TaPDIL1Aα*, *TaPDIL1Aβ*, *TaPDIL1Aγ*, *TaPDIL1Aδ*, and *TaPDIL1B*), group II (*TaPDIL2*), group III (*TaPDIL3A*), group IV (*TaPDIL4D*), and group V (*TaPDIL5A*), were cloned. The expression levels of recombinant TaPDIL1Aα, TaPDIL1B, TaPDIL2, TaPDIL3A, TaPDIL4D, and TaPDIL5A in *Escherichia coli* were established from the cloned cDNAs. All recombinant proteins were expressed in soluble forms and purified. Aside from TaPDIL3A, the recombinant proteins exhibited oxidative refolding activity on reduced and denatured ribonuclease A. Five groups of PDI family proteins were distributed throughout wheat caryopses, and expression levels of these proteins were higher during grain filling than in the late stage of maturing. Localization of these proteins in the ER was confirmed by fluorescent immunostaining of the immature caryopses. In mature grains, the five groups of PDI family proteins remained in the aleurone cells and the protein matrix of the starchy endosperm.

Conclusions: High expression of PDI family proteins during grain filling in the starchy endosperm suggest that these proteins play an important role in forming intramolecular disulfide bonds in seed storage proteins. In addition, these PDI family proteins that remain in the aleurone layers of mature grains likely assist in folding newly synthesized hydrolytic enzymes during germination.

Keywords: Protein disulfide isomerase, Oxidative refolding, Starchy endosperm, Aleurone cell, Endoplasmic reticulum, Bread wheat

Background

The major seed proteins of bread wheat (*Triticum aestivum* L.) include gliadins and glutenins. These storage proteins are synthesized in the rough endoplasmic reticulum (ER) of starchy endosperm cells and accumulate in two kinds of protein bodies derived from the ER and protein storage vacuoles [1-3]. Many of the wheat seed storage proteins have intramolecular disulfide bonds [4]. For example, gliadins have three to four intramolecular disulfide bonds in the C-terminal domain. In the case of γ-gliadins, formation of intramolecular disulfide bonds in the ER has been demonstrated to be essential for transport to the Golgi apparatus and deposition into protein bodies [5-8]. Therefore, mutations in cysteine residues or the reduction of disulfide bonds result in precipitation into insoluble aggregates in the ER [6-8].

Generally, the formation of disulfide bonds in proteins synthesized in the rough ER occurs mainly via dithiol/disulfide transfer reactions catalyzed by protein disulfide isomerase (PDI) (EC 5.3.4.1) and PDI-related proteins in eukaryotes [9]. PDI has two thioredoxin domains that contain the redox active site CGHC (**a** and **a'**) and two inactive domains (**b** and **b'**) [10]. Other PDI family members contain one or more thioredoxin domains [11]. In dicotyledonous *Arabidopsis*, a set of 22 orthologs of

* Correspondence: urade@kais.kyoto-u.ac.jp
Division of Agronomy and Horticultural Science, Graduate School of Agriculture, Kyoto University, Uji, Kyoto 611-0011, Japan

known PDI family proteins was discovered by a genome-wide search, and these orthologs were separated into phylogenetic groups I-X [12].

The physiological functions and biochemical properties of several plant PDI family members have been studied. AtPDIL5, an *Arabidopsis* ortholog of group I PDI family proteins, is expressed in endothelial cells of developing seeds and traffics together with the cysteine proteases RD21 and CP43 from the ER through the Golgi to vacuoles [13]. Studies of an AtPDIL5-null mutant revealed that AtPDIL5 is required for proper seed development and regulates the timing of programmed cell death by chaperoning and inhibiting cysteine proteases and serving as a redox-sensitive protease regulator during their trafficking to vacuoles before endothelial cells undergo programmed cell death. AtPDIL2-1, an *Arabidopsis* ortholog of group IV PDI family proteins, has been shown to act in maternal sporophytic tissues to affect embryo sac development [14]. A truncated AtPDIL2-1 mutant has been demonstrated to function as a gain-of-function mutant in sporophytic tissues and to affect ovule structure and impede embryo sac development, thereby disrupting pollen tube guidance.

In *Oldenlandia affinis*, a coffee family (Rubiaceae) plant, Oa PDI, an *Oldenlandia* ortholog of group I PDI family proteins, has been shown to be involved in the biosynthesis of the knotted circular proteins termed cyclotides [15]. In addition, Oa PDI dramatically enhances the oxidative folding of kalata B1 at physiological pH *in vitro*.

GmPDIL-1, GmPDIL-2, GmPDIL-3, GmPDIS-1 and –2, and GmPDIM, soybean orthologs of groups I-V PDI family proteins have been identified, and the recombinant proteins of GmPDIL-1, GmPDIL-2, GmPDIS-1, GmPDIS-2 and GmPDIM, but not GmPDIL-3, have been demonstrated to possess oxidative refolding activities [16-19]. In addition, GmPDIL-1, GmPDIL-2, GmPDIS-1, and GmPDIM have been shown to be involved in the folding of the soybean seed storage proteins proglycinin and β-conglycinin in the ER of cotyledon cells.

In monocotyledonous rice, grains of the mutant *eps*, which lack the rice ortholog (PDIL1-1) of group I PDI proteins, fail to generate the normal prolamin-containing protein bodies-I and accumulate the 57-kD proglutelin polypeptide in aggregate with the prolamin polypeptides via intermolecular disulfide bonds in small ER-derived protein bodies of uniform size (0.5 m in diameter) [20]. In addition, PDIL1-1 is asymmetrically distributed within the cortical cisternal ER, and this ortholog is essential for the maturation of proglutelin only when its rate of synthesis significantly exceeds its export from the ER. These findings suggest that rice PDIL1-1 helps retain proglutelin in the cisternal ER lumen until it attains competence for ER export [21]. Furthermore, analysis of the T-DNA insertion mutant revealed that rice PDIL1-1 deficiency causes a chalky phenotype, thick aleurone layer, lower protein content, and higher free sugar content in grains than the wild-type rice protein, suggesting that rice PDIL1-1 is involved in regulatory activities for various proteins that are essential for the synthesis of grain components [22]. Rice PDIL2;3, a rice ortholog of group V PDI family proteins, has been shown to be efficiently targeted to the surface of protein bodies-I in a redox active site-dependent manner and to play an important role in the accumulation of Cys-rich 10-kD prolamin (crP10) in the core of PB-I [23]. Complementation experiments using *eps* have indicated that rice PDIL2;3 and PDIL1;1 are not functionally redundant for disulfide bond formation in structurally diverse storage proteins and that these proteins play distinct roles in protein body development.

In wheat, the importance of PDI family proteins in seed storage protein folding and accumulation has been long predicted. Oxidative refolding activity was first reported in the embryo of bread wheat in 1978 [24], and activity in the endosperm was subsequently reported [25,26]. In 1995, a cDNA encoding a typical PDI was cloned. In addition, a 60-kD glycoprotein, which had a partial amino acid sequence homologous to the amino acid sequence predicted from the cDNA, was purified from grains of bread wheat by Shimoni *et al.* [27,28]. Since then, many cDNAs encoding wheat PDI family proteins have been identified [29-37]. Furthermore, d'Aloisio *et al.* reported the cloning of cDNAs encoding one typical PDI categorized into group I and eight PDI family proteins categorized into groups II-VIII [34]. These investigators also found that genes encoding these PDI family proteins are located in chromosome regions syntenic to those in rice [36]. Quantitative analysis of mRNAs transcribed from these genes revealed that these genes were constitutively expressed in all tissues examined but were characterized by different expression profiles [34]. Based on mRNA expression data, PDI family proteins (especially in groups I-V) have been proposed to play essential roles in grain development; however, the enzymatic activities of these individual proteins have not been characterized. In this study, we report the cloning of cDNAs encoding bread wheat PDI family proteins of groups I-V (TaPDIL1A, TaPDIL1B, TaPDIL2, TaPDIL3A, TaPDIL4D, and TaPDIL5A) and characterization of the enzymatic properties of their recombinant proteins. In addition, this report describes the expression and subcellular localization of these PDI family proteins in developing and mature caryopses.

Results and discussion

Cloning of wheat PDI family genes

In this study, we harvested mRNAs of bread wheat PDI family protein orthologs from groups I, II, III, IV, and V [12] from the immature caryopses and performed PCR using primer sets designed from their nucleotide sequences as reported previously [34]. Next, we generated and cloned the cDNAs encoding these PDI orthologs. We obtained five clones categorized into group I (*TaPDIL1Aα* [DDBJ: AB933341], *TaPDIL1Aβ* [DDBJ:AB933345], *TaPDIL1Aγ* [DDBJ:AB933342], *TaPDIL1Aδ* [DDBJ:AB933343], and *TaPDIL1B* [DDBJ:AB933344]) and one clone categorized into group II (*TaPDIL2,* [DDBJ:AB933346]), group III (*TaPDIL3A*, [DDBJ:AB933347]), group IV (*TaPDIL4D*, [DDBJ:AB933348]) and group V (*TaPDIL5A*, [DDBJ: AB933349]) (Table 1). Among these clones, *TaPDIL1Aβ*, *TaPDIL1Aγ*, *TaPDIL1Aδ*, and *TaPDIL2* have novel nucleotide sequences. The domain structures predicted from the amino acid sequences encoded by these cDNAs are shown in Figure 1. PDI family proteins categorized into group I are members of the representative eukaryotic PDI, which possesses an N-terminal signal sequence, two thioredoxin-like motifs with a CGHC active site (**a** and **a'** domains), two putative thioredoxin-folded domains without active site (**b** and **b'** domains), and a C-terminal KDEL sequence that functions as an ER retention signal [38,39]. In earlier studies [33,34], three genes (*TaPDIL1-1 [CPDI4A], TaPDIL1-2,* and *TaPDIL1-3*) encoding group I PDI family proteins were shown to be located on chromosome arms 4AL, 4BS, and 4DS, respectively. Recently, the wheat genome draft sequences were deposited in the databases of the wheat portal (http://wheat-urgi.versailles.inra.fr/Seq-Repository/) [40]. Nucleotide sequences of the cDNAs of *TaPDIL1Aα, TaPDIL1Aβ, TaPDIL1Aγ,* and *TaPDIL1Aδ* were coincident with those of the *TaPDIL1-1* gene and contigs on chromosome 4AL. Alignments of the nucleotide sequences of these cDNA with the genomic nucleotide sequence of *TaPDIL1A* [GenBank:AJ868102] suggested that they are produced by intron retention-type alternative splicing of the first exon (Figure 2) [41]. *TaPDIL1Aα, TaPDIL1Aβ, TaPDIL1Aγ,* and *TaPDIL1Aδ* cDNAs encoded proteins of 515, 501, 487, and 485 amino acids, respectively (Table 2, Additional file 1: Figure S1). TaPDIL1Aβ, TaPDIL1Aγ, and TaPDIL1Aδ lack the amino acid sequences Pro21-A34, Ala15-Leu42, and Ala13-Leu42 found in TaPDIL1Aα, respectively. The putative signal peptides of TaPDIL1Aγ and TaPDIL1Aδ (Met1-Ala18 and Met1-Ala16, respectively) were shorter than those of TaPDIL1Aα and TaPDIL1Aβ (Met1-Ala25 and Met1-Ala26, respectively). The effects of such alterations in the signal peptide on TaPDIL1A protein targeting to the ER are unclear. The nucleotide sequence of *TaPDIL1B* cDNA was matched to that of GPDIA-4B, the *TaPDIL1-2* genomic gene [GenBank: AJ868103] [33,34], and contigs on chromosome 4BS. TaPDIL1Aα, −1β, −1γ, and -1δ and TaPDIL1B have a conserved arginine involved in the regulation of the active site redox potential and a conserved glutamic acid that facilitates the release of the active site from a mixed disulfide with substrate in human PDI [42-44] in both the **a** and **a'** domains (Table 2 and Additional files 1: Figure S1 and Additional file: 2: Figure S2).

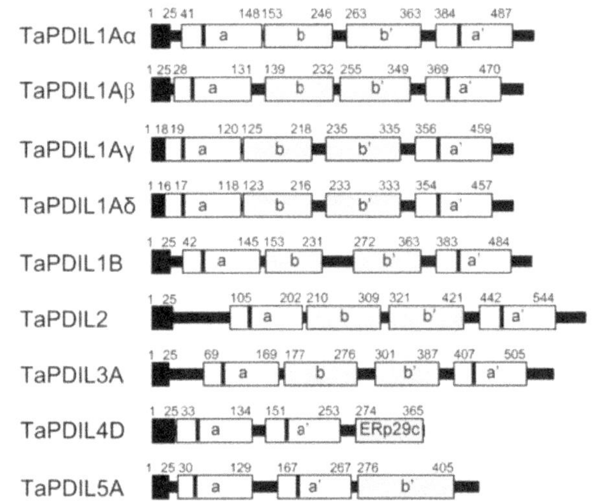

Figure 1 Wheat PDI family protein domain structures deduced from the cDNA clones. The boxes indicate the domain boundaries predicted by an NCBI conserved domain search. Black boxes in domains a and a' represent the CXXS/C motif. The N-terminal black box represents a signal peptide that was predicted using SignalP-4.0 euk software.

Table 1 Wheat PDI family proteins cloned in this study

Clone	ORF (nt)	Accession number	Clone name reported previously	Accession number
TaPDIL1Aα	1545	AB933341	CPDI4A*, TaPDIL1-1a**, TaPDIL1-1c*	AJ868105
TaPDIL1Aβ	1503	AB933345	-	-
TaPDIL1Aγ	1461	AB933342	-	-
TaPDIL1Aδ	1455	AB933343	-	-
TaPDIL1B	1536	AB933344	CPDI4B*, TaPDIL1-1b**	AJ868106
TaPDIL2	1755	AB933346	-	-
TaPDIL3A	1623	AB933347	TaPDIL3-1**	FN555317
TaPDIL4D	1101	AB933348	TaPDIL4-1**	FN555318
TaPDIL5A	1320	AB933349	TaPDIL5-1a**	FN555320

*reference 33, **reference 34.

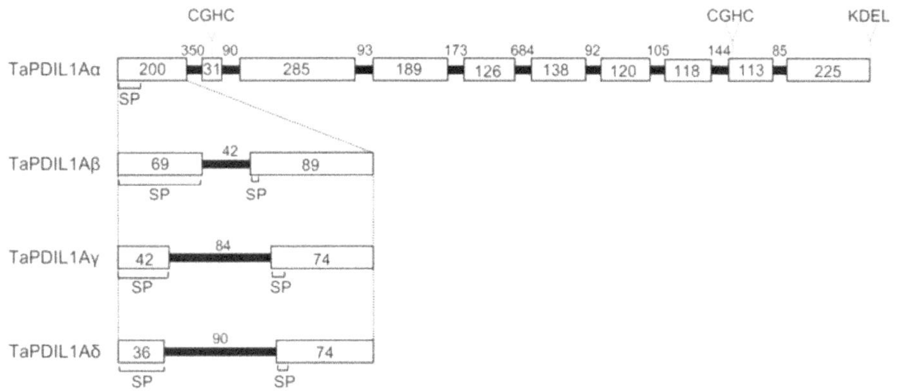

Figure 2 Splicing sites in *TaPDIL1Aα*, *TaPDIL1Aβ*, *TaPDIL1Aγ*, and *TaPDIL1Aδ*. Open boxes indicate exons, and solid black lines denote introns. The numbers indicate the size of each exon and intron (bp). The positions of the signal peptide (SP), the two CGHC motifs, and the C-terminal KDEL sequence are indicated.

The *TaPDIL2* cDNA encoded a 585- amino acid protein (Figure 1 and Table 2), and the nucleotide sequence of *TaPDIL2* cDNAs was 99% identical to that of *TaPDIL2-1* [34]. No contig sequences matched with the sequence of *TaPDIL2* cDNA found in the wheat genome draft sequences. The putative amino acid sequence of TaPDIL2 was also 99% identical to that of TaPDIL2-1 (Additional file 3: Figure S3). TaPDIL2 was categorized as a group II PDI family protein that possesses an N-terminal signal sequence, aspartic acid-rich flanking region, two thioredoxin-like motifs with a CGHC active site (**a** and **a'** domains), two putative thioredoxin-folded domains without active sites (**b** and **b'** domains), and a C-terminal KDEL sequence. TaPDIL2 also has a conserved arginine and a conserved glutamic acid in both the **a** and **a'** domains (Table 2 and Additional file 4: Figure S4).

The *TaPDIL3A* cDNA encoded a 541-amino acid protein (Table 2), and the nucleotide sequence was matched to contigs on chromosome 7AS. TaPDIL3A was categorized as a group III PDI family protein and possesses the same domain structure as TaPDIL1A (Figure 1). However, the sequences of the active sites in the **a** and **a'** domains (CERS and CVDC) differ from the representative motif CGHC. In addition, TaPDIL3A lacks the conserved glutamic acid in the **a** domain, as well as the conserved arginine in the **a** and **a'** domains (Additional file 5: Figure S5).

The nucleotide sequence of the *TaPDIL4D* cDNA was matched to that of contigs on chromosome 1AS.

Table 2 Characteristics of the wheat PDI family proteins cloned in this study

Name	Total amino acid residues	Signal peptide*	Molecular weight	pI	Conserved active site	Consensus N-glycosylation site(s)	Conserved arginine(s)	Conserved charge pairs	ER retention signal
TaPDIL1Aα	515	1–25	56579.13	4.99	C68GHC71, C412GHC415	N283	R136, R475	E62/K96, E406/K439	KDEL
TaPDIL1Aβ	501	1–26	55256.73	5.07	C54GHC57, C398GHC401	N269	R122, R461	E48/K82, E392/K425	KDEL
TaPDIL1Aγ	487	1–18	53977.23	5.10	C40GHC43, C384GHC387	N255	R108, R447	E34/K68, E378/K411	KDEL
TaPDIL1Aδ	485	1–16	53679.83	5.10	C38GHC41, C382GHC385	N253	R106, R445	E32/K66, E376/K399	KDEL
TaPDIL1B	512	1–25	56425.02	5.03	C68GHC71, C412GHC415	N283	R136, R475	E62/K96, E406/K439	KDEL
TaPDIL2	585	1–28	63588.76	4.61	C129GHC132, C470GHC473	N109, N212	R136, R535	E123/K157, E464/K497	KDEL
TaPDIL3A	541	1–23	59594.01	4.95	C96ERS99, C435VDC438	N150	R160	E429K462	KDEL
TaPDIL4D	367	1–30	40260.95	6.17	C60GHC63, C179GHC182	-	R125, R244	E54/K87, E173/K211	-
TaPDIL5A	440	1–22	47207.89	5.30	C57GHC60, C194GHC197	N164, N170	R119, R257	E51/K89, E188/K226	NDEL

*Signal peptides were predicted using SignalP 4.1 Server by the Technical University of Denmark.

TaPDIL4D was categorized as a group IV PDI family protein, which is unique to plants and possesses an N-terminal signal peptide, two thioredoxin-like motifs with a CGHC active site (**a** and **a'** domains), and a domain homologous to the C-terminal domain of mammalian ERp29 (Figure 1 and Table 2) [45].

TaPDIL5A was categorized as a group V PDI family protein. This sequence encodes a protein of 440 amino acids that is homologous to mammalian P5 [46], and the nucleotide sequence of *TaPDIL5A* cDNA was matched to that of contigs on chromosome 5AL. TaPDIL5A possesses an N-terminal signal peptide, two thioredoxin-like motifs with a CGHC active site (**a** and **a'** domains), a putative thioredoxin-folded **b** domain, and a C-terminal NDEL sequence similar to KDEL (Figure 1 and Table 2). TaPDIL4D and TaPDIL5A contain a conserved arginine and a conserved glutamic acid in both the **a** and **a'** domains (Table 2 and Additional files 6: Figure S6 and Additional files 7: Figure S7).

Expression and characterization of recombinant wheat PDI family proteins

To investigate the enzymatic properties of wheat PDI family proteins, recombinant proteins without the putative N-terminal signal peptide were prepared from cDNAs of *TaPDIL1Aα*, *TaPDIL1B*, *TaPDIL2*, *TaPDIL3A*, *TaPDIL4D*, and *TaPDIL5A* using an *Escherichia coli* (*E. coli*) expression system, and then the proteins were purified (Figure 3A-F). All recombinant proteins were expressed as soluble proteins (Figure 3A-F) and eluted in a monomeric form from a

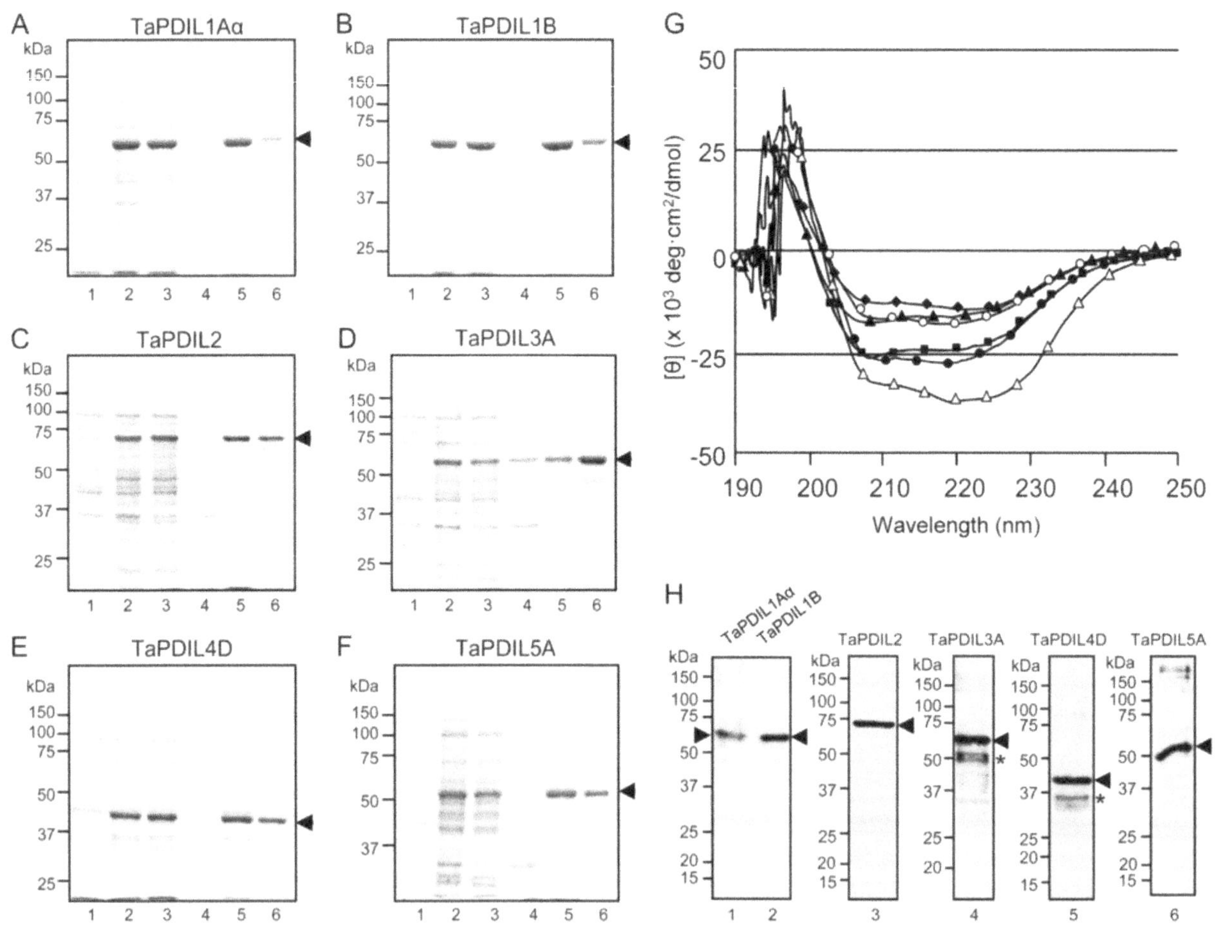

Figure 3 Expression of recombinant wheat PDI family proteins. Recombinant TaPDIL1Aα **(A)**, TaPDIL1B **(B)**, TaPDIL2 **(C)**, TaPDIL3A **(D)**, TaPDL4D **(E)**, and TaPDIL5A **(F)** were expressed in *E. coli.* cells transformed with each expression plasmid and incubated with 0.4 mM IPTG at 30°C for 0 (lane 1) or 20 h (lane 2). Soluble (lane 3) and insoluble (lane 4) fractions were separated from these cells after 20 h by centrifugation after sonication. Each recombinant PDI family protein (arrowheads) was purified by His-tag column chromatography (lane 5) followed by gel filtration chromatography (lane 6). Proteins in each sample were separated by SDS-PAGE and stained with Coomassie Brilliant Blue. **(G)** The CD spectra of the purified recombinant TaPDIL1Aα (●), TaPDIL1B (○), TaPDIL2 (▲), TaPDIL3A (■), TaPDL4D (♦), and TaPDIL5A (Δ) were determined at a concentration of 0.3 mg/ml. **(H)** The purified recombinant TaPDIL1Aα (lane 1), TaPDIL1B (lane 2), TaPDIL2 (lane 3), TaPDIL3A (lane 4), TaPDIL4D (lane 5), or TaPDIL5A (lane 6) was detected by western blot analysis with anti-TaPDIL1Aα serum (lanes 1 and 2), anti-TaPDIL2 serum (lane 3), anti-TaPDIL3A serum (lane 4), anti-TaPDIL4D serum (lane 5), or anti-TaPDIL5A serum (lane 6). Asterisks indicate degradation products.

gel filtration column (data not shown). All recombinant TaPDIL family proteins had circular dichroism (CD) spectra typical of well-folded α/β-type proteins (Figure 3G).

The oxidative refolding activities of these recombinant proteins were analyzed with reduced and denatured ribonuclease A (RNaseA) as a substrate. Both TaPDIL1Aα and TaPDIL1B exhibited higher activities than the other recombinant PDI family proteins (Table 3). Furthermore, the activities of TaPDIL1Aα and TaPDIL1B were approximately 1.5-fold higher than that of the recombinant soybean ortholog GmPDIL-1 [17] despite 63% share identity in amino acid sequence (Additional file 2: Figure S2). The activity of TaPDIL2 was nearly identical to that of soybean ortholog GmPDIL-2 [17], which has an amino acid sequence with 62% shared identity (Additional file 4: Figure S4). The recombinant TaPDIL3A had no oxidative refolding activity similar to soybean GmPDIL3 [19]. The lack of oxidative refolding activity of TaPDIL3A and GmPDIL-3 is thought to stem from their atypical active site motifs and the lack of a conserved arginine that is necessary for active site activity as described above. Furthermore, the activity of TaPDIL4D was approximately three- to four-fold higher than those of soybean orthologs, GmPDIS-1 and GmPDIS-2 [16], even though 75% and 74%, respectively, of the amino acid sequences are identical (Additional file 6: Figure S6). Finally, TaPDIL5A exhibited approximately four-fold greater activity than that of the soybean ortholog GmPDIM [18], which shares a 79% identical amino acid sequence (Additional file 7: Figure S7).

Expression of PDI family proteins in wheat caryopses

We generated antiserum using recombinant TaPDIL1Aα, TaPDIL2, TaPDIL3A, TaPDIL4D, or TaPDIL5A as an antigen. Each antiserum cross-reacted with the recombinant protein used for the immunization (Figure 3H). The antiserum against TaPDIL1Aα also cross-reacted with recombinant TaPDIL1B. The expression of each PDI family protein in wheat caryopses was confirmed by western blot

Table 3 Oxidative refolding activity of recombinant wheat PDI family proteins

Wheat PDI family protein	Activity (mmol/min/mol)	Soybean PDI ortholog	Activity (mmol/min/mol)
TaPDIL1Aα	736 ± 4	GmPDIL-1	460[b]
TaPDIL1B	665 ± 49		
TaPDIL2	287 ± 20	GmPDIL-2	259[b]
TaPDIL3A	N.D.	GmPDIL-3	N.D.[d]
TaPDIL4D	179 ± 7	GmPDIS-1	66[a]
		GmPDIS-2	43[a]
TaPDIL5A	175 ± 33	GmPDIM	45[c]

[a, b, c, d]Values are quoted from references 16, 17, 18, and 19, respectively.
Oxidative refolding activity was assayed using reduced and denatured RNaseA as a substrate.

analyses with these sera. The sera against TaPDIL1Aα, TaPDIL2, TaPDIL3A, TaPDIL4D, and TaPDIL5A recognized a 60-, 70-, 69-, 40-, and 47-kDa band, respectively, on gels containing proteins extracted from caryopses of bread wheat (Figure 4A). Because anti-TaPDIL1Aα serum cross-reacted with recombinant TaPDIL1B, a highly conserved protein categorized as group I PDI family proteins, it is likely that splicing variants of TaPDIL1Aα, TaPDIL1B, and TaPDIL1-3 [34] may all be detected by western blot analysis with this serum. Likewise, the sera against TaPDIL2, TaPDIL3A, TaPDIL4D, and TaPDIL5A were presumed to cross-react with their homolog categorized into group II, III, IV, or V, respectively. Thus, the proteins that cross-reacted with sera against TaPDIL1Aα, TaPDIL2, TaPDIL3A, TaPDIL4D, or TaPDIL5A were referred to as TaPDIL1, TaPDIL2, TaPDIL3, TaPDIL4, or TaPDIL5, respectively.

The sizes of the bands that cross-reacted with anti-TaPDIL4D and anti-TaPDIL5A sera approached the molecular sizes calculated from the amino acid sequences of TaPDL4D and TaPDIL5A, respectively, in the absence of the signal peptide. The sizes of bands that cross-reacted with the sera against TaPDIL1Aα, TaPDIL2, and TaPDIL3A were larger than the calculated molecular sizes based on the amino acid sequences of these PDI family proteins without the signal peptide, but these proteins contain one or two consensus N-glycosylation sites, suggesting that these PDI family proteins contain N-glycan(s). When caryopses proteins were digested with endoglycosidase H or F, the masses of the bands that were recognized by the sera against TaPDIL1Aα, TaPDIL2, and PDIL3A became smaller (Figure 4B), supporting the notion of N-glycosylation of these proteins in wheat caryopses. Because endoglycosidase H specifically cleaves high mannose-type oligosaccharides but not complex oligosaccharides from glycoproteins, these results indicate that the oligosaccharides attached to TaPDIL1, TaPDIL2, and TaPDIL3 are high mannose-type oligosaccharides. In agreement with these findings, TaPDIL1 purified from wheat was previously shown to be a glycoprotein [28]. In addition to PDI family proteins, we examined wheat orthologs of calreticulin and ER oxidoreductin 1 (Ero1) using western blot analyses. Calreticulin acts cooperatively with the human PDI family protein ER-60/ERp57 to fold monoglycosylated N-glycoproteins [47-50]. Ero1 is a PDI-activating enzyme located in the ER [51-54]. The active centers of PDI family proteins oxidize cysteine residues in substrate proteins and become a reduced form. PDI family proteins require other oxidizing molecules such as Ero1 for reoxidation of their active center cysteine residues. The sera against calreticulin and Ero1 cross-reacted with a 52-kDa band and 63-kDa band, respectively (Figure 4A). These band sizes were similar to those calculated from the putative amino acid sequences of

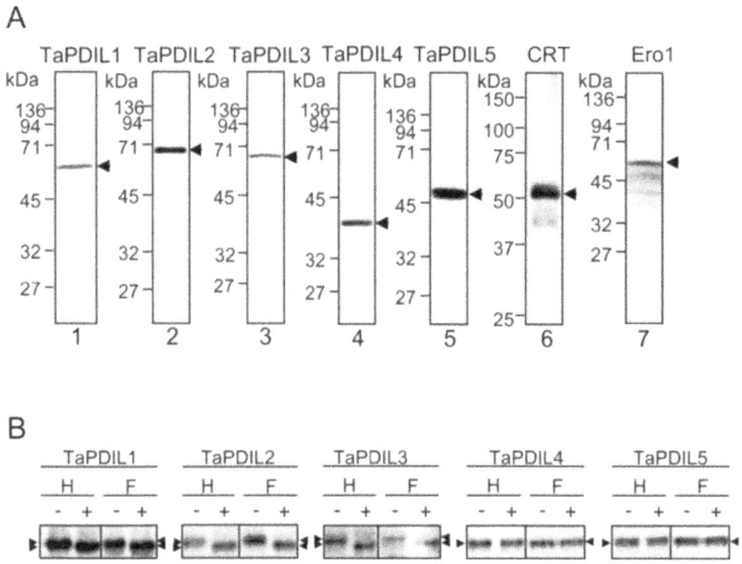

Figure 4 Expression of wheat PDI family proteins in wheat caryopses. (A) Proteins (25 μg) extracted from the caryopses at 25 (lanes 1, 3) or 15 (lanes 2, 4-7) dap were analyzed by western blot with serum against TaPDIL1Aα (lane 1), TaPDIL2 (lane 2), TaPDIL3A (lane 3), TaPDIL4D (lane 4), TaPDIL5A (lane 5), calreticulin (lane 6), or Ero1 (lane 7). **(B)** TaPDIL1, TaPDIL2, and TaPDIL3 are high mannose-type N-glycosylated proteins in wheat caryopses. The proteins extracted from the caryopses were treated with (+) or without (–) endoglycosidase H (H) or endoglycosidase F (F). The proteins (20 μg for TaPDIL1 and TaPDIL2; 30 μg for TaPDIL3, TaPDIL4, and TaPDIL5) were separated by SDS-PAGE and immunostained with serum against TaPDIL1Aα, TaPDIL2, TaPDIL3A, TaPDIL4D, or TaPDIL5A.

the wheat orthologs of calreticulin (50,114; [EMS59406]) and Ero1 (65,619; [EMS58683]) [55].

The expression levels of the PDI family proteins, calreticulin, and Ero1 in the caryopses during maturation were determined by western blot analyses (Figure 5A). In addition, the levels of each of the PDI family proteins in bread wheat caryopses and flour were semi-quantitatively assessed by determining their band intensities on western blots (Figure 5B and Table 4). Levels of TaPDIL1 in the caryopses were highest at 10–15 dpa, which corresponds to the time of greatest synthesis of the seed storage proteins such as gliadins and glutenins (Figure 5A, Additional file 8: Figure S8). The levels of TaPDIL2, TaPDIL4, and TaPDIL5 in caryopses were also highest at 5–15 dpa in the maturing period. The level of TaPDIL3 in the caryopses was highest at 5 dpa and then decreased with time. The content of TaPDIL1 was the highest among PDI family proteins (Table 4). The expression of *TaPDIL1* mRNA in

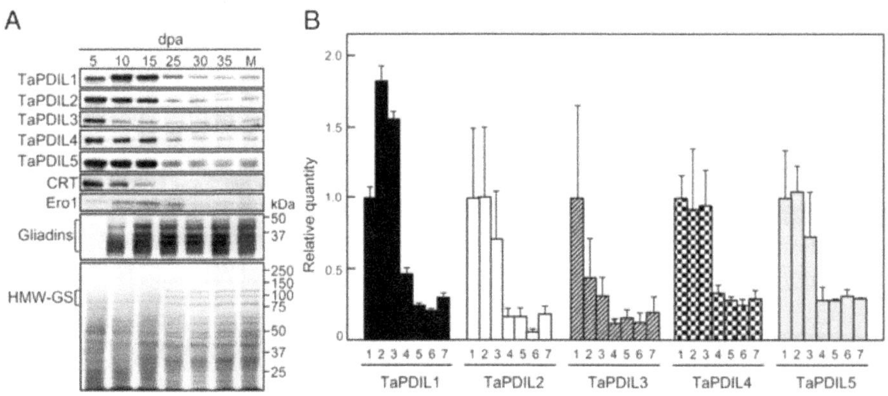

Figure 5 Expression of wheat PDI family proteins in wheat caryopses during maturation. (A) Proteins (25 μg) extracted from caryopses at 5 (lane 1), 10 (lane 2), 15 (lane 3), 25 (lane 4), 30 (lane 5), and 35 dpa (lane 6) as well as mature grain (M) were analyzed by western blot as described in Figure 4A. For detection of gliadins, anti-gliadin serum was used. Total proteins were stained with Coomassie Brilliant Blue (CBB). HMW-GS indicates high molecular glutenin subunit. **(B)** Levels of TaPDIL1, TaPDIL2, TaPDIL3, TaPDIL4, and TaPDIL5 in caryopses at 5 (1), 10 (2), 15 (3), 25 (4), 30 (5) and 35 dpa (6) and in mature grain (7) were estimated from the band intensities on western blots in A. Values were calculated as a ratio to the value obtained at 5 dpa. Data represent the mean ± standard error of three experiments.

Table 4 PDI family proteins in wheat caryopses and flour

Wheat PDI family protein	Content (pmol/mg protein)							
	5 dpa	10 dpa	15 dpa	25 dpa	30 dpa	35 dpa	Mature	flour
TaPDIL1	15.0	27.4	23.2	6.9	3.4	2.9	4.3	7.3
TaPDIL2	3.9	3.8	2.7	0.5	0.6	0.4	0.6	1.1
TaPDIL3	0.7	0.3	0.2	0.1	0.1	0.1	0.1	0.1
TaPDIL4	12.7	11.6	12.1	4.1	3.2	3.0	3.5	1.5
TaPDIL5	1.5	1.3	1.2	0.3	0.2	0.2	0.3	0.3

Contents by weight of PDI family proteins were estimated from the band intensities on western blot analysis of the caryopses (Figure 4B) and flour (Haruyokoi) with recombinant TaPDIL1Aα, TaPDIL2, TaPDIL3A, TaPDIL4D, and TaPDIL5A used as standard proteins for determination of TaPDIL1, TaPDIL2, TaPDIL3, TaPDIL4, and TaPDIL5, respectively. Molar quantities of TaPDIL1, TaPDIL2, TaPDIL3, TaPDIL4, and TaPDIL5 included in 1 mg of caryopses protein or flour protein were calculated with each molecular weight of TaPDIL1Aα, TaPDIL2, TaPDIL3A, TaPDIL4D, and TaPDIL5A (as shown in Table 2), respectively.

the developing caryopses has been reported to be the highest of those of PDI family proteins [34]. The expression of wheat Ero1 was highest at 15 dpa (Figure 5A). Wheat Ero1 may activate wheat PDI family proteins, but whether plant Ero1 activates plant PDI family proteins *in vivo* remains unproven. Knockdown of rice Ero1 (OsERO1) revealed that native disulfide bond formation in proglutelins depends on an electron transfer pathway involving OsERO1 [56]. Wheat calreticulin was also highly expressed during the period of grain filling (Figure 5A). Levels of all PDI family proteins, calreticulin, and Ero1 were drastically diminished in the late maturing period (25–35 dpa; Figure 5A and B) when grain weights decreased due to desiccation (Additional file 8: Figure S8); however, low levels of PDI family proteins remained even in the mature grains. Calreticulin and Ero1 were not detected in the 25–35 dpa caryopses or in the mature grains.

Localization of PDI family proteins in wheat caryopses

Localization of each PDI family protein in the immature wheat caryopses at 10 and 20 dpa was investigated by immunofluorescence microscopy. The antisera for PDI family proteins were confirmed to specifically react with their respective native recombinant PDI family protein (Additional file 9: Figure S9). The aleurone cells, pericarp, and starchy endosperm of the caryopsis were labeled with the sera raised against TaPDIL1Aα, TaPDIL2, TaPDIL3A, TaPDIL4D, or TaPDIL5A (Figure 6). These distributions suggest that these PDI family proteins are required for folding nascent proteins that are synthesized *de novo* in these tissues during grain development.

The TaPDIL1 group proteins, TaPDIL2, TaPDIL3A, and TaPDIL5A contain putative signal peptides as well as the C-terminal ER retention signal. Therefore, these PDI family proteins were expected to localize to the ER. In addition, the ER-localization of TaPDIL4D, which has an N-terminal signal peptide but no C-terminal ER retention signal, was also predicted because soybean orthologs GmPDIS-1 and GmPDIS-2 localize to the ER [16]. The ER-localization of all PDI family proteins in the starchy endosperm was indicated by their network-like distribution in the cytoplasm and co-localization with calreticulin, a well-known ER luminal protein (Figure 7A-E). In the starchy endosperm, peripheral regions of A-type starch granules were strongly immunostained with anti-PDI family and anti-calreticulin sera, suggesting that the ER was densely packed around the starch granules. The existence of all wheat PDI family proteins in the ER of the starchy endosperm cells during grain filling suggests that these factors play important roles in the folding of storage proteins such as gliadins and glutenins, which are synthesized and folded in the ER. γ-Gliadin has been shown to require PDI family proteins for folding by *in vitro* translational experiment [5]. In the aleurone cells, members of all five groups of PDI family proteins were labeled along with calreticulin in the periphery of aleurone grains and network structure (Figure 7F-J), suggesting localization of these proteins to the ER. In aleurone grains, storage proteins accumulate during seed development [57]. These PDI family proteins may assist in the folding of storage proteins synthesized in the ER of the aleurone cells, and the resultant folded storage proteins accumulate in the aleurone grains. However, the specific details of the transport of storage proteins from the ER to aleurone grains remains unclear. In addition, the five groups of PDI family proteins were also detected at the boundaries of aleurone cells. The localization mechanism and physiological roles of those PDI family proteins at the aleurone cell boundary remain to be elucidated.

In the mature grains, the members of the five groups of PDI family proteins were labeled in the aleurone cells and starchy endosperm (Figure 8A-E). In the aleurone cells of the mature wheat grain, the ER has been found in two regions: the periphery of the aleurone grain-spherosome complex and the cytoplasm as spread network structures [58]. The periphery of the aleurone grain-spherosome complex was mainly labeled with antisera to the five anti-PDI family proteins and calreticulin (Figure 8F-J). Based on these findings, we predict that the PDI family proteins that localized in the aleurone cells of the mature grain accumulated prior to grain formation and function in folding newly synthesized hydrolytic enzymes in the ER during

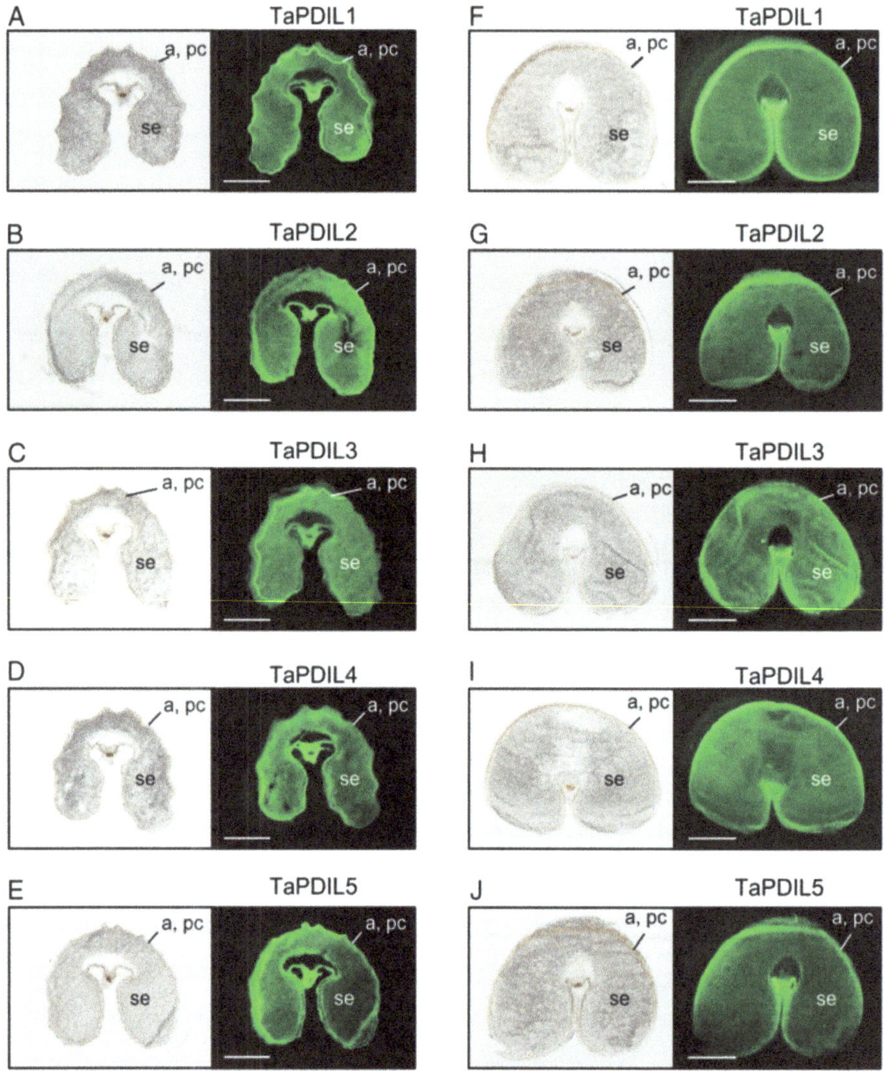

Figure 6 Distribution of wheat PDI family proteins in immature wheat caryopses. Cross sections of caryopses at 10 dpa **(A-E)** and 20 dpa **(F-J)** were immunostained with serum against TaPDIL1Aα **(A, F)**, TaPDIL2 **(B, G)**, TaPDIL3A **(C, H)**, TaPDIL4D **(D, I)**, or TaPDIL5A **(E, J)**. Specimens were observed with a stereomicroscope SZX16. Fluorescent images are shown on the right panels in **(A-J)**. Visible light images collected simultaneously are shown on the left in **A-J**. se, starchy endosperm; a, aleurone layer; pc, pericarp. Scale bar = 1 mm.

germination. In mature cereal grain endosperm, only aleurone cells are alive [59]. When cereal grains imbibe water, α-amylases and cysteine proteinases, which may contain intramolecular disulfide bonds [60,61], are synthesized in the ER of the aleurone cells in response to gibberellic acid secretion into the endosperm from the embryo via the scutellum [62,63]. These hydrolytic enzymes are secreted to the starchy endosperm to break down starch and protein that accumulated in the starchy endosperm to supply carbon and nitrogen used in germination. Whereas the secretion of α-amylases and cysteine proteases from the aleurone layers increased in response to gibberellic acid treatment for 24 h, the level of the barley TaPDIL-1 ortholog was not affected by gibberellic acid treatment [64], suggesting that the barley TaPDIL-1 ortholog that remains in the aleurone cell of the wheat grain may help fold the newly synthesized hydrolytic enzymes.

During the late stage of maturation, the starchy endosperm cells undergo programmed cell death, and the protein bodies fuse to the protein matrix [65]. The analyzed members of the five groups of PDI family proteins were distributed throughout the fused protein matrix (Figure 8K-O). Because wheat flour arises from milled starchy endosperm, the five groups of PDI family proteins must be present in flour (Table 4). We previously detected oxidative refolding activity in an extract from wheat flour, and its activity was inhibited by the PDI inhibitor bacitracin [66]. Therefore, PDI family proteins may be active in both the starchy endosperm of the dry mature grain and flour. It is unclear whether the PDI

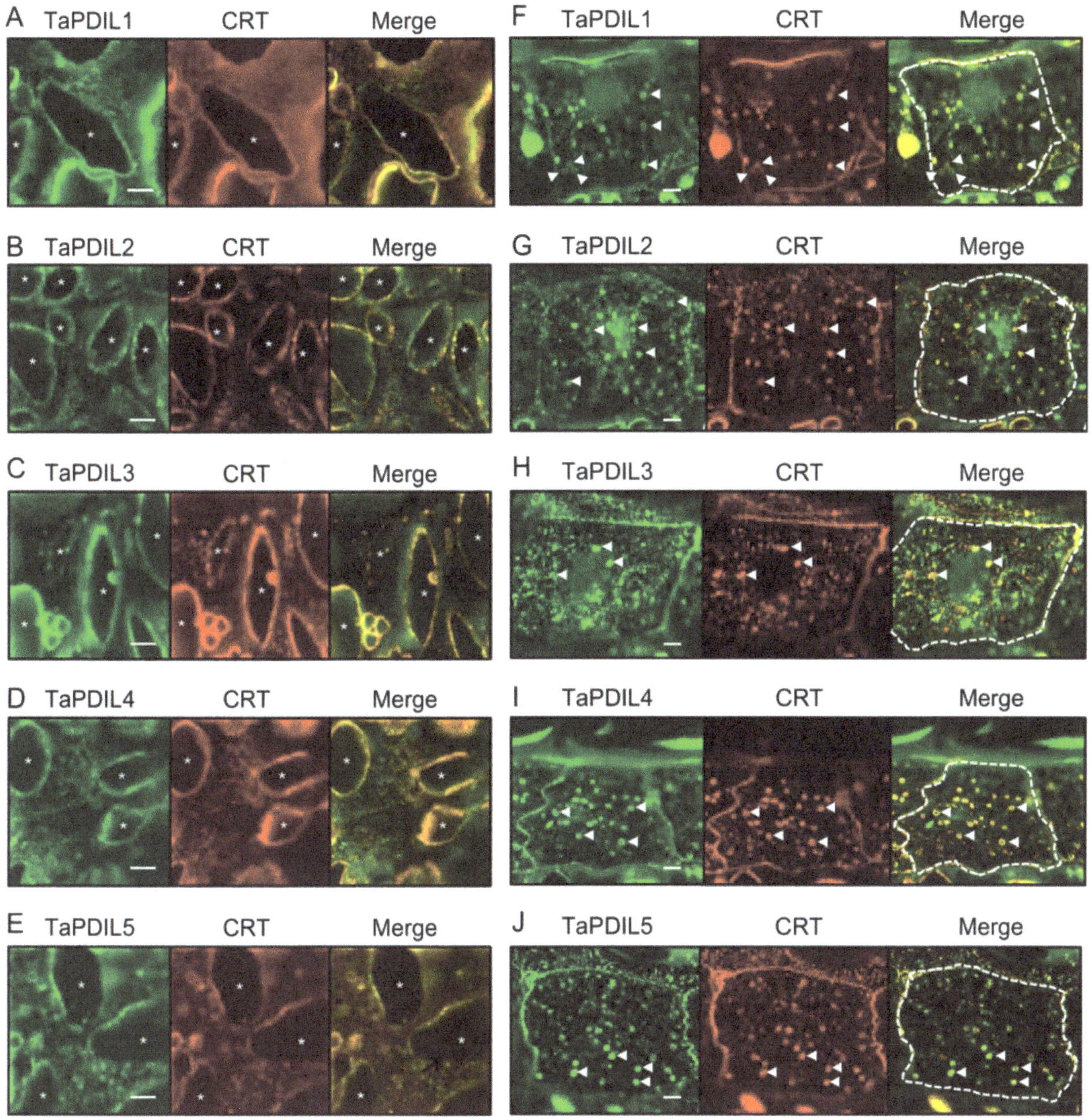

Figure 7 Subcellular distribution of wheat PDI family proteins in immature starchy endosperm and aleurone cells. Cross sections of caryopses at 10 dpa were immunostained with a combination of sera against TaPDIL1Aα (**A** and **F**, green), TaPDIL2 (**B** and **G**, green), TaPDIL3A (**C** and **H**, green), TaPDIL4D (**D** and **I**, green), or TaPDIL5A (**E** and **J**, green) and calreticulin (CRT, red). Specimens were observed with a Confocal Imaging System FV1200. Merged images of red and green are also shown. Endosperm sections and aleurone cells are shown in the left and right panels, respectively. Asterisks indicate A-type starch granules. Arrowheads indicate aleurone grains. White dotted lines show the profile of one aleurone cell. Scale bar = 5 μm.

family proteins that remain in the starchy endosperm of the wheat grain play a physiological role during germination. Further studies are needed to identify the substrate proteins associated with each PDI family protein *in vivo* and to determine the effects of knockdown of these factors on wheat phenotypes, such as seed storage protein accumulation and plant body growth. Furthermore, PDI family proteins may affect the quality of food products. We previously demonstrated that PDI family proteins play a role in retaining glutenin macropolymers during dough mixing, and these polymers are the most important factor for dough's tensile strength [66]. We semi-quantitatively analyzed the levels of the five PDI family protein groups in a wheat flour sample (Table 4) and found that among the five groups of PDI family proteins, TaPDIL1 levels were the highest. Because the oxidative refolding activities of

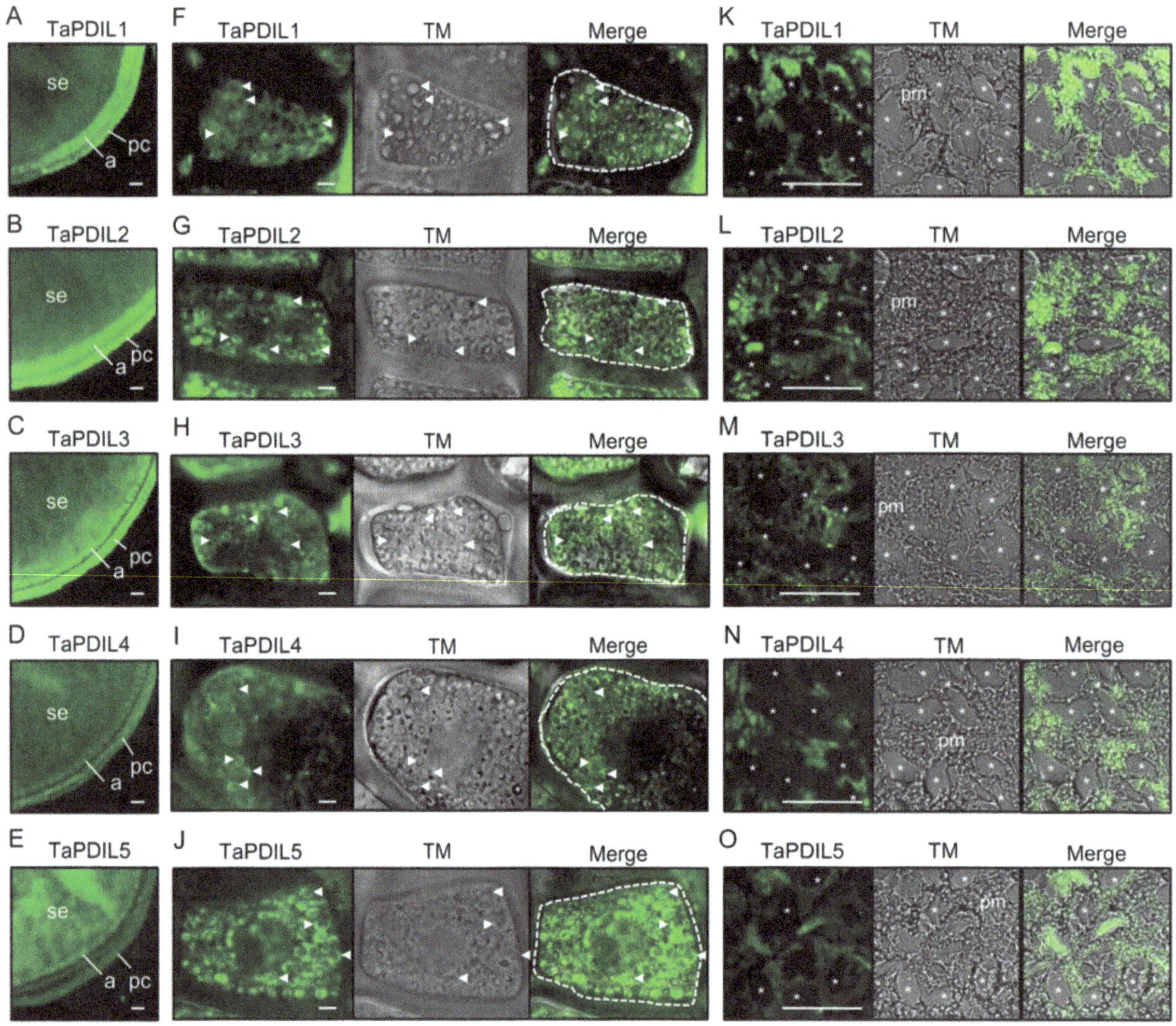

Figure 8 Distribution of wheat PDI family proteins in the mature aleurone cells and the protein matrix of wheat grains. Cross sections of mature grains were immunostained with sera against TaPDIL1Aα **(A, F, K)**, TaPDIL2 **(B, G, L)**, TaPDIL3A **(C, H, M)**, TaPDIL4D **(D, I, N)**, or TaPDIL5A **(E, J, O)**. Specimens were observed with a fluorescence microscope BZ-9000 **(A-E, K-O)** and a Confocal Imaging System FV1200 **(F-J)**. Visible light images (gray) collected simultaneously are shown in the middle panels **(F-O)**. Merged fluorescent and visible light images are also shown in the right panels **(F-O)**. a, aleurone layer; pc, pericarp; pm, protein matrix; se, starchy endosperm. Arrowheads indicate aleurone grains. White dotted lines show the profile of one aleurone cell. Asterisks indicate A-type starch granules. Scale bars in **A-E** = 250 μm. Scale bars in **F-J** = 5 μm. Scale bars in **K-O** = 50 μm.

recombinant TaPDIL1A and TaPDIL1B were the strongest of the five recombinant PDI family proteins, TaPDIL1 may primarily function in retaining glutenin macropolymers during dough mixing. Therefore, PDI family proteins may be an important contributing factor to the quality wheat flour, and they may serve as targets for genetic manipulation in the future.

Conclusions

In eukaryotes, PDI and PDI family proteins function in multiple essential capacities such as oxidative folding of nascent proteins, molecular chaperoning, antigen presentation, degradation of abnormal proteins, and redox signaling. Most information about such functions has been obtained from mammalian cell studies. The physiological functions and functional mechanisms of plant PDI family proteins remain largely unknown. Here, we demonstrated that four of the five groups of wheat PDI family proteins exhibit oxidative refolding activity on reduced and denatured RNaseA. All five groups of PDI family proteins were highly expressed throughout the caryopses and in particular, in the starchy endosperm during grain filling. As expected from their primary structures, these proteins localized to the ER. These results suggest that PDI family proteins play an important role in folding seed storage proteins synthesized in the

ER. In mature grain, the PDI family proteins from all five groups remain in the ER surrounding the aleurone grains of the aleurone cells and the protein matrix of the starchy endosperm. Because many of the hydrolytic enzymes are synthesized and folded in the ER of the aleurone cells and secreted to the starchy endosperm to break down starch and storage proteins during the early stage of germination, these PDI family proteins may function in folding nascent hydrolytic enzymes. In conclusion, our work provides significant basic information about the physiological functions of these PDI family proteins during the wheat life cycle. In the future, it will be necessary to identify the *in vivo* substrate proteins and determine the effects of selective silencing of PDI family genes on wheat phenotypes with respect to seed storage protein folding, transport, and accumulation; seed germination; and plant body growth.

Methods

Plant material

Bread wheat plants were grown in an open field at Kyoto University. The developing caryopses were collected 147 days after seeding bread wheat plants (*Triticum aestivum* cv. Yumeshiho) for isolation of mRNA to clone wheat PDI family cDNAs. The developing caryopses from 5–35 dpa were collected from *Triticum aestivum* cv. Haruyokoi at 5-day intervals for western blot analysis and immunohistochemistry. The caryopses were frozen in liquid nitrogen and stored at −80°C immediately after harvest.

cDNA cloning

The cloning of TaPDIL1, TaPDIL2, TaPDIL3, TaPDIL4, and TaPDIL5 cDNAs was performed using RT-PCR. Total RNA was isolated using the Sepasol RNA I (Nacalai Tesque, Kyoto, Japan) reagent according to the manufacturer's protocol. The amplification of cDNA from total RNA was performed with a Prime Script II RTase (TaKaRa Bio Inc., Shiga, Japan) using the following oligonucleotide primers: 5'-CATGGCGATCTCCAAGGTC-3' and 5'-CTC TACCTGTCTGCTGCTAGC-3' for TaPDIL1 (GenBank: AJ277379), 5'-TGGACTGAATTCGGCGGATCCATTTC CACTCC CACTTCCCCCAACG-3' and 5'-CATCCCCT GGTTTCGGCGTCGGCTC-3' for TaPDIL2 (GenBank: FN555316), 5'-TGGACTGAATTCCCCTCTCCTAGATC TCGGAGGAGGAGCGC-3' and 5'-TGGACTGAATTCA TGGCTACTGCGTAACCGTGACCAACCCCTAC-3' for TaPDIL3 (GenBank:FN555317), 5'-TGGACTCTCGAGG TGCAAGAAGAACAGGTGCCAACCG-3' and 5'-TGGA CTCTCGAGCCGCTAAACTTTCACTGCCATCTCTCT GATCTC-3' for TaPDIL4 (GenBank:FN555318), and 5'-TGGACTAAGCTTCCGGCTTCCAGAAATTTTTCA ACGACGC-3' and 5'-TGGACTAAGCTTCCACCTTG CACATCAGAGCTTTCTCCCAC-3' for TaPDIL5 (GenBank:FN555320). Nested PCRs were performed with the DNA fragments amplified by the RT-PCR as a template using the following oligonucleotide primers: 5'-ATGGC GATCTCCAAGGTCTGGATC-3' and 5'- CTGCTGCTA GCAAGACTGATGC-3' for TaPDIL1, 5'-CCATGGCGG CGATGCCGATGC-3' and 5'-GGCTCCTACTTGTTGT CAATGGTG-3' for TaPDIL2, 5'-GGAGGAGCGCGAT GAGGGCGACG-3' and 5'-CATTTTCCTTCAACGCGG CCAGC-3' for TaPDIL3, 5'-GGTCA CCCGAGCTCG CAG-3' and 5'-CTTCACT TCTCTCTTGTGGC-3' for TaPDIL4, and 5'-TGGACTAAGCTTCCGGCTTCCAGA AATTTTTCAACGACGC-3' and 5'-CACATCAGAGCA AGTGAAGC-3' for TaPDIL5. The amplified TaPDIL1, TaPDIL2, and TaPDIL3 DNA fragments were subcloned into pCR Blant II-TOPO (Invitrogen, Carlsbad, CA) and transformed in *E. coli* DH5. The amplified TaPDIL4 and TaPDIL5 DNA fragments were subcloned into T-Vector pMD20 (TaKaRa Bio Inc.) and transformed into *E. coli* DH5 cells. The inserts in the plasmid vectors were sequenced using the fluorescence dideoxy chain termination method with an ABI PRISM® 3100-Avant Genetic Analyzer (Applied Biosystems, Foster City, CA).

Construction of His-tagged expression plasmids

Expression plasmids encoding mature His-tagged TaPDIL1Aα, TaPDIL1B, TaPDIL2, TaPDIL3A, TaPDIL4D, and TaPDIL5A without the putative signal peptide were constructed. Briefly, the DNA fragment was amplified from the cDNA of TaPDIL1Aα, TaPDIL1B, TaPDIL2, TaPDIL3A, TaPDIL4D, and TaPDIL5A by PCR using the following oligonucleotide primers: 5'-GACGACGACAA GATG GAGGAGGCCGCCGCCGCCGAGGAG-3' and 5'-GAGGAGAAGCCCGGTCAGAGCTCGTCCTTCAG AGGCTC-3' for TaPDIL1Aα and TaPDIL1B, 5'-GACG ACGACAAGATGGCAGTCCCCACCTCCAACCCGG-3' and 5'-GAGGAGAAGCCCGGTCACAACTCGTCCTTC GGGTTCGAAC-3' for TaPDIL2, 5'-GACGACGACAA GATG GCGAAGCTCGATCTGGACGAGGTG-3' and 5'-GAGGAGAAGCCCGGTCATAGCTCATCCTTGACA TTGTC-3' for TaPDIL3A, 5'-GACGACGACAAGATGG ACGAGGTGCTTGCCCTCACGGAG-3' and 5'-GAGG AGAAGCCCGGTCAAGAGGAGAAGGCTGAAAGAA TG-3' for TaPDIL4D, and 5'-GACGACGACAAGATGC TCTACTCCGCCGGCTCCCCGGTC-3' and 5'-GAGG AGAAGCCCGGTCACAACTCGTCGTTGGGCGCAGA G-3' for TaPDIL5A. The amplified DNA fragment was subcloned into the ligation-independent cloning site of the pET46Ek/LIC vector (EMD Biosciences, Inc., San Diego, CA). The recombinant proteins have a His-tag linked to the amino terminus.

Expression and purification of recombinant wheat PDI family proteins

E. coli BL21(DE3) cells were transformed with the expression vectors as described above. The expression of recombinant proteins was induced by the addition of 0.4 mM isopropyl thiogalactoside at 30°C for 20 h. All expressed recombinant proteins were soluble in *E. coli.* The cells from 1 L culture broth were collected by centrifugation, washed twice by suspending in phosphate-buffered saline and centrifugation, disrupted by sonication in 8 mL of 20 mM Tris (hydroxymethyl) aminomethane (Tris)-HCl buffer (pH 8) containing 5 mM imidazole and 0.5 M NaCl (binding buffer), and then centrifuged at 10,000 × g for 30 min at 4°C. The supernatant was filtered through a Millex Syringe-driven filter 33 mm (Millipore Corporation, Billerica, MA) and applied to a column packed with Ni Sepharose 6 Fast Flow (GE Healthcare, Piscataway, NJ). After washing the column with binding buffer containing 60 mM imidazole, recombinant proteins were eluted with binding buffer containing 1 M imidazole, concentrated with a Vivaspin 20 (GE Healthcare), and then subjected to gel filtration chromatography on a TSK gel G3000SW column (Tosoh, Tokyo, Japan) equilibrated with 20 mM Tris–HCl buffer (pH 7.4) containing 0.15 M NaCl and 10% glycerol. Recombinant proteins that eluted in the inside volume fractions were collected. The concentrations of purified recombinant proteins were determined from their absorbance at 280 nm using the molar extinction coefficients calculated with the modified method of Gill and von Hippel [67]. Extinction coefficients of 43,820, 42,330, 40,840, 33,265, 36,620, and 54,275 M^{-1} cm^{-1} were used for recombinant TaPDIL1Aα, TaPDIL1B, TaPDIL2, TaPDIL3A, TaPDIL4D, and TaPDIL5A, respectively.

The CD spectra of recombinant proteins in 20 mM Tris–HCl buffer (pH 7.4) containing 150 mM NaCl and 10% glycerol were obtained using a J-720 spectropolarimeter (JASCO Corp., Tokyo, Japan) in a 1-mm path-length cell with a scan speed of 20 nm/min at 14°C.

RNaseA refolding assay

Oxidative refolding activity was assayed by measuring the RNase activity produced through the regeneration of the active form from reduced and denatured RNaseA that was prepared as described previously by Creighton [68]. Each reaction mixture comprised 180 mM [4-(2-hydroxyethyl)-1-piperazinyl]ethanesulfonic acid (pH 7.5), 150 mM NaCl, 2 mM $CaCl_2$, 0.5 mM glutathione disulfide, 2 mM glutathione, 1 mg/mL reduced RNaseA, and 0.25 μM recombinant wheat PDI family proteins. The reaction mixture was incubated at 25°C. An aliquot (16 μl) of the reaction mixture was removed, and RNaseA activity was measured spectrophotometrically at 284 nm with cytidine2':3'-cyclic monophosphate as the substrate [69]. Reactivation of reduced RNaseA in the absence of recombinant protein was subtracted from reactivation in the presence of recombinant wheat PDI family proteins.

Antibodies

Antisera were prepared by Japan Bio Serum (Hiroshima, Japan). Purified recombinant TaPDIL1Aα, TaPDIL2, TaPDIL3A, TaPDIL4D, and TaPDIL5A were emulsified with Freund's complete adjuvant and intradermally injected into female guinea pigs. Anti-calreticulin and anti-Ero1 guinea pig sera were prepared with recombinant soybean calreticulin and soybean Ero1 expressed in *E. coli* as described previously [16]. Anti-gliadin rabbit serum was purchased from Sigma-Aldrich. Inc. (St. Louis, MO).

Western blot analysis

Wheat caryopses that had been frozen in liquid nitrogen were ground into fine powders with a Multi-beads Shocker (Yasui Kikai. Osaka, Japan) [70]. Proteins were extracted from ground caryopses by boiling for 5 min in sodium dodecyl sulfate -polyacrylamide gel electrophoresis (SDS-PAGE) sample buffer [71] containing a 1% cocktail of protease inhibitors (Sigma-Aldrich). Debris in the sample was removed by centrifugation at 5000 × *g* for 20 min. The concentrations of proteins in the samples were measured with a protein assay kit (RC DC protein assay, Bio-Rad Laboratories, Hercules, CA). To cleave the N-glycan modifications of the proteins by endoglycosidase H (Sigma-Aldrich) or endoglycosidase F (Sigma-Aldrich), proteins were extracted from the caryopses in 2% SDS, 0.1 M Tris–HCl (pH 8.6), 1% Nonidet P-40. The extract was diluted 20-fold for glycosidase H in 0.1 M Tris–HCl (pH 8.6)/1% Nonidet P-40 or for glycosidase F in 50 mM potassium phosphate buffer (pH 5.5) containing 0.25% 2-mercaptoethanol. Proteins (0.4 mg) were treated with 10 mU endoglycosidase H or 125 mU endoglycosidase F at 37°C for 16 h. Proteins were subjected to SDS-PAGE [71] and blotted onto polyvinylidene difluoride membranes. The proteins were then immunostained with the specific guinea pig or rabbit antisera as primary antibodies and horseradish peroxidase-conjugated anti-guinea pig IgG serum (Promega Corporation, Madison, WI) or anti-rabbit IgG serum (Santa Cruz Biotechnology, Inc., Dallas, TX) as secondary antibodies. The bands were visualized using the Western Lightning Plus-ECL Enhanced Chemiluminescence Substrate (Perkin Elmer Life Sciences, Boston, MA).

Semi-quantitative assay of wheat PDI family proteins in caryopses

Proteins extracted from the developing caryopses were separated along with known amounts of recombinant PDI family proteins on the same gel by SDS-PAGE and detected by western blot analysis as described above. The amount of each PDI family protein was calculated from the band intensities of the sample and the corresponding recombinant protein. Recombinant TaPDIL1Aα, TaPDIL2, TaPDIL3A, TaPDIL4D, and TaPDIL5A were used as standards for the semi-quantitative analysis of TaPDIL1, TaPDIL2, TaPDIL3, TaPDIL4, and TaPDIL5, respectively.

Microscopic observation

Caryopses from developing and mature grains were cut and fixed with 4% paraformaldehyde in 80 mM piperazine-1,4-bis (2-ethanesulfonic acid) at pH 6.5, 5 mM ethylene glycol tetraacetic acid, and 2 mM $MgCl_2$ for 2 h at room temperature. The fixed caryopses were dehydrated with a series of 50, 50, 70, 70, 80, 90, 95, and 99.5% ethanol for 40 min each at room temperature. The dehydrated caryopses were embedded in Historesin (Leica Microsystems, Wetzlar, Germany) and sliced into sections with a rotary microtome PR-50 (Yamato Kohki Industrial Co., Ltd., Saitama, Japan). The sections were mounted on glass slides and stained with guinea pig antisera against TaPDIL1Aα, TaPDIL2, TaPDIL3A, TaPDIL4D, and TaPDIL5A and subsequently with a secondary antibody, which was Cy3-conjugated anti-guinea pig IgG goat serum (Chemicon International, Temecula, CA). For detection of calreticulin, specimens were stained with an anti-calreticulin rabbit serum followed by a biotin-anti-rabbit IgG goat serum (Cortex Biochem, San Leandro, CA) and incubation with Cy5-streptavidin (GE Healthcare Bio-Sciences Corp.). The specimens were examined on a stereomicroscope SZX16 equipped with an SZX2-FGFP filter and a DP73 camera (Olympus Co., Tokyo, Japan), a Confocal Imaging System FV1200 (Olympus), and a fluorescence microscope BZ-9000 (Keyence Corp., Osaka, Japan).

Additional files

Additional file 1: Figure S1. Alignment of amino acid sequences of TaPDIL1Aα, TaPDIL1Aβ, TaPDIL1Aγ, TaPDIL1Aδ, and TaPDIL1B.

Additional file 2: Figure S2. Alignment of amino acid sequences of TaPDIL1Aα and soybean GmPDIL-1.

Additional file 3: Figure S3. Alignment of amino acid sequences of TaPDIL2 and TaPDIL2-1.

Additional file 4: Figure S4. Alignment of amino acid sequences of TaPDIL2 and soybean GmPDIL-2.

Additional file 5: Figure S5. Alignment of amino acid sequences of TaPDIL3A and soybean GmPDIL-3a.

Additional file 6: Figure S6. Alignment of amino acid sequences of TaPDIL4D, soybean GmPDIS-1, and soybean GmPDIS-2.

Additional file 7: Figure S7. Alignment of amino acid sequences of TaPDIL5A and soybean GmPDIM.

Additional file 8: Figure S8. Caryopsis weight after anthesis.

Additional file 9: Figure S9. Dot blot analysis of recombinant wheat PDI family proteins.

Abbreviations

CD: Circular dichroism; dpa: Days post anthesis; ER: Endoplasmic reticulum; Ero1: ER oxidoreductin 1; *Escherichia coli*: *E. coli*; PDI: protein disulfide isomerase; RNaseA: Ribonuclease A; SDS-PAGE: Sodium dodecyl sulfate-polyacrylamide gel electrophoresis; Tris: Tris (hydroxymethyl) aminomethane.

Competing interests

The authors declare that they have no competing interests.

Authors' contributions

RU designed experiments and drafted the manuscript. SK, YK, and TM performed sample preparation and experimental procedures for the characterization of the PDI family genes. SK performed experiments to assess recombinant PDI family protein expression and enzyme assays. SK and YH performed sample preparation, western blot analysis, and microscope analysis. SK and RU performed data analyses. All the authors discussed the results. RU provided financial support for the study. All authors reviewed, discussed, and approved the final manuscript.

Acknowledgements

We thank Dr. Tatsuhiko Shiraiwa and Dr. Koki Homma of Kyoto University for providing the Yumeshiho wheat grains. We thank Dr. Koichi Nagasawa, Dr. Miwako Ito, and Dr. Tadashi Tabiki for providing the Haruyokoi wheat flour. This work was supported in part by grants from the Takano Foundation.

References

1. Levanony H, Rubin R, Altschuler Y, Galili G. Evidence for a novel route of wheat storage proteins to vacuoles. J Cell Biol. 1992;119:1117–28.
2. Galili G, Shimoni Y, GioriniSilfen S, Levanony H, Altschuler Y, Shani N. Wheat storage proteins: Assembly, transport and deposition in protein bodies. Plant Physiol Biochem. 1996;34:245–52.
3. Tosi P, Parker M, Gritsch CS, Carzaniga R, Martin B, Shewry PR. Trafficking of storage proteins in developing grain of wheat. J Exp Bot. 2009;60:979–91.
4. Shewry PR, Tatham AS. Disulphide bonds in wheat gluten proteins. J Cereal Sci. 1997;25:207–27.
5. Bulleid NJ, Freedman RB. Defective co-translational formation of disulphide bonds in protein disulphide-isomerase-deficient microsomes. Nature. 1988;335:649–51.
6. Orsi A, Sparvoli F, Ceriotti A. Role of individual disulfide bonds in the structural maturation of a low molecular weight glutenin subunit. J Biol Chem. 2001;276:32322–9.
7. Altschuler Y, Galili G. Role of conserved cysteines of a wheat gliadin in its transport and assembly into protein bodies in xenopus oocytes. J Biol Chem. 1994;269:6677–82.
8. Shimoni Y, Galili G. Intramolecular disulfide bonds between conserved cysteines in wheat gliadins control their deposition into protein bodies. J Biol Chem. 1996;271:18869–74.
9. Hatahet F, Ruddock LW. Protein disulfide isomerase: A critical evaluation of its function in disulfide bond formation. Antioxid Redox Signal. 2009;11:2807–50.
10. Darby NJ, Kemmink J, Creighton TE. Identifying and characterizing a structural domain of protein disulfide isomerase. Biochemistry. 1996;35:10517–28.
11. Lu J, Holmgren A. The thioredoxin superfamily in oxidative protein folding. Antioxid Redox Signal. 2014;21:457–70.
12. Houston NL, Fan C, Xiang QY, Schulze JM, Jung R, Boston RS. Phylogenetic analyses identify 10 classes of the protein disulfide isomerase family in plants, including single-domain protein disulfide isomerase-related proteins. Plant Physiol. 2005;137:762–78.
13. Ondzighi CA, Christopher DA, Cho EJ, Chang S-C, L. Staehelina LA. Arabidopsis protein disulfide isomerase-5 inhibits cysteine proteases during trafficking to vacuoles before programmed cell death of the endothelium in developing seeds. Plant Cell. 2000;20:2205–20.
14. Wang H, Boavida LC, Ron M, McCormick S. Truncation of a protein disulfide isomerase, pdil2-1, delays embryo sac maturation and disrupts pollen tube guidance in *Arabidopsis thaliana*. Plant Cell. 2008;20:3300–11.
15. Gruber CW, Čemažar M, Clark RJ RJ, Horibe T, Renda RF, Anderson MA, et al. A novel plant protein-disulfide isomerase involved in the oxidative folding of cystine knot defense proteins. J Biol Chem. 2007;282:20435–46.
16. Wadahama H, Kamauchi S, Ishimoto M, Kawada T, Urade R. Protein disulfide isomerase family proteins involved in soybean protein biogenesis. FEBS J. 2007;274:687–703.
17. Kamauchi S, Wadahama H, Iwasaki K, Nakamoto Y, Nishizawa K, Ishimoto M, et al. Molecular cloning and characterization of two soybean protein

disulfide isomerases as molecular chaperones for seed storage proteins. FEBS J. 2008;275:2644–58.
18. Wadahama H, Kamauchi S, Nakamoto Y, Nishizawa K, Ishimoto M, Kawada T, et al. A novel plant protein disulfide isomerase family homologous to animal P5 – molecular cloning and characterization as a functional protein for folding of soybean seed-storage proteins. FEBS J. 2008;275:399–410.
19. Iwasaki K, Kamauchi S, Wadahama H, Ishimoto M, Kawada T, Urade R. Molecular cloning and characterization of soybean protein disulfide isomerase family proteins with nonclassic active center motifs. FEBS J. 2009;276:4130–41.
20. Takemoto Y, Coughlan SJ, Okita TW, Satoh H, Ogawa M, Kumamaru T. The rice mutant esp2 greatly accumulates the glutelin precursor and deletes the protein disulfide isomerase. Plant Physiol. 2002;128:1212–22.
21. Satoh-Cruz M, Crofts AJ, Takemoto-Kuno Y, Sugino A, Washida H, Crofts N, et al. Protein disulfide isomerase Like 1–1 participates in the maturation of proglutelin within the endoplasmic reticulum in rice endosperm. Plant Cell Physiol. 2005;51:1581–93.
22. Kim YJ, Yeu SY, Park BS, Koh H-J, Song JT, Seo HS. Protein disulfide isomerase-like protein 1–1 controls endosperm development through regulation of the amount and composition of seed proteins in rice. PLOS ONE. 2012;7:e44493.
23. Onda Y, Nagamine A, Sakurai M, Kumamaru T, Ogawa M, Kawagoe Y. Distinct roles of protein disulfide isomerase and P5 sulfhydryl oxidoreductases in multiple pathways for oxidation of structurally diverse storage proteins in rice. Plant Cell. 2011;2:210–23.
24. Grynberg A, Nicolas J, Drapron R. Some characteristics of protein disulfide isomerase (E.C.5.3.4.1) from wheat (*Triticum vulgare*) embryo. Biochimie. 1978;60:547–51.
25. Roden LT, Miflin BJ, Freedman RB. Protein disulfide-isomerase is located in the endoplasmic-reticulum of developing wheat endosperm. FEBS Letts. 1982;138:121–4.
26. DE Azevedo GMV, Brockway BE, Freedman RB, Greenwell P, Roden LT. Protein disulfide-isomerase of mature and developing wheat. Biochem Soc Trans. 1984;12:1043.
27. Shimoni Y, Segal C, Zhu X-z, Galili C. Nucleotide sequence of a wheat cDNA encoding protein disulfide isomerase. Plant Physiol. 1995;107:281.
28. Shimoni Y, Zhu X-z, Levanony H, Segal G, Galili C. Purification, characterization, and Intracellular localization of glycosylated protein disulfide isomerase from wheat grains. Plant Physiol. 1995;108:327–35.
29. Cia M, Dominici L, Tanzarella OA, Porceddu E. Chromosomal assignment of gene sequences coding for protein disulphide isomerase (PDI) in wheat. Theor Appl Genet. 1999;98:405–10.
30. Johnson JC, Clarke BC, Bhave M. Isolation and characterisation of cDNAs encoding protein disulphide isomerases and cyclophilins in wheat. J Cereal Sci. 2001;34:159–71.
31. Ciafi M, Paolacci AR, Dominici L, Tanzarella OA, Porceddu E. Molecular characterization of gene sequences coding for protein disulide isomerase (PDI) in durum wheat (*Triticum turgidum* ssp. durum). Gene. 2001;265:147–56.
32. Johnson JC, Bhave M. Molecular characterization of the protein disulphide isomerase genes of wheat. Plant Sci. 2004;167:397–410.
33. Ciaffi M, Paolacci AR, D'Aloisio E, Tanzarella OA, Porceddu E. Cloning and characterization of wheat PDI (protein disulfide isomerase) homoeologous genes and promoter sequences. Gene. 2006;366:209–18.
34. d'Aloisio E, Paolacci AR, Dhanapal AP, Tanzarella OA, Porceddu E, Ciaffi M. The protein disulfide isomerase gene family in bread wheat (*T. aestivum* L.). BMC Plant Biol. 2010;10:101.
35. Dhanapal AP, Ciaffi M, Porceddu E, d'Aloisio E. Protein disulphide isomerase promoter sequence analysis of *Triticum urartu*, *Aegilops speltoides* and *Aegilops tauschii*. Plant Genet Resour: Char Util. 2011;9:338–41.
36. d'Aloisio E, Paolacci AR, Dhanapal AP, Tanzarella OA, Porceddu E, Ciaffi M. Protein disulphide isomerase family in bread wheat (*Triticum aestivum* L.): genomic structure, synteny conservation and phylogenetic analysis. Plant Genet Resour: Char Util. 2011;9:342–6.
37. Wu H, Dorse S, Bhave M. In silico identification and analysis of the protein disulphide isomerases in wheat and rice. Biologia. 2012;67:48–60.
38. Munro S, Pelham HR. An Hsp70-like protein in the ER: identity with the 78 kd glucose-regulated protein and immunoglobulin heavy chain binding protein. Cell. 1986;46:291–300.
39. Munro S, Pelham HR. A C-terminal signal prevents secretion of luminal ER proteins. Cell. 1987;48:899–907.
40. The International Wheat Genome Sequencing Consortium (IWGSC): A chromosome-based draft sequence of the hexaploid bread wheat (*Triticum aestivum*) genome. Science. 2014, 345: 1251788-1–1251788-11.
41. Barbazuk WB, Fu Y, McGinnis KM. Genome-wide analyses of alternative splicing in plants: Opportunities and challenges. Genome Res. 2008;18:1381–92.
42. Lappi AK, Lensink MF, Alanen HI, Salo KE, Lobell M, Juffer AH, et al. A conserved arginine plays a role in the catalytic cycle of the protein disulphide isomerases. J Mol Biol. 2004;335:283–95.
43. Ellgaard L, Ruddock LW. The human protein disulphide isomerase family: substrate interactions and functional properties. EMBO Rep. 2005;6:28–32.
44. LeMaster DM, Springer PA, Unkefer CJ. The role of the buried aspartate of Escherichia coli thioredoxin in the activation of the mixed disulfide intermediate. J Biol Chem. 1997;1997(272):29998–30001.
45. Demmer J, Zhou CM, Hubbard MJ. Molecular cloning of ERp29, a novel and widely expressed resident of the endoplasmic reticulum. FEBS Lett. 1997;402:145–50.
46. Kikuchi M, Doi E, Tsujimoto I, Horibe T, Tsujimoto Y. Functional analysis of human P5, a protein disulfide isomerase homologue. J Biochem. 2002;132:451–5.
47. Zapun A, Darby NJ, Tessier DC, Michalak M, Bergeron JJ, Thomas DY. Enhanced catalysis of ribonuclease B folding by the interaction of calnexin or calreticulin with ERp57. J Biol Chem. 1998;273:6009–12.
48. Oliver JD, Roderick HL, Llewellyn DH, High S. ERp57 functions as a subunit of specific complexes formed with the ER lectins calreticulin and calnexin. Mol Biol Cell. 1999;10:2573–82.
49. Molinari M, Helenius A. Glycoproteins form mixed disulphides with oxidoreductases during folding in living cells. Nature. 1999;402:90–3.
50. Urade R, Okudo H, Kato H, Moriyama T, Arakaki Y. ER-60 domains responsible for interaction with calnexin and calreticulin. Biochemistry. 2004;43:8858–68.
51. Frand A, Kaiser C. The ERO1 gene of yeast is required for oxidation of protein dithiols in the endoplasmic reticulum. Mol Cell. 1998;1:161–70.
52. Pollard M, Travers K, Weissman J. Ero1p: A novel and ubiquitous protein with an essential role in oxidative protein folding in the endoplasmic reticulum. Mol Cell. 1998;1:171–82.
53. Cabibbo A, Pagan M, Fabbri M, Rocchi M, Farmery M, Bulleid N, et al. ERO1-L, a human protein that favors disulfide bond formation in the endoplasmic reticulum. J Biol Chem. 2000;275:4827–33.
54. Araki K, Inaba K. Structure, mechanism, and evolution of Ero1 family enzymes. Antioxid Redox Signal. 2012;16:790–9.
55. Ling HQ, Zhao S, Liu D, Wang J, Sun H, Zhang C, et al. Draft genome of the wheat A-genome progenitor *Triticum urartu*. Nature. 2013;496:87–90.
56. Onda Y, Kumamaru T, Kawagoe Y. ER membrane-localized oxidoreductase Ero1 is required for disulfide bond formation in the rice endosperm. Proc Natl Acad Sci U S A. 2009;106:14156–61.
57. Zheng Y, Wang Z. Protein accumulation in aleurone cells, sub-aleurone cells and the center starch endosperm of cereals. Plant Cell Rep. 2015;33:1607–15.
58. Buckhout TJ, Gripshover BM, Morre DJ. Endoplasmic reticulum formation during germination of wheat seeds. A quantitative electron microscope study. Plant Physiol. 1981;68:1319–22.
59. Bradbury D, MacMasters MM, Cull IM. Structure of the mature wheat kernel. III. Microscopic structure of the endosperm of the hard red winter wheat. Cereal Chem. 1956;33:361–73.
60. Tkachuk R R, Kruger JE. Wheat alpha-amylases. II. Physical characterization. Cereal Chem. 1974;51:508–28.
61. Kiyosaki T, Asakura T, Matsumoto I, Tamura T, Terauchi K, Funaki J, et al. Wheat cysteine proteases triticain alpha, beta and gamma exhibit mutually distinct responses to gibberellin in germinating seeds. J Plant Physiol. 2009;166:101–6.
62. Mirbahar RB, Laidman DL. Gibberellic acid-stimulated a-amylase secretion and phospholipid metabolism in wheat aleurone tissue. Biochem J. 1982;208:93–100.
63. Kiyosaki T, Matsumoto I, Asakura T, Funaki J, Kuroda M, Misaka T, et al. Gliadainm a gibberellin-inducible cysteine proteinase occurring in germinationg seeds of wheat, *Triticum aestivum* L., specifically digests gliadin and is regulatd by intrinsic cystatins. FEBS J. 2007;274:1908–17.
64. Barba-Espín G, Dedvisitsakul P, Hägglund P, Svensson B, Finnie C. Gibberellic acid-induced aleurone layers responding to heat shock or tunicamycin provide insight into the N-glycoproteome, protein secretion, and endoplasmic reticulum stress. Plant Physiol. 2014;164:951–65.
65. Loussert C, Popineau Y, Mangavel C. Protein bodies ontogeny and localization of prolamin components in the developing endosperm of wheat caryopses. J Cereal Sci. 2008;47:445–56.
66. Koh A, Nishimura K, Urade R. Relationship between endogenous protein disulfide isomerase family proteins and glutenin macropolymer. J Agric Food Chem. 2010;58:12970–5.

67. Pace CN, Vajdos F, Fee L, Grimsley G, Gray T. How to measure and predict the molar absorption coefficient of a protein. Protein Sci. 1995;4:2411–23.
68. Creighton TE. Kinetics of refolding of reduced ribonuclease. J Mol Biol. 1977;113:329–41.
69. Lyles MM, Gilbert HF. Catalysis of the oxidative folding of ribonuclease A by protein disulfide isomerase: dependence of the rate on the composition of the redox buffer. Biochemistry. 1991;30:613–9.
70. Akiyama H, Watanabe T, Wakabayashi K, Nakade S, Yasui S, Sakata K, et al. Quantitative detection system for maize sample containing combined-trait genetically modified maize. Anal Chem. 2005;77:7421–8.
71. Laemmli UK. Cleavage of structural proteins during the assembly of the head of bacteriophage T4. Nature. 1970;227:680–5.

Metabolic patterns associated with the seasonal rhythm of seed survival after dehydration in germinated seeds of *Schismus arabicus*

Bing Bai[1,2], David Toubiana[1], Tanya Gendler[1], Asfaw Degu[1], Yitzchak Gutterman[1] and Aaron Fait[1*]

Abstract

Background: Seed of *Shismus arabicus*, a desert annual, display a seasonal tolerance to dehydration. The occurrence of a metabolic seasonal rhythm and its relation with the fluctuations in seed dehydration tolerance was investigated.

Results: Dry seeds metabolism was the least affected by the season, while the metabolism of germinated and dehydrated seeds exhibit distinct seasonal patterns. Negative associations exist between amino acids, sugars and TCA cycle intermediates and seed survival, while positive relations exist with seed germination. In contrast, associations between the level of secondary metabolites identified in the dehydrated seeds and survival percentage were evenly distributed in positive and negative values, suggesting a functional role of these metabolites in the establishment of seed dehydration tolerance.

Conclusion: Our results indicate the occurrence of metabolic biorhythms in germinating and dehydrating seeds associated with seasonal changes in germination and, more pronouncedly, in seed dehydration tolerance. Increased biosynthesis of protective compounds (polyphenols) in dehydrating seeds during the winter season at the expenses of central metabolites likely contributes to the respective enhanced dehydration tolerance monitored.

Keywords: Seed germination, Seed survival, Dehydration, Metabolomics, Annual rhythm

Background

Seed germination is affected by the genetic background and the environmental conditions during seed development and post-dispersal [1,2]. The main factor regulating seed germination is the availability of water, which initiates the seed metabolism through water uptake by rehydrating membranes and oxygenating the inner parts of the seed [3]. Other determinants, such as day length [4], temperature [5] and osmoticum [6] also can modulate seed germination. Germination in arid environments exposes germinated seeds to unpredictable rainfall and prolonged period of drought. Hence, mechanisms for regulation of germination and tolerance to dehydration have evolved determining the degree and timing for germination, a trait, which likely evolved in conjunction with seed survival following dehydration [7]. A significant amount of knowledge has been accumulated on the annual periodicity of the germinability of stored seeds [8-11]. *Arabidopsis* seeds were reported to follow an annual dormancy cycling by an altering sensitivity to the environmental stimulus such as temperature, light and nitrite in different seasons [12,13]. In weed seeds annual dormancy cycles are linked to a continuum of physiological changes possibly related to changes in membrane properties [14] such as the fluidity and membrane protein conformation [15], likely promoting gas exchange in the inner parts of the seed altering its redox state. Reactive oxygen species and nitric oxide were recently suggested to be involved in the regulation of dormancy [16]. Whilst annual periodicity in dormancy of seeds received much attention, more elusive phenomena were shown to be seasonal dependent. For example, *Digitalis purpurea* L. seeds germinated over a period of 13 months under controlled condition were shown to vary in their content of sterol at the same germination stage [17]. The seasonal regulation

* Correspondence: fait@bgu.ac.il
[1]Ben-Gurion University of the Negev, Jacob Blaustein Institutes for Desert Research, French Associates Institute for Agriculture and Biotechnology of Drylands, Midreshet Ben-Gurion 84990, Israel
Full list of author information is available at the end of the article

of sterol was found to correlate with annual cycle in germination likely for the purpose of membrane stabilization and protection during cold winter. More recently a three year study demonstrated the occurrence of annual periodicity in dehydration tolerance of germinated seeds [18]. *Schismus arabicus* Nees (Poaceae), a common fodder in Negev desert, germinated uniformly throughout the year at 80-100%; however the percentage of surviving seed to controlled dehydration experiments varied depending on the season.

Dehydration response in plants involves all levels of cellular activity [19] including metabolic reorganization [20]. For example, the biosynthesis of sugars and polyphenols play a significant role in protein and membrane protection against the effect of dehydration; trehalose, raffinose, galactinol and umbelliferose can promote the formation of protective glass matrix [21-24]; flavonoids can provide a chemical barrier by decreasing permeability to moisture [25] limiting damage during storage [26]; tocopherols - lipophilic antioxidants can limit non-enzymatic lipid oxidation during seed dehydration, storage, and early germination stages [27,28]. Recently metabolite profiling showed the induction of energy metabolism and the biosynthesis of specialized antioxidant as possibly linked with increased germination following dehydration of imbibed Arabidopsis seeds [29].

The aim of the present study was to explore the metabolic basis of seasonal periodicity in seed germination and survival following dehydration in *Shismus arabicus*.

Methods

Schismus arabicus Nees caryopses (seeds) were collected in April 2005 from a natural habitat near Sede Boker in the Negev (34°46′E 30°51′N; 460 m a.s.l). The caryopses were separated and stored in glass vials, placed into brown paper bags and stored at 40°C in darkness controlled with thermostat (Environette, Lab-Line, Illinois, USA) as described earlier [18]. In the current set of experiments only caryopses of the size 350–425 μm were used, which showed to have the highest germination rates and percentage of germination [30].

Seed germination, dehydration and seed survival measurements

Germination and dehydration experiments were conducted exactly as described in [18]. The experiment started in June 2010 lasting 12 months until May 2011. Briefly, caryopses were germinated in four replicates of 50 caryopses each on wetted (1.5 ml) Fisher No. 1 filter paper vertically positioned under in a vial 55 mm high and 33 mm in diameter. 1.5 ml of distilled water was placed at the bottom of each vial, and the vials were closed and placed at 25°C in darkness. After 24 h of wetting, the average percentage germination was determined. After 24 hours imbibition, the germinated seeds with radicle length of about 0.2-0.3 mm measured by microscope (Olympus SZ61, with scale) were transferred to dry 5 cm diameter Petri dishes and allowed to dry at 25 ± 1°C and 10–15% relative humidity (RH), measured by a thermo-hygrograph throughout the sets of experiments. Following 180 min dehydrated germinated seeds were stored in the same conditions for 21 days. After the period of dry storage, the filter papers with the dehydrated seeds were placed on petri dishes and re-wetted with 1.5 ml water. The closed petri dishes were stored first in darkness at 15°C for 48 h, and then at 15°C under low light of 100 μmol m^{-2} s^{-1}. Seeds were scored as "survived" when both root and coleoptile elongation continued after 21-d rehydration (Additional file 1c).

Extraction for the identification and quantification of metabolites

50 dry caryopses, germinated seeds and dehydrated seeds per replicate were extracted for parallel metabolite profiling as described in [31]. Seeds were homogenized using previously cooled mortar and pestle with liquid nitrogen and extracted in a pre-chilled methanol:chloroform:water extraction solution (1:2.5:1 v/v) for 30 min at 4°C shaking. Standards, i.e. 0.2 mg/ml ribitol, 1 mg/ml ampicillin in water, 1 mg/ml corticosterone in methanol and 5 mg/ml heptadecanoic acid in chloroform, were subsequently added. After centrifugation at 2,200 g, the remaining pellet was extracted in a second step with 500 μl methanol/chloroform. The extracts were combined and 500 μl of water was added to the supernatant to separate the chloroform phase from the water/methanol phase. The latter was used for metabolite analysis via GC-MS DSQII (Thermo-Fisher ltd.) and UPLC-Xevo-QTOF-MSMS (Waters ltd) exactly as described in [29]. A volume of 200 μl of water/methanol extract was reduced to dryness in vacuum. Residues were derivatized and analyzed via an established GC-MS based method adapted to seeds [32]. GC-MS data were processed by Xcalibur® and normalized by the internal standard ribitol. The UPLC raw data were recorded with the aid of MassLynx version 4.1 software (Waters ltd). Metabolites were identified by using MassLynx software and searched against metabolite database Chemspider (http://www.chemspider.com/). The quantification of the compounds is based on the relative peak response area of each mass signal after pareto scaling in the chromatograms and normalized to the tissue DW.

Statistical analysis

The significance between the germination percentage of the caryopses and percentage of seeds that survived was tested by one-way ANOVA following arcsin transformation. Principal component analysis (PCA), t-test and ANOVA were implemented using the software TMEV [33]. The term significant is used in the text for p-values lower than 0.05 ($p < 0.05$).

Network analysis

The coordinated behavior of metabolites can be delineated using graph theory, where the nodes represent metabolites and the relationship between them is demonstrated as edges. The generation of the graphs was based on the correlation analysis of all metabolites and the two physiological traits (germination and survival percentage). Prior to correlation analysis, each metabolite was normalized by its respective mean calculated across the time-point measurements. Physiological traits were arcsin transformed. In addition each component (metabolites and physiological traits) of the dataset was pareto-scaled. Normal distribution was tested across all time-points by employing a Shapiro-Wilk test. In most cases (dry seed network = 74.0%, germination network = 92.2%, dehydration network = 79.2%) the assumption of normal distribution was violated. Thus, the non-parametric Spearman's rank correlation was chosen over the parametric Pearson correlation to compute correlation coefficients.

To reconstruct a network capturing coordinated changes in metabolic and physiology profiles, first the corresponding p-value threshold Spearman rank correlation coefficient ensuring a q-value of 0.05 was determined. Second, the adequate correlation coefficient threshold was chosen by assessing four different network properties, i.e. average node degree, clustering coefficient, network density, and diameter. For a full description on these network properties the reader is referred to [34]. The correlation coefficient, at which the network displayed a robust behavior across a range of p-values in all four properties, was chosen as the threshold for network construction. Subsequently, the network was analyzed for communities by employing the walk trap community algorithm [35]. The significance of the communities with more than nine nodes was tested by performing a Wilcoxon signed rank test. The test was performed by assessing the degree of node-connectivity [34] of the isolated community as compared to the nodes of the community still embedded in the network of which all community specific edges have been subtracted.

All computations for network visualizations were generated in the R environment. The software Cytoscape [36] version 2.8.3 was used for network visualization itself. Network properties and communities were computed by using the igraph R package (Additional file 2).

Results

Seed germination and recovery after rehydration

Germination and seed survival percentage during the 2010–2011 experiment closely reflected the values measured in a previous three-year study conducted on seeds from a different harvest [18]. In detail, germination percentage of *S. arabicus* caryopses from June to September 2010 was above 90% with a decreasing trend in November to 76.5 ± 2.2% (Figure 1). From December 2010 to April 2011, the percentage of germination was kept at about 90%, followed by a significant drop in May to 66.0 ± 4.3%. This relatively stable germination percentage contrasted drastically the changes scored for seed survival following dehydration. Seed survival dropped to 0% in July 2010 and peaked during the month of January 2011 with 100% recovery in all four independent preparations with 50 seeds each (Figure 1). The seasonal survival displayed similar patterns to the meteorological data as indicated in (Figure 1 and Additional file 3), obtained from the meteorological station at the Jacob Blaustein Institutes for Desert Research of the Ben Gurion University of the Negev (http://www.bgu.ac.il/BIDR/research/phys/meteorology/). Soil and air temperature, global radiation and rainfall are all following a seasonal cycle, characterized by the high temperature, more intensive radiation and complete lack of rainfall during the summer and by colder temperatures, reduced radiation and low rainfall during the winter (Additional file 3).

Seed germination and dehydration are characterized by season-specific metabolite profiles

Metabolite profiles of central metabolism of dry, germinated and dehydrated seeds were generated (Figure 2). Principal Component Analysis (PCA) was conducted on the year-long metabolite dataset to investigate the relative impact of single metabolites and seasonal changes on metabolic shift (Figure 3 and Additional file 4). In the dry seeds, monthly metabolic changes were minor, leading to no visible separation between monthly profiles

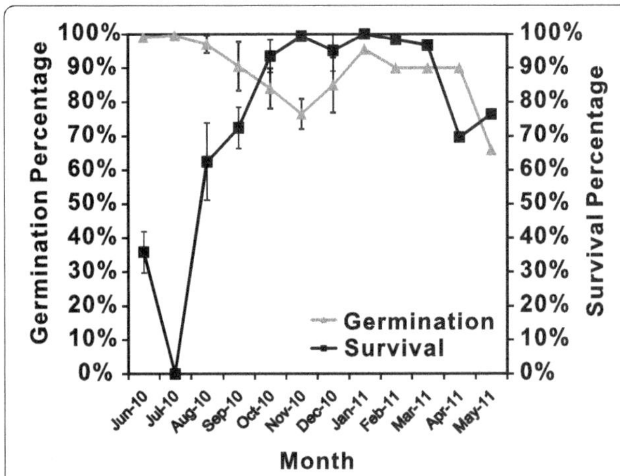

Figure 1 Monthly changes in seed germination percentage and seed survival. Seed germination percentage was measured following 24 hours of imbibition; survival percentage was measured following a three-week dehydration of the germinated seeds.

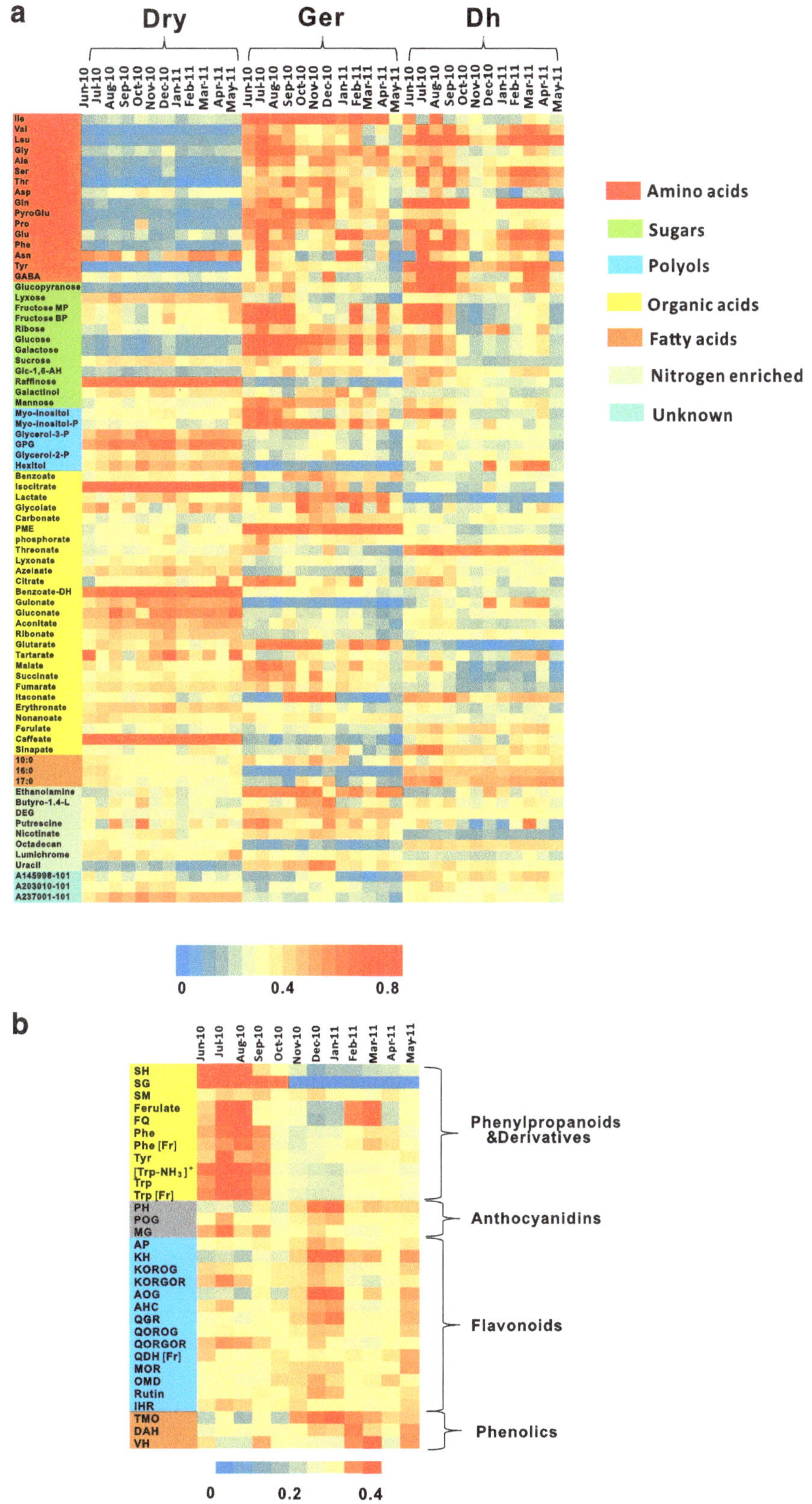

Figure 2 Heatmap visualization of primary metabolic profile of dry seeds (Dry), germinated (Ger) and dehydrated seeds (Dh) (a) and secondary metabolic profile of dehydrated seeds (b) through the year. The value of each metabolite entry was scaled between 0–1 as indicated by the color scale in the heatmap.

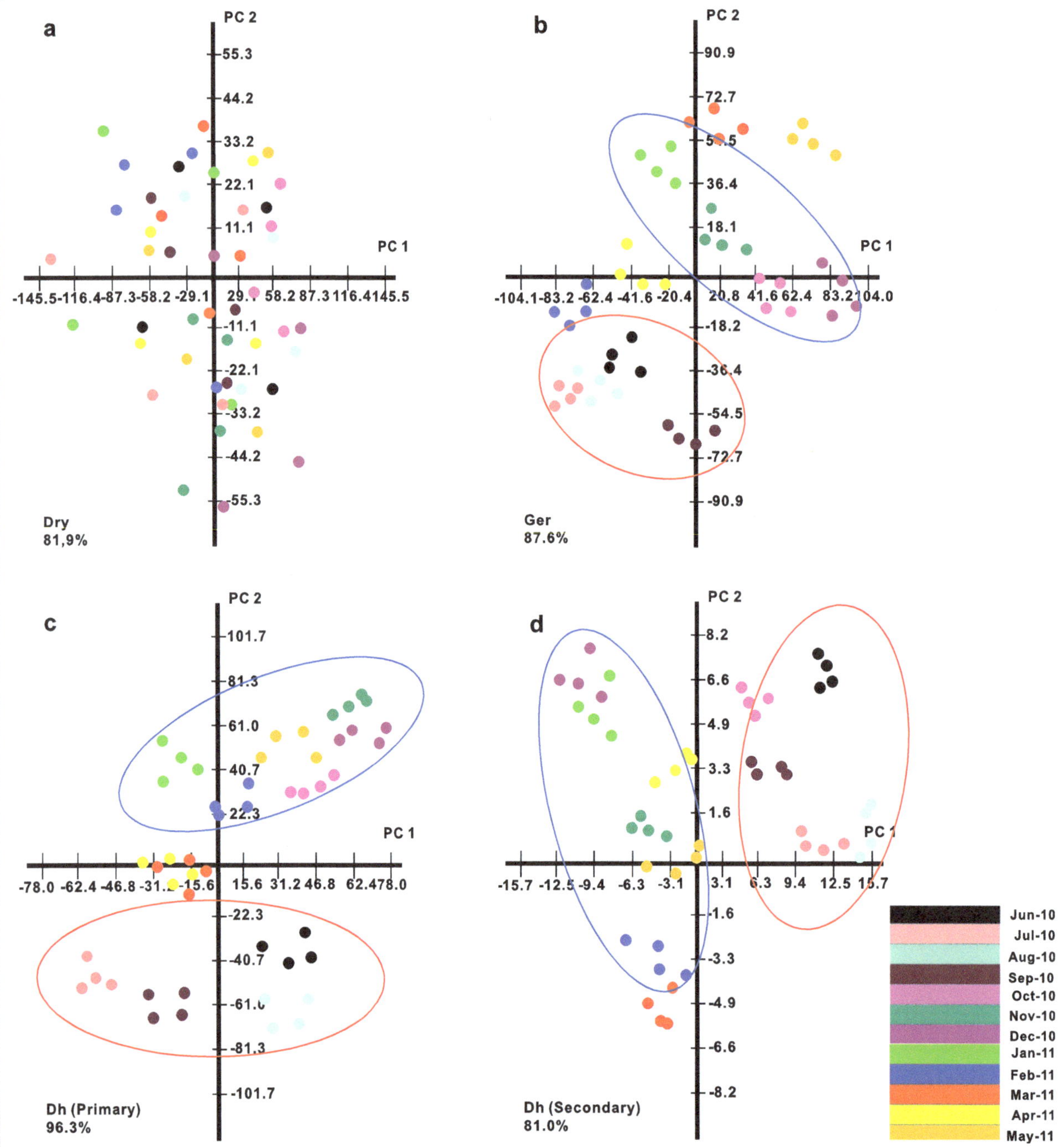

Figure 3 Principal Component Analysis (PCA) of seasonal effect on primary metabolite content of dry seeds (a), germinated seeds (b) and dehydrated seeds (c) and on secondary metabolites content of dehydrated seeds (d) through the year. The percentage of total variation explained by the first two principal components are shown. The separation of summer and winter is shown in red and blue ellipses.

(Figure 3a and Additional file 5). In contrast, germination on a monthly resolution was associated by changes in the metabolic profiles of the germinated seeds across the year. For example, notable is the separation on the 1st component of the "summer" samples (June-September) from the winter samples (October-January) (Figure 3b). Monthly dehydrated seeds also displayed a seasonal metabolite profiles in a similar but more accentuated manner than germinated seeds. Summer (June, July, August and September 2010) and winter months (November and December 2010, January and February 2011) could be distinguished on PC2 (Figure 3c). Same seasonal separation following dehydration was shown for secondary metabolites by PCA (Figure 3d).

Metabolic changes during *Schismus* seed germination and dehydration

First we investigated the common changes in metabolite profiles during germination and dehydration, independent of the season. Central and specialized metabolites followed generally conserved patterns during germination and dehydration (Figure 2 and Additional file 6a and b). As expected, seed germination was characterized by enhanced carbon - nitrogen metabolism featured by the accumulation of glucose, glucopyranose and fructose at the expenses of sucrose, the accumulation of glutamine, pyroglutamate and proline together with most amino acids, with the exception of asparagine. The decrease of aconitate and isocitrate associated to the TCA cycle was coupled to the accumulation of 2-oxoglutarate, succinate and malate. Free fatty acids and associated glycerol derivatives dropped in content in germinated seeds compared to dry ones, suggesting a role in supporting early germinative processes. Cell wall associated metabolites glucoronate, gulonate and lyxose decreased sharply during seed germination, in contrast to a 23 fold change (FC) accumulation of glucose, a 17 FC accumulation of galactose and a two fold increase of mannose, suggesting an expected repartitioning of C metabolism. Raffinose drastically decreased in content upon imbibition validating its role as transient C storage molecule as suggested earlier [37,38]. The shikimate derived and precursors of the phenylpropanoid pathway, phenylalanine (5.8 FC) and tyrosine (9.8 FC) accumulated in the germinated seeds, but not so the derived caffeate and ferulate (Additional files 5 and 6a).

Dehydration of germinated seeds resulted in attenuated effect on the abundance of most intermediates of the central metabolism (Additional file 5 and 6b). Notable was the accumulation of the non-proteinogenic amino acid GABA (1.8 FC), hydrophobic branched chain amino acids glutamine (12 FC), valine (1.6 FC), leucine (2.8 FC), and serine (2.2 FC) and shikimate derived phenylalanine (1.6 FC) and tyrosine (3.0 FC). Also ascorbate precursor galactose and derived threonate, raffinose and pentose phosphate pathway intermediates gluconate, gulonate and lyxonate accumulated during dehydration. Dehydration induced the activation of the phenylpropanoid pathway reflected by the accumulation of precursor amino acids and representative phenylpropanoids sinapate, caffeate and ferulate (Figure 2 and Additional files 5 and 6b).

Seasonal impact on germinated and dehydrated seeds

Germination percentage in the summer and winter were relatively stable (Figure 1), while the metabolism of germinated seeds displayed a seasonal pattern (Additional files 6c and 7). In the summer, a significant ($p < 0.05$) higher abundance was observed in glycolysis intermediates glucose, fructose and intermediates associated with energy production succinate, fumarate and malate in TCA cycles as compared with the seeds germinated in the winter months. The latter were instead characterized by relatively higher itaconate, tartarate, glycolate, glycerol derivatives and identified fatty acids, and also by significantly higher level of intermediates of the pentose phosphate pathway, in contrast to 1/4 the content of myo-inositol. Seeds germinated in the winter showed also accumulation of phenolic compounds ferulate, sinapate and "protective" sugars such as galactinol and to a lesser extent raffinose and Additional file 5.

Seasonal changes significantly affected the seed survival percentage following dehydration, which was measured at an average of 42.7% in the summer compared with 98.2% in the winter (Additional file 8).

Metabolite profiles on dehydrated seeds during the season (Additional file 6d) generally followed a similar pattern to the one observed in germinated seeds. Hence the identified shifts in metabolism in germinated seeds from summer to winter might be functional to seed survival upon dehydration. In the summer, the carbon pool, particularly of the sugars and TCA cycle intermediates was greater than in the winter, suggesting a lower carbon partitioning rate (Additional file 6d). Dehydrated seeds showed a seasonal trend in the free pool of amino acids being greater in the summer. Similarly to central metabolism, secondary metabolites showed seasonal specific profiles (Figure 2b, Additional files 6d and 7). Ten phenylpropanoids were accumulated mainly in the summer together with the two aromatic amino acids phenylalanine and tyrosine (Figure 2b). However, down-stream phenylpropanoid derived phenolic compound flavonoids (thirteen out of fourteen) and anthocyanins (all three) detected were higher in winter season, suggesting a seasonal dependent regulation of the biochemical steps linking the higher and lower portion of the phenylpropanoid pathway.

Interestingly, putatively identified 1-O-sinapoyl-β-D-glucose was detectable following dehydration only during the summer from June to October, months characterized by the lowest seed dehydration tolerance.

Network analysis sheds insights into the relation between germination, survival percentage and metabolism

In an attempt to understand the coordinated metabolic shifts characterizing dry seeds, germinated seeds and dehydrated seeds across the year and in order to identify key metabolites associated with seed tolerance to dehydration, we employed correlation-based network analysis (CNA). Within the CN we included the relationships between the physiological traits (germination and survival percentage) and metabolites.

The metabolic network of the dried seed (Additional file 9a) is composed of four main communities, of which communities 1 and 3 are significant in respect to the community affiliation (Comm. 1 – p = 1.58e-09, Comm. 2 – p = 0.821, Comm. 3 – p = 0.009, Comm. 4 – p = 0.085). Comm. 1 incorporates all represented compound classes and is characterized by a high degree of positive correlations and homogenous nodal degree (number of links per node). The most profound feature of this network is the few relations with the physiological traits tested. The germination percentage correlates solely, and negatively, to gulonate while, the seed survival correlates negatively to the germination percentage, a feature suggesting that overall in periods of low germinability the seeds that do germinate have a high probability of tolerating the dehydration process. On the other hand in periods of high germinability the percentage of survival is not a prominent feature.

The metabolic network associated with germination (Additional file 9b) also displays four main communities. However, the structure of the communities reveals notable differences. Community 3 and 4 of the dry-seed metabolite network are not present in the germinated network emphasizing the shifts that undergo amino acid and energy related metabolism from the dry seed to germination. The current view reveals two main communities, which integrate a similar number of nodes (Comm. 1 = 21 nodes and Comm. 2 = 23 nodes), including most amino acids and sugars present in the network. Furthermore, Comm. 1 integrates the two physiological traits indicating that the profiles tested for each correlates significantly to the metabolic profiles measured during germination. Taken together, these data and the chronological order of events suggest that the processes during germination are primary factors affecting seed tolerance to dehydration.

The dehydration metabolite network (Additional file 9c) revealed the occurrence of two main communities. The most salient community (p-value of 1.82e-14) shown is the densely intra-connected community 1, incorporating 40 of the 71 nodes and exemplifying the highly coordinated shift across the year taking place in dehydrated seeds. This community entails the entire array of compound classes represented, as well as the two physiological traits, seed germination and seed survival, suggesting that the germination percentage and seed dehydration tolerance across the year are in part the expression of a defined and coordinated shifts in the metabolic phenotype.

To further investigate the relationship of survival and germination with the metabolite profiles changes during the year, isolated subgraphs integrating solely the adjacent nodes were generated (Figure 4). The commonalities between germination metabolism and dehydration metabolism within the subgraphs were highlighted by dashed lines. The germination percentage correlates positively to all compounds connected, whilst- generally - survival percentage correlates negatively to the same compounds, i.e. the sugars fructose, glucose, and the cell wall associated sugar galactose as well as the sugar alcohol myo-inositol in both networks. Specifically, when dehydrated seeds are low in intermediates of the glycolysis and TCA cycle, the corresponding survival percentage is relatively high; a similar trend is observed for the amino acids for the shikimate pathway, and stress related Pro, GABA and Ala. Outstanding is the positive relation between the content of ethanolamine in the dehydrated seeds and their survival. When the yearly metabolite profiles of the dehydrated seeds was subjected to network analysis, metabolite 1-O-sinapoyl-β-D-glucose highly correlated with another putatively identified compound sinapic acid hexose (SH, r = 0.85, p = 0.0004) and with its precursor sinapate(r = 0.77, p = 0.0033), however low correlation was found with the down-stream metabolite sinapoyl malate (SM, r = 0.49, p = 0.1034). It also displayed the strongest negative connection with seed survival and also negatively correlated with other metabolite cluster such as flavonoid and anthocyanidin, especially four nodes pelargonidin hexose (PH), Apigenin-7-O-glucoside (AOG), kaempferol hexose (KH) and pentanoic acid (TMO), which are strongly positively interconnected. In contrast to the central metabolites, several of the secondary metabolites in the dehydrated seeds positively related with their survival percentage, TMO, KH, AOG and PH.

Discussion

Seed germination displaying annual rhythm was the focus of several studies and shown to be at least in part under endogenous regulators [39]. However a significant gap in knowledge exists in respect to the metabolic processes associated with germination rhythms. Here we presented the first evidence of seasonal metabolic fluctuations in germinated seeds and dehydrated seeds and its relation to germination and seed survival following dehydration. In spite of relatively stable germination across the year, the number of seeds of *S. arabicus* that could survive three weeks dehydration was largely affected by the annual periodicity of dehydration treatment. Namely seeds survived dehydration in the winter season from October to March with almost 100% survival, in contrast to the low survival in the summer from April to September and consistent with previous reports. Neither germination, nor seed survival were found to be associated with the metabolism of the dry seed, which displayed a uniform profile throughout the year. These results suggest a distinct regulation mechanism of seasonal changes in dehydration tolerance of *S. arabicus* compared with dormancy cycling in *Arabidopsis* dry seeds, which is attributed to the integration of the molecular physiological state with changes in sensitivity to

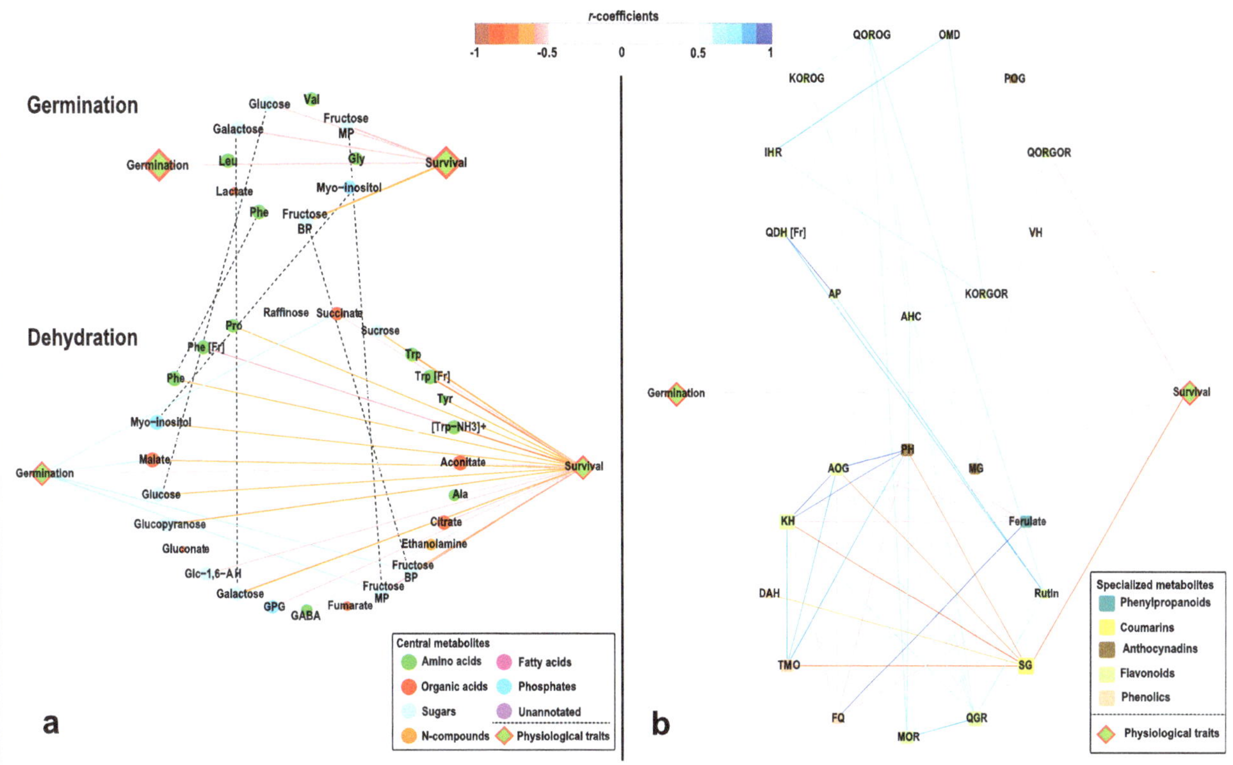

Figure 4 Integrated network for the association of central (a) and specialized metabolites (b) with seed germination and seed survival percentage. Nodes in the network are color-coded according to their compound classes and shaped according to their specificity (elliptical = central metabolism, rounded rectangles = specialized metabolism, diamond shaped = physiological traits). Relative sizes of nodes correspond to their degree of connectivity. The Spearman rank correlation was employed to compute all pairwise correlations between metabolites across the timeline. Solely significant correlations were chosen to be depicted. A significance level of $q < 0.05$ and an r-value of >0.5 were considered to be significant.

the environment [8,12,40]. The endogenous rhythm of *S. arabicus* is presumably set already in advance during seed development. An endogenous molecular clock is likely stored in the form of mRNAs and epigenetic phenomena [41,42]. During seed imbibition, water intake activates the cellular metabolism by enhancing enzymes activities associated storage reserve mobilization, hormone mobilization and seed respiration [43,44]. These processes are reflected in seeds by decreased sucrose content and concurrent increase of glycolysis and TCA intermediates and fatty acids, and by the conversion of Asn to Asp, as shown in this study and as previously reported in *Arabidopsis* [38]. During seed germination, also the flavonoid biosynthetic pathway is induced as shown by the accumulation of shikimate derived Phe and Tyr precursors of the phenylpropanoid pathway, and the reduction of caffeate and ferulate could suggest for enhanced integration within downstream processes. Upon dehydration, stress related metabolic processes are induced [45,46] including the accumulation of GABA, branched chain amino acids Val and Leu, raffinose and galactose and phenylpropanoids [47].

Seasonal rhythm affects stress related metabolism linking to seed germination and seed survival

Seed germination and seed survival are negatively related (Figure 1 and Figure 4a) suggesting that during periods of low germination, those seeds that do germinate will eventually tolerate dehydration. Is germination the period of priming for seasonal changes in seed tolerance? Metabolite profiling and network analysis suggest that, for central metabolism, namely this might be the case. The experiments were conducted in controlled conditions, where water, temperature and light during germination were at constant levels throughout the season. Nevertheless, germinated seeds in the summer show a very different metabolite profile compared to the winter, characterized by increased amino acids content, accumulation of primary sugars, TCA cycle and shikimate intermediates. A characteristic of the winter, imbibed seeds (and better germinating) was a general lower content of metabolites throughout the profile, except for the accumulation of itaconate, and glycolate. These results might suggest a higher metabolic turnover.

When gradually dehydrated, seeds accumulated amino acids, sugars and fatty acids particularly in the summer likely dedicated to the formation of a glassy matrix to counter the loss of water [48,49]. Network analysis could differentiate those metabolites jointly associated with trends in germination and dehydration tolerance from those specific to dehydration and likely more relevant to seed survival (Figure 4), e.g. sucrose, TCA cycle intermediates and ethanolamine accumulation. The annual rhythm of seed survival was also associated with the accumulation of phenylpropanoid precursors of the shikimate pathway in the summer and of downstream compounds, including kaempferol, quercetin and their derivatives in the winter. These flavonoids have been long recorded to be involved in wounding response, pathogen plant interaction and provide protection from irradiation and UV [50-54]. These results suggest for an enhanced capacity of the winter-seeds to repartition the C pool and accumulate protective compounds, especially in the polyphenol group, thus reducing detrimental cellular damages. A compound, tentatively identified as 1-O-sinapoyl-β-D-glucose, was present at detectable level from June to October only, i.e. the summer period. Its significant correlation with the precursor sinapate and low correlation with the down-stream product sinapoyl malate indicates that the turnover of sinapate could be a potential marker for dehydration tolerance.

No significant differences were detected in membrane permeability between seasons (Additional file 10). Hence we can safely conclude that the differences encountered are not due to impaired oxygen diffusion in the inner parts of the seed during the summer or differences in the membrane stability. Nevertheless we cannot exclude the occurrence of other processes within the dry seed that might affect the seasonal differences in dehydration tolerance.

Conclusion

By employing seeds of a desert annual, *Schismus arabicus*, metabolic profiling of dry seeds, germinated and dehydrated seeds revealed metabolic features closely associated with the documented annual rhythm of seed survival. Overall metabolite profiling and network analysis show that metabolic processes during germination seem to characterize the degree of seed dehydration tolerance. In the summer, the accumulation of central metabolites during germination, likely from a lower turnover, and a lower content of "protective" compounds could contribute to the lower tolerance of the seed to dehydration. The existence of inhibitory compounds accumulating during the summer, e.g. 1-O-sinapoyl-β-D-glucose, should be further investigated. In addition, future studies shall investigate the regulatory processes involved in the metabolic and physiological patterns here characterized including the occurrence of associated epigenetic phenomena during seed development.

Availability of supporting data

The data sets supporting the results of this article are included within the article and its additional files.

Additional files

Additional file 1: *Schismus arabicus* Nees caryopses (seeds) were collected in April 2005 from a natural habitat near Sede Boker in the Negev and the seeds around 425 mm were selected for the experiment. Dry seeds were aligned on the filter paper (a) followed by 24 hours imbibition (b). Then the germinated seeds were subjected to controlled drying for 21 days and rehydrated by reapplying water. Following rehydration, seeds were scored as viable based on their ability to reestablish root, coleoptile and a continuation of coleoptile elongation (c).

Additional file 2: Network properties. Listed are network properties, corresponding to the networks in Figure 4 and Additional file 9, used to determine the significant edges shown in the networks.

Additional files 3: Meteorological data in the Sede Boqer area between 2010 and 2011. The data were obtained and summarized from the meteorological station at the Institutes for Desert Research Midreshet Ben Gurion, Sede Boqer.

Additional file 4: Metabolite loading of dry seeds (DRY), germinated seeds (GER) and dehydrated seeds (DH) in PCA plot (Figure 3). The loading value of first three principal components are shown.

Additional file 5: t-test (p=0.05) of detected metabolites in different developmental stages and with seasonal effect. Values following each metabolite are p value, −log10(−p) and false discovery rate (FDR) of each compared group which are germinated seed compared with dry seeds (GER&DRY), dehydrated seeds compared with germinated seeds (DH&GER), and their respective fold changes under seasonal effect represented by the ratio between winter and summer during germination (WIN&SUM GER) and dehydration (WIN&SUM DH).

Additional file 6: Schematic view of metabolites enrichment during seed germination (a) and dehydration (b) and the temporal distribution of metabolites of imbibed seeds (c) and dehydrated seeds (d) in different seasons. Jun, Jul, Aug and Sep were selected as the representative of summer months and Nov, Dec, Jan and Feb were selected as the representatives of winter month. Standard paired t-test was used to compare the metabolite content in germinated seeds with dry seed and dehydrated seeds with germinated seeds in each month and standard unpaired t-test was performed to compare the metabolites significantly enriched in each season. Red and blue circles represent increase or decrease, respectively, in metabolite abundance during seed germination (a) or dehydration (b), in summer (c) and winter (d), p=0.05.

Additional file 7: Fold changes of detected metabolites in different developmental stages and with seasonal effect. Value for each metabolite mean fold change (FC) of three replicates of each compared group which are germinated seed compared with dry seeds (GER&DRY), dehydrated seeds compared with germinated seeds (DH&GER), and their respective fold changes under seasonal effect represented by the ratio between winter and summer during germination (WIN&SUM GER) and dehydration (WIN&SUM DH).

Additional file 8: The average germination and seed survival percentage following dehydration in summer and winter. *p=0.05, **p=0.01.

Additional file 9: Network visualization of metabolites as analyzed on dry *Shismus arabicus* seeds (a), germinated seeds (b), dehydrated seeds (c). Metabolites are clustered according to the walktrap community algorithm. Positive correlations are denoted as blue edges, negative correlations are denoted as red edges. The sizes of the nodes represent the relative degree of connectivity, The widths of edges in the network correspond to the relative magnitude of correlation estimated.

Additional file 10: The dry seed leakage conductivity during the summer and winter months.

Abbreviations

DEG: Diethylenglycol; Benzoate DH: Benzoic acid, 3, 4-dihydroxy PME, Phosphoratemonomethyl ester; PyroGlu: Pyroglutamate; GPG: Glycerophosphoglycerol; SH: Sinapic acid hexose; SG: 1-O-sinapoyl-β-D-glucose; SM: Sinapoyl malate; Phe: Phenylalanine; Phe [Fr]: Phenylalanine Fragment; Tyr: Tyrosine; Trp: Tryptophan; Trp [Fr]: Tryptophan Fragment; PH: Pelargonidin hexose; POG: Peonidin 3-O-glucoside; MG: Malvidin-3-glucoside; AP: Artonin P; KH: Kaempferol hexose; KOROG: Kaempferolerol-3-O-rutinoside-7-O-glucoside; KORGOR: Kaempferol-3-O-a-L-rhamnopyranosyl(1,2)-b-D-glucopyranoside-7-O-a-L-rhamnopyranoside; AOG: Apigenin-7-O-glucoside; AHC: Apigenin-C-hexoside; QGR: Quercetinrcetin-glucose-rhamnose; QOROG: Quercetin 3-O-rutinoside-7-O-glucoside; QORGOR: Quercetin-3-O-a-L-rhamnopyranosyl (1,2)-b-D-glucopyranoside-7-O-a-L rhamnopyranoside; QDH [Fr]: Quercetin-deoxyhexoside-hexoside fragment; MOR: 3-Methylquercetin 3-O-rutinoside; OMD: O-methylquercetin-deoxyhexoside; IHR: Isorhamnetin-Hex-Rha; TMO: (S)-2-(3-(4-hydroxyphenethoxy)-4-nitrobenzamido)-5(methylthio) pentanoic acid; FQ: Feruloylquinic acid; DAH: Dihydroxybenzoic acid hexoside; VH: Dihydroxy-methyl-benzoic acid hexoside (vanillic acid hexoside).

Competing interests

The authors declare that they have no competing interests.

Authors' contributions

Authors, who have made substantial contributions to conception, design of experiments: BB, AF. Acquisition of data, analysis and interpretation of data: BB, AD and AF. Authors who have contributed to performing experiments: BB, TG and AF. Authors who have been involved in drafting the manuscript: BB, DT and AF. Authors who have revised it critically: BB, AF, DT and IG. Authors who have given final approval of the version to be published: all. Authors who agree to be accountable for all aspects of the work in ensuring that questions related to the accuracy or integrity of any part of the work are appropriately investigated and resolved: all. All authors read and approved the final manuscript.

Acknowledgements

We would like to thank Noga Sikron for the assistence in metabolic analysis. Special thanks go to the support from Albert Katz International School, Ben Gurion University. The Koshland and Goldinger foundation and the Pearlstein foundation are acknowledged for their financial support.

Author details

[1]Ben-Gurion University of the Negev, Jacob Blaustein Institutes for Desert Research, French Associates Institute for Agriculture and Biotechnology of Drylands, Midreshet Ben-Gurion 84990, Israel. [2]Current address: Department of Molecular Plant Physiology, Utrecht University, Utrecht 3584 CH, The Netherlands.

References

1. Galpaz N, Reymond M. Natural Variation in Arabidopsis thaliana Revealed a Genetic Network Controlling Germination Under Salt Stress. PLoS ONE. 2010;5(12):e15198. doi:10.1371/journal.pone.0015198.
2. Joosen RV, Arends D, Li Y, Willems LA, Keurentjes JJ, Ligterink W, et al. Identifying genotype-by-environment interactions in the metabolism of germinating arabidopsis seeds using generalized genetical genomics. Plant Physiol. 2013;162(2):553–66.
3. Bradford KJ. A water relations analysis of seed germination rates. Plant Physiol. 1990;94(2):840–9.
4. Densmore R. Effect of day length on germination of seeds collected in Alaska. Am J Bot. 1997;84(2):274.
5. Carter CT, Brown LS, Ungar IA. Effect of temperature regimes on germination of dimorphic seeds of Atriplex prostrata. Biol Plantarum. 2003;47(2):269–72.
6. Negm FB, Smith OE. Effects of ethylene and carbon dioxide on the germination of osmotically inhibited lettuce seed. Plant Physiol. 1978;62(4):473–6.
7. Zhu Y, Dong M, Huang Z. Adaptation strategies of seed germination and seedling growth to sand dune environment. Ying Yong Sheng Tai Xue Bao. 2006;17(1):137–42.
8. Bouwmeester HJ, Karssen CM. Seasonal periodicity in germination of seeds of Chenopodium-Album L. Ann Bot-London. 1993;72(5):463–73.
9. Tokumasu S, Kakihara F. Seasonal germination periodicity of imbibed dormant seeds of rape (Brassica-Napus L). Sci Hortic-Amsterdam. 1990;42(1–2):1–7.
10. Young JA, Evans RA, Raguse CA, Larson JR. Germinable seeds and periodicity of germination in annual grasslands. Hilgardia. 1981;49(2):1–37.
11. Stoller EW, Wax LM. Periodicity of germination and emergence of some annual weeds. Weed Sci. 1973;21(6):574–80.
12. Footitt S, Clay HA, Dent K, Finch-Savage WE. Environment sensing in spring-dispersed seeds of a winter annual Arabidopsis influences the regulation of dormancy to align germination potential with seasonal changes. New Phytol. 2014;202(3):929–39.
13. Footitt S, Huang Z, Clay HA, Mead A, Finch-Savage WE. Temperature, light and nitrate sensing coordinate Arabidopsis seed dormancy cycling, resulting in winter and summer annual phenotypes. Plant J. 2013;74(6):1003–15.
14. Baskin JM, Baskin CC. Annual dormancy cycles in buried weed seeds: a continuum. BioScien. 1985;35(8):492.
15. Hilhorst HWM. The regulation of secondary dormancy. The membrane hypothesis revisite. Seed Sci Res. 1998;8(2):77–90.
16. Arc E, Sechet J, Corbineau F, Rajjou L, Marion-Poll A. ABA crosstalk with ethylene and nitric oxide in seed dormancy and germination. Front Plant Sci. 2013;4:63.
17. Jacobsohn MK, Jacobsohn GM. Annual variation in the sterol content of digitalis purpurea L. seedlings. Plant Physiol. 1976;58(4):541–3.
18. Rachmilevitch S, Gutterman Y, Gendler T. Survival of Schismus arabicus seedlings exposed to desiccation depends on annual periodicity. Planta. 2010;231(6):1475–82.
19. Bartels D. Desiccation tolerance studied in the resurrection plant Craterostigma plantagineum. Integr Comp Biol. 2005;45(5):696–701.
20. Yobi A, Wone BWM, Xu WX, Alexander DC, Guo LN, Ryals JA, et al. Comparative metabolic profiling between desiccation-sensitive and desiccation-tolerant species of Selaginella reveals insights into the resurrection trait. Plant J. 2012;72(6):983–99.
21. Buitink J, Hemminga MA, Hoekstra FA. Is there a role for oligosaccharides in seed longevity? An assessment of intracellular glass stability. Plant Physiol. 2000;122(4):1217–24.
22. Crowe JH, Crowe LM, Chapman D. Preservation of Membranes in Anhydrobiotic Organisms - the Role of Trehalose. Science. 1984;223(4637):701–3.
23. Wolkers WF, Walker NJ, Tablin F, Crowe JH. Human platelets loaded with trehalose survive freeze-drying. Cryobiology. 2001;42(2):79–87.
24. Whittaker A, Martinelli T, Bochicchio A, Vazzana C, Farrant J. Comparison of sucrose metabolism during the rehydration of desiccation-tolerant and desiccation-sensitve leaf material of Sporobolus stapfianus. Physiol Plant. 2004;122(1):11–20.
25. Debeaujon I, Peeters AJ, Leon-Kloosterziel KM, Koornneef M. The TRANSPARENT TESTA12 gene of Arabidopsis encodes a multidrug secondary transporter-like protein required for flavonoid sequestration in vacuoles of the seed coat endothelium. Plant Cell. 2001;13(4):853–71.
26. Sagasser M, Lu GH, Hahlbrock K, Weisshaar B. A-thaliana TRANSPARENT TESTA 1 is involved in seed coat development and defines the WIP subfamily of plant zinc finger proteins. Gene Dev. 2002;16(1):138–49.
27. Sattler SE, Gilliland LU, Magallanes-Lundback M, Pollard M, DellaPenna D. Vitamin E is essential for seed longevity, and for preventing lipid peroxidation during germination. Plant Cell. 2004;16(6):1419–32.
28. Horvath G, Wessjohann L, Bigirimana J, Jansen M, Guisez Y, Caubergs R, et al. Differential distribution of tocopherols and tocotrienols in photosynthetic and non-photosynthetic tissues. Phytochemistry. 2006;67(12):1185–95.
29. Bai B, Sikron N, Gendler T, Kazachkova Y, Barak S, Grafi G, et al. Ecotypic Variability in the Metabolic Response of Seeds to Diurnal Hydration-Dehydration Cycles and its Relationship to Seed Vigor. Plant Cell Physiol. 2012;53(1):38–52.

30. Gutterman Y. Drought tolerance of the dehydrated root of Schismus arabicus seedlings and regrowth after rehydration, affected by caryopsis size and duration of dehydration. Israel J Plant Sci. 2001;49(2):123–8.
31. Weckwerth W, Wenzel K, Fiehn O. Process for the integrated extraction identification, and quantification of metabolites, proteins and RNA to reveal their co-regulation in biochemical networks. Proteomics. 2004;4(1):78–83.
32. Lisec J, Schauer N, Kopka J, Willmitzer L, Fernie AR. Gas chromatography mass spectrometry-based metabolite profiling in plants. Nat Protoc. 2006;1(1):387–96.
33. Saeed AI, Sharov V, White J, Li J, Liang W, Bhagabati N, et al. TM4: a free, open-source system for microarray data management and analysis. Biotechniques. 2003;34(2):374–8.
34. Toubiana D, Fernie AR, Nikoloski Z, Fait A. Network analysis: tackling complex data to study plant metabolism. Trends Biotechnol. 2013;31(1):29–36.
35. Pons P, Latapy M. Computing communities in large networks using random walks. LNCS. 2005;3733:284–93.
36. Shannon P, Markiel A, Ozier O, Baliga NS, Wang JT, Ramage D, et al. Cytoscape: a software environment for integrated models of biomolecular interaction networks. Genome Res. 2003;13(11):2498–504.
37. Baud S, Dubreucq B, Miquel M, Rochat C, Lepiniec L. Storage reserve accumulation in Arabidopsis: metabolic and developmental control of seed filling. Arabidopsis Book. 2008;6:e0113.
38. Fait A, Angelovici R, Less H, Ohad I, Urbanczyk-Wochniak E, Fernie AR, et al. Arabidopsis seed development and germination is associated with temporally distinct metabolic switches. Plant Physiol. 2006;142(3):839–54.
39. Rawat MMS, Thapliyal RC. Endogenous rhythm in seed germination of Dendrocalamus strictus. Seed Sci Technol. 2003;31(1):21–7.
40. Bouwmeester HJ, Karssen CM. Annual changes in dormancy and germination in seeds of sisymbrium-officinale (L) scop. New Phytol. 1993;124(1):179–91.
41. Nakabayashi K, Okamoto M, Koshiba T, Kamiya Y, Nambara E. Genome-wide profiling of stored mRNA in Arabidopsis thaliana seed germination: epigenetic and genetic regulation of transcription in seed. Plant J. 2005;41(5):697–709.
42. Holdsworth MJ, Finch-Savage WE, Grappin P, Job D. Post-genomics dissection of seed dormancy and germination. Trends Plant Sci. 2008;13(1):7–13.
43. Alencar NL, Innecco R, Gomes-Filho E, Gallao MI, Alvarez-Pizarro JC, Prisco JT, et al. Seed reserve composition and mobilization during germination and early seedling establishment of Cereus jamacaru D.C. ssp. jamacaru (Cactaceae). An Acad Bras Cienc. 2012;84(3):823–32.
44. Tozzi HH, Takaki M. Histochemical analysis of seed reserve mobilization in Passiflora edulis Sims fo. flavicarpa O. Deg. (yellow passion fruit) during germination. Braz J Biol. 2011;71(3):701–8.
45. Job C, Rajjou L, Lovigny Y, Belghazi M, Job D. Patterns of protein oxidation in Arabidopsis seeds and during germination. Plant Physiol. 2005;138(2):790–802.
46. Barba-Espin G, Diaz-Vivancos P, Job D, Belghazi M, Job C, Hernandez JA. Understanding the role of H2O2 during pea seed germination: a combined proteomic and hormone profiling approach. Plant Cell Environ. 2011;34(11):1907–19.
47. Dixon RA, Paiva NL. Stress-Induced Phenylpropanoid Metabolism. Plant Cell. 1995;7(7):1085–97.
48. Angelovici R, Galili G, Fernie AR, Fait A. Seed desiccation: a bridge between maturation and germination. Trends Plant Sci. 2010;15(4):211–8.
49. Taji T, Ohsumi C, Iuchi S, Seki M, Kasuga M, Kobayashi M, et al. Important roles of drought- and cold-inducible genes for galactinol synthase in stress tolerance in Arabidopsis thaliana. Plant J. 2002;29(4):417–26.
50. Sasaki K, Takahashi T. A flavonoid from Brassica rapa flower as the UV-absorbing nectar guide. Phytochemistry. 2002;61(3):339–43.
51. Ryan KG, Swinny EE, Markham KR, Winefield C. Flavonoid gene expression and UV photoprotection in transgenic and mutant Petunia leaves. Phytochemistry. 2002;59(1):23–32.
52. Li J, Ou-Lee TM, Raba R, Amundson RG, Last RL. Arabidopsis Flavonoid Mutants Are Hypersensitive to UV-B Irradiation. Plant Cell. 1993;5(2):171–9.
53. Pasold S, Siegel I, Seidel C, Ludwig-Muller J. Flavonoid accumulation in Arabidopsis thaliana root galls caused by the obligate biotrophic pathogen Plasmodiophora brassicae. Mol Plant Pathol. 2010;11(4):545–62.
54. Mizuno H, Yazawa T, Kasuga S, Sawada Y, Ogata J, Ando T, et al. Expression level of a flavonoid 3′-hydroxylase gene determines pathogen-induced color variation in sorghum. BMC Res Notes. 2014;7:761.

Functional characterization of the diatom cyclin-dependent kinase A2 as a mitotic regulator reveals plant-like properties in a non-green lineage

Marie JJ Huysman[1,2,3,4], Atsuko Tanaka[4,5], Chris Bowler[4], Wim Vyverman[3] and Lieven De Veylder[1,2*]

Abstract

Background: Cyclin-dependent kinases (CDKs) are crucial regulators of cell cycle progression in eukaryotes. The diatom *CDKA2* was originally assigned to the classical A-type CDKs, but its cell cycle phase-specific transcription at the G2-to-M phase transition is typical for plant-specific B-type CDKs.

Results: Here, we report the functional characterization of CDKA2 from the diatom *Phaeodactylum tricornutum*. Through a yeast two-hybrid library screen, CDKA2 was found to interact with the G2/M-specific CDK scaffolding factor CKS1. Localization of CDKA2 was found to be nuclear in interphase cells, while in cells undergoing cytokinesis, the signal extended to the cell division plane. In addition, overexpression of *CDKA2* induced an overall reduction in the cell growth rate. Expression analysis of cell cycle marker genes in the overexpression lines indicates that this growth reduction is primarily due to a prolongation of the mitotic phase.

Conclusions: Our study indicates a role for CDKA2 during cell division in diatoms. The functional characterization of a CDK with clear CDKB properties in a non-green organism questions whether the current definition of B-type CDKs being plant-specific might need revision.

Keywords: Cyclin-dependent kinase, CDKA2, Cell cycle, Cell division, Cytokinesis, Diatom, Mitosis, *Phaeodactylum tricornutum*

Background

In eukaryotes, control of cell cycle progression is driven by an evolutionarily conserved family of serine/threonine kinases, the cyclin-dependent kinases (CDKs). CDKs form functional heterodimers with regulatory cyclin subunits [1,2]. Together, CDKs and their cyclin partners control cell cycle progression at the G1 (Gap 1)-to-S (Synthesis) and the G2 (Gap 2)-to-M (Mitosis) phase transitions through the phosphorylation of target proteins involved in DNA replication and mitosis, respectively [3]. The activity of CDK/cyclin complexes is regulated at multiple levels, including the interaction with inhibitors or scaffolding proteins, and phosphoregulation of the CDK subunit.

In contrast to fission yeast (*Schizosaccharomyces pombe*) and budding yeast (*Saccharomyces cerevisiae*), in which only a single CDK (Cdc2/Cdc28) controls the cell cycle [4,5], animals and plants possess multiple CDKs [1,6]. The most conserved cell cycle regulators possess a typical PSTAIRE cyclin-binding motif (Cdc2/Cdc28 in yeast, Cdk1/Cdk2 in animals and A-type CDKs in plants). In addition, a class of CDKs specific to plants has been shown to control the cell cycle, being called the B-type CDKs [7-10]. B-type CDKs in higher plants possess a variant of the PSTAIRE motif, either PPTALRE (CDKB1) or PPTTLRE (CDKB2) [6] and, unlike the A-type CDKs that are required for both the G1-to-S and G2-to-M phase transition, they only play a role at the G2/M boundary [11].

* Correspondence: lieven.deveylder@psb.vib-ugent.be
[1]Department of Plant Systems Biology, Flanders Institute for Biotechnology (VIB), 9052 Ghent, Belgium
[2]Department of Plant Systems Biology, VIB, and Bioinformatics, Ghent University, 9052 Ghent, Belgium
Full list of author information is available at the end of the article

With the recent advances in sequencing techniques, more genomes have become available, including several from different algal groups [12]. This wealth of new data makes it possible to study B-type CDK evolution by comparative genomics. CDKB-like sequences have been identified in different algal species, including the green algae *Ostreococcus tauri* [13], *Chlamydomonas reinhardtii* [14], *Micromonas sp.* and *Micromonas pusilla*, the red alga *Cyanidioschyzon merolae* [15], and also the brown alga *Ectocarpus siliculosus* [16]. Remarkably, the CDKB-like sequences of *O. tauri* and *C. reinhardtii* have been reported to represent functional homologs of A-type CDKs, mainly by their ability to complement *cdc28* temperature-sensitive yeast mutants [15,17], indicating that B-type CDKs might originate from a duplication and subsequent specification of the A-type CDKs [18]. However, to date, their functionality to complement higher plant B-type CDKs has not been investigated.

Phaeodactylum tricornutum is a unicellular marine diatom belonging to the heterokont (or stramenopile) lineage [19]. This diatom multiplies by binary division and, unlike most other diatoms, it lacks a sexual phase during its life cycle, rendering this diatom a perfect model species to study vegetative reproduction [20-23]. Furthermore, due to the presence of a light-dependent phase during its cell cycle [21,24], the cell division process in *P. tricornutum* can easily be synchronized by implementation of alternating light/dark cycles [21]. Phylogenetic analysis of all CDKs identified in *P. tricornutum* revealed the presence of two A-type CDKs, of which CDKA1 shows the classical PSTAIRE motif, while CDKA2 shows the divergent PSTALRE cyclin-binding motif [21]. The latter deviates only by one amino acid from the CDKA (PSTAIRE) and CDKB (P[P/S]T[A/T]LRE) hallmarks. Such a PSTALRE motif is also present in the *Dictyostelium discoideum* CDC2 homolog (DdCDK1) [25], the *C. merolae* CDKA protein [15], the *O. tauri* CDKB protein [13] and the *E. siliculosus* CDKA2 protein [16]. Moreover, unlike typical CDKAs, transcription of the *P. tricornutum CDKA2* gene is cell cycle-regulated and shows a peak of transcription at the G2-to-M phase transition [21]. However, based on its current phylogenetic position and its transcription pattern, it is impossible to define whether *CDKA2* represents a functional ortholog of A- or B-type CDKs.

The main objective of this study was to functionally elucidate the role of CDKA2 during the cell cycle in *P. tricornutum*. Overexpression of *CDKA2* resulted in a delayed growth phenotype, mainly at the G2/M phase of the cell cycle. Moreover, microscopic analysis of cells expressing a fluorescently-tagged version of CDKA2 showed a relocalization of CDKA2 from the nucleus to the cell division plane just before cytokinesis. Together, these data suggest a role for CDKA2 during mitosis in diatoms.

Results

Phylogenetic analysis of CDKA2

Considering the ambiguous classification of CDKA2, its phylogenetic position among the CDK family members was reanalyzed (Figure 1). The recent release of genome data from several stramenopile algae, including the diatom *Fragilariopsis cylindrus*, the brown alga *E. siliculosus* [26] and the eustigmatophyte alga *Nannochloropsis gaditana* [27], allowed us to include CDK sequences of these species in the multiple sequence alignment. In the updated phylogenetic tree, the diatom CDKA2 sequences group together with CDKA2 and CDKB-like sequences from the other stramenopile groups (*E. siliculosus*, *N. gaditana* and *Phytophthora sojae*) as a well-supported clade that may represent a stramenopile-specific cluster. Although this cluster is clearly distinct from the plant B-type CDK cluster, it does not belong to the archetypical A-type CDK cluster either. Based on this updated phylogenetic analysis, it therefore remains impossible to determine whether CDKA2 is more related to the A-type or B-type CDKs (Figure 1).

Cell cycle phase-dependent expression of *CDKA2*

Previous work suggested that *CDKA2* transcription is cell cycle-regulated and shows a peak at the G2-to-M phase transition [21]. However, the experimental setup used in that study did not allow assessing expression levels beyond the metaphase point, since a microtubule depolymerizing agent, nocodazole, was added to the cells to increase the proportion of cells at mitosis. Here, we further analyzed the temporal expression of *CDKA1* and *CDKA2* during the cell cycle in synchronized *P. tricornutum* cells in the absence of nocodazole (Figure 2). In contrast to *CDKA1*, which displays no reproducible temporal expression profile during the cell cycle, *CDKA2* levels start accumulating during the G1/S phase and reach a maximum peak at the G2/M phase, coinciding with the peak expression of the mitotic marker *cyclin B1* (*CYCB1*).

CDKA2 interacts specifically with CKS1

To identify putative interactors of the *P. tricornutum* CDKA proteins, yeast two-hybrid (Y2H) cDNA library screens were conducted using the full-length CDKA1 and CDKA2 sequences fused to the GAL4 DNA-binding domain as bait. Using CDKA2 as bait, four possible interacting proteins were detected (Additional file 1: Table S1). However, only one of these interactors (Cdc Kinase Subunit 1, CKS1) could be confirmed (Additional file 2: Figure S1). Previously, we have shown that *CKS1* is predominantly transcribed during the G2/M phases, coinciding with the expression pattern of *CDKA2* [21]. Using CDKA1 as bait, 84 clones were isolated, representing 12 different putative interacting proteins (Additional file 1: Table S1). Of these, three belong

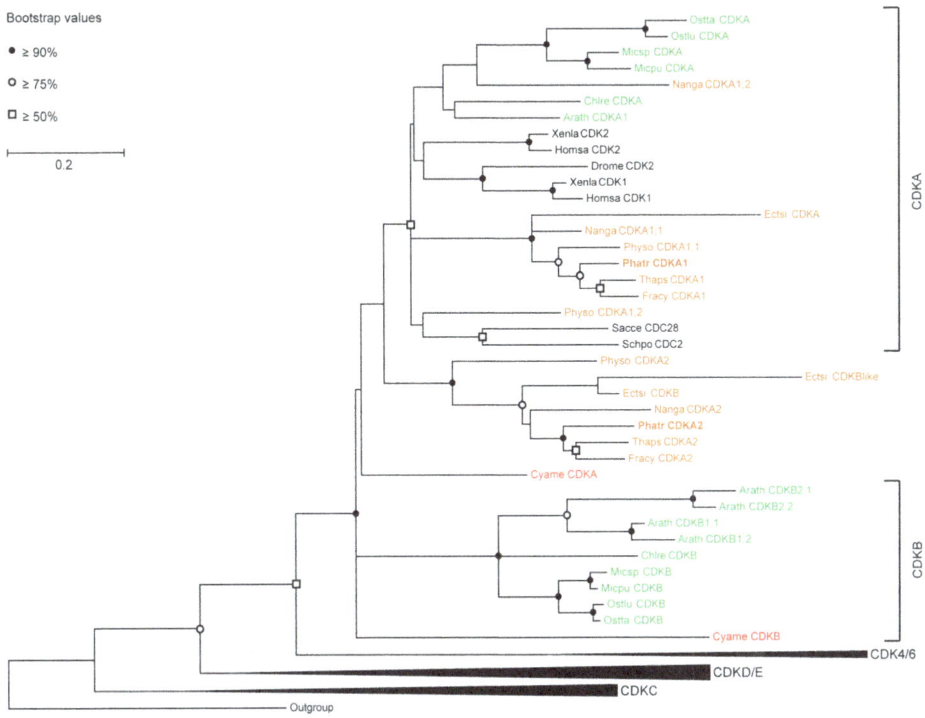

Figure 1 Phylogenetic analysis of the cyclin-dependent kinases of *P. tricornutum*. Maximum-likelihood tree (MEGA5.1, 1000 replicates) of the CDK family. The *P. tricornutum* sequences are shown in bold. Stramenopile sequences are shown in brown, higher plant and green algal sequences are indicated in green, red algal sequences in red and animal or yeast sequences in black. Abbreviations: Arath, *Arabidopsis thaliana*; Chlre, *Chlamydomonas reinhardtii*; Cyame, *Cyanidioschyzon merolae*; Drome, *Drosophila melanogaster*; Ectsi, *Ectocarpus siliculosus*; Fracy, *Fragilariopsis cylindrus*; Homsa, *Homo sapiens*; Micpu, *Micromonas pusilla*; Micsp, *Micromonas sp.*; Nanga, *Nannochloropsis gaditana*; Ostlu, *Ostreococcus lucimarinus*; Ostta, *Ostreococcus tauri*; Phatr, *Phaeodactylum tricornutum*; Physo, *Phytophthora sojae*; Sacce, *Saccharomyces cerevisiae*; Schpo, *Schizosaccharomyces pombe*; Thaps, *Thalassiosira pseudonana* and Xenla, *Xenopus laevis*. The outgroup is represented by human CDK10.

to the cyclin family (cyclin P1 (CYCP1), cyclin P6 (CYCP6) and diatom-specific cyclin 7 (dsCYC7)) [21].

To test the binding specificity of the two baits, pairwise Y2H assays were performed by co-transforming them in yeast with the prey proteins CKS1, CYCP1, CYCP6, dsCYC7 and dsCYC2, which was previously described as an interactor of CDKA1 [28]. Interestingly, CDKA1 was not able to bind CKS1 in the Y2H assay, whereas CDKA2 did not interact with any of the cyclins tested (Figure 3), indicating that the interactions picked up by Y2H library screens are specific for each CDK.

Overexpression of *CDKA2* reduces the cell growth rate by interfering with G2/M phase progression

To study the function of CDKA2 during the cell cycle of *P. tricornutum*, the effect of *CDKA2* overexpression on cell cycle progression was determined. Transgenic lines were generated that overexpress *CDKA2* fused at its C-terminus to *yellow fluorescent protein* (*YFP*) (*CDKA2-YFP*) under control of the *fucoxanthin chlorophyll binding protein B* (*fcpB*) promoter. Overexpression of *CDKA2-YFP* was evaluated by transcript analysis of *CDKA2* (Figure 4a) and *YFP* (Figure 4b) using real-time quantitative PCR (RT qPCR) in wild-type (WT) and transgenic cells. Two lines expressed an approximately two- to three-fold higher level of *CDKA2* transcripts compared to WT cells (CDKA2-YFP A4 and CDKA2-YFP B1, respectively) (Figure 4a) due to the overexpression of the *CDKA2-YFP* fusion (Figure 4b), while another line (CDKA2-YFP B9) showed no increase of *CDKA2* (Figure 4a). In contrast to the CDKA2-YFP A4 and B1 lines, no expression of *YFP* could be detected for the CDKA2-YFP B9 line (Figure 4b). This line was therefore used as an internal transformation control in downstream analyses.

To determine if *CDKA2* overexpression alters cell cycle dynamics, a growth analysis was performed. Optical density determination, which can be used as an estimate of cell number, was performed in the morning during nine consecutive days, in lines grown under constant light conditions. Cells expressing increased levels of *CDKA2-YFP* (CDKA2-YFP A4 and CDKA2-YFP B1) showed longer generation times compared to the control cells (WT and CDKA2-YFP B9) (Figure 4c). To determine which cell cycle phase was affected, we measured the transcript levels of *CYCP6*, *E2F1* and *CYCB1*

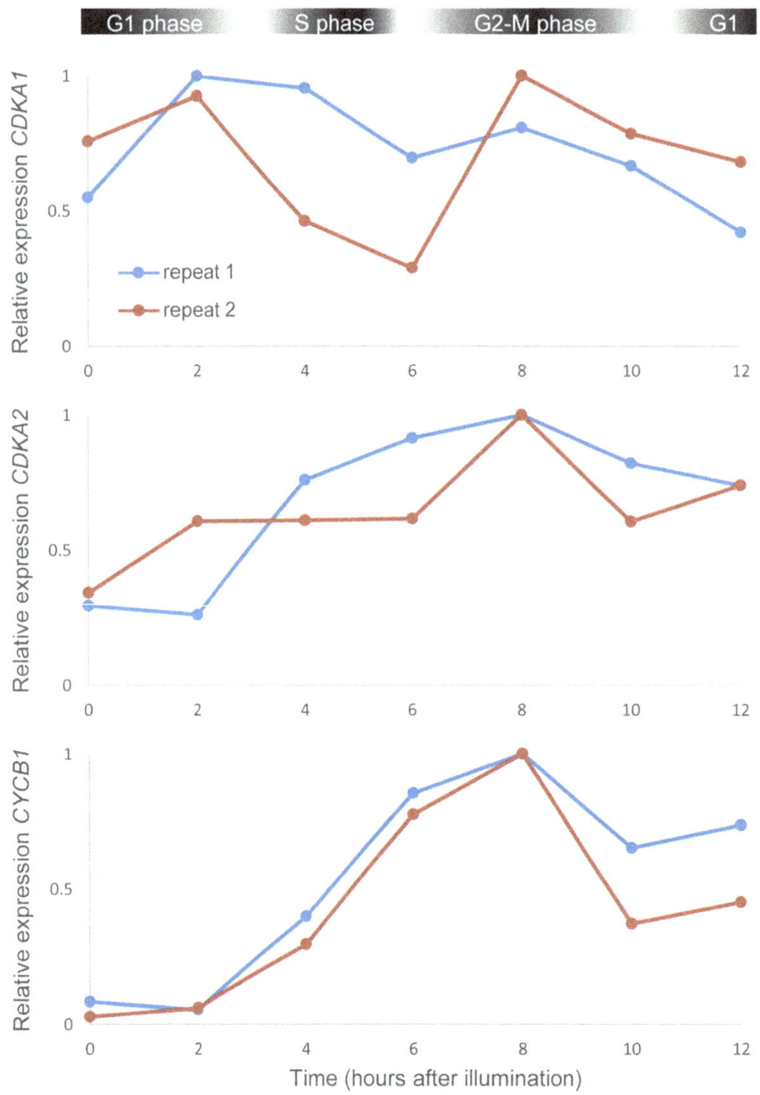

Figure 2 Transcript expression profiles of *CDKA1* and *CDKA2* during the cell cycle in wild-type *P. tricornutum* cells. Transcript levels were measured by nCounter analysis, normalized using four reference genes (*EF1a*, *histone H4*, *RPS*, and *UBI-4*) and rescaled to the maximum expression value (=1). Approximate cell cycle phase timing for this sample series was reported previously [22] and is indicated at the top.

in exponentially growing asynchronous *CDKA2*-overexpressing and control cells. These genes represent diatom cell cycle marker genes specific for the G1, S and G2/M phases, respectively [21,28]. In both *CDKA2* overexpression lines, we detected higher transcript levels of the mitotic marker *CYCB1*, and slightly lower *E2F1* levels compared to those of the control cells (Figure 4d), indicating that *CDKA2*-overexpressing cells spend more time at the G2/M phase.

CDKA2 relocalizes from the nucleus to the division plane before cytokinesis

To explore the subcellular localization of CDKA2 during the cell cycle, we monitored YFP fluorescence in the *CDKA2-YFP* overexpression cells. To avoid aberrant localization patterns due to the overexpression of *CDKA2*, we selected the CDKA2-YFP A4 line for this analysis, which shows only about two-fold higher expression levels of *CDKA2* (Figure 4a). For microscopic observations, cells were synchronized to enrich for cells at the different phases of the cell cycle, including interphase and mitosis. Confocal laser-scanning microscopy (CLSM) revealed a predominant nuclear and weak cytosolic YFP fluorescence in interphase cells that contain undivided or divided translocating chloroplasts (Figure 5a). This predominant nuclear localization of the YFP signal was confirmed by co-localization of Hoechst33342 DNA staining (Additional file 3: Figure S2). Remarkably, in cells with fully translocated daughter chloroplasts, the YFP fluorescence extended to the cell division plane between the two

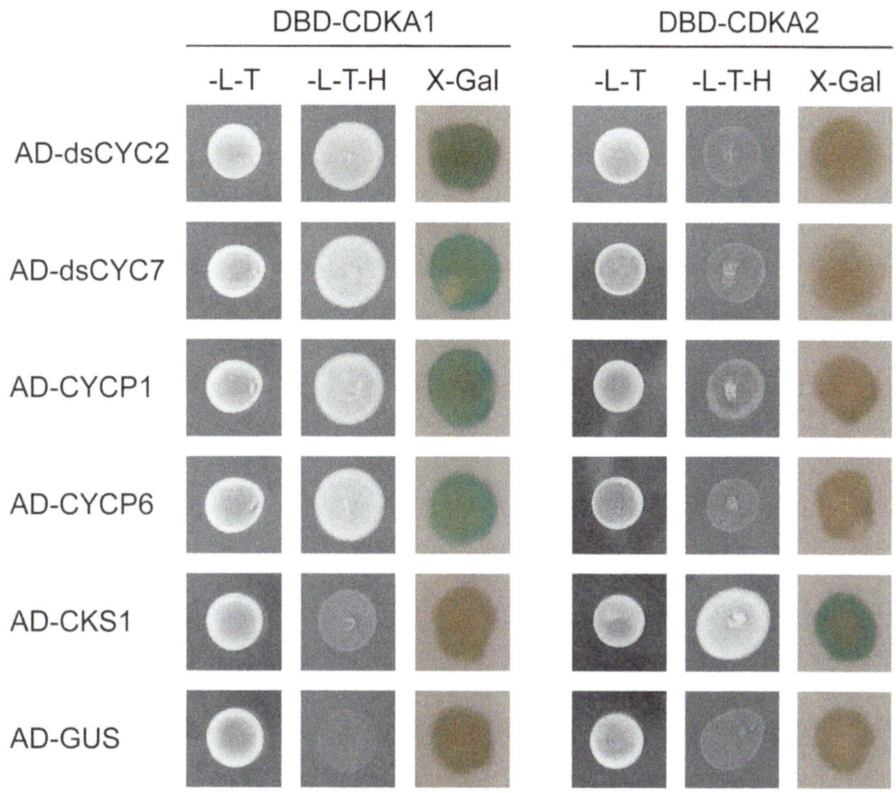

Figure 3 Interactions with CDKA1 and CDKA2 proteins using pairwise Y2H co-transformation assays. Yeast PJ694-alpha cells were co-transformed with bait (DBD) and prey (AD) plasmid as indicated. Co-transformation was analyzed on medium lacking leucine and tryptophan (-L-T). Co-transformants were tested for their ability to activate the histidine marker gene by assessing yeast growth on medium lacking leucine, tryptophan and histidine (-L-T-H) and for their ability to activate the *LacZ* reporter gene (X-Gal). As a negative control, the *GUS* gene was used. For each combination, three independent colonies were screened, of which one is shown.

daughter chloroplasts and resembled dot-like signals (Figure 5b). Three-dimensional reconstruction of the Z-stack of confocal images clearly demonstrated the localization of CDKA2-YFP fluorescence to a plate-like structure (Additional files 4: Movie S1 and 5: Movie S2). Following division, the YFP signal is localized in the nucleus of both daughter cells.

Discussion

In this study, we provide experimental evidence elucidating the role of CDKA2 in the diatom *P. tricornutum*, a CDK previously assigned to the A-type family, but displaying a PSTALRE motif, a motif in-between those of the conserved A-type and plant-specific B-type CDKs [6,21]. In contrast to classical A-type CDKs, transcription of diatom *CDKA2* fluctuates during the cell cycle, increasing from the G1/S phase onward and showing a peak in its expression at mitosis, thus resembling the expression pattern of B-type CDKs [29,30] and suggesting a role for CDKA2 in the control of mitosis. Re-examination of the phylogenetic position of *CDKA2*, by including closely related organisms, assigned it to a stramenopile cluster of CDKA/B-like sequences. The phylogenetic tree indicates a rapid radiation of CDK types near the base of the CDKA and CDKB lineage, possibly representing a *CDK* gene diversification phase during the early stages of eukaryotic evolution. The hypothesis that CDKA2 may function as a mitotic regulator is further supported by the localization of CDKA2-YFP at the cell division plane in pre-cytokinetic cells. This localization pattern suggests that CDKA2 may play a role in the positioning of the cleavage furrow or the formation of cytokinetic structures, or in the recruitment of one or various components to the cell division plane. In mammals, the activity of PRC1, a microtubule (MT) binding and bundling protein responsible for MT stabilization during cytokinesis, is regulated through phosphorylation by CDKs [31]. Also in plant cells, MT-dependent association of CDKs with mitotic structures and the functional involvement of CDKs in the organization of specific MT arrays have been reported [32-36].

CDKA2 was found to interact with CKS1, whose gene was previously reported to be transcribed mainly at the G2/M phase in synchronized cells [21]. The interaction

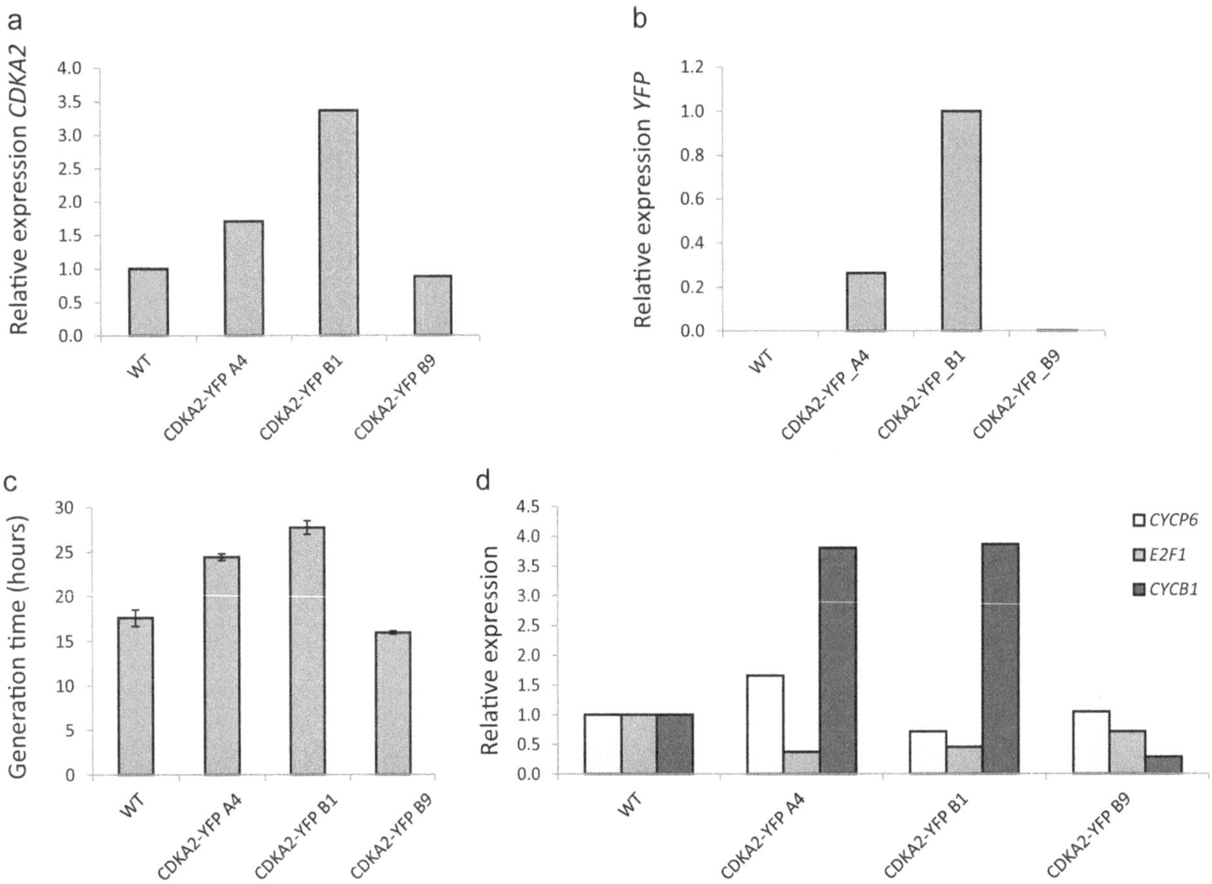

Figure 4 Effect of *CDKA2-YFP* overexpression on cell cycle progression in *P. tricornutum*. **(a)** Real-time qPCR analysis of *CDKA2* transcript levels in WT and transgenic lines. **(b)** Real-time qPCR analysis of *YFP* transcript levels in WT and transgenic lines. **(c)** Generation times of overexpression (A4 and B1) and control lines (WT and B9) grown under constant light. Error bars represent standard errors of the mean of three biological replicates. **(d)** Real-time qPCR analysis of different cell cycle marker genes in overexpression (A4 and B1) and control lines (WT and B9).

with CKS1 was not surprising, since CKS1 is a member of a conserved family of small proteins that are believed to act as docking factors that mediate the interaction of CDKs with regulatory proteins and putative substrates [37]. CKS1 might therefore represent a good bait to identify putative substrates of CDKA2 in further studies, which might help us to determine the precise function and action of CDKA2 during diatom mitosis. Interestingly, CKS1 was found not to bind CDKA1, resembling the situation in animal cells, in which CKS proteins have been reported to bind to Cdk2 complexes, but not to the G1/S-specific Cdk4/6 complexes [37-39].

In plants, overexpression of wild-type *CDK* genes generally does not trigger a phenotype [7,8,40]. However, recently, the overexpression of *CDKA1* in tomato fruit was reported to cause an increase in cell division [41], while the overexpression of *CDKB1* or *CDKB2* reduces the cell division rate in tomato fruit pericarp [42]. In addition, overexpression of *CDKB2;1* and *CDKB2;2* in *Arabidopsis* reduces cell cycle progression in the meristem due to regulatory defects at the G2-to-M transition [43]. Therefore, overexpression of B2-type *CDK*s appears to induce a consistent cell cycle phenotype, suggesting a universal role for these CDKs during cell cycle regulation. Here, we observed major growth defects in *P. tricornutum* cells overexpressing *CDKA2*. Most prominently, cell generation time almost doubled, despite only a slight increase in *CDKA2* transcript levels. Analysis of the expression of cell cycle marker genes demonstrated that this delay most probably results from a lengthening of the G2/M phases, as *CYCB1* levels were clearly elevated in *CDKA2*-overexpressing cells. Several hypotheses could explain the *CDKA2* overexpression phenotype. First, the overexpression construct could compete with endogenous *CDK* genes for rate-limiting interacting proteins (e.g., mitotic cyclins or CKS1). On the other hand, the mitotic arrest observed in the *CDKA2*-overexpressing cells might result from aberrant levels and timing of CDKA2 activity. In eukaryotes, exit from mitosis strictly depends on the downregulation of CDK activity, which is

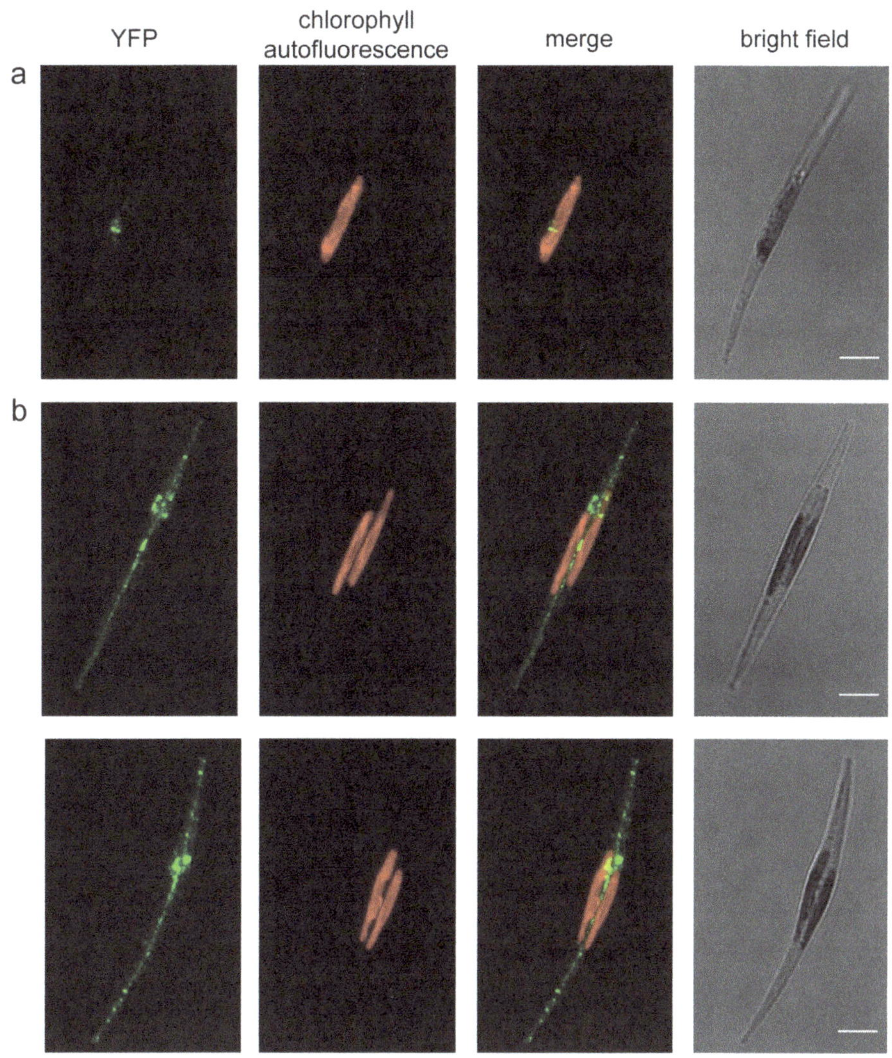

Figure 5 Localization of CDKA2-YFP in *P. tricornutum*. Confocal images of *CDKA2-YFP* overexpressing cells during different stages of the cell cycle. Maximum intensity Z-projections of CLSM analyses acquired with a Leica SP5 device are shown. For the bright field panels, a representative single plane is shown. **(a)** Interphase cell with undivided chloroplast. **(b)** Cells during cytokinesis with divided chloroplast and nucleus. At the time of cytokinesis, the signal was no longer prominent in the nucleus, but it was also targeted to the plane of division. The YFP signal is indicated in green and chlorophyll autofluorescence from the chloroplast in red. Scale bars represent 5 μm.

initiated by the destruction of mitotic cyclins by the anaphase promoting complex/cyclosome (APC/C) [44,45]. In a previous study, we have identified all essential components of this ubiquitin ligase complex in *P. tricornutum* and reported the cell cycle phase-specific transcription of its activators CDC20 and CDH1 [22], suggesting that a similar control mechanism for mitotic exit by a decline in CDK activity might exist in diatoms.

In yeast and metazoans, CDC25 phosphatases are known to activate CDKs by opposing the activity of the WEE1/MYT1/MIK1 family of inhibitory kinases [46]. Phosphorylation of conserved Thr (T14) and/or Tyr (Y15) residues results in CDK inactivation upon activation of the cell cycle checkpoints caused by triggers that should stop the cell cycle, such as DNA damage or mitotic defects. Dephosphorylation of these residues by CDC25 renders the CDK/cyclin complex active and hence stimulates cell cycle progression [47]. Although Tyr phosphorylation in plants is important to arrest the cell cycle under stress conditions, it does not seem to be crucial for G2-to-M progression [40,48-50]. Interestingly, all organisms in which B-type CDKs have been identified appear to lack a functional CDC25 phosphatase [13,51]. The only exception is *O. tauri*, which contains both a CDC25 phosphatase and a B-type-like CDK. However, functional analysis of the *O. tauri* CDKB indicated some CDKA features, including the ability to complement a yeast *cdc28* mutant, suggesting that this protein is rather atypical and has not yet achieved all the functional properties of higher plant B-type CDKs [17]. Because of an intriguing number of parallels in transcriptional, biochemical and functional properties of mammalian CDC25 and plant B-type CDKs, it was suggested that the CDC25-mediated regulatory mechanisms might have been replaced in plants by a mechanism governed by the plant-specific B-type CDKs [18]. Despite the presence of almost all regulatory components of the eukaryotic cell cycle in *P. tricornutum*, diatoms also lack a clear CDC25 phosphatase homolog [21], which may explain the need for a CDKB-like CDK, such as CDKA2, controlling the G2-to-M progression.

Conclusions

In this study we addressed the ambiguous nature of CDKA2 in *P. tricornutum*. CDKA2 was originally assigned to the A-type CDKs, but displays some typical characteristics of the plant-specific B-type CDKs, including cell cycle phase-dependent transcription at the G2-to-M transition. Subcellular localization of CDKA2 at the cell division plane during cytokinesis and its interaction with the G2/M expressed cell cycle regulator CKS1, point to a function for CDKA2 during mitosis. In addition, *CDKA2* overexpression resulted in the prolongation of the mitotic phase and an increase in cell cycle duration, demonstrating its role as a mitotic regulator. This is the first functional characterization of a CDK with clear CDKB properties in a non-green lineage, indicating that, while B-type CDKs form a clear plant-specific clade at the phylogenetic level, functional orthologs can be identified in other eukaryotic groups.

Methods

Phylogenetic analysis

Sequence data was retrieved from the NCBI RefSeq database (*A. thaliana, C. merolae, C. reinhardtii, D. melanogaster, H. sapiens, N. gaditana, P. sojae, S. cerevisiae, S. pombe* and *X. laevis*), through the JGI portal (*P. tricornutum* and *T. pseuodonana*) or the pico-PLAZA database (*E. siliculosus, F. cylindrus, M. pusilla, M. sp., O. tauri* and *O. lucimarinus*) [12,52]. Multiple alignments based on CDK amino acid sequences were generated with MUSCLE [53] and then manually improved, yielding 226 amino acid positions (Additional file 6: Figure S3). To define subclasses within the gene families, phylogenetic trees were built that included reference CDK sequences from animals, yeast, higher plants, stramenopiles and several algal species. MEGA5.1 was used to construct the maximum-likelihood tree using the rtREV + G + I model. To test the significance of the nodes, bootstrap analysis was applied using 1,000 replicates.

Diatom culture conditions

Phaeodactylum tricornutum (Pt1 8.6; accession numbers CCAP 1055/1 and CCMP2561) was grown in f/2 medium without silica (f/2-Si) [54] made with autoclaved filtered sea water. For routine cultivation, cells were grown at 18-20°C in a 12-h light/12-h dark regime and 70–100 μmol photons m^{-2} s^{-1}. Liquid cultures were shaken at 100 rpm. For biolistic transformation, *P. tricornutum* cells were grown on solid f/2-Si medium containing 1% Select agar (Sigma).

nCounter analysis

The sample series that was used to address the cell cycle phase-dependent transcription of *CDKA1* and *CDKA2* in wild-type *P. tricornutum* cells was described elsewhere [22,28]. Synchronization, RNA extraction and Nanostring nCounter analysis (Nanostring Technologies) were performed as described before [22,28]. Briefly, transcript levels were measured in multiplexed reactions using the nCounter analysis system by the VIB MicroArrays Facility (www.microarrays.be) as described before [55] and normalized using four reference genes (*EF1a, histone H4, RPS,* and *UBI-4*). Additional file 7: Table S2 gives an overview of the nCounter probe pairs used in this study. The analysis was done using two biological replicates.

Yeast two-hybrid analysis

Yeast two-hybrid bait plasmids were generated through recombinational GATEWAY cloning (Invitrogen). To generate the bait ENTRY clones, full-length open reading frames of the *P. tricornutum CDKA1* and *CDKA2* were isolated and cloned into the pENTR-D-TOPO vector as described elsewhere [28]. The obtained ENTRY clones were recombined in the pDEST32 (bait) vector (Invitrogen) by attL × attR recombination, resulting in translational fusions between the proteins and the GAL4 DNA-binding domains. Plasmids encoding the bait constructs were transformed in the yeast strain PJ694-alpha (MATa; trp1-901, leu2-3,112, ura3-52, his3-200, gal4D, gal80D, LYS2::GAL1-HIS3, GAL2-ADE2, met2GAL7-lacZ) by the LiAc method [56] to generate bait strains. Y2H library screens were performed using a custom-made Y2H cDNA library (Invitrogen) described before [28]. To this end, the respective bait strain was transformed with 50 μg of prey plasmids derived from the Y2H cDNA library according to the protocol described in the Yeast Protocol Handbook (Clontech). For each Y2H library screen at least 10^6 transformants were screened. Putative positive Y2H interactions were selected on synthetic dextrose (SD) plates lacking Leu, Trp and His. Growing colonies were streaked on SD medium lacking Leu, Trp and His, and plasmid was purified from yeast patches using the Zymoprep I Yeast Plasmid Minipreparation Kit (Zymo Research) according to the manufacturers' instructions. Yeast plasmid was used as a template in a PCR reaction with primers flanking the gateway cloning site of pDEST22 (pDEST22_Fw: TATAACGCGTTTGGAATCACT and pDE

ST22_Rv: AGCCGACAACCTTGATTGGAGAC), and the obtained PCR product was sequenced and blasted against the *P. tricornutum* genome database (http://genome.jgi-psf.org/Phatr2/Phatr2.home.html) to identify the putative interactor protein. To confirm the protein-protein interactions of interest, the respective prey and bait plasmids were co-transformed in the yeast strain PJ694-alpha, and the interactions were retested on medium lacking Leu, Trp and His and by X-Gal testing.

Generation of the *CDKA2* overexpression construct

The full-length sequence of *CDKA2* was isolated and amplified by PCR using the CDKA2_Fw (CACCATGGAACGTTACCATAAGATAGAAAAG) and CDKA2-C_Rv (GATGTTTTCCTTATCCAAGTCATCA) primers designed to allow C-terminal fusion. The purified fragment was cloned into the pENTR-D-TOPO vector (Invitrogen) using the directional TOPO cloning strategy and the obtained ENTRY clone was subsequently recombined into pDEST-C-EYFP, a diatom adapted destination vector for C-terminal fusion with the YFP fluorescent marker under the control of the *fcpB* promoter [57], using the Gateway attL × attR recombination reaction (Invitrogen).

Biolistic transformation

The CDKA2-YFP expression construct was introduced into *P. tricornutum* by microparticle bombardment as previously described [58]. As a selection marker, the pAF6 plasmid was co-transformed with the overexpression construct to confer resistance to phleomycin [57,58]. Transformants were initially selected based on their ability to grow on medium containing phleomycin (100 μg/ml final concentration). Individual resistant colonies were both restreaked on f/2-Si agar plates and grown in liquid f/2-Si medium without antibiotics for further analysis.

Real-time quantitative PCR

Cells were grown in continuous light conditions to desynchronize the cells. For RNA extraction, 5 x 10^7 exponentially growing cells were collected by centrifugation (15 minutes at 3,000 rpm, 4°C), fast frozen in liquid nitrogen and stored at −70°C. Cell lysis and RNA extraction was performed using TriReagent (Molecular Research Center, Inc., Cincinnati, OH, USA) according to the manufacturer's instructions. Contaminating genomic DNA was removed by DNaseI treatment (Promega). To assess RNA concentration and purity, spectrophotometry was used (NaNodrop ND-1000, Wilmington, DE). Total RNA was reverse transcribed using iScript reverse transcriptase (Roche). Finally, 10 ng of cDNA was used as template in each qPCR reaction.

Samples in triplicate were amplified on the Lightcycler 480 platform with the Lightcycler 480 SYBR Green I Master mix (Roche Applied Science), in the presence of 0.5 μM gene-specific primers (YFP_Fw: TGCTTCGCCCGCTACCC and YFP_Rv: ATGTTGCCGTCCTCCTTGAAG; E2F1_Fw: CCCTAAGCGGCGGATTTACG; E2F1_Rv: AAGCGACGAGCCAAGAAGAAGC; other primers see [21]). The cycling conditions were 10 min polymerase activation at 95°C and 45 cycles at 95°C for 10 s, 58°C for 15 s and 72°C for 15 s. Amplicon dissociation curves were recorded after cycle 45 by heating from 65°C to 95°C. Data were analyzed using the qbase$^+$ software package (Biogazelle) using the stably expressed *EF1a* and *TubA* as normalization genes [57].

Growth analysis

To monitor growth, cells were grown at constant illumination in a 24-well plate (Falcon), in a total volume of 1 ml, over a time period of nine days. Absorbances of the cultures were measured at 405 nm using the VICTOR^3 Multilabel Plate Reader (Perkin-Elmer) each day in the morning. Obtained growth curves of triplicate cultures were LN(2)-transformed and average generation times were calculated by determination of the derivative of the values between the points of maximal slope (exponential growth phase).

Microscopic analysis

Images were obtained with a confocal laser-scanning microscope Leica SP5 using a HCX PL APO CS 63.0x1.40 OIL objective, and a Zeiss LSM710 equipped with a C-Apochromat 63x/1.20 W Korr M27 objective. Using the Leica SP5, chlorophyll autofluorescence and YFP fluorescence were excited at 514 nm and detected at 617–687 nm and 520–560 nm, respectively. Using the Zeiss LSM710, chlorophyll autofluorescence and YFP fluorescence were excited at 514 nm and detected at 630–730 nm and 520–550 nm, respectively. Nuclear DNA was stained using the dye Hoechst 33342 (life technologies) at a final concentration of 5 μg / ml and stained cells were visualized by illumination at 405 nm and detection at 410–190 nm. Confocal images were processed further using either the ImageJ 1.45 (Leica images) or LSM Browser 4.2 (Zeiss images) software. 3D projections of the Z-stacks were created using the 3D Project tool (brightest-point projection) in ImageJ 1.45.

Availability of supporting data

All relevant supporting data can be found within the supplementary files accompanying to this article. Phylogenetic data supporting the results of this article are available in the TreeBASE repository, http://purl.org/phylo/treebase/phylows/study/TB2:S17171.

Additional files

Additional file 1: Table S1. List of putative CDKA1 and CDKA2 interactor proteins identified by Y2H cDNA library screening.

Additional file 2: Figure S1. Confirmation of the candidate CDKA2-interactors identified by a Y2H library screen using pairwise Y2H co-transformation assays. Yeast PJ694-alpha cells were co-transformed with bait (DBD) and prey (AD) plasmid as indicated. Co-transformation was analyzed on medium lacking leucine and tryptophan (-L-T). Co-transformants were tested for their ability to activate the histidine marker gene by assessing yeast growth on medium lacking leucine, tryptophan and histidine (-L-T-H) and for their ability to activate the *LacZ* reporter gene (X-Gal). As a negative control, the *GUS* gene was used. For each combination three independent colonies were screened, of which one is shown. Note that one clone (unknown1) was identified as a general false-positive in the Y2H assay, since it was able to activate itself on medium lacking histidine.

Additional file 3: Figure S2. Confirmation of the main nuclear localization of CDKA2-YFP. Confocal laser-scanning image (acquired with a Zeiss LSM710) of a *CDKA2-YFP*-expressing cell during interphase. Green, YFP signal; Red, chlorophyll autofluorescence; Blue, nuclear Hoechst 33342 staining. Scale bars represent 5 μm.

Additional file 4: Movie S1. Three-dimensional reconstruction of the Z-stack of confocal images of a *CDKA2-YFP*-expressing cell shown in Figure 5b. The YFP signal is indicated in green.

Additional file 5: Movie S2. Three-dimensional reconstruction of the Z-stack of confocal images of a *CDKA2-YFP*-expressing cell shown in Figure 5b. The YFP signal is indicated in green and chlorophyll autofluorescence is indicated in red.

Additional file 6: Figure S3. Multiple sequence alignment of the CDK sequences (generated by MUSCLE) used to build the phylogenetic tree shown in Figure 1.

Additional file 7: Table S2. Overview of the nCounter code set probe pairs.

Abbreviations

APC/C: Anaphase promoting complex / cyclosome; CDK: Cyclin-dependent kinase; CLSM: Confocal laser-scanning microscopy; Y2H: Yeast two-hybrid; YFP: Yellow fluorescent protein.

Competing interests

The authors declare that they have no competing interests.

Authors' contributions

MJJH, CB, WV and LDV conceived and designed the study. MJJH performed the phylogenetic analysis, generated and analyzed transgenic lines, participated in the microscopic analysis, performed Y2H assays and drafted the manuscript. AT participated in the microscopic analysis. All authors read and approved the final manuscript.

Acknowledgements

The authors wish to thank Annick Bleys for help in preparing the manuscript. This work was supported by grants of the Research Foundation Flanders (G.0288.13) and Ghent University (BOF project GOA 01G01715). MJJH acknowledges the European Molecular Biology Organization (EMBO) for a short-term fellowship (ASTF 93-2011) to visit CB's group at Institut de Biologie de l'Ecole Normale Supérieure (IBENS) Paris (France). CB additionally acknowledges the ERC "Diatomite" Advanced Grant Award, and the French Government 'Investissements d'Avenir' programmes MEMO LIFE (ANR-10-LABX-54) and PSL* Research University (ANR-11-IDEX-0001-02).

Author details

[1]Department of Plant Systems Biology, Flanders Institute for Biotechnology (VIB), 9052 Ghent, Belgium. [2]Department of Plant Systems Biology, VIB, and Bioinformatics, Ghent University, 9052 Ghent, Belgium. [3]Protistology and Aquatic Ecology, Department of Biology, Ghent University, 9000 Ghent, Belgium. [4]Environmental and Evolutionary Genomics Section, Institut de Biologie de l'Ecole Normale Supérieure, Centre National de la Recherche Scientifique, Unité Mixte de Recherche 8186, Institut National de la Santé et de la Recherche Médicale U1024, Ecole Normale Supérieure, 75230 Paris, Cedex 05, France. [5]Current address: Muroran Marine Station, Field Science Center for Northern Biosphere, Hokkaido University, Muroran, Hokkaido 051-0013, Japan.

References

1. Morgan DO. Cyclin-dependent kinases: engines, clocks, and microprocessors. Annu Rev Cell Dev Biol. 1997;13:261–91.
2. Inzé D, De Veylder L. Cell cycle regulation in plant development. Annu Rev Genet. 2006;40:77–105.
3. Murray AW. Recycling the cell cycle: Cyclins revisited. Cell. 2004;116:221–34.
4. Hartwell LH, Culotti J, Pringle JR, Reid BJ. Genetic control of the cell division cycle in yeast. Science. 1974;183:46–51.
5. Nurse P, Thuriaux P. Regulatory genes controlling mitosis in the fission yeast *Schizosaccharomyces pombe*. Genetics. 1980;96:627–37.
6. Joubès J, Chevalier C, Dudits D, Heberle-Bors E, Inzé D, Umeda M, et al. CDK-related protein kinases in plants. Plant Mol Biol. 2000;43:607–20.
7. Porceddu A, Stals H, Reichheld J-P, Segers G, De Veylder L, Barrôco R, et al. A plant-specific cyclin-dependent kinase is involved in the control of G_2/M progression in plants. J Biol Chem. 2001;276:36354–60.
8. Boudolf V, Barrôco R, de Almeida Engler J, Verkest A, Beeckman T, Naudts M, et al. B1-type cyclin-dependent kinases are essential for the formation of stomatal complexes in *Arabidopsis thaliana*. Plant Cell. 2004;16:945–55.
9. Boudolf V, Lammens T, Boruc J, Van Leene J, Van Den Daele H, Maes S, et al. CDKB1;1 forms a functional complex with CYCA2;3 to suppress endocycle onset. Plant Physiol. 2009;150:1482–93.
10. Lee J, Das A, Yamaguchi M, Hashimoto J, Tsutsumi N, Uchimiya H, et al. Cell cycle function of a rice B2-type cyclin interacting with a B-type cyclin-dependent kinase. Plant J. 2003;34:417–25.
11. Mironov VV, De Veylder L, Van Montagu M, Inzé D. Cyclin-dependent kinases and cell division in plants - The nexus. Plant Cell. 1999;11:509–22.
12. Vandepoele K, Van Bel M, Richard G, Van Landeghem S, Verhelst B, Moreau H, et al. pico-PLAZA, a genome database of microbial photosynthetic eukaryotes. Environ Microbiol. 2013;15:2147–53.
13. Robbens S, Khadaroo B, Camasses A, Derelle E, Ferraz C, Inzé D, et al. Genome-wide analysis of core cell cycle genes in the unicellular green alga *Ostreococcus tauri*. Mol Biol Evol. 2005;22:589–97.
14. Bisova K, Krylov DM, Umen JG. Genome-wide annotation and expression profiling of cell cycle regulatory genes in *Chlamydomonas reinhardtii*. Plant Physiol. 2005;137:475–91.
15. Čížková M, Pichová A, Vitová M, Hlavová M, Hendrychová J, Umysová D, et al. CDKA and CDKB kinases from *Chlamydomonas reinhardtii* are able to complement *cdc28* temperature-sensitive mutants of *Saccharomyces cerevisiae*. Protoplasma. 2008;232:183–91.
16. Bothwell JH, Marie D, Peters AF, Cock JM, Coelho SM. Role of endoreduplication and apomeiosis during parthenogenetic reproduction in the model brown alga *Ectocarpus*. New Phytol. 2010;188:111–21.
17. Corellou F, Camasses A, Ligat L, Peaucellier G, Bouget F-Y. Atypical regulation of a green lineage-specific B-type cyclin-dependent kinase. Plant Physiol. 2005;138:1627–36.
18. Boudolf V, Inzé D, De Veylder L. What if higher plants lack a CDC25 phosphatase? Trends Plant Sci. 2006;11:474–9.
19. De Martino A, Meichenin A, Shi J, Pan KH, Bowler C. Genetic and phenotypic characterization of *Phaeodactylum tricornutum* (Bacillariophyceae) accessions. J Phycol. 2007;43:992–1009.
20. De Martino A, Amato A, Bowler C. Mitosis in diatoms: rediscovering an old model for cell division. Bioessays. 2009;31:874–84.
21. Huysman MJJ, Martens C, Vandepoele K, Gillard J, Rayko E, Heijde M, et al. Genome-wide analysis of the diatom cell cycle unveils a novel type of cyclins involved in environmental signaling. Genome Biol. 2010;11:R17.
22. Huysman MJJ, Martens C, Vyverman W, De Veylder L. Protein degradation during the diatom cell cycle: Annotation and transcriptional analysis of SCF and APC/C ubiquitin ligase genes in *Phaeodactylum tricornutum*. Mar Genomics. 2014;14:39–46.
23. Huysman MJJ, Vyverman W, De Veylder L. Molecular regulation of the diatom cell cycle. J Exp Bot. 2014;65:2573–84.
24. Brzezinski MA, Olson RJ, Chisholm SW. Silicon availability and cell-cycle progression in marine diatoms. Mar Ecol Prog Ser. 1990;67:83–96.
25. Michaelis C, Weeks G. Isolation and characterization of a *cdc2* cDNA from *Dictyostelium discoideum*. Biochim Biophys Acta. 1992;1132:35–42.
26. Cock JM, Sterck L, Rouze P, Scornet D, Allen AE, Amoutzias G, et al. The Ectocarpus genome and the independent evolution of multicellularity in brown algae. Nature. 2010;465:617–21.

27. Radakovits R, Jinkerson RE, Fuerstenberg SI, Tae H, Settlage RE, Boore JL, et al. Draft genome sequence and genetic transformation of the oleaginous alga *Nannochloropsis gaditana*. Nat Commun. 2013;3:686.
28. Huysman MJJ, Fortunato AE, Matthijs M, Schellenberger Costa B, Vanderhaeghen R, Van den Daele H, et al. AUREOCHROME1a-mediated induction of the diatom-specific cyclin *dsCYC2* controls the onset of cell division in diatoms (*Phaeodactylum tricornutum*). Plant Cell. 2013;25:215–28.
29. Sorrell DA, Menges M, Healy JMS, Deveaux Y, Amano C, Su Y, et al. Cell cycle regulation of cyclin-dependent kinases in tobacco cultivar bright yellow-2 cells. Plant Physiol. 2001;126:1214–23.
30. Menges M, Hennig L, Gruissem W, Murray JA. Cell cycle-regulated gene expression in Arabidopsis. J Biol Chem. 2002;277:41987–2002.
31. Jiang W, Jimenez G, Wells NJ, Hope TJ, Wahl GM, Hunter T, et al. PRC1: a human mitotic spindle-associated CDK substrate protein required for cytokinesis. Mol Cell. 1998;2:877–85.
32. Weingartner M, Binarova P, Drykova D, Schweighofer A, David J-P, Heberle-Bors E, et al. Dynamic recruitment of Cdc2 to specific microtubule structures during mitosis. Plant Cell. 2001;13:1929–43.
33. Stals H, Bauwens S, Traas J, Van Montagu M, Engler G, Inzé D. Plant CDC2 is not only targeted to the pre-prophase band, but also co-localizes with the spindle, phragmoplast, and chromosomes. FEBS Lett. 1997;418:229–34.
34. Mineyuki Y, Yamashita M, Nagahama Y. $p34^{cdc2}$ kinase homologue in the preprophase band. 1991;162:182–6.
35. Colasanti J, Cho S-O, Wick S, Sundaresan V. Localization of the functional $p34^{cdc2}$ homolog of maize in root tip and stomatal complex cells: Association with predicted division sites. Plant Cell. 1993;5:1101–11.
36. Ayaydin F, Vissi E, Mészáros T, Miskolczi P, Kovács I, Fehér A, et al. Inhibition of serine/threonine-specific protein phosphatases causes premature activation of cdc2MsF kinase at G2/M transition and early mitotic microtubule organisation in alfalfa. Plant J. 2000;23:85–96.
37. Pines J. Cell cycle: Reaching for a role for the Cks proteins. Curr Biol. 1996;6:1399–402.
38. Azzi L, Meijer L, Reed SI, Pidikiti R, Tung HYL. Interaction between the Cell-Cycle-Control Proteins P34cdc2 and P9ckshs2 - Evidence for 2 Cooperative Binding Domains in P9ckshs2. Eur J Biochem. 1992;203:353–60.
39. Azzi L, Meijer L, Ostvold AC, Lew J, Wang JH. Purification of a 15-Kda Cdk4-Binding and Cdk5-Binding Protein. J Biol Chem. 1994;269:13279–88.
40. Hemerly A, Engler JD, Bergounioux C, Van Montagu M, Engler G, Inzé D, et al. Dominant negative mutants of the Cdc2 kinase uncouple cell division from iterative plant development. EMBO J. 1995;14:3925–36.
41. Czerednik A, Busscher M, Angenent GC, de Maagd RA. The cell size distribution of tomato fruit can be changed by overexpression of *CDKA1*. Plant Biotechnol J. 2015;13:259–68.
42. Czerednik A, Busscher M, Bielen BAM, Wolters-Arts M, de Maagd RA, Angenent GC. Regulation of tomato fruit pericarp development by an interplay between *CDKB* and *CDKA1* cell cycle genes. J Exp Bot. 2012;63:2605–17.
43. Andersen SU, Buechel S, Zhao Z, Ljung K, Novák O, Busch W, et al. Requirement of B2-type cyclin-dependent kinases for meristem integrity in *Arabidopsis thaliana*. Plant Cell. 2008;20:88–100.
44. Weingartner M, Criqui M-C, Mészáros T, Binarova P, Schmit A-C, Helfer A, et al. Expression of a nondegradable cyclin B1 affects plant development and leads to endomitosis by inhibiting the formation of a phragmoplast. Plant Cell. 2004;16:643–57.
45. Zachariae W, Nasmyth K. Whose end is destruction: cell division and the anaphase-promoting complex. Genes Dev. 1999;13:2039–58.
46. Perry JA, Kornbluth S. Cdc25 and Wee1: analogous opposites? Cell Div. 2007;2:12.
47. Lew DJ, Kornbluth S. Regulatory roles of cyclin dependent kinase phosphorylation in cell cycle control. Curr Opin Cell Biol. 1996;8:795–804.
48. De Schutter K, Joubès J, Cools T, Verkest A, Corellou F, Babiychuk E, et al. *Arabidopsis* WEE1 kinase controls cell cycle arrest in response to activation of the DNA integrity checkpoint. Plant Cell. 2007;19:211–25.
49. Zhang KR, Diederich L, John PCL. The cytokinin requirement for cell division in cultured *Nicotiana plumbaginifolia* cells can be satisfied by yeast cdc25 protein tyrosine phosphatase. Implications for mechanisms of cytokinin response and plant development. Plant Physiol. 2005;137:308–16.
50. Schuppler U, He P-H, John PCL, Munns R. Effect of water stress on cell division and cdc2-like cell cycle kinase activity in Wheat leaves. Plant Physiol. 1998;117:667–78.
51. Khadaroo B, Robbens S, Ferraz C, Derelle E, Eychenié S, Cooke R, et al. The first green lineage cdc25 dual-specificity phosphatase. Cell Cycle. 2004;3:513–8.
52. Grigoriev IV, Nordberg H, Shabalov I, Aerts A, Cantor M, Goodstein D, et al. The Genome Portal of the Department of Energy Joint Genome Institute. Nucleic Acids Res. 2012;40:D26–32.
53. Edgar RC. MUSCLE: multiple sequence alignment with high accuracy and high throughput. Nucleic Acids Res. 2004;32:1792–7.
54. Guillard RRL. Culture of phytoplankton for feeding marine invertebrates. In: Smith WL, Canley MH, editors. Culture of marine invertebrate animals. New York: Plenum Press; 1975. p. 29–60.
55. Geiss GK, Bumgarner RE, Birditt B, Dahl T, Dowidar N, Dunaway DL, et al. Direct multiplexed measurement of gene expression with color-coded probe pairs. Nat Biotechnol. 2008;26:317–25.
56. Gietz D, St Jean A, Woods RA, Schiestl RH. Improved method for high efficiency transformation of intact yeast cells. Nucleic Acids Res. 1992;20:1425.
57. Siaut M, Heijde M, Mangogna M, Montsant A, Coesel S, Allen A, et al. Molecular toolbox for studying diatom biology in *Phaeodactylum tricornutum*. Gene. 2007;406:23–35.
58. Falciatore A, Casotti R, Leblanc C, Abrescia C, Bowler C. Transformation of nonselectable reporter genes in marine diatoms. Marine Biotechnol. 1999;1:239–51.

6

The transcriptome of *Utricularia vulgaris*, a rootless plant with minimalist genome, reveals extreme alternative splicing and only moderate sequence similarity with *Utricularia gibba*

Jiří Bárta[1], James D Stone[2,3], Jiří Pech[1], Dagmara Sirová[1], Lubomír Adamec[4], Matthew A Campbell[5] and Helena Štorchová[2*]

Abstract

Background: The species of *Utricularia* attract attention not only owing to their carnivorous lifestyle, but also due to an elevated substitution rate and a dynamic evolution of genome size leading to its dramatic reduction. To better understand the evolutionary dynamics of genome size and content as well as the great physiological plasticity in this mostly aquatic carnivorous genus, we analyzed the transcriptome of *Utricularia vulgaris*, a temperate species with well characterized physiology and ecology. We compared its transcriptome, namely gene content and overall transcript profile, with a previously described transcriptome of *Utricularia gibba*, a congener possessing one of the smallest angiosperm genomes.

Results: We sequenced a normalized cDNA library prepared from total RNA extracted from shoots of *U. vulgaris* including leaves and traps, cultivated under sterile or outdoor conditions. 454 pyrosequencing resulted in more than 1,400,000 reads which were assembled into 41,407 isotigs in 19,522 isogroups. We observed high transcript variation in several isogroups explained by multiple loci and/or alternative splicing. The comparison of *U. vulgaris* and *U. gibba* transcriptomes revealed a similar distribution of GO categories among expressed genes, despite the differences in transcriptome preparation. We also found a strong correspondence in the presence or absence of root-associated genes between the *U. vulgaris* transcriptome and *U. gibba* genome, which indicated that the loss of some root-specific genes had occurred before the divergence of the two rootless species.

Conclusions: The species-rich genus *Utricularia* offers a unique opportunity to study adaptations related to the environment and carnivorous habit and also evolutionary processes responsible for considerable genome reduction. We show that a transcriptome may approximate the genome for gene content or gene duplication estimation. Our study is the first comparison of two global sequence data sets in *Utricularia*.

Keywords: Transcriptome, Root-associated genes, Alternative splicing, *Utricularia vulgaris*

Background

Members of the rootless genus *Utricularia* (Lentibulariaceae) are the most versatile and cosmopolitan among carnivorous plants, exhibiting great morphological and ecophysiological plasticity [1-3]. Approximately 50 species of *Utricularia* are aquatic or amphibious, growing in standing, nutrient-poor humic waters. While their ecology and carnivorous habit have been researched previously [3], increasing attention has been given to the peculiarities of *Utricularia* genomes - miniature size in many species within the family [4,5], highly increased nucleotide substitution rates across the genomes of all three cellular compartments: mitochondrial, plastid, and nuclear [6-9], and to the extremely dynamic evolution of genome size at the level of species or even single populations [4,10]. *Utricularia gibba* possesses one of the smallest haploid

* Correspondence: storchova@ueb.cas.cz
[2]Institute of Experimental Botany CAS, Rozvojová 263 6- Lysolaje, Praha 16502, Czech Republic
Full list of author information is available at the end of the article

angiosperm genomes known, approximately one-half that of *Arabidopsis thaliana*, with chromosomes of bacterial size [4,5]. *U. gibba* was the subject of the first broad survey of nuclear gene transcripts in *Utricularia* species [11,12], revealing several interesting aspects of their physiology and morphology. Supporting physiological data, the global transcript analysis revealed specific expression patterns of genes involved in respiration, DNA repair, ROS detoxification, and nutrient uptake in different plant tissues. The sequencing and analysis of the *U. gibba* genome [13] additionally revealed a compressed genome architecture with highly reduced intergenic regions and nearly free of retrotransposons.

Utricularia spp. have since been identified as prime candidates for further research on the complexities of plant ecophysiology associated with carnivory, metagenomic surveys of trap microbial communities, novel plant nitrogen/nutrient utilization pathways, the ecology of prey attraction, whole-plant and trap comparative development, and the evolution of a minimalist angiosperm genome [3,5,14-20]. *Utricularia gibba*, however, is not a good candidate species for many ecological and physiological experiments due to its minute size and extremely small traps. We have therefore chosen the ecologically well-characterized temperate *Utricularia vulgaris* [3,16-18] as our model for a broad transcriptome analysis. Its ecophysiology is subtly but meaningfully distinct from that of *U. gibba*, offering the possibility for a comprehensive comparison of genome-wide expression patterns between the two species.

In this study, we report the results of 454 GS-FLX Titanium sequencing of a polyA-selected and normalized cDNA library from *U. vulgaris*, derived from a pooled sample of multiple tissue types, including functional annotation of expressed gene content. We compared this transcriptome to the *U. gibba* transcriptome [12] and showed that, despite different methods of preparation and tissue composition, the overall gene expression pattern and gene distribution among GO categories were very similar between the two species. We also analysed several cases of alternative splicing (AS) in the *U. vulgaris* transcriptome, including a gene for which this post-transcriptional process has not been investigated in any plant species.

Although any transcriptome should be viewed as incomplete, it may serve as an acceptable proxy for the genome in a species without complete genomic information, such as *U. vulgaris*, provided that it is prepared from multiple tissues and various environmental conditions [21,22]. We demonstrate the usefulness of the *U. vulgaris* transcriptome for the identification of gene losses and duplications during the course of evolution of the genus *Utricularia*.

Results

Transcriptome assembly

In total, 1,405,703 reads were generated by 454 pyrosequencing of the *U. vulgaris* normalized cDNA library, 1,389,835 of them passed built-in quality filtering. 91.5% of the initial, raw reads were assembled by Newbler 2.7 and produced 19,522 isogroups containing 41,407 isotigs, roughly corresponding to the individual transcripts. In addition, 64,188 singletons longer than 100 nt were obtained. Isotigs and singletons were combined together into a unique transcript (UT_U.vulgaris) data set representing the *U. vulgaris* transcriptome. To facilitate the comparison between our data and the *U. gibba* transcriptome published by [12], raw reads of *U. gibba* were downloaded from DNA Data Bank of Japan (DDBJ) under the submission SRA029151 and assembled by Newbler 2.7 using the same parameters as adopted for the *U. vulgaris* transcriptome (UT_U.gibba). Table 1 compares the transcriptome assemblies of the two species. Our *U. vulgaris* data set consisted of nearly twice as many raw reads, a higher proportion of which assembled into contigs, than the *U. gibba* dataset. The *U. vulgaris* assembly also resulted in a higher number of isogroups and much higher (about three fold) number of isotigs. Furthermore, our *U. vulgaris* assembly produced only 64,188 singletons compared to the 99,900 singletons remaining after *de novo U. gibba* assembly. The *U. vulgaris* data set contained about 2.1 isotigs per isogroup, whereas only 1.2 isotigs per isogroup, on average, were found in the *U. gibba* assembly. The much higher number of isotigs in the *U. vulgaris* transcriptome, both relative (per isogroup) and absolute, was at least partly caused by the method of cDNA library preparation. Our *U. vulgaris* cDNA library was normalized, increasing the number of rare transcripts represented by isotigs.

Transcriptome annotation

39,006 *U. vulgaris* isotigs (96%) gave significant BLAST hit against the NCBI nr protein database (BLASTX algorithm, e-value cutoff 10^{-5}). These sequences were further annotated using the BLAST2GO annotation pipeline. 30,392 isotigs (73% of total isotigs) were successfully annotated. 9,794 isotigs (33% of annotated isotigs) were assigned with enzyme codes (E.C.). The average level of annotations in GO hierarchy was 5,868. The total number of assigned Gene Ontology terms was 212,122 (Table 2).

Of the total 58,363 *U. vulgaris* singletons, 23,212 (40%) gave a significant BLAST hit against the NCBI nr protein database under the same parameters as used for isotigs. 14,536 singletons (25% of total singletons) were successfully annotated and 4,121 singletons (28% of annotated singletons) were assigned with E.C. The average level of annotations in GO hierarchy was 5,791. The total number of assigned Gene Ontology terms was 90,735. The much lower proportion of *U. vulgaris* singletons yielding

Table 1 Transcriptomes comparison

	Utricularia vulgaris	*Utricularia gibba*
Biological source of RNA	Shoots, cultivated under sterile conditions	Shoots and flowers natural conditions
cDNA library preparation	Oligo dT enrichment normalized library	Oligo dT enrichment without normalization
inputFile NumReads	1 405 703	817 792
numAlignedReads	1 271 732; 91.5% of total	707 880; 86.56% of total
Number of Assembled	1 151 110	638 520
Number of Partially assembled	120 448	69 288
Number of Singletons	64 188	99 900
Number Outlier (e.g. chimeras)	23 169	9967
Number of TooShort Reads	30 086	6
Number of Isogroups	19 522	13 591
Avg Contig Count	2.6	1.4
Number With One Contig	11 168; 57.2%	11 676; 85.9%
Avg Isotig Count	2.1	1.2
Number With One Isotig	11 213	11 705
Number of Isotigs	41 407	16 465
Avg Contig Count	3.4	1.6
Largest Contig Count	18	13
Number With One Contig	11 168	11 676
Avg Isotig Size	1 514	705
N50 Isotig Size	1905	839
Largest Isotig Size	12 895	3 326
Number Of Large Contigs	22 001	7 763
Avg Large Contig Size	962	819
Number Of all Contigs	51 270	19 120

U. vulgaris and *U. gibba* transcriptomes assembled by Newbler 2.7.

significant BLAST hits, compared with the isotigs, may be due to their short sizes and also due to the presence of transcripts derived from microbes without any NCBI record.

The results of the GO annotations of the UT_U.gibba transcriptome are given in Table 2. The proportion of annotated isotigs is a bit lower and the proportion of annotated singletons is a bit higher in *U. gibba* than in *U. vulgaris*. This difference results from a higher amount of unassembled reads in UT_U.gibba relative to UT_U. vulgaris. The proportion of isotigs with an assigned E.C. was also higher in *U. vulgaris* than in *U. gibba*.

Table 2 GO Annotation summary

	Isotigs		Singletons	
	U.vulgaris	*U.gibba*	*U.vulgaris*	*U.gibba*
NCBI nr	96%	89%	40%	47%
Number of GO terms	212 122	95 741	90 735	264 218
GO annotation	73%	75%	25%	39%
Biological process	54%	52%	54%	53%
Molecular function	13%	12%	14%	13%
Cellular component	34%	36%	32%	34%
E.C. assignment	33%	28%	25%	33%

Values in % indicate the percentage of sequences/groups with one or more significant blast hits/ annotations based on an e-value cut-off of 10^{-5}.

Despite of the differences in cultivation conditions, plant tissues used for RNA extraction, cDNA library preparation and assembly parameters, the general partition of isotigs into basic KEGG categories was very similar between the two *Utricularia* species (Figure 1). "Catalytic activity" and "Binding" were the prevalent categories among Molecular Function. "Cell" and "Organelle" dominated in the Cellular Component section. Abundant categories "Metabolic Process" and "Cellular" were followed by slightly less numerous categories "Response to stimulus" and "Biological regulation". The high representation of the "Single-organism process" category appeared due to co-existing microbes.

We summarized the results of our *U. vulgaris* transcriptome assembly and annotation and created a web-accessible database (http://utricularia.prf.jcu.cz/index.php) which can be easily searched by BLAST or annotation.

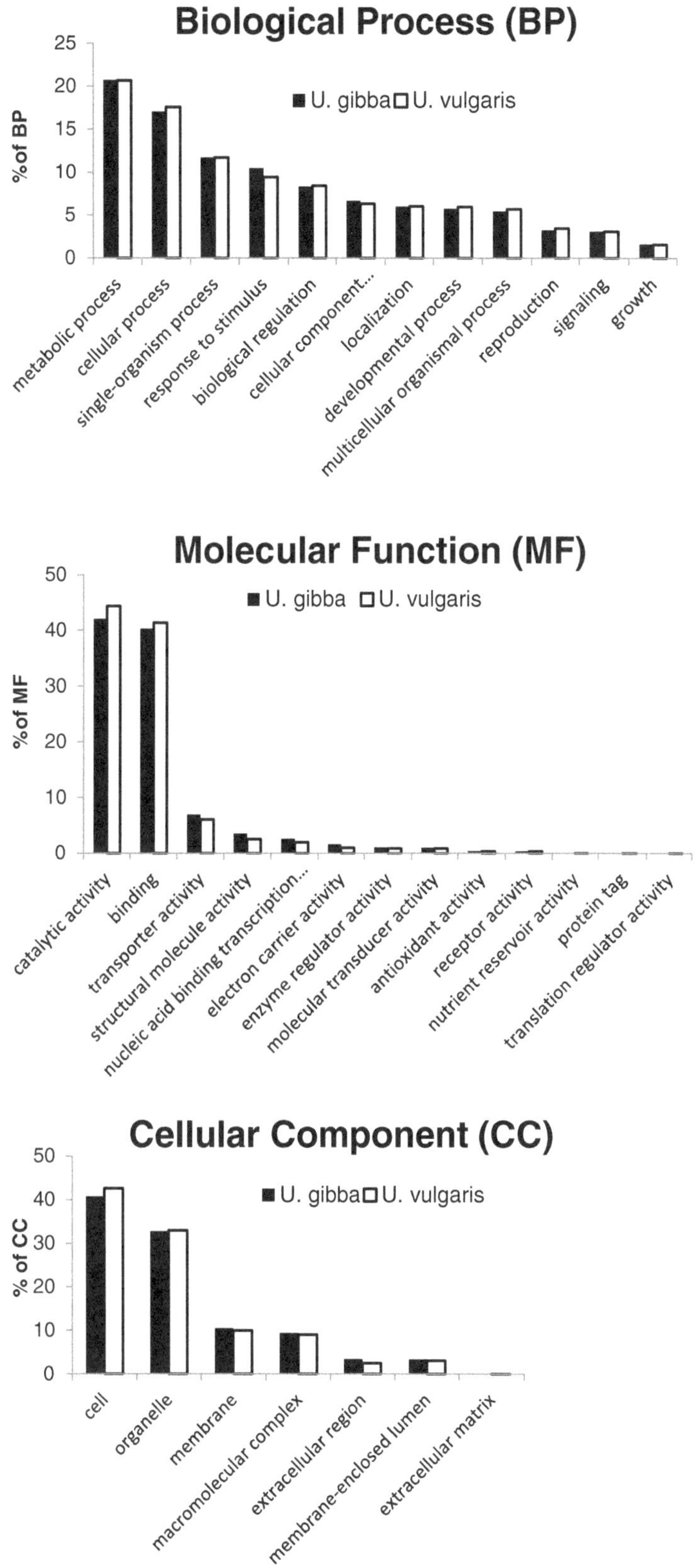

Figure 1 Distribution of GO categories. The comparison of the distribution of unique transcripts (isotigs and singletons) between *U. gibba* and *U. vulgaris* transcriptomes in three main GO categories.

The composition of transcriptomes

More than 99% of the isotigs with significant hits were assigned by MEGAN to plants (Streptophytes) in both *U. vulgaris* and *U. gibba*. All remaining isotigs (38 in *U. vulgaris* and 87 in *U. gibba*) belonged to Fungi, Metazoa, unicellular eukaryotes, and prokaryotes (Figure 2). The taxonomic diversity of singletons was much higher: 5.3% and 10.6% of singletons with significant hits were assigned outside the Streptophytes in *U. vulgaris*, and *U. gibba*, respectively (Additional file 1). The non-plant sequences were mostly derived from microbial commensals, as well as a minor fraction from animal (fish, worm) RNA contamination. The very low proportion of prokaryotic sequences was due to the polyA+ RNA used to prepare cDNA. As prokaryotic mRNAs rarely contain polyA+ tails, they were mostly eliminated. The proportion of non-plant transcripts is probably higher among singletons, because many of them may not have produced statistically significant hits due to incomplete microbial records in public databases. The abundance of microbe-derived transcripts was higher in the *U. gibba* transcriptome prepared only from plants grown under natural conditions and colonized with microbes. In contrast, the *U. vulgaris* transcriptome was constructed from a pooled RNA sample prepared from the plants cultivated under both sterile and non-sterile conditions.

Large isogroups and alternative splicing

The isogroups containing numerous isotigs may include transcripts derived from several or many loci, e.g. retroposons, or from transcripts undergoing AS [23]. The *U. vulgaris* transcriptome contained six isogroups with > 100 isotigs, 23 isogroups with > 45 isotigs, and 332 isogroups with > 10 isotigs. The largest, isogroup 00018 in *U.vulgaris*, included 480 isotigs derived from various members of a large *BETA GLUCOSIDASE* gene family.

In contrast, the *U. gibba* assembly contained zero isogroups with > 100 isotigs, only two isogroups with > 45 isotigs, and 17 isogroups with > 10 isotigs (Additional file 2). The main reason for such a high difference in the number of large isogroups with many isotigs between the UT_U. vulgaris and UT_U.gibba transcriptome assemblies seems to be the method of cDNA preparation. Normalization of the cDNA library led to the enrichment of rare transcripts in *U. vulgaris*, including alternatively spliced mRNAs. The largest isogroup in the UT_U.gibba assembly, which was generated without a cDNA normalization step, contained only 68 isotigs, representing transcripts coding for the small subunit of Rubisco, the most abundant protein on Earth. When read counts are extremely high, as in the case of Rubisco, sequencing errors occur in multiple reads which are then assembled into separate, artifactual contigs. Some large isogroups in *U. gibba* also represented transcripts derived from multiple loci-e.g. isogroup 00005 (*KETOACYL COA SYNTHASE* family) [24] or the isogroups 00002 and 00012, which gave no hits in BLAST searches of NCBI databases, but yielded multiple hits against the *U. gibba* genome draft (CoGe-id36222).

Alternative splicing appears to be the main reason for the transcript abundance and diversity in many of the

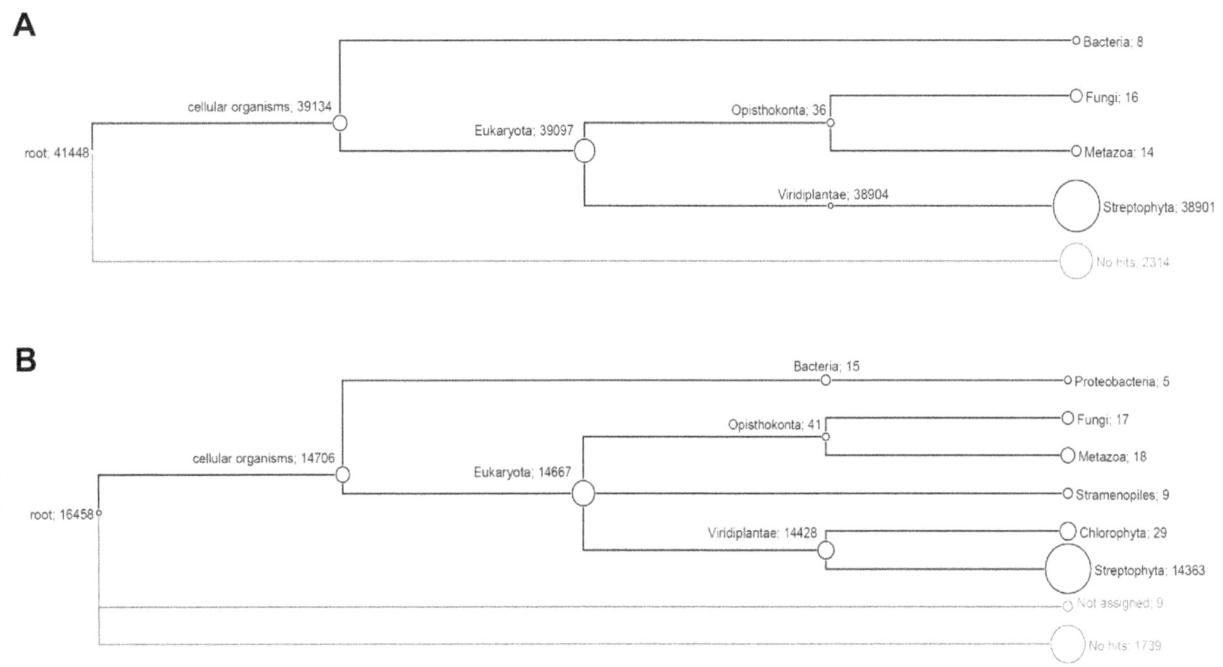

Figure 2 Taxonomic assignment. Dendrogram showing number of MEGAN assigned *U. vulgaris* **(A)** and *U. gibba* **(B)** isotigs.

largest isogroups in *U. vulgaris* and in *U. gibba*. These isotigs contain contigs corresponding to *Arabidopsis* exons and also numerous contigs which may be assigned to introns based on their position between two exons. Additional file 2 compares the 23 and 17 largest isogroups of *U. vulgaris* and *U. gibba*, respectively. They represent various genes or gene families belonging to similar structure and function categories.

Only one large isogroup (00008) appears to be the same in both *Utricularia* species. It encodes a family of ATP dependent RNA helicases. Its *Arabidopsis* homologs (At5g11170, At5g11200) are involved in a wide range of RNA metabolism including pre-mRNA splicing, mRNA transport, turnover, translation initiation etc. [25,26]. They undergo AS, as documented by genome-wide analysis of transcript variants [27]. Five contigs of the isogroup 00008 in *U. vulgaris* match *Arabidopsis* exons, suggesting extensive AS of transcripts derived from at least two related genes. The more than four fold higher isotig count of the 00008 isogroup in *U. vulgaris* than in *U. gibba* may again reflect a significant enrichment in rare transcripts due to cDNA normalization of *U. vulgaris* transcriptome, or reflect a lower extent of AS in *U. gibba*. Three other isogroups of *U. vulgaris* could participate in the control of AS, including the isogroup 00013, encoding a homolog of *AFC2* protein kinase, which underwent extreme AS (producing multiple splice variants from the same primary transcript) in *Arabidopsis* [27]. The remaining large isogroups with AS code for membrane proteins with multiple domains, proteins involved in protein degradation, or fulfilling regulatory functions. Two large isogroups (00020, 00061) were assigned to retroposons in the *U. vulgaris* transcriptome. No large isogroup corresponding to transposons or retroposons was found in the *U. gibba* transcriptome, however three single isotigs were.

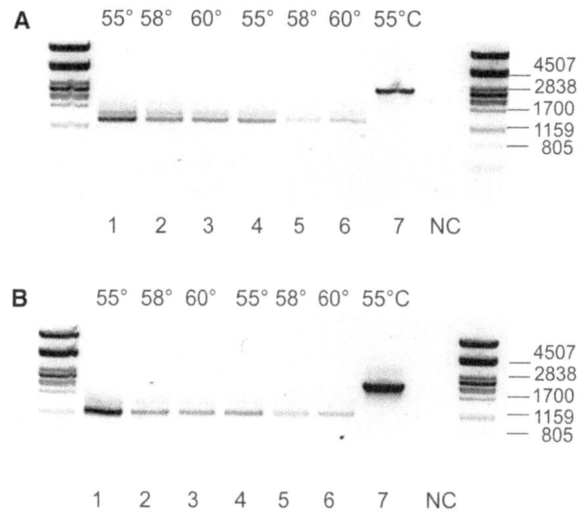

Figure 3 PCR amplification with the *LCB-1-P lyase* specific primers. An agarose gel (1.2%) electrophoresis of PCR fragments amplified from the gene encoding LCB-1-P lyase (isogroup 000007) in *U. vulgaris* with different templates. 1–6: cDNA prepared from RNA extracted from two different plant individuals, 7: genomic DNA. NC: negative control with water instead of DNA. **(A)** PCR with exon-specific primers UV405_F1 and UV405_R1. **(B)**. PCR with intron-specific primer UV405_F2 and exon-specific primer UV405_R1. Annealing temperature is indicated above the lanes. Standard of molecular weights is shown on the both sides of the gel.

Two isogroups with extreme alternative splicing

We selected two isogroups of *U. vulgaris* with very high isotig counts for more detailed analysis. After aligning all 277 isotigs of the isogroup 00007, we found that all of them were derived from the same locus, because only one sequence variant (contig) corresponded to each exon of the homologous *Arabidopsis* gene, At1g27980, coding for sphingoid long-chain base 1-phosphate lyase (LCB-1-P lyase) (Additional file 3). We assigned eight contigs to eight introns based on a comparison with the homologous *Arabidopsis* gene. The retention of variable numbers of introns was responsible for the observed extreme AS in this isogroup. Only one isotig 00648 contained the correct ORF with genetic information for a functional protein. To confirm AS experimentally, we designed primers targeted to exon 6 or intron 6 (forward) and exon 15 (reverse) and ran PCR (Figure 3). The size of PCR fragment generated from genomic DNA (2.4 kb) with exon-specific primers UV405_F1 and UV405_R1 agreed with the expected size of this genomic region (2,353 bp). The amplification of cDNA produced a strong band (1.3 kb) corresponding to correctly spliced mRNA with no introns (1,377 bp) and several weak upper bands most likely derived from partially spliced mRNA with retained introns. The primers spanning from intron 6 to exon 15 (UV405_F2 and UV405_R1) produced a PCR fragment from genomic DNA as well as one strong band (1.1 kb) and a few weaker ones from cDNA. The strong band amplified from cDNA provided evidence for intron 6 retention, because no amplification with this primer pair could occur if only correctly spliced mRNA were present in the transcript pool.

To achieve the correct assembly of alternative transcripts in a species without reference genome is very difficult. It becomes even more challenging if multiple similar paralogous genes are transcribed and alternatively spliced. In such cases, chimeric misassembled contigs are frequently generated [28]. The isogroup 00006 homologous to the *ETHYLENE INSENSITIVE 2 EIN2* gene (At5g03280) in *Arabidopsis* is an example of the mixture of alternatively spliced transcripts derived from at least two loci. We identified contigs corresponding to the exons and introns of the *EIN2* gene. Several putative exons existed in two sequence variants and occurred in chimeric isotigs. We confirmed the occurrence of two *EIN2* gene copies by direct sequencing of *U. vulgaris* genomic DNA. We designed two primer pairs UV304_F1, R1 and UV308_F1, R1 (Additional file 4) and amplified and sequenced a part of exon 7 from both *EIN2* paralogs. The alignment (1,360 bp) of *U. vulgaris* sequences with *EIN2* homologs across angiosperms was used in phylogenetic analysis to generate MP and ML trees (Figure 4). The trees constructed by both methods showed the same topology and confirmed a relatively recent duplication of *EIN2*, preceding the divergence of *U. vulgaris* and *U. gibba*. We found only one *EIN2* homolog in the *U. gibba* genome (CoGe-id36222).

The ratio of non-synonymous and synonymous substitutions (Ka/Ks) in the pairwise comparison between *EIN2-1* and *EIN2-2* of *U. vulgaris* was 0.41, between *EIN2* of *U. vulgaris* and *EIN2* of *U. gibba* was 0.44 (for both *U. vulgaris EIN2* paralogs). The data suggest no variation in evolutionary constraints.

Putative orthologs between *U. vulgaris* and *U. gibba*

We performed a reciprocal BLAST hit search to identify putative orthologs between the UT_U.vulgaris transcriptome and a 19475-mRNA database derived from the genomic draft of *U. gibba*, which represents an *in silico* transcriptome of this species. We chose the *U. gibba* transcriptome derived from a genomic draft, because it supposedly represented more complete set of transcripts than the experimental transcriptome UT_U. gibba.

We identified 12,267 putative orthologous pairs, 10,600 of them contained *U. vulgaris* isotigs and the remaining 1,667 pairs contained *U. vulgaris* singletons. The orthologs represented about 42.9% of all genes annotated in the *U. gibba* genomic draft. We ordered putative orthologous pairs between *U. gibba* and *U. vulgaris* according to the sequence similarity of the regions aligned by BLAST. The distribution of orthologs assigned to individual similarity classes is shown in Figure 5. Most orthologous pairs

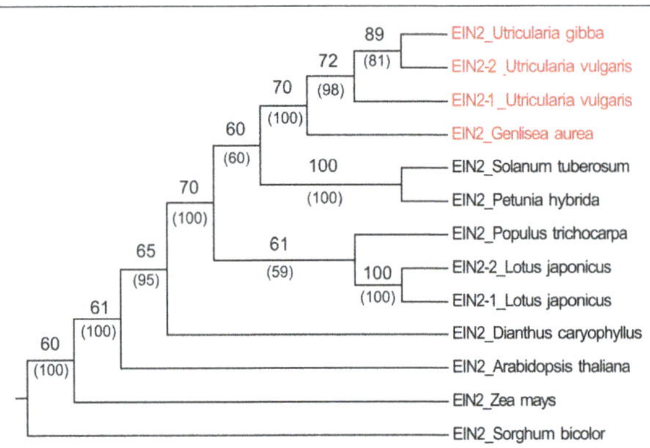

Figure 4 Phylogenetic analyses of *EIN2*. MP and ML tree constructed from the alignment of partial *EIN2* sequences across angiosperms exhibited the same topology. Bootstrap supports calculated from 1000 pseudoreplicates are shown above branches (MP) or below branches in parentheses (ML).

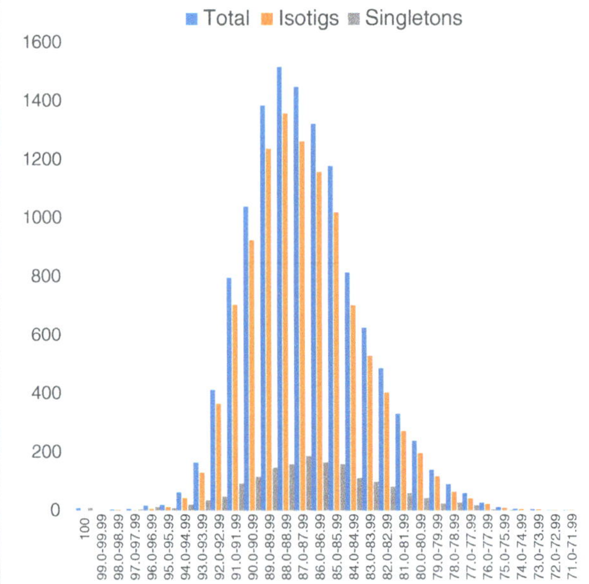

Figure 5 Ortholog similarity distribution. The distribution of putative orthologous pairs between *U. gibba* and *U. vulgaris* according to their sequence similarity. Each bar represents the number of pairs of a given similarity spanning 1% interval. Blue–all orthologous pairs, orange–pairs with *U. vulgaris* isotigs, grey–pairs with *U. vulgaris* singletons.

exhibited a sequence similarity of 85%-90%, whether or not they included *U. vulgaris* isotigs or singletons. Singletons are much shorter than an average isotig (1,514 bp; Table 1), thus they often represent incomplete transcripts. Their sequence similarity depends on whether they are derived from a more or less conserved part of the gene, it does not reflect the similarity across an entire gene. For this reason, we performed the following analyses of the most conserved orthologs with the pairs containing only *U. vulgaris* isotigs, not singletons.

Because the overall sequence similarity of putative *U. vulgaris-U. gibba* orthologs was rather low (median 87%), we investigated which GO categories were enriched among the most conserved orthologous pairs. GO enrichment (AgriGo) [29] analysis of the most conserved orthologs (with similarity higher than 93%) against all orthologs identified 36 significantly enriched GO categories (Additional file 5). They belonged to the genes encoding proteins conserved across all angiosperms (ribosomal proteins, tubulins, small GTP-binding proteins, mitochondrial respiratory chain proteins, etc.). Their proportion in respective similarity classes of putative orthologs increased with increasing sequence similarity (Figure 6). Detailed inspection of GO categories enriched among highly conserved orthologs between *U. gibba* and *U. vulgaris* revealed genes which were less similar to their *Arabidopsis* counterparts than the rest of the highly conserved orthologs, namely *MYOSIN XI B* (homolog of At1g04160) and

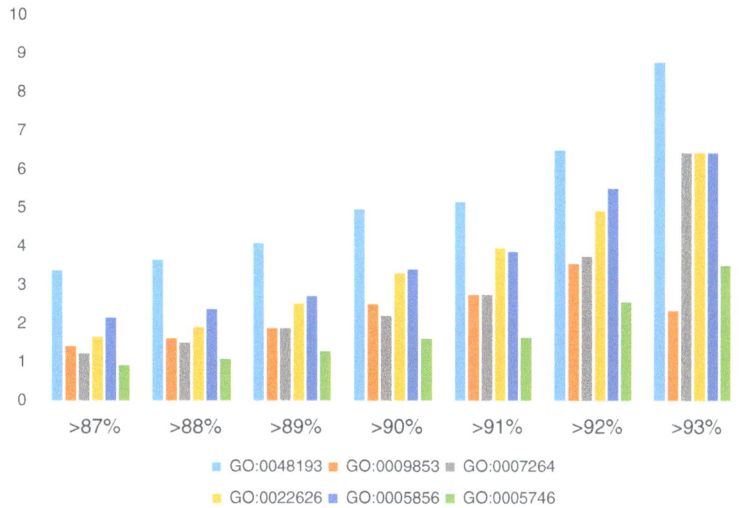

Figure 6 GO enrichment. The enrichment of particular GO categories (in % of total GO categories) in the subsets of orthologous pairs between *U. vulgaris* and *U. gibba* with ascending sequence similarity. GO:0048193 Golgi vesicle transport; GO:0009853 Photorespiration GO: 0007264 Small GTPase mediated signal transduction; GO: 00022626 Cytosolic ribosomes; GO:0005856 Cytoskeleton; GO: 0005746 Mitochondrial respiratory chain.

TIP GROWTH DEFECTIVE 1 (TIP1) (a homolog of At5g20350). Interestingly, both genes play a role in root hair development in *Arabidopsis*. As neither *U. gibba* nor *U. vulgaris* produce roots, it is probable that the two genes have gained a novel or modified functions in *Utricularia*, explaining why their sequences are highly similar between both *Utricularia* species, but less similar to *Arabidopsis* homologs.

Root-specific genes in rootless *Utricularia*

As *Utricularia vulgaris* does not form roots, some of the genes involved in root development and function might have been lost. Ibarra-Laclette et al. [13] published a list of the genes associated with root in *A. thaliana*, but not found in the genome of *U. gibba*. They include *MYB* transcription factors, MADS box genes, cell-wall-associated kinases, nitrate transporter etc. (Ibarra Laclette et al. [13]). We selected the root-associated genes absent in *U. gibba*, supplemented additional genes exclusively or predominantly expressed in the roots of *A. thaliana* and checked their presence among putative orthologs identified by BLASTX-TBLASTN reciprocal BLAST search between *A. thaliana* protein data set (TAIR10_prot_20101214) and the UT_U.vulgaris transcriptome. We found a strong correspondence between the absence of particular genes in the *U. gibba* genome [13] and their absence in the *U. vulgaris* transcriptome (Additional file 6). Moreover, we did not find the counterparts of additional *Arabidopsis* root-associated genes in the *U. vulgaris* transcriptome, notably transcription factors involved in root hair (e.g. *ROOT HAIR DEFECTIVE 6-RHD6*, *WEREWOLF–WER*) or root cap development (e.g. *BEARSKIN 1–BRN1*, *FEZ*, *SOMBRERO–SMB*). These genes were also missing in the *U. gibba* genome (CoGe-id36222). On the other hand, some genes such as *AUXIN RESPONSE FACTOR 4* (*ARF4*) or *INDOLE-3-ACETIC ACID INDUCIBLE* (*IAA*) genes found putative orthologs in both the *U. vulgaris* transcriptome and the *U. gibba* genome (Additional file 6). Six of 13 *U. vulgaris* root-associated genes (isotigs or single reads) under study contained complete ORFs, the rest of isotigs represented partial sequences, which reflected an overall incompleteness of experimental transcriptomes.

Because the transcriptomes are incomplete, the absence of an ortholog in the transcriptome, by itself, is not the evidence of its absence in the corresponding genome. Thus, we may not exclude the possibility that the absence of any respective gene in the *U. vulgaris* transcriptome was caused by its low or missing transcription. However, the coincident absence of 48 root-associated genes and concordant presence of 11 root-associated genes in the *U. vulgaris* transcriptome and the *U. gibba* genome suggest that, at least in the case of root genes, the UT_U. vulgaris transcriptome reflects the gene content of the *U. vulgaris* genome.

Interestingly, two copies of the gene *AUXIN RESPONSE FACTOR 16* (*ARF16*), duplicated in the *U. gibba* genome [13], were also found in the *U. vulgaris* transcriptome. The full agreement between the sets of root-associated genes lost in *U. gibba* and *U. vulgaris* and the concordance of the genes duplicated in both species support the notion that deletions and duplications of genes involved in root-associated genes occurred before the divergence of the two *Utricularia* species.

Discussion

Transcriptome comparison

Recent progress in next generation sequencing makes it possible to sequence the genomes and transcriptomes of non-model plants to an unprecedented extent. The 1000 plants (one KP or 1KP) initiative (https://sites.google.com/a/ualberta.ca/onekp) is just one example of current efforts. The genomic draft of *U. gibba* [13] has attracted attention because it represents one of the smallest genomes in the plant kingdom and opened the possibility to study the mechanisms responsible for genome contraction in plants. The availability of a sequenced genome and an experimental transcriptome of *U. gibba* generated by 454 pyrosequencing from various organs [12], made *U. gibba* a suitable species for comparative transcriptomics in *Utricularia*. We chose a temperate congener *U. vulgaris* as a counterpart for comparison. Both species share an aquatic carnivorous life style, lack of roots, and display rapid apical shoot growth, but exhibit partly distinct ecophysiology (turion formation in *U. vulgaris*, possible terrestrial life in *U. gibba*; see [1].

We utilized the two kinds of transcriptomes available for *U. gibba* in our comparative studies. The *in silico* transcriptome derived from the genomic draft is assumed to be more complete than the experimental transcriptome, which is known to lack transcripts due to low or missing gene expression [21]. However, an *in silico* transcriptome is only as good and complete as the annotation of the genome of interest. It may also erroneously assign virtual transcript to a pseudogene which is not expressed. In contrast, an experimental transcriptome is comprised of real transcripts able to capture alternatively spliced transcripts derived from the same gene. Considering advantages and disadvantages of the two kinds of transcriptomes, we decided to prefer the *in silico* transcriptome for the identification of putative orthologs between *U. gibba* and *U. vulgaris*. In contrast, experimental transcriptome of *U. gibba* [12] was used to compare expressed gene categories and also to identify alternatively spliced transcripts.

The transcriptome of *U. gibba* [12] was prepared from inflorescences in addition to submersed parts of plants, but, unlike *U. vulgaris* transcriptome, it did not include sterile plants. Another distinction was the application of cDNA normalization to the construction of the *U. vulgaris* transcriptome but not the *U. gibba* transcriptome. Despite these differences, the proportions of annotated GO categories were very similar (Figure 1).

The distribution of GO categories in *Utricularia* was also in line with previously published transcriptomic data from carnivorous species of *Sarracenia* [30]. This study used only a quarter of the data of the *U. vulgaris* transcriptome, without pooling various tissues or developmental stages. Despite methodological differences, the proportions of GO categories were very similar between *U. vulgaris* and *Sarracenia*. Only the categories "Response to stimulus" and "Biological regulation" were much higher in *Utricularia* than recorded for *Sarracenia*. This difference may reflect distinct life styles of both carnivorous genera. Whereas *Sarracenia* is a robust slowly growing terrestrial perennial plant, *Utricularia* is a fast growing aquatic plant which has to cope with sudden changes of environment (nutrient level, salinity, streaming or even temporary desiccation). Alternatively, the impact of very different data sets cannot be excluded. In this case, it would affect only the two GO categories, which appears unlikely.

Examples of alternative splicing

Normalization of cDNA is recommended for the study of AS, because it increases the proportion of rare mRNAs, often represented by alternatively spliced transcripts. This approach revealed that 61.2% of intron-containing genes were alternatively spliced in *Arabidopsis* [27]. We cannot directly compare the extent of AS in *U. vulgaris* and *Arabidopsis*, because missing genomic information in *U. vulgaris* prevents the accurate assignment of splice variants. However, 23 largest isogroups in *U. vulgaris* (Additional file 2) matched *Arabidopsis* homologs with more than 3 splice variants belonging to 25% of the genes with the highest level of AS in *Arabidopsis* [27], which suggests that a similar set of genes is highly alternatively spliced in *U. vulgaris* and *Arabidopsis*.

The gene encoding ATP dependent RNA helicase (isogroup 00008) was shown to be alternatively spliced not only in *U. vulgaris*, but also in *U. gibba*. It participates in the control of mRNA splicing and export in *Arabidopsis* [25,26] and its expression is regulated by AS in this plant. It is therefore possible that the paralogs encoding ATP dependent RNA helicase (isogroup 00008) play similar roles in *Utricularia*.

We also documented extreme AS in the isogroup 00007 in *U. vulgaris* homologous to the gene for LCB-1-P lyase in *Arabidopsis*. The function of LCB-1-P lyase or sphingosine-1-phosphate (SPH-1-P) lyase in plants is not fully understood. Ng et al. [31], Coursol S et al. [32] described the role of SPH-1-P as a lipid messenger in guard cell abscisic acid (ABA) response. Nishikawa et al. [33] showed that LCB-1-P was degraded by LCB-1-P lyase (encoded by *AtDPL1*, At1g27980), which was located in the endoplasmic reticulum. LCB-1-P lyase regulates LCB-1-P content and through this activity participates in stomata closure and dehydration stress response in *Arabidopsis* [33,34].

Our *U. vulgaris* transcriptome was generated from submersed plant organs which did not develop stomata. However, we have verified that above-water flower stems of *U. vulgaris* contain stomata (Adamec, unpublished results). It is therefore possible that transcripts with retained introns represent a pool from which functional transcripts may be readily formed by additional splicing when LCB-1-P lyase becomes necessary. This protein may be needed when the submersed plant body continues to grow above water. A similar regulatory role of intron retention was observed, for example, in the fern *Marsilea vestita*, where transcripts with retained introns were stored in spores and spliced after germination [35].

The expression of the gene for LCB-1-P lyase in submersed plant organs lacking stomata may also suggest that it fulfills a distinct function in *Utricularia*, not associated with stomata. As SPH-1-P affects ion channels in guard cell protoplasts [32], we speculate that this lipid messenger may have a role in water pumping regulation in *Utricularia* traps, which is associated with potassium channels in trap bifid glands [36]. Finally, it is possible that extreme AS in the isogroup 00007 coding for LCB-1-P lyase does not have any regulatory function and represents an error in a complex splicing process. To our knowledge, AS of transcripts encoding LCB-1-P lyase has not been studied in any plant species. We cannot determine whether this also occurs in *U. gibba*, because non-normalized transcriptomes have only limited potential to detect AS.

Duplication of the *EIN2* gene in *U. vulgaris*

We found a duplication of *EIN2* gene in *U. vulgaris* transcriptome. This gene is essential for ethylene signaling and occurs in a single copy in many plant species and its duplication is rare among angiosperms [37,38]. Two *EIN2* paralogs undergoing accelerated evolution were recently identified in *Lotus japonicus* [39]. They regulate not only a response to ethylene, but also nodulation in the course of symbiosis with rhizobia. The two *EIN2* paralogs in *Utricularia vulgaris* may be important for the interaction between plants and microbes, similar to the role of the two *EIN2* genes in the symbiosis between leguminous plants and bacteria [39]. We speculate that the duplication of the *EIN2* gene occurred early in the evolution of the genus *Utricularia* and might have been associated with the transition to a carnivorous life-style. Subsequently, one copy was lost in *U. gibba*. Similar Ka/Ks ratios in pairwise comparisons of *Utricularia EIN2* genes do not indicate any shift in function. However, it should be emphasized, that only parts of the *U. vulgaris EIN2* genes, confirmed by Sanger sequencing, were analyzed. The examination of additional species of Lentibulariaceae regarding *EIN2* multiplication will shed light on the evolution and function of this important gene.

Low sequence similarity between putative *U. gibba-U. vulgaris* orthologs

The median value of sequence similarity among 12,267 putative orthologous pairs (measured as high-scoring segment pair of BLAST alignments) between the two *Utricularia* species was only 87% (Figure 5), much lower than for example a median ortholog similarity between two *Corylus* species (98%)- [40] or between chimpanzee and human (about 93.7%- [41]) which belong to different genera. *U. vulgaris* and *U. gibba* are classified in the same generic section *Utricularia* [1,5], although they are not sister species. The reason for the high divergence between *U. gibba* and *U. vulgaris* appears to be a high substitution rate, described in Lentibulariaceae as one of the characteristics of the plant carnivorous syndrome [11,42]. However, these two species still displayed very high sequence similarity in orthologs encoding ribosomal proteins, components of respiratory chain and cytoskeletal proteins, which were ultraconserved across the plant kingdom. We also observed a high conservation of some genes involved in root-associated function, which were generally less conserved among angiosperms. This could be explained by a functional shift shared by the two rootless aquatic *Utricularia* species.

The loss of root-associated genes

We found a perfect coincidence between the absence of root-associated genes in the *U. gibba* genome [13] and the absence of their counterparts in the *U. vulgaris* transcriptome. The correspondence between both *Utricularia* species was also observed in additional root-associated genes, not specifically analyzed by [13] (Additional file 6). Although the absence of particular genes in our transcriptome may be due to intrinsic incompleteness of any experimental transcriptomic data, the high coincidence between genomic and transcriptomic gene occurrences in two species suggests that many root-associated genes are indeed missing in the *U. vulgaris* genome. It is probable that the loss of root-associated genes had occurred already in the ancestor of *U. gibba* and *U. vulgaris*. The comparison of the presence or absence of root-associated genes in additional *Utricularia* species will be very useful for understanding the adaptation to an aquatic rootless carnivorous life-style.

Besides gene losses, gene duplications could be also very informative regarding the evolution and consequences of aquatic carnivory in plants. For example, the duplication of *ARF16* in *U. gibba* [13] was also observed in the *U. vulgaris* transcriptome. In contrast, the *EIN2* duplication event was unique for *U. vulgaris*.

Conclusions

Our study is the first example of comparative transcriptomics in the species-rich genus *Utricularia*. We compared the transcriptome of *U. vulgaris* with the previously published transcriptome of *U. gibba* [12] and confirmed a general similarity of their expression profiles. Both *Utricularia* species displayed some distinctions from the *Sarracenia* transcriptome [30] due to the aquatic life-style different from a terrestrial carnivory. Average sequence similarity of the putative orthologs between *U. vulgaris* and *U. gibba* was lower than 90%, which could be caused by elevated substitution rate in this genus [11]. A strong correspondence in presence or absence of root-associated genes between *U. vulgaris* and *U. gibba* suggests that the loss of some root-specific genes had occurred before the two rootless species separated. Future transcriptome and genome comparisons over numerous *Utricularia* species will certainly deepen our understanding of plant evolution and adaptation.

Methods

Plant material

We utilized *U. vulgaris* from three sources for RNA extraction: (1) sterile plants from a meristem tissue culture (courtesy of Ing. Kamil Pásek, Ostrava-Poruba, Czech Republic), growing in a half-strength Gamborg B5 liquid medium with 500 mg l^{-1} KNO_3, microelements, vitamins, and 2.5% sucrose, but without other organic substances [43]; (2) sterile plants transplanted axenically from tissue culture into an aerated mineral medium [44], with reduced amounts of phosphorus and nitrogen where plants were grown in 16/8 h light/dark regime of fluorescent light at 23°C for 4 weeks, attaining the length of approximately 20 cm; and (3) plants with colonized traps, approximately 35 cm in length, cultivated in a 750 litre outdoor plastic container closely simulating natural dystrophic conditions [45], with the litter of robust sedges (*Carex* spp.) used as a substrate [45] and partial shading (usually 20-50% of incident irradiance). Plants cultivated outdoor were occasionally fed fine zooplankton prey to support growth. No experiments on humans or animals, or cloning experiments were carried out. All activity complied with ethics rules.

Tissue for RNA extraction was harvested in the middle of the dark period to reduce the amount of Rubisco-coding mRNA in samples, frozen immediately in liquid nitrogen, and stored at-70°C until further processing.

RNA extraction, cDNA library construction, and 454 sequencing

Total RNA was extracted from entire shoots, including leaves and traps, from all three sources described above. Inflorescences were not used in this study. Approximately 800 mg of fresh weight (2 individuals) of whole shoots in three replicates from each source were ground in liquid nitrogen using a mortar and pestle and RNA was extracted using RNeasy Plant Midi Kit (Qiagen Inc., Valencia, CA, USA). The results of the parallel extractions were pooled in equal proportions into a single sample and the quality and quantity of the extracted RNA were verified by gel electrophoresis and by spectrophotometer, measuring the 230/260 ratio with Biophotometer (Eppendorf, Germany).

A pooled RNA sample was provided to GATC Biotech (Konstanz, Germany) for the construction of a cDNA library and subsequent sequencing. Library construction involved DNAselection for polyadenylated (polyA+) transcripts to enrich for protein-coding mRNAs, DNase I treatment and normalization through denaturation/reassociation of cDNA, to improve the representation of low-copy transcripts and thereby maximize gene discovery. The resulting library was sequenced with a full picotiter plate on a 454 GS-FLX sequencer with Titanium reagents (Roche Applied Science, Indianapolis, IN, USA), using standard 454 protocols.

Transcriptome assembly and annotation

Raw reads obtained by 454 pyrosequencing were submitted to European Nucleotide Archive under the study accession number PRJEB8057 (http://www.ebi.ac.uk/ena/data/view/PRJEB8057).

We assembled our raw reads with Newbler 2.7 using the following command line settings:–cdna,–it 500,–ig 1000,–icc 200. The subsequent annotation steps excluded singletons that were <100 bp in length. Annotation of isotigs (contigs assembled by Newbler and roughly corresponding to individual transcripts) and singletons (transcripts represented by single reads) was done using the BLAST2GO pipeline [46] with the following parameters: Annotation rule cutoff 55, e-value 10^{-6}, Hit-HSP overlap 0, GO weight 5.

MEGAN v 5.7.0 [47] with default parameters was used to get the taxonomical annotation of isotigs and singletons. MEGAN uses BLASTX output and LCA (Lowest Common Ancestor) classifier based on the actual NCBI taxonomy to obtain the taxonomical annotations.

Search for orthologs

The orthologous transcripts between *U. vulgaris* and *U. gibba* were identified by means of reciprocal BLAST search [48]. We used the transcriptome of *U. vulgaris* containing both isotigs and the singletons which passed quality filtering (min phred score 25, min length 100, max ambiguities 0, max homopolymer 8). As the *U. gibba* transcriptome published by Ibarra-Laclette et al. [12] might have been incomplete, we downloaded the database 19475-mRNA, constructed from the complete genome of *U. gibba*, from CoGe website (https://genomevolution.org/CoGe/-id36222). We generated local databases and used BLASTN, threshold e–10, to retrieve the best hits in reciprocal BLAST searches. Reciprocal BLAST outputs were compared by RBH (reciprocal blast hit) script and putative orthologs which appeared as the same pair in both BLAST outputs were identified. The similarity of orthologs in % was based on high-scoring segment pair (HSP) of BLAST alignments.

The orthologs between *U.vulgaris* and *Arabidopsis* (TAIR10_prot_20101214) were searched in a similar way, with BLASTX (threshold e–10) and TBLASTN (threshold e–10) instead of BLASTN, which means that predicted protein sequences were aligned.

The analysis of large isogroups

We selected the isogroups containing more than 45 isotigs in *U. vulgaris* and more than 10 isotigs in *U. gibba* and found their *Arabidopsis* counterparts using the BLASTX search in TAIR (www.arabidopsis.org). We assigned their functions based on TAIR annotations. We aligned all isotigs from two selected isogroups of *U. vulgaris*, each displaying an extremely high number of isotigs, by MUSCLE implemented in Geneious 7.1.5. We identified introns retained in the transcripts with the help of Layout output from Newbler 2.7 and by comparison with gene structures in *Arabidopsis* (TAIR).

GO enrichment analysis

We performed the analysis of enrichment of particular GO categories among putative *U. gibba–U. vulgaris* orthologs by means of AgriGO (http://bioinfo.cau.edu.cn/agriGO) [29]. We input the list of annotated putative orthologs as a *customized annotated reference* and the list of orthologs with specific similarity as *a customized annotation.* We applied Yekutieli multi-test adjustment, significance level 0.05 and a minimum number of mapping entries of 4.

RT PCR confirmation

To verify transcript sequences inferred from the 454 assembly, we collected *U. vulgaris* plants grown outdoor and isolated total RNA by RNeasy Plant Mini Kit (Qiagen). RNA was treated with DNase I (DNA-free, Ambion, TX, USA) to eliminate a DNA contamination. The treatment was repeated until no amplification was observed in PCR reaction with RNA as a template. One μg of RNA and oligo dT primers (500 ng) were heated 5 min at 65°C, chilled on ice and mixed with Transcriptor buffer (Roche Scientific, Mannheim, Germany), 0.5 μl of Protector RNase Inhibitor (Roche Scientific), 2 μl of 10 mM dNTPs and 10 units of Transcriptor Reverse Transcriptase (Roche Scientific). The first strand of cDNA was synthesized at 55°C for 30 min and then amplified with the primers (Additional file 4) targeted to isotig 00405 (isogroup 00007). PCR was run in a T Gradient cycler (Biometra, Goettingen, Germany) with Taq polymerase (Promega, Madison, WI, USA) under the following conditions: initial denaturation 2 min at 94°C; 36 cycles consisting of 40 s at 93°C, 45 s at given annealing temperature and 2 min at 72°C; final extension 10 min at 72°C.

PCR confirmation

Genomic DNA from *U. vulgaris* plants grown outdoor was isolated by the sorbitol extraction [49]. PCR with the primers (Additional file 4) targeted to isotigs 00304 or 00306 (isogroup 00006) was run as described above at annealing temperature 60°C. PCR products were Sanger sequenced and partial genomic sequences corresponding to two paralogs of *EIN2* in *U. vulgaris* were deposited in GenBank under the numbers KP279653 and KP279654.

Phylogenetic analyses

We downloaded *EIN2* sequences of *U. gibba* (Sc00341_g15851 from CoGe-id36222), *Genlisea aurea* (EPS70151), *Solanum tuberosum* (XM006354273), *Petunia hybrida* (AY353249), *Populus trichocarpa* (XM002326149), *Lotus japonicus* (chr5.CM1729.230 and chr1. CM0012.1100), *Dianthus caryophyllus* (HQ441183), *Arabidopsis thaliana* (NM12040), *Zea mays* (AY359584), and *Sorghum bicolor* (XM002457067). We aligned these angiosperm EIN2 sequences with partial protein sequences EIN2-1 and EIN2-2 from *U. vulgaris* and created a 1,361 amino acid long gapped-alignment in Geneious 7.1.5. We then adopted a maximum parsimony (MP) approach with stepwise addition to the starting tree, no topology constraints, tree bisection-reconnection (TBR) swap, and 1000 bootstrap pseudoreplicates to construct a 60% consensus tree in PAUP*4 v10 [50]. Gaps were treated as missing characters. To confirm phylogenetic positions of the *EIN2* paralogs in *Utricularia* we performed Maximum Likelihood (ML) analysis in PAUP*4 v10 [50] of the same nucleotide alignment under HKY85 evolution model, using the same tree construction methods as described above, 1000 bootstrap pseudoreplicates.

Availability of supporting data

We summarized the results of our *U. vulgaris* transcriptome assembly and annotation and created a web-accessible database (http://utricularia.prf.jcu.cz/index.php) which can be easily searched by BLAST or annotation. Raw reads obtained by 454 pyrosequencing were submitted to European Nucleotide Archive under the study accession number PRJEB8057 (http://www.ebi.ac.uk/ena/data/view/PRJEB8057). The sequences of *EIN2* paralogs were deposited in GenBank under the numbers KP279653 and KP279654. The data sets undelying phylogenetic *EIN2* studies was submitted to TreeBase under the ID 17158. The sequences of *U. gibba* were obtained from the study of Ibarra-Laclette et al. [12] (doi:10.1186/1471-2229-11-101).

Additional files

Additional file 1: Dendogram showing number of MEGAN assigned *U. vulgaris* (A) and *U. gibba* (B) singletons.

Additional file 2: Isogroups with numerous isotigs in *U. vulgaris* and *U. gibba*.

Additional file 3: Layout of the isotigs comprising the isogroup 00007. Exons are marked in yellow, contigs corresponding to exons and introns are blue.

Additional file 4: PCR primers used in this study.

Additional file 5: GO categories significantly enriched among *U. gibba*–*U. vulgaris* putative orthologs with sequence similarity higher than 93%.

Additional file 6: Root-associated genes in *U. vulgaris* and *U. gibba*.

Abbreviations

ARF: *AUXIN RESPONSE FACTOR*; AS: Alternative splicing; *BRN*: *BEARSKIN*; DDBJ: DNA Data Bank of Japan; *EIN*: *ETHYLENE INSENSITIVE* GO, Gene Onthology; HSP: high scoring segment pair; *IAA*: *INDOLE-3-ACETIC ACID INDUCIBLE*; LCB-1-P: long-chain base 1-phosphate; MP: maximum parsimony; RBH: reciprocal blast hit; *RHD*: *ROOT HAIR DEFECTIVE*; TBR: tree bisection-reconnection; *SBM*: *SOMBRERO*; SPH-1-P: sphingosine-1-phosphate; *TIP*: *TIP GROWTH DEFECTIVE*; *WER*: *WEREWOLF*.

Competing interests

The authors declare that they have no competing interests.

Authors' contributions

JB extracted RNA, performed transcriptome assembly and annotation, and contributed to manuscript preparation. JDS and MAC provided troubleshooting and advice with sequence analyses and contributed to manuscript preparation JP wrote custom scripts and provided computer assistance. DS contributed to experimental design, performed plant cultivation, collected plant material, and contributed to manuscript preparation. LA contributed to experimental design, optimized the composition of cultivation media, and contributed to manuscript preparation. HS optimized RNA extraction, designed data analyses, performed phylogenetic analysis, GO enrichment and AS analysis, set up a major part of the manuscript. All authors read and approved the final manuscript.

Acknowledgements

The authors are grateful to Jaroslav Vrba and Jakub Borovec for their valuable contribution to plant cultivation, and to Daniel B Sloan for advice on web site construction and transcriptome assembly. They are also thankful to Naoki Takebayashi for bioinformatics advice, Kamil Pásek for providing axenic tissue cultures of *U. vulgaris*, and Kateřina Haškovcová for excellent technical assistance. Access to computing and storage facilities owned by parties and projects contributing to the National Grid Infrastructure MetaCentrum, provided under the programme 'Projects of Large Infrastructure for Research is greatly acknowledged. This work was funded by the grant of the Grant Agency of the Czech Republic P504/11/0783 to LA, and further supported by Czech Research Programme No. RVO67985939 to LA, and by the project "Integration of the experimental and population biology using new methods of interdisciplinary issues-the way to excellence with young scientists," Reg.No.: CZ.1.07/2.3.00/30.0048, funded by the European Social Fund (ESF) and the state budget of Czech Republic through the Operational Program Education for Competitiveness (OPEC)" to JDS and HS.

Author details

[1]Faculty of Science, University of South Bohemia, Branišovská 31, České Budějovice 37005, Czech Republic. [2]Institute of Experimental Botany CAS, Rozvojová 263 6- Lysolaje, Praha 16502, Czech Republic. [3]Institute of Botany CAS, Zámek 1, Průhonice 25243, Czech Republic. [4]Institute of Botany CAS, Section of Plant Ecology, Dukelská 135, Treboň 37982, Czech Republic. [5]Institute of Fundamental Sciences, Massey University, Private Bag 11222, Palmerston North 4442, New Zealand.

References

1. Taylor P. The Genus *Utricularia*: A Taxonomic Monograph. London: Kew Bulletin; 1989. Additional Series XIV.
2. Guisande C, Granado-Lorencio C, Andrade-Sossa C, Duque SR. Bladderworts. Funct Plant Sci Biotechnol. 2007;1:58–68.
3. Adamec L. Ecophysiological look at plant carnivory: Why are plants carnivorous? In: Seckbach J, Dubinski Z, editors. All Flesh is Grass. Plant-Animal Interrelationships. Cellular Origin, Life in Extreme Habitats and Astrobiology. Volume 16. Dordrecht - Heidelberg - London - New York: Springer Science + Business Media B. V; 2011. p. 455–89.
4. Greilhuber J, Borsch T, Müller K, Worberg A, Porembski S, Barthlott W. Smallest angiosperm genomes found in Lentibulariaceae, with chromosomes of bacterial size. Plant Biol. 2006;8:770–7.
5. Veleba A, Bureš P, Adamec L, Šmarda P, Lipnerová I, Horová L. Genome size and genomic GC content evolution in the miniature genome-sized family Lentibulariaceae. New Phytol. 2014;203:22–8.
6. Jobson RW, Albert VA. Molecular rates parallel diversification contrasts between carnivorous plant sister lineages. Cladistics. 2002;18:127–36.
7. Müller K, Borsch T, Legendre L, Porembski S, Theisen I, Barthlott W. Evolution of carnivory in Lentibulariaceae and the Lamiales. Plant Biol. 2004;6:477–90.
8. Müller K, Borsch T, Legendre L, Porembski S, Barthlott W. Recent progress in understanding the evolution of carnivorous Lentibulariaceae (Lamiales). Plant Biol. 2006;8:748–57.
9. Müller K, Borsch T. Phylogenetics of *Utricularia* (Lentibulariaceae) and molecular evolution of the trnK intron in a lineage with high substitutional rates. Plant Syst Evol. 2005;250:39–67.
10. Albert VA, Jobson RW, Michael TP, Taylor DJ. The carnivorous bladderwort (*Utricularia*, Lentibulariaceae): a system inflates. J Exp Bot. 2010;61:5–9.
11. Ibarra-Laclette E, Albert VA, Herrera-Estrella A, Herrera-Estrella L. Is GC bias in the nuclear genome of the carnivorous plant *Utricularia* driven by ROS-based mutation and biased gene conversion? Plant Signal Behav. 2011;6:1631–4.
12. Ibarra-Laclette E, Albert VA, Perez-Torres CA, Zamudio-Hernandez F, de Ortega-Estrada MJ, Herrera-Estrella A, et al. Transcriptomics and molecular evolutionary rate analysis of the bladderwort (*Utricularia*), a carnivorous plant with a minimal genome. BMC Plant Biol. 2011;11:101.
13. Ibarra-Laclette E, Lyons E, Hernández-Guzmán G, Pérez-Torres CA, Carretero-Paulet L, Chang TH, et al. Architecture and evolution of a minute plant genome. Nature. 2013;498:94–8.
14. Peroutka M, Adlassnig W, Volgger M, Lendl T, Url WG, Lichtscheidl IK. *Utricularia*: a vegetarian carnivorous plant? Algae as prey of bladderwort in oligotrophic bogs. Plant Ecol. 2008;199:153–62.
15. Sirová D, Borovec J, Černá B, Rejmánková E, Adamec L, Vrba J. Microbial community development in the traps of aquatic *Utricularia* species. Aquat Bot. 2009;90:129–36.
16. Sirová D, Borovec J, Šantrůčková H, Šantrůček J, Vrba J, Adamec L. *Utricularia* carnivory revisited: Plants supply photosynthetic carbon to traps. J Exp Bot. 2010;61:99–103.
17. Sirová D, Borovec J, Picek T, Adamec L, Nedbalová L, Vrba J. Ecological implications of organic carbon dynamics in the traps of aquatic carnivorous *Utricularia* plants. Funct Plant Biol. 2011;38:583–93.
18. Sirová D, Šantrůček J, Adamec L, Bárta J, Borovec J, Pech J, et al. Dinitrogen fixation associated with shoots of aquatic carnivorous plants: is it ecologically important? Ann Bot. 2014;114:125–33.
19. Adamec L. Different reutilization of mineral nutrients in senescent leaves of aquatic and terrestrial carnivorous *Utricularia* species. Aquat Bot. 2014;119:1–6.
20. Vincent O, Weisskopf C, Poppinga S, Masselter T, Speck T, Joyeux M, et al. Ultra-fast underwater suction traps. Proc R Soc B. 2011;278:2909–14.
21. Martin JA, Wang Z. Next-generation transcriptome assembly. Nat Rev Genet. 2011;12:671–82.
22. Ge XX, Chen HW, Wang HL, Shi AP, Liu KF. De novo assembly and annotation of *Salvia splendens* transcriptome using the Illumina platform. PLoS One. 2014;9:e87693.
23. Reddy ASN, Marquez Y, Kalyna M, Barta A. Complexity of the alternative splicing landscape in plants. Plant Cell. 2013;25:3657–83.
24. Joubes J, Raffaele S, Bourdenx B, Garcia C, Laroche-Traineau J, Moreau P, et al. The VLCFA elongase gene family in *Arabidopsis thaliana*: phylogenetic analysis, 3D modelling and expression profiling. Plant Mol Biol. 2008;67:547–66.
25. Kammel C, Thomaier M, Sorensen BB, Schubert T, Laengst G, Grasser M, et al. Arabidopsis DEAD-Box RNA Helicase UAP56 interacts with both RNA and DNA as well as with mRNA export factors. PLoS One. 2013;8:e60644.
26. Owttrim GW. RNA helicases and abiotic stress. Nucl Acids Res. 2006;34:3220–30.
27. Marquez Y, Brown JWS, Simpson C, Barta A, Kalyna M. Transcriptome survey reveals increased complexity of the alternative splicing landscape in *Arabidopsis*. Genome Res. 2012;22:1184–95.

28. Yang Y, Smith S. Optimizing *de novo* assembly of short-read RNA-seq data for phylogenomics. BMC Genomics. 2013;14:328.
29. Du Z, Zhou X, Ling Y, Zhang Z, Su Z. agriGO: a GO analysis toolkit for the agricultural community. Nucl Acids Res. 2010;38:W64–70.
30. Srivastava A, Rogers WL, Breton CM, Cai L, Malmberg RL. Transcriptome analysis of *Sarracenia*, an insectivorous plant. DNA Res. 2011;18:253–61.
31. Ng CKY, Carr K, McAinsh MR, Powell B, Hetherington AM. Drought-induced guard cell signal transduction involves sphingosine-1-phosphate. Nature. 2001;410:596–9.
32. Coursol S, Fan LM, Le Stunff H, Spiegel S, Gilroy S, Assmann SM. Sphingolipid signalling in *Arabidopsis* guard cells involves heterotrimeric G proteins. Nature. 2003;423:651–4.
33. Nishikawa M, Hosokawa K, Ishiguro M, Minamioka H, Tamura K, Hara-Nishimura I, et al. Degradation of sphingoid long-chain base 1-phosphates (LCB-1Ps): Functional characterization and expression of AtDPL1 encoding LCB-1P lyase involved in the dehydration stress response in *Arabidopsis*. Plant Cell Physiol. 2008;49:1758–63.
34. Nakagawa N, Kato M, Takahashi Y, Shimazaki K, Tamura K, Tokuji Y, et al. Degradation of long-chain base 1-phosphate (LCBP) in Arabidopsis: functional characterization of LCBP phosphatase involved in the dehydration stress response. J Plant Res. 2012;125:439–49.
35. Boothby TC, Zipper RS, van der Weele CM, Wolniak SM. Removal of retained introns regulates translation in the rapidly developing gametophyte of *Marsilea vestita*. Dev Cell. 2013;24:517–29.
36. Sydenham PH, Findlay GP. The rapid movement of the bladder of *Utricularia* sp. Aust J Biol Sci. 1972;26:1115–26.
37. Alonso JM, Hirayama T, Roman G, Nourizadeh S, Ecker JR. EIN2, a bifunctional transducer of ethylene and stress responses in *Arabidopsis*. Science. 1999;284:2148–52.
38. Shibuya K, Barry KG, Ciardi JA, Loucas HM, Underwood BA, Nourizadeh S, et al. The central role of PhEIN2 in ethylene responses throughout plant development in petunia. Plant Phys. 2004;136:2900–12.
39. Miyata K, Kawaguchi M, Nakagawa T. Two distinct EIN2 genes cooperatively regulate ethylene signaling in *Lotus japonicas*. Plant Cell Phys. 2013;54:1469–77.
40. Ma H, Lu ZQ, Liu BB, Qiu Q, Liu JQ. Transcriptome analyses of a Chinese hazelnut species *Corylus mandshurica*. BMC Plant Biol. 2013;13:152.
41. Nielsen R, Bustamante C, Clark AG, Glanowski S, Sackton TB, Hubisz MJ, et al. A scan for positively selected genes in the genomes of humans and chimpanzees. PLoS Biol. 2005;3:e170.
42. Wicke SK, Schaeferhof B, DePamphilis CW, Muller KF. Disproportional plastome-wide increase of substitution rates and relaxed purifying selection in genes of carnivorous Lentibulariaceae. Mol Biol Evol. 2014;31:529–45.
43. Adamec L, Pásek K. Photosynthetic CO_2 affinity of aquatic carnivorous plants growing under nearly-natural conditions and *in vitro*. Carniv Plant Newslett (Fullerton). 2009;38:107–13.
44. Sato C, Schnoor JL, McDonald DB, Huey J. Test medium for the growth of *Nitrosomonas europaea*. Appl Environ Microbiol. 1985;49:1101–7.
45. Adamec L. How to grow *Aldrovanda vesiculosa* outdoors. Carniv Plant Newslett (Fullerton). 1997;26:85–8.
46. Conesa A, Gotz S, Garcia-Gomez JM, Terol J, Talon M, Robles M. BLAST2GO: a universal tool for annotation, visualization and analysis in functional genomics research. Bioinformatics. 2005;21:3674–6.
47. Huson DH, Auch AF, Qi J, Schuster SC. MEGAN analysis of metagenomic data. Genome Res. 2007;17:377–86.
48. Gabaldon T, Koonin EV. Functional and evolutionary implications of gene orthology. Nat Rev Genet. 2013;14:360–6.
49. Štorchová H, Hrdličková R, Chrtek Jr J, Tetera M, Fitze D, Fehrer J. An improved method of DNA isolation from plants collected in the field and conserved in saturated NaCl/CTAB solution. Taxon. 2000;49:79–84.
50. Swofford DL. PAUP*: Phylogenetic analysis using parsimony (*and other methods). Version 4. Sunderland, Massachusetts: Sinauer Associates; 2003.

7

Proteomic and metabolomic analyses provide insight into production of volatile and non-volatile flavor components in mandarin hybrid fruit

Qibin Yu[1], Anne Plotto[2], Elizabeth A Baldwin[2], Jinhe Bai[2], Ming Huang[1], Yuan Yu[1], Harvinder S Dhaliwal[3] and Frederick G Gmitter Jr[1*]

Abstract

Background: Although many of the volatile constituents of flavor and aroma in citrus have been identified, the knowledge of molecular mechanisms and regulation of volatile production are very limited. Our aim was to understand mechanisms of flavor volatile production and regulation in mandarin fruit.

Result: Fruits of two mandarin hybrids, Temple and Murcott with contrasting volatile and non- volatile profiles, were collected at three developmental stages. A combination of methods, including the isobaric tags for relative and absolute quantification (iTRAQ), quantitative real-time polymerase chain reaction, gas chromatography, and high-performance liquid chromatography, was used to identify proteins, measure gene expression levels, volatiles, sugars, organic acids and carotenoids. Two thirds of differentially expressed proteins were identified in the pathways of glycolysis, citric acid cycle, amino acid, sugar and starch metabolism. An enzyme encoding valencene synthase gene (*Cstps1*) was more abundant in Temple than in Murcott. Valencene accounted for 9.4% of total volatile content in Temple, whereas no valencene was detected in Murcott fruit. Murcott expression of *Cstps1* is severely reduced.

Conclusion: We showed that the diversion of valencene and other sesquiterpenes into the terpenoid pathway together with high production of apocarotenoid volatiles might have resulted in the lower concentration of carotenoids in Temple fruit.

Keywords: Apocarotenoid volatiles, Carotenoids, Sesquiterpene synthase, Citrus, Gene expression

Background

Fruit volatiles are essential components of fruit flavor, have defense mechanisms against biotic and abiotic stresses, and contribute to various physiological and ecological functions during plant development [1]. Flavor in mandarin fruit is the result of a combination of sugars (glucose, sucrose and fructose), acids (citric and malic), flavonoids, limonoids, and volatile compounds [2]. Volatiles in mandarin fruit belong to several chemical families such as terpenes, hydrocarbons, aldehydes, esters, alcohols, ketones and sulfur compounds [3]. Terpenoids play a central role in generating the chemical diversity, and accounted for 85–95% of volatiles in tangerine fruit [4]. Most volatiles are derived from a diverse set of non-volatile precursors, simple or complex molecules including amino acids, fatty acids, carbohydrates and carotenoids, which can be grouped into four biosynthetic classes: terpenoids, fatty acids, branched-chain amino acids and aromatic amino acids such as phenylalanine [5]. Virtually all of these precursors are essential human nutrients [6].

Breeding for improvement of fruit flavor is a very challenging task when using classical breeding methods due to the difficulty of scoring and quantifying such a complex trait. The presence of a single volatile molecule, even at a relatively high level, does not mean that it contributes to either flavor or liking [7]. To complicate matters further, some volatiles can also impact the perception of sweetness and *vice versa* [8]. So far, we still do not really understand how all of these volatiles and

* Correspondence: fgmitter@ufl.edu
[1]University of Florida, Institute of Food and Agricultural Sciences, Citrus Research and Education Center, Lake Alfred, FL 33850, USA
Full list of author information is available at the end of the article

non-volatiles are integrated into the unique flavor perception of a fruit. For breeding programs, screening for the large range of flavor chemicals is not practically possible. Therefore, it is important to characterize the molecular mechanisms and regulation of flavor in order to understand the complexity of this trait. Knowledge of biosynthetic pathways of fruit flavor compounds and regulatory mechanisms will lead to efficient breeding strategies, such as to identify markers that track flavor-associated chemicals.

Several studies in tomato, peach, strawberry and banana have been performed, identifying and characterizing the most important genes and encoded enzymes involved in aroma-related volatiles [9-14], however, very few studies have been carried out in citrus [15]. Although volatile constituents of flavor and aroma have been identified in tangerine [3,4,16], research on the mechanisms of regulation or modulation, especially in citrus, is very limited. Progress in gene isolation related to volatile production has been impeded by the lack of information concerning plant secondary metabolism, with flavor-associated volatiles [17]. Even for some of the most important metabolites, pathways for synthesis have only recently been established or remain to be established [18]. An integrated approach, including metabolomics, genomics, transcriptomics and proteomics, and determining fundamental metabolism, can make an important contribution toward this goal [2,19-22].

In the present study, we selected contrasting volatile and non-volatile profiles between two mandarin hybrids: Murcott and Temple. The two hybrids have similar genetic backgrounds due to having the same general parentage of mandarin and sweet orange, although their exact origins are unknown [23]. Despite that, both of these cultivars have good fruit flavor, although previous studies indicate that Temple is much richer in volatiles than Murcott, especially in sesquiterpenes and esters [4]. In addition to a comparison of volatile and non-volatile (sugars, acids, and carotenoids) compounds, and the interrelationships of these chemical components, a comparative iTRAQ (isobaric tags for relative and absolute quantification) proteome analysis was used to identify qualitative and quantitative differences in the proteome between the two hybrids at three levels of maturity. iTRAQ is a powerful approach, using isotope labeling coupled with multidimensional liquid chromatography and tandem mass spectrometry (MS), thereby enabling sensitive assessment and quantification of protein levels [24-26]. This analysis helped to better understand the pathways and genes controlling synthesis of flavor volatiles during mandarin hybrid fruit maturation, and to identify enzymes and genes involved in their biosynthesis pathways, especially concerning the terpenoid biosynthesis pathway.

Results

Differences in sugar, organic acid and carotenoid content between Murcott and Temple

Fruits of Temple and Murcott were different in flesh color (Figure 1). There were differences for sugars, organic acids and carotenoids between Temple and Murcott at the three maturity stages. Among sugars, only sucrose and total sugars were higher in Murcott than Temple at stage 3, and total soluble solids content (SSC) at stage 1 and 3. However, no differences were found in fructose and glucose. Among acids, Temple was higher than Murcott for citric acid at stage1, malic acid and titratable acidity (TA) at stage 1 and 2, and ascorbic acid at all three stages, respectively. The pH values for Temple were significantly lower at stage 2. Overall, ascorbic acid was 21 times higher in Temple than Murcott. SSC/titratable acidity (TA) was lower in Temple at stage 1 and 2. SSC/TA is an indicator of maturity in citrus, and no differences were found between the two cultivars in stage 3. All carotenoids, except α-carotene for stage 2 and 3 and lutein for stage 1, were significantly higher in Murcott than in Temple (Figure 2).

Differences in aroma volatiles between Murcott and Temple

A total of 121 volatile compounds were detected by gas chromatography-mass spectrometry (GC-MS), with 108 compounds in Temple and 60 compounds in Murcott, respectively (Additional file 1: Table S1). Only 48 volatiles were found in both Temple and Murcott. There were 46 volatiles unique to Temple, in addition to 14 unknown compounds, whereas 12 volatiles were found only in Murcott (Table 1). The sum of total relative peak areas (peak area of compounds divided by peak area of internal standard) was twice as high in Temple than in Murcott, 21.9 for Temple, 11.5 for Murcott, respectively (Table 2). Terpenoid-related compounds contributed more than 85 and 95% of the total volatiles in Temple and Murcott respectively, also the volatile profile was markedly different. Valencene accounted for 9.4% of the total profile in Temple, whereas no valencene nor nootkatone was detected in Murcott. Sesquiterpenes were 0.15% and 3.10% and esters were 0.38% and 7.16% in Murcott and Temple, respectively. We found seven carotenoid-derived volatiles in Temple: nerol, neral, geranial, neryl acetate, α-ionone, geranyl acetone, and β-ionone. In contrast, only two of these, neryl acetate and geranyl acetone, were found in Murcott. D-limonene was the most abundant volatile compound which accounted for 80.8% and 64.4% of the volatile profile in Murcott and Temple, respectively. Murcott had two branched aldehydes, 3-methyl pentanal and 4-methyl hexanal, which were lacking in Temple. However, Temple had one branched alcohol, 3-methyl-1-butanol, and one branched ester, ethyl 2-methylbutyrate, likely to have been derived

Figure 1 Cross section of Temple and Murcott mandarin hybrid fruit.

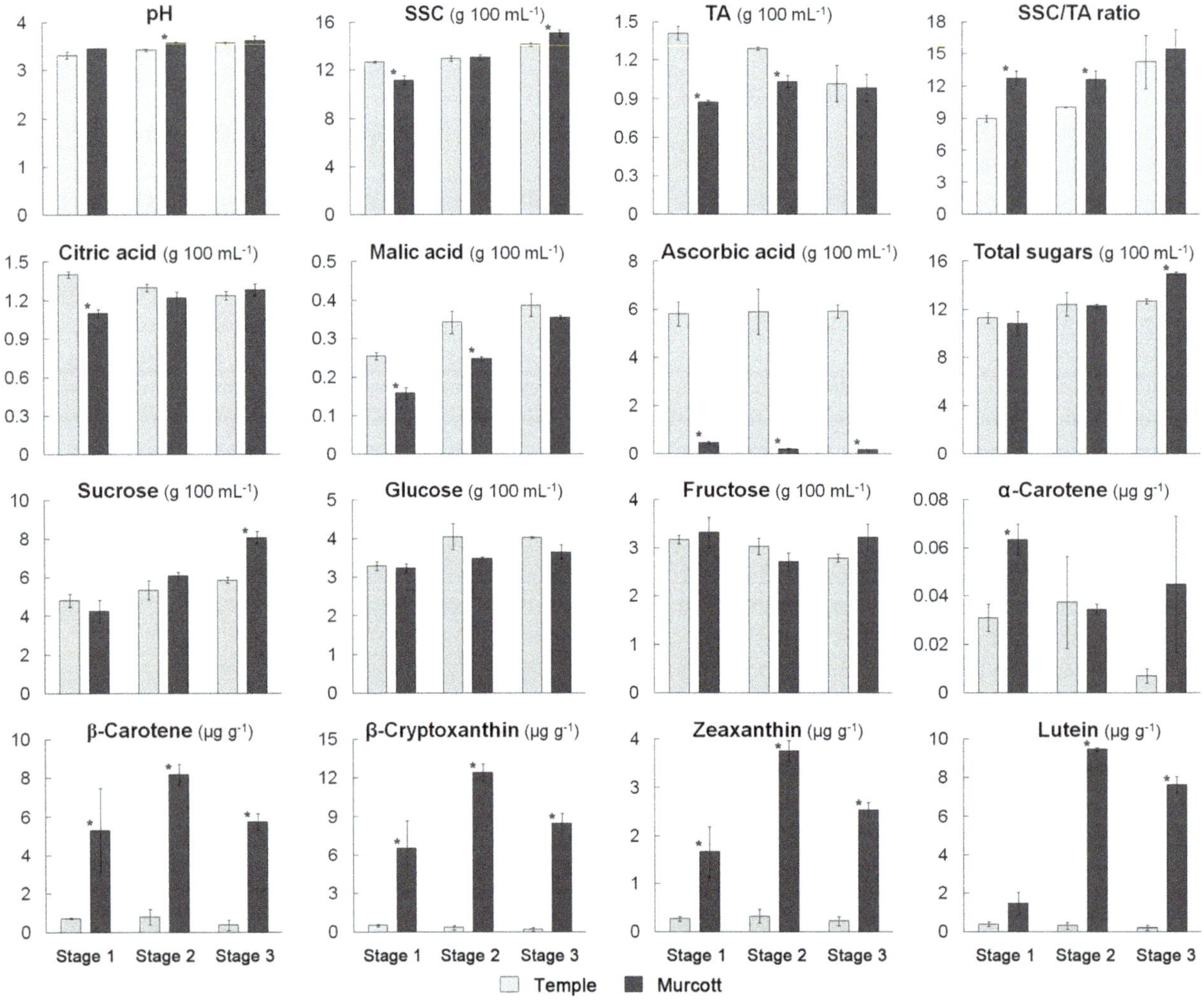

Figure 2 Sugar, organic acid and carotenoid content in Temple and Murcott mandarin hybrid fruit at three developmental stages (stage 1: 22-Dec-2008; stage 2: 30-Jan-2009; and stage 3: 11-Mar-2009). Student's T-test was used to determine the statistical significance of the differences between mean values for Temple and Murcott at the same developmental stage; standard error bars are provided. *: significant difference ($P < 0.05$); SSC: soluble solids content; TA: titratable acidity.

Table 1 Volatiles in Temple and Murcott mandarin hybrid fruit arranged by chemical class

Temple only	Murcott only	Both
Monoterpenes	**Monoterpenes**	**Monoterpenes**
Isoterpinolene	β-Pinene	α-Thujene
3-Carene	(+)-4-Carene	α-Pinene
2-Carene	**Aldehydes**	Sabinene
3-Methyl-4-methylenebicyclo [3.2.1]oct-2-ene	Butanal	β-Myrcene
	3-Methyl pentanal	α-Phellandrene
Sesquiterpenes	4-Methyl hexanal	γ-Terpinene
β-Elemene	p-Menth-1-en-9-al	p-Cymene
β-Cubebene	p-Menth-1-en-9-al isomer	d-Limonene
β-Humulene	**Ester**	β-Phellandrene
α-Caryophyllene	Ethyl acetate	γ-Terpiene
α-Selinene	**Ether**	p-Mentha-3, 8-diene
γ-Selinene	Ethyl ether	Terpinolene
Valencene	**Hydrocarbons**	**Sesquiterpenes**
Aromadendrene	(E,E)-2,6-dimethyl-1,3,5,7-octatetraene	α-Cubebene
Calamenene		Copaene
(−)-α-Panasinsen	**Furans**	Caryophyllene
Eremophilene	2-n-Butyl furan	δ-Cadinene
Eudesma-3,7-diene	2-Pentyl furan	**Aldehydes**
4,11-Selinadiene		Acetaldehyde
Aldehydes		Propanal
(E)-2-Pentenal		Pentanal
Geranial (carotenoid)		Hexanal
Neral (carotenoid)		Heptanal
Ketones		Octanal
Acetone		Nonanal
Nootkatone		Decanal
α-Ionone (carotenoid)		(E)-2-Hexenal
β-Ionone (carotenoid)		(E)-2-Heptenal
Alcohols		(E)-2-Octenal
1-Hexanol		(E)-2-Nonenal
3-Methylbutanol		(E)-2-Decenal
(Z)-p-Mentha-2,8-dien-1-ol		Perillaldehyde
β-Terpineol		**Ketones**
Nerol (carotenoid)		1-Pentene-3-one
Esters		3-Pentanone
Ethyl butanoate		4-Heptanone
Ethyl 2-butenoate		d-Carvone
Ethyl 2-methylbutanoate		Dihydrocarvone
Ethyl pentanoate		Geranyl acetone (carotenoid)
Ethyl hexanoate		**Alcohols**
Ethyl-3-hydroxyhexanoate		Ethyl alcohol
Ethyl octanoate		1-Penten-3-ol
Propyl butanoate		Linalool
Methyl butanoate		Terpinen-4-ol
Methyl hexanoate		α-Terpineol
Hexyl acetate		**Esters**
Linalool acetate		Octyl acetate
Terpinyl acetate		Citronellol acetate
Ether		Neryl acetate (carotenoid)
1,8-Cineole		**Hydrocarbons**
Hydrocarbons		1,3-Pentadiene
(E)-2,6-Dimethyl-2, 6-octadiene		(Z)-2,6-Dimethyl-2, 6-octadiene
1,5-Dimethyl-cyclooctadiene		(+/−)-4-Acetyl-1-methylcyclohexene
Furan		
2-Ethyl furan		**Furan**
		2-Methyl furan

Carotenoid-derived volatiles are in parentheses.

from the branched alcohol, whereas Murcott did not have these compounds (Table 2).

Differentially expressed proteins in Temple versus Murcott

We identified 280 differentially expressed proteins in Temple versus Murcott (Additional file 1: Table S2). Of these identified proteins, 92 were significantly differentially expressed in juice sacs at the three ripening stages (fold change > 1.5, $P < 0.05$) (Table 3). We found 42, 54 and 45 expressed proteins in ripening stage 1, stage 2 and stage 3, respectively. There were 22 proteins in common between stage 1 and 2, 24 between stage 2 and 3, whereas only 9 proteins in common were identified between stage 1 and 3. Five proteins were present across all three stages: hypothetical protein (gi|225442225), superoxide dismutase (SOD) (gi|77417715), phospholipase D alpha (gi|169160465), plastid-lipid-associated protein (gi|62900641), and UDP-glucosyltransferase family 1 protein (gi|242199340). All proteins were more highly expressed in Murcott than Temple in stage 2, whereas most proteins were more highly expressed in Temple than Murcott in stage 1. In stage 3, 13 proteins were up-regulated versus 32 down-regulated in Temple versus Murcott. We found several important proteins involved in volatile production. Phospholipase D alpha (gi|169160465), a key enzyme involved in membrane deterioration which produces precursors to aliphatic alcohols and aldehydes,

Table 2 Content of major volatile classes in Temple and Murcott mandarin hybrid fruit

Chemical class	Murcott	Temple	P value
Aliphatic alcohols	0.045 ± 0.021	0.094 ± 0.043	0.356
Branched alcohols	n. d.	0.002 ± 0.001	
Aliphatic aldehydes	0.910 ± 0.257	0.755 ± 0.138	0.442
Branched aldehydes	0.005 ± 0.002	n. d.	
Aliphatic esters	0.044 ± 0.017	1.561 ± 0.246	0.000
Branched esters	n. d.	0.006 ± 0.001	
Aliphatic ketones	0.014 ± 0.001	0.019 ± 0.002	0.001
d-Limonene	9.266 ± 1.203	14.03 ± 2.317	0.110
Monoterpenes except d-Limonene	0.937 ± 0.141	1.323 ± 0.217	0.191
Valencene	n. d.	2.053 ± 0.367	
Sesquiterpenes except Valencene	0.017 ± 0.004	0.677 ± 0.004	0.000
Terpene alcohols	0.123 ± 0.013	0.720 ± 0.144	0.007
Terpene aldehydes	0.013 ± 0.003	0.026 ± 0.005	0.035
Terpene esters	0.011 ± 0.011	0.061 ± 0.010	0.004
Terpene ketones	0.057 ± 0.011	0.061 ± 0.010	0.764
Ethers	n. d.	0.348 ± 0.073	
Furans	0.022 ± 0.004	n. d.	
Other hydrocarbon	n. d.	0.149 ± 0.031	
Other	0.005 ± 0.001	0.007 ± 0.001	0.390
Total	11.47 ± 1.51	21.90 ± 3.000	0.030

Total ion current of target compound was divided by that of internal standard, 3-hexanone.

was up-regulated in Temple versus Murcott at stage 1, but not stage 2 and 3. The Family1 glycotranferases might affect biosynthesis and accumulation of glycosides that bind volatile terpenoids. Isopentenyl diphosphate Delta-isomerase I (gi|6225526) isomerizes isopentenyl diphosphate (IPP) to its isomer dimethylallyl diphosphate (DMAPP) and was up-regulated in Murcott versus Temple at ripening stage 2. Valencene synthase (gi|33316389) was the protein that was the most different between the two cultivars, being 25 times higher in Temple than in Murcott at ripening stage 3. Several proteins from the glycolysis pathway were identified: triosephosphate isomerase (gi|77540216), a triosphosphate isomerase-like protein (gi|76573375), and pyruvate decarboxylase (gi|17225598). All were only expressed in ripening stage 3, and were higher in Murcott than in Temple. A citrate synthase precursor (gi|624676) was found in ripening stage 1, up-regulated in Temple in comparison with Murcott. In addition to citrus synthase, malate dehydrogenase (gi|27462762) and isocitrate dehydrogenase (gi|5764653) of the tricarboxylic acid (TCA) cycle were also found and downregulated in Temple versus Murcott. Glutamate decarboxylase (gi|70609690) and aspartate aminotransferase (gi|255551036), involved in glutamate synthesis, were also identified.

Gene annotation was conducted using the Blast2GO program for all 92 identified proteins. The biological interpretation was further completed by assigning them to metabolic pathways using Kyoto Encyclopedia of Genes and Genomes (KEGG) annotation. KEGG analysis assigned the 46 differentially expressed proteins to 48 metabolic pathways (Additional file 1: Table S3). Most biosynthetic pathways identified were glycolysis, citric acid cycle, sugar synthesis, amino acid synthesis and terpene synthesis. Additional file 2: Figure S1 shows the distributions of GO terms (2nd level GO terms) according to biological processes, cellular components and molecular function. Most differentially expressed proteins were predicted to be involved in carbohydrate, amino acid, and lipid metabolism as well as in energy production. We found 10 enzymes involved in the glycolysis pathway and 16 enzymes involved in different amino acid pathways (Table 4; Additional file 1: Table S3).

Discussion

In this study, two thirds of differentially expressed proteins were identified in the pathways of glycolysis and TCA as well as amino acid, sugar and starch metabolism (Tables 3 and 4). This is understandable, because the upstream precursors for most volatiles come from carbohydrate metabolism, mainly through sugar and starch metabolism through the glycolysis pathway, which is important for providing the carbon skeleton and toward the different branches that lead to the aforementioned volatiles. Most organic acids, amino acids, terpenes and fatty acids are produced from glycolysis and TCA. For amino acids, the carbon skeletons are derived from 3-phosphoglycerate, phosphoenolpyruvate or pyruvate generated in glycolysis, or from 2-oxoglutarate and oxaloacetate generated in TCA [20]. Terpenoids are enzymatically synthesized de novo from acetyl CoA and pyruvate provided by the carbohydrate pools in plastids and the cytoplasm [27].

The differences in protein expression between Temple and Murcott were due to the different ripening patterns of these two hybrids. Temple is a middle-late variety whereas Murcott is a very late variety; however in Florida citrus production conditions, and depending on season, Temple and Murcott maturity times may overlap. These differences in time of maturity might explain proteins being more highly expressed in Temple than Murcott in stage 1, whereas all proteins were more highly expressed in Murcott than Temple in stage 2, and mixed protein expression levels were seen in stage 3. Feng *et al.* [28] found that glutamate decarboxylase (gi|70609690) was one of two proteins likely associated with carbohydrate and acid metabolism in the ripening

Table 3 Differentially expressed proteins in fruit flesh of Temple (Te) versus Murcott (Mu) mandarin hybrid fruit

Accession	Name	Species	iTRAQ ratio fold change					
			Stage 1		Stage 2		Stage 3	
			Te/Mu	P value	Te/Mu	P value	Te/Mu	P value
gi|11596186	cystatin-like protein	*Citrus x paradisi*	4.041	0.036	0.647	0.002		
gi|118061963	extracellular solute-binding protein, family 5	*Roseiflexus castenholzii* DSM 13941					0.483	0.047
gi|119367477	putative H-type thioredoxin	*Citrus cv.* Shiranuhi	10.782	0.001	0.404	0.001		
gi|119367479	putative cyclophilin	*Citrus cv.* Shiranuhi			0.588	0.037	2.347	0.002
gi|121485004	cytosolic phosphoglycerate kinase	*Helianthus annuus*	5.535	0.002				
gi|124360080	Galactose mutarotase-like	*Medicago truncatula*					1.724	0.003
gi|125546170	hypothetical protein OsI_14032	*Oryza sativa* Indica Group			0.561	0.014		
gi|14031067	dehydrin COR15	*Citrus x paradisi*					2.806	0.000
gi|147809484	hypothetical protein	*Vitis vinifera*			0.608	0.022	0.696	0.065
gi|147836508	hypothetical protein	*Vitis vinifera*					1.630	0.024
gi|147853192	hypothetical protein	*Vitis vinifera*					1.803	0.018
gi|15219028	26.5 kDa class I small heat shock protein-like	*Arabidopsis thaliana*					0.491	0.008
gi|15235730	phosphoenolpyruvate carboxykinase (ATP), putative/PEP carboxykinase, putative/PEPCK, putative	*Arabidopsis thaliana*					1.899	0.034
gi|159471948	U2 snRNP auxiliary factor, large subunit	*Chlamydomonas reinhardtii*			0.255	0.044		
gi|166850556	CTRSFT1-like protein	*Poncirus trifoliata*	3.261	0.011	0.237	0.005		
gi|169160465	phospholipase D alpha	*Citrus sinensis*	4.060	0.000	0.240	0.000	0.573	0.000
gi|17225598	pyruvate decarboxylase	*Fragaria x ananassa*					0.286	0.012
gi|183579873	chitinase	*Citrus unshiu*	1.534	0.012				
gi|192912988	40S ribosomal protein S4	*Elaeis guineensis*	1.601	0.049				
gi|218202932	14-3-3 protein	*Dimocarpus longan*			0.227	0.016		
gi|221327587	ascorbate peroxidase	*Citrus maxima*	4.863	0.000	0.180	0.049		
gi|2213425	hypothetical protein	*Citrus x paradisi*	0.627	0.000	0.524	0.001		
gi|223949137	unknown	*Zea mays*	5.116	0.003				
gi|224069008	predicted protein	*Populus trichocarpa*	6.992	0.001				
gi|224099429	predicted protein	*Populus trichocarpa*			0.587	0.014	0.316	0.002
gi|224109966	predicted protein	*Populus trichocarpa*			0.476	0.040		
gi|224127346	predicted protein	*Populus trichocarpa*			0.156	0.007	0.641	0.043
gi|224128794	predicted protein	*Populus trichocarpa*			0.298	0.007	0.382	0.022
gi|224135985	predicted protein	*Populus trichocarpa*			0.248	0.006	0.366	0.021
gi|225424861	PREDICTED: hypothetical protein isoform 2	*Vitis vinifera*					0.536	0.040
gi|225425914	PREDICTED: hypothetical protein	*Vitis vinifera*			0.429	0.002	0.425	0.010
gi|225439785	PREDICTED: hypothetical protein	*Vitis vinifera*			0.441	0.007	0.658	0.023
gi|225441981	PREDICTED: hypothetical protein	*Vitis vinifera*			0.304	0.002	0.568	0.007

Table 3 Differentially expressed proteins in fruit flesh of Temple (Te) versus Murcott (Mu) mandarin hybrid fruit *(Continued)*

gi|225442225	PREDICTED: hypothetical protein	*Vitis vinifera*	9.896	0.015	0.576	0.010	0.571	0.002
gi|225451968	PREDICTED: similar to mangrin	*Vitis vinifera*	4.507	0.040	0.263	0.095		
gi|231586	ATP synthase subunit beta	*Hevea brasiliensis*			0.134	0.004	0.555	0.007
gi|242199340	UDP-glucosyltransferase family 1 protein	*Citrus sinensis*	7.535	0.002	0.394	0.008	0.539	0.030
gi|255539613	phosphoglucomutase, putative	*Ricinus communis*			0.142	0.020		
gi|255543156	conserved hypothetical protein	*Ricinus communis*	7.967	0.000				
gi|255544686	eukaryotic translation elongation factor, putative	*Ricinus communis*			0.424	0.006	0.323	0.008
gi|255550111	heat-shock protein, putative	*Ricinus communis*	3.788	0.043				
gi|255551036	aspartate aminotransferase, putative	*Ricinus communis*					0.599	0.037
gi|255561582	Patellin-3, putative	*Ricinus communis*			0.588	0.017		
gi|255571742	peptidase, putative	*Ricinus communis*			0.275	0.004		
gi|255586766	monodehydroascorbate reductase, putative	*Ricinus communis*			0.429	0.003	0.493	0.001
gi|255641409	unknown	*Glycine max*			0.645	0.021		
gi|255642211	unknown	*Glycine max*	0.521	0.011	0.121	0.001		
gi|255644696	unknown	*Glycine max*	5.914	0.002				
gi|257659867	unnamed protein product	*Linum usitatissimum*			0.329	0.235	0.368	0.047
gi|257675725	unnamed protein product	*Zea mays*	3.832	0.019				
gi|257690969	unnamed protein product	*Citrus sinensis*			0.384	0.002		
gi|257712573	unnamed protein product	*Brassica napus*	9.086	0.011			0.664	0.006
gi|257720002	unnamed protein product	*Glycine max*			0.551	0.001	0.387	0.007
gi|257726687	unnamed protein product	*Zea mays*	1.650	0.035	0.387	0.001		
gi|27462762	malate dehydrogenase	*Lupinus albus*					0.305	0.003
gi|29124973	unknown	*Populus tremuloides*					2.039	0.031
gi|33316389	valencene synthase	*Citrus sinensis*					25.730	0.022
gi|33325127	eukaryotic translation initiation factor 5A isoform VI	*Hevea brasiliensis*					1.914	0.039
gi|33340236	copper/zinc superoxide dismutase	*Citrus limon*	3.706	0.001	0.638	0.004		
gi|37524017	COR15	*Citrus clementina x Citrus reticulata*	10.311	0.006			2.382	0.010
gi|3790102	pyrophosphate-dependent phosphofructokinase alpha subunit	*Citrus x paradisi*	1.724	0.025			0.554	0.011

Table 3 Differentially expressed proteins in fruit flesh of Temple (Te) versus Murcott (Mu) mandarin hybrid fruit *(Continued)*

gi|40646744	mitochondrial citrate synthase precursor	*Citrus junos*			0.201	0.032	0.553	0.018
gi|4580920	vacuole-associated annexin VCaB42	*Nicotiana tabacum*			0.209	0.046	0.330	0.007
gi|4704605	glycine-rich RNA-binding protein	*Picea glauca*	4.452	0.009				
gi|530207	heat shock protein	*Glycine max*	4.177	0.045				
gi|544437	Probable phospholipid hydroperoxide glutathione peroxidase	*Citrus sinensis*					3.140	0.039
gi|5764653	NADP-isocitrate dehydrogenase	*Citrus limon*			0.430	0.006	0.437	0.003
gi|6094476	Thiazole biosynthetic enzyme	*Citrus sinensis*			0.228	0.007		
gi|6166140	Elongation factor 1-delta 1	*Oryza sativa* Japonica Group	7.427	0.045	0.654	0.028		
gi|6225526	Isopentenyl-diphosphate Delta-isomerase I	*Clarkia breweri*			0.562	0.033		
gi|624674	heat shock protein	*Citrus maxima*						
gi|624676	citrate synthase precursor	*Citrus maxima*	2.731	0.020				
gi|62900641	Plastid-lipid-associated protein	*Citrus unshiu*	6.082	0.002	0.289	0.000	0.662	0.022
gi|63333659	beta-1,3-glucanase class III	*Citrus clementina x Citrus reticulata*			0.493	0.141	2.712	0.000
gi|6518112	H + −ATPase catalytic subunit	*Citrus unshiu*	4.754	0.017	0.598	0.007		
gi|6682841	sucrose synthase	*Citrus unshiu*	3.194	0.025	0.632	0.009		
gi|6682843	sucrose synthase	*Citrus unshiu*			0.144	0.008	0.575	0.024
gi|7024451	glycine-rich RNA-binding protein	*Citrus unshiu*	1.886	0.531				
gi|70609690	glutamate decarboxylase	*Citrus sinensis*	3.588	0.025	0.643	0.043		
gi|7269241	UDPglucose 4-epimerase-like protein	*Arabidopsis thaliana*	0.424	0.011	0.158	0.004		
gi|74486744	translation elongation factor 1A-9	*Gossypium hirsutum*	4.923	0.008				
gi|76573375	triosphosphate isomerase-like protein	*Solanum tuberosum*					0.311	0.000
gi|77417715	SOD	*Citrus maxima*	0.638	0.017	0.118	0.010	0.322	0.013
gi|77540216	triosephosphate isomerase	*Glycine max*					0.514	0.022
gi|77744899	temperature-induced lipocalin	*Citrus sinensis*	4.028	0.018	0.548	0.016		
gi|82623427	glyceraldehyde 3-phosphate dehydrogenase-like	*Solanum tuberosum*			0.661	0.297		
gi|862480	valosin-containing protein	*Glycine max*	1.510	0.029	0.374	0.010		
gi|870794	polyubiquitin	*Arabidopsis thaliana*	4.534	0.005				
gi|90820120	UDP-glucose pyrophosphorylase	*Cucumis melo*	7.835	0.028				

Table 3 Differentially expressed proteins in fruit flesh of Temple (Te) versus Murcott (Mu) mandarin hybrid fruit *(Continued)*

gi|9082317	actin	*Helianthus annuus*	3.959	0.051	0.527	0.001		
gi|9280626	UDP-glucose pyrophosphorylase	*Astragalus membranaceus*	9.821	0.002			1.626	0.022
gi|9757974	polyubiquitin	*Arabidopsis thaliana*			0.585	0.011		

The P value was selected from the most significant one among three biological replications. Additional file 1: Table S2 has the result from all three biological replications. Stage 1 was on December 22, 2008, Stage 2 was on January 30, 2009, and Stage 3 was on March 11, 2009.

fruit. In our study, this protein is expressed more in Temple at stage 1, but less in stage 2 than Murcott. This might also explain the differences in levels of volatiles, sugar, organic acids in different stages between Temple and Murcott.

Sugar, TCA and glycolysis biosynthesis

Sucrose is the major sugar translocated in the plant, the major photo-assimilate stored in the plant, and can be degraded by cell wall sucrose synthase to glucose and fructose. Glucose can be converted into pyruvate, generating small amounts of adenosine triphosphate (ATP) and nicotinamide adenine dinucleotide reduced form (NADH) via the glycolysis pathway. Glucose phosphomutase (gi|255539613, EC 5.4.2.2) was down-regulated in Temple in stage 2, and is an enzyme responsible for the conversion of D-glucose 1-phosphate into D-glucose 6-phosphate. Sucrose synthase (gi|6682841/gi|6682843, EC 2.4.1.13) catalyzes the degradation of sucrose into UDP-glucose and fructose, up-regulated in Temple at stage 1 and down-regulated in stage 2 and 3. The high expression of sucrose synthase in Murcott stage 2 might partially explain why Murcott had higher sucrose than Temple (Figure 2). Sucrose, in turn, is derived from hexose phosphates through UDP-glucose pyrophosphorylase, (gi|90820120, gi|9280626, EC 2.7.7.9). The glycolysis biosynthesis is a central pathway that produces important precursor metabolites: six-carbon compounds of glucose-6P and fructose-6P and three-carbon compounds of glycerone-P, glyceraldehyde-3P, glycerate-3P, phosphoenolpyruvate, and pyruvate. Acetyl-CoA and another important precursor metabolite are produced by oxidative decarboxylation of pyruvate. The reaction, mediated by phosphofructokinase (gi|3790102, EC 2.7.1.11), is one of the key control points of glycolysis in plants. This reaction catalyzes the interconversion of fructose-6-phosphate and fructose-1, 6-bisphosphate.

Citric acid is the main organic acid in citrus fruit juice. Yun *et al.* [29] found citric acid comprised up to 90% of the total organic acid content throughout the entire postharvest period. Citrate may be utilized by three major metabolic pathways for sugar production, amino acid synthesis, and acetyl-CoA metabolism. 2-Oxoglutarate can be then metabolized to an amino acid such as glutamate. Six enzymes acting in the TCA cycle were identified in our study including: pyruvate decarboxylase (gi|17225598, EC 4.1.1.1), malate dehydrogenase (gi|27462762, EC 1.1.1.37), isocitrate dehydrogenase (NADP+) (gi|5764653, EC 1.1.1.42), dihydrolipoyllysine-residue acetyltransferase (gi|225442225, EC 2.3.1.12), citrate synthase (gi|624676, EC 2.3.3.1) and phosphoenolpyruvate (PEP) carboxykinase (gi|15235730, EC 4.1.1.49). The pyruvate decarboxylase enzyme, down-regulated in Temple, links the TCA cycle to glycolysis. Plant cells can convert PEP to malate via oxaloacetate in reactions catalyzed by PEP carboxykinase (gi|15235730, EC 4.1.1.49) and malate dehydrogenase (gi|27462762, EC 1.1.1.37) [1]. Citrate can be produced by condensation of oxaloacetate and acetyl-CoA, catalyzed by citrate synthase which was up-regulated in Temple in stage 2. Citrate synthase is the rate-limiting enzyme of the TCA cycle [29]. The result might explain the higher citric acid content in Temple than Murcott. The oxidative decarboxylation of isocitrate into 2-oxoglutarate is mediated by the action of isocitrate dehydrogenase. The last step of the TCA pathway is the interconversion of malate to oxaloacetate utilizing nicotinamide adenine dinucleotide oxidized form (NAD+) /NADH and is catalyzed by malate dehydrogenase. In general, however, the changes of enzymes in the TCA cycle and glycolysis cannot fully explain the difference of organic acid and sugar contents in Temple compared to Murcott. Katz *et al.* [21] indicated that changes in metabolite amounts in fruit do not always correlate well with protein expression levels, reflecting the complication of regulated pathway outputs.

Amino acids, oxidization, ascorbate-glutathione cycle

KEGG pathway analysis conducted by Blast2GO indicated that seven enzymes are involved in the glutathione metabolic pathway (Table 4). In plants, glutathione is crucial for biotic and abiotic stress management. It is a pivotal component of the glutathione-ascorbate cycle, a system that reduces poisonous hydrogen peroxide. Pan *et al.* [30] found that expression levels of five antioxidative enzymes (catalase, peroxidase, ascorbate peroxidase, glutathione reductase and superoxide dismutase) were altered in a mutant orange "Hong Anliu" which has a high level of lycopene, and implied a regulatory role of oxidative stress on carotenogenesis. In our study, the protein expression of L-ascorbate peroxidase (gi|221327587, EC 1.11.1.11), phospholipid-hydroperoxide glutathione

Table 4 KEGG assigned differentially expressed proteins between Temple and Murcott mandarin hybrid fruit in metabolic pathways

KEGG pathway	Pathway	Enzyme number
Carbohydrate metabolism	Amino sugar and nucleotide sugar metabolism	ec:2.7.7.9, ec:3.2.1.14, ec:5.1.3.2,ec:5.4.2.2
	Ascorbate and aldarate metabolism	ec:1.10.3.3, ec:1.11.1.11, ec:1.6.5.4
	Butanoate metabolism	ec:4.1.1.15
	Tricarboxylic acid cycle (TCA)	ec:1.1.1.37, ec:1.1.1.42, ec:2.3.1.12, ec:2.3.3.1, ec:4.1.1.49
	Fructose and mannose metabolism	ec:2.7.1.11, ec:2.7.1.90, ec:4.1.2.13,ec:5.3.1.1
	Galactose metabolism	ec:2.7.1.11, ec:2.7.7.9, ec:5.1.3.2, ec:5.4.2.2
	Glycerophospholipid metabolism	ec:3.1.4.4
	Glycolysis/Gluconeogenesis	ec:1.2.1.12, ec:2.3.1.12, ec:2.7.1.11, ec:2.7.2.3, ec:4.1.1.1, ec:4.1.1.49, ec:4.1.2.13, ec:5.1.3.3, ec:5.3.1.1, ec:5.4.2.2
	Glyoxylate and dicarboxylate metabolism	ec:1.1.1.37, ec:1.11.1.6, ec:2.3.3.1
	Pentose and glucuronate interconversions	ec:2.7.7.9, ec:3.1.1.11
	Pentose phosphate pathway	ec:1.1.1.49, ec:2.7.1.11, ec:4.1.2.13, ec:5.4.2.2
	Pyruvate metabolism	ec:1.1.1.37, ec:2.3.1.12, ec:4.1.1.49, ec:4.4.1.5
Amino acid metabolism	Alanine, aspartate and glutamate metabolism	ec:2.6.1.1, ec:2.6.1.2, ec:4.1.1.15
	Arginine and proline metabolism	ec:2.6.1.1, ec:3.5.3.1
	beta-Alanine metabolism	ec:4.1.1.15
	Cysteine and methionine metabolism	ec:2.6.1.1
	Glutathione metabolism	ec:1.1.1.42, ec:1.1.1.49, ec:1.11.1.11, ec:1.11.1.12, ec:1.11.1.15, ec:1.11.1.9, ec:2.5.1.18
	Phenylalanine metabolism	ec:1.11.1.7,ec:2.6.1.1
	Phenylalanine, tyrosine and tryptophan biosynthesis	ec:2.6.1.1
	Taurine and hypotaurine metabolism	ec:4.1.1.15
	Tryptophan metabolism	ec:1.11.1.6
	Tyrosine metabolism	ec:2.6.1.1
	Valine, leucine and isoleucine degradation	ec:2.3.1.168
Other secondary metabolites	Isoquinoline alkaloid biosynthesis	ec:2.6.1.1
	Novobiocin biosynthesis	ec:2.6.1.1
	Tropane, piperidine and pyridine alkaloid biosynthesis	ec:1.11.1.6
	Streptomycin biosynthesis	ec:5.4.2.2
Energy metabolism	Carbon fixation in photosynthetic organisms	ec:1.1.1.37, ec:2.6.1.1, ec:2.6.1.2, ec:2.7.2.3, ec:4.1.1.49, ec:4.1.2.13, ec:5.3.1.1
	Carbon fixation pathways in prokaryotes	ec:1.1.1.37, ec:1.1.1.42
	Inositol phosphate metabolism	ec:5.3.1.1
	Methane metabolism	ec:1.1.1.37, ec:1.11.1.6, ec:1.11.1.7, ec:2.7.1.11, ec:4.1.2.13
	Oxidative phosphorylation	ec:3.6.3.6
Lipid metabolism	alpha-Linolenic acid metabolism	ec:5.3.99.6
	Arachidonic acid metabolism	ec:1.11.1.9
	Ether lipid metabolism	ec:3.1.4.4
	Primary bile acid biosynthesis	ec:1.3.1.3
	Steroid degradation	ec:1.1.1.145
	Steroid hormone biosynthesis	ec:1.1.1.145, ec:1.3.1.3
Metabolism of terpenoids and polyketides	Terpenoid backbone biosynthesis	ec:5.3.3.2

Table 4 KEGG assigned differentially expressed proteins between Temple and Murcott mandarin hybrid fruit in metabolic pathways *(Continued)*

Nucleotide metabolism	Arginine and proline metabolism	ec:3.5.3.11
	Cysteine and methionine metabolism	ec:4.4.1.14
	Purine metabolism	ec:3.6.1.3, ec:5.4.2.2
Xenobiotics biodegradation and metabolism	Chlorocyclohexane and chlorobenzene degradation	ec:3.1.1.45
	Drug metabolism - cytochrome P450	ec:2.5.1.18
	Fluorobenzoate degradation	ec:3.1.1.45
	Metabolism of xenobiotics by cytochrome P450	ec:2.5.1.18
	Toluene degradation	ec:3.1.1.45

peroxidase (gi|544437, EC 1.11.1.12), superoxide dismutase (SOD) (gi|77417715), and monodehydroascorbate reductase (gi|255586766, EC 1.6.5.4), were mixed (Table 3). SOD and monodehydroascorbate reductase had lower expression in Temple, whereas, other proteins were higher in stage 1 and 3, and lower in stage 2 (Table 3). We could not define a clear relationship between antioxidative enzyme activity and the amount of carotenoids. The discrepancy is likely due to other regulatory pathways, since there are many steps involved in the biosynthesis pathways that are tightly regulated [31]. Liu *et al.* [32] found glutamate decarboxylase is an enzyme catalyzing the conversion of L-glutamate to γ-aminobutyric acid, and suggested that it is possible that glutamate decarboxylase (gi|70609690) could participate in regulating the cytosolic pH.

Volatile biosynthesis

All terpenoids derive from the common building units isopentenyl diphosphate (IPP) and its isomer dimethylallyl diphosphate (DMADP). Both IPP and DMADP are synthesized via two parallel pathways, the mevalonate (MVA) pathway, which is active in the cytosol, and the methylerythritol 4-phosphate (MEP) pathway, which is active in the plastids. In this study, IPP isomerase (gi|6225526) upregulated in Murcott relative to Temple, catalyzes isomerization between IPP and dimethylallyl diphosphate (Table 3). Aharoni *et al.* [33] found that the pool of IPP in the plastids might affect the formation of sesquiterpenes in the cytosol given that transport of isoprenoid precursors is known to occur from the plastids to the cytosol. A valencene synthase (gi|33316389) expression explains the difference in valencene content between Temple and Murcott. Sharon-Asa *et al.* [15] isolated and characterized the valencene synthase gene (*Cstps1*) and reported that valencene accumulates during the ripening of Valencia orange fruits together with *Cstps1*. Results from the current work agreed with their study (Additional file 2: Figure S2-A). In order to validate the result, real-time PCR showed that the gene expression of *Cstps1* was found to be over 217 and 2720 times higher in Temple than in Murcott on Dec 22, 2008 and March 11, 2009, respectively (Figure 3). Murcott expression of *Cstps1* gene is very severely reduced.

Non-volatile sugar conjugates constitute a large pool of precursors for many of the important flavor volatiles. Enzymes synthesizing and hydrolyzing these sugar conjugates are likely to influence the volatile profiles. Family 1 glycosyltransferases (gi|242199340), often referred to as UDP glycosyltransferases, is the largest in the plant kingdom [34], which catalyze the transfer of a glycosyl moiety from UDP-sugars to a wide range of acceptor molecules. Glycosyltransferase might affect biosynthesis and accumulation of glycosides of volatile terpenoids.

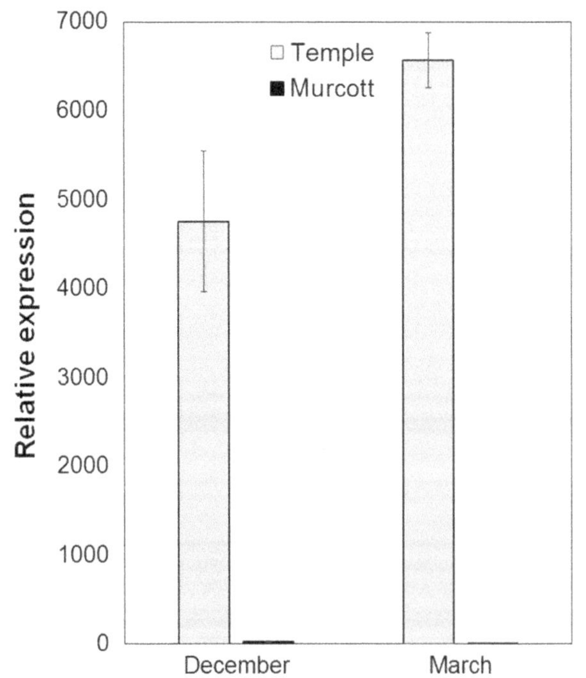

Figure 3 QRT-PCR validation of the expression profiles of *Cstsp1* genes at two time points. Results were expressed relative to the value of the expression of Murcott *Cstps1* in March.

Fan *et al.* [35] identified three putative terpenoid UDP-glycosyltransferase (UGT) genes in sweet orange. The different expression of glycotranferase family 1 in three stages of fruit ripening in Temple might explain the difference in terpenoid volatile levels compared with Murcott.

Fatty acids play a major role in ester volatile synthesis. We have identified the phospholipid D (gi|169160465, EC 3.1.4.4) in all three ripening stages. Oke *et al.* [36] found that the transgenic tomato fruits with an antisense phospholipase D (PLD) showed improved red color, lycopene content, and results suggest that a reduction in PLD activity may lead to increased membrane stability and preservation of membrane compartmentalization that can have positive quality impacts for transgenic fruit and their products. We did not find major enzyme differences downstream, such as the lipoxygenase (LOX) pathway, which comprises the action of phospholipase, lipoxygenase, and hydroxyperoxide. The lipid-derived volatiles represent the bulk of aroma volatiles in tomato and are generated by the lipoxygenase (LOX) pathway [37]. In addition, pyruvate decarboxylase (gi|17225598) is believed to be involved in the pathway that provide aldehydes and alcohols for ester synthesis [38].

Correlation between valencene/sesquiterpenes accumulation and total carotenoids

It is generally recognized that the cytosolic MVA pathway is responsible for the synthesis of sesquiterpenes, phytosterols and ubiquinone, whereas monoterpenes, gibberellins, abscisic acid, carotenoids and the prenyl moiety of chlorophylls, plastoquinone and tocopherol are produced in plastids via the MEP pathway [27,39]. Although the subcellular compartmentation of MVA and MEP pathways allows them to operate independently, metabolic "crosstalk" between the two pathways was prevalent, particularly in the direction of plastids to cytosol [5] (Figure 4). Prenyltransferase condenses dimethylallyl diphosphate with two IPP molecules to produce FPP or three IPP to geranylgeranyl diphosphate (GGPP). In this study, Temple, had lower carotenoids but higher number of apocarotenoid volatiles than Murcott (Additional file 1: Table S1). Davidovich-Rikanati *et al.* [11] indicated that a transgenic tomato expressing a monoterpene synthesis gene resulted in lighter color in comparison with wild type tomatoes. Because GGPP is the precursor of the carotenoids, the activity of valencene synthase (*Cstps1*) converting FPP to valencene could be one of the limiting steps for carotenoid production in Temple (Figure 4). The important flavor volatile genes are those that encode enzymes responsible for synthesis of the end products and those encoding factors that regulate pathway output [18]. Valencene synthase (*Cstps1*) is the protein for synthesis of the end product, valencene. Klee *et al.* [18] indicated that all of the apocarotenoid volatile QTLs identified to date are associated with carotenoid biosynthetic enzymes, and substrate availability rather than enzyme synthesis appears to be limiting apocarotenoid volatiles. Our study indicated that the high concentration of carotenoids in Murcott might be due to its lack of valencene synthase activity (Figure 3; Additional file 2: Figure S2-B) as well as less sesquiterpenes and other carotenoid derived volatiles

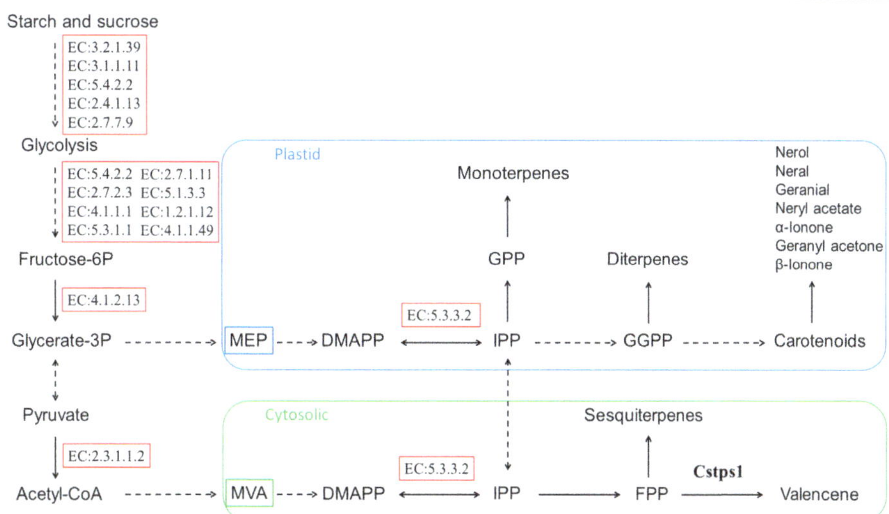

Figure 4 Summary of metabolic pathways leading to terpenoid-associated volatile synthesis. The differently expressed KEGG enzymes between Temple and Murcott mandarin hybrid fruit are in red boxes. The second metabolites are presented in yellow boxes. Pathway names are presented in the blue box. In most cases, arrows indicate multiple enzyme reactions. Abbreviations: MEP, 2-C-methyl-D-erythritol 4-phosphate; MVA, mevalonate; IPP, isopentenyl diphosphate; DMAPP, dimethyl-allyl diphosphate; GPP, geranyl diphosphate; FPP, farnesyl diphosphate; GGPP, geranylgeranyl diphosphate; Cstps1,valencene synthase.

(Additional file 1: Table S1), compared with Temple. In tomato and watermelon, studies have indicated that carotenoid pigmentation patterns have profound effects on apocarotenoid volatile compositions [40,41]. By comparison with Murcott, our results suggest that the diversion of high valencene and other sesquiterpenes into the terpenoid pathway together with high production of apocarotenoid volatiles might have resulted in the lower concentration of carotenoids in Temple.

Conclusions

Two thirds of differently expressed proteins were identified in the pathway of glycolysis and TCA, as well as amino acid, sugar and starch metabolism. This highlights the importance of these metabolic pathways for providing the carbon skeleton of the upstream precursors for most volatiles. Total carotenoids were significantly higher and apocarotenoid volatiles lower in Murcott than in Temple. It appears that high concentrations of apocarotenoid volatile compounds may result in low concentrations of carotenoids in Temple. In addition, we found that valencene synthase (*Cstps1*) was severely reduced in Murcott, and consequently, no valencene was detected in Murcott fruit during development, while substantial amounts were present in Temple. Further study is needed to confirm if there is a relationship between carotenoid concentrations and apocarotenoid volatile compounds, sesquiterpenes such as valencene, in citrus fruit. Improving fruit flavor is a challenging task using classic breeding methods because of the difficulty in scoring and quantifying such a complex trait. An increased understanding of biosynthetic pathways for fruit flavor compounds and corresponding regulatory mechanisms will lead to more efficient breeding strategies to improve flavor.

Methods

Plant material

Fruit of Murcott and Temple cultivars were collected on three harvest dates (designated as Stage 1, 2, and 3 respectively): 22 December 2008, 30 January 2009, and 11 March 2009 from groves at the University of Florida, Citrus Research and Education Center (UF-CREC) (Figure 1). These trees were grown under the same environmental conditions of soil, irrigation and illumination. Fruit maturity for Murcott and Temple was determined based on previous results [4], and three years of measurements of volatiles and non-volatiles at different stages amoung 14 mandarin hybrids including Temple and Murcott. Sample fruits were also selected based on fruit of similar size, color, and flavor by experienced breeders. Both Temple and Murcott have the same rootstock, Cleopatra mandarin, and are grown in the center part of field. In total, 20 fruits were collected randomly around the tree, 10 fruits for protein and 10 fruits for volatile compound identification, respectively. Three to four fruits were bulked as biological replications for proteome analysis.

Sugars, organic acids and carotenoids analysis

The measurement of sugars and acids was based on the method described by Baldwin *et al.* [42]. For titratable acidity (TA) and soluble solids content (SSC), TA was determined by titrating to pH8.2 with 0.1 M NaOH using an autotitrator (Mettler Toledo DL50, Columbus, OH) and SSC using a refractometer (Atago PR-101, Tokyo, Japan). Individual sugar and acid analysis was performed via high performance liquid chromatography (HPLC). Approximately 40 g of juice was extracted using 70 mL of an 80% ethanol/deionized water solution. The mixture was boiled for 15 min, cooled, and filtered (Whatman #4 filter paper, Batavia, IL). The filtered solution was brought to 100 mL with 80% ethanol. A total of 10 mL of the filtered solution was then passed through a C^{18} Sep-Pak (Waters/Millipore), followed by a 0.45 μm Millipore (Siemens-Millipore, Shrewbury, MA) filter. Individual sugars analysis was performed by HPLC with a refractive index detector (Perkin Elmer, Norwalk, Conn) equipped with a Waters Sugar Pak column [43-45]; The mobile phase was 10^{-4} M ethylenediaminetetraacetic acid disodium calcium salt (CaEDTA) (0.5 mL min^{-1} flow rate at 90°C). All results are expressed as g 100 mL^{-1} juice. Organic acids, including ascorbic acid, were analyzed using a Perkin-Elmer Series 200 auto sampler (Waltham, MA), a Spectra System P4000 pump, and a Spectra System UV 6000 LP detector (Thermo Fisher Scientific, Waltham, MA). Acids were separated on an AltechOA1000 Prevail organic acid column with a flow rate of 0.2 mL min^{-1} at 35°C and a mobile phase of 0.01 N H_2SO_4 [42,46]. The injection volume was 20 μL.

Carotenoids in the pellet and supernatant were analyzed using HPLC. Juice samples (30 mL) were centrifuged at 10,000 × g for 15 min. The pellet extracts were collected by dissolving pellets in acetone. Both pellet extracts and supernatants were individually filtered through a 0.45 μm filter into amber vials and stored at −20°C until injected into an HPLC (20 μL loop) equipped with an YMC carotenoid column (YMC Co. Ltd., Komatsu City, Japan). Elution conditions included a three-solvent gradient composed initially of water/methanol/methyl tertbutyl ether (4/81/15, v/v/v), and changed to linear gradients of 4/6/90 (v/v/v) by 60 min at a flow rate of 1 mL min^{-1}, at 30°C. Compounds were detected using a photo diode array (PDA) detector scanning 200–700 nm at 5 nm increments, identified using standards (Sigma, Carotenoid Nature) and quantified using absorbance measurements. Values for pellet extracts and supernatants were then added together for each sample.

Volatile compound identification

Sample preparation for volatile and aroma identification used the same methods as previously described [4]. Briefly, Temple and Murcott samples were juice composites of 10 fruits with 2 replications of 5 fruits. The fruit were washed, rinsed and gently juiced manually using a table-top manual juicer (model 3183; Oster, Rye, NY, USA) to avoid potential peel components (peel oil) entering the juice. Juice samples (2.5 mL) were placed in 20 mL glass vials (Gerstel, Inc., Baltimore, MD, USA) along with saturated sodium chloride solution (2.5 mL) to help drive volatiles into the headspace and inhibit any potential enzymatic activity. An internal standard (3-hexanone, 1 ppm) was added to juice samples. The vials were capped and stored at −20°C until analyzed. The extraction of aroma volatiles was performed using solid-phase microextraction (SPME) with an MPS-2 auto sampler (Gerstel). The vials were incubated at 40°C for 30 min and volatile compounds were identified by comparison of their mass spectra with library entries (NIST/EPA/NIH Mass Spectral Library, version 2.0; National Institute of Standards and Technology, Gaithersburg, MA, USA), as well as by comparing retention indices (RIs) with published RIs on both columns. Volatiles were semi-quantified by dividing peak area with the peak area of the internal standard.

Statistical analysis of volatile and non-volatile compounds

Two pooled samples from ten fruits were used for each harvesting time. All calculations were based on means of harvesting time. The differences of volatile and non-volatile compounds between Temple and Murcott were examined by an analysis of variance using the PROC GLM procedure of the SAS 9.4 statistical software package (http://www.sas.com).

Protein extraction

Protein extraction was modified based on the following description [21]. Briefly, the juice sacs were ground in homogenization buffer containing 0.5 M MOPS-KOH pH 8.5, 1.5% PVPP, 7.5 mM EDTA, 2 mM DTT, 0.1 mM PMSF, and 0.1% (v/v) protease inhibitor cocktail (Sigma, St. Louis, MO, USA). The homogenates were filtered through four layers of cheesecloth and centrifuged at 1500 × g for 20 min to eliminate cellular debris and nuclei. The pellet was discarded and the supernatant was centrifuged at 12000 × g for 20 min at 4°C. Soluble protein was precipitated in ammonium sulfate (85%) and collected by centrifugation at 12000 × *g*. The pellets were resuspended in a buffer containing 10 mM KH_2PO_4 and 0.5 mM DTT and desalted with a PD-10 column (Amersham Bioscience, GE Healthcare, Piscataway, NJ, USA) according to manufacturer's instruction. Protein concentration was determined using the Bio-Rad Bradford protein assay (Bio-Rad, Hercules, CA. USA). One hundred µg protein from each sample was precipitated in 80% cold acetone at −20°C overnight, centrifuged at 18,000 rpm for 20 min at 4°C, and washed once with 80% cold acetone.

iTRAQ Labeling and data analysis

In total, 18 samples were labeled and analyzed (2 cultivars × 3 maturity levels × 3 replications). Three to four fruits were pooled with 100 µg protein as one replication. iTRAQ labeling and data analysis were performed as a service by the Interdisciplinary Center for Biotechnology Research (ICBR) Proteomic Core facility at the University of Florida (Gainesville, FL, USA). For protein digestion, iTRAQ labeling and cation exchange were done according to the company's protocols and described by Zhu *et al.* [47]. Briefly, the MS/MS data were analyzed by a thorough search considering biological modifications against the NCBI subset of green plants fasta database (downloaded on November, 2010) using the Paragon™ Algorithm of PROTEINPILOT v3.0 software suite (Applied Biosystems). For relative quantification of proteins, only MS/MS spectra unique to a particular protein and for which the sum of the signal-to-noise ratio for all of the peak pairs was greater than 9 were used for quantification (Applied Biosystems). To be identified as being differentially expressed, a protein had to be quantified with at least three spectra, a $p < 0.05$, and a ratio -fold change of at least 2 in more than two independent experiments (i.e. at least six peptides). Protein identities were confirmed using BLAST at the NCBI. Gene ontology analysis of identified proteins was carried out using Blast2GO [48]. The biological interpretation of the differentially expressed proteins was further completed by assigning them to metabolic pathways using Kyoto Encyclopedia of Genes and Genomes (KEGG) annotation. For proteins identified more than once, only the most significant identified protein was selected. In addition, functional classification of total identified proteins was analyzed by Blast2Go with default parameters (https://blast2go.com).

RNA extraction and quantitative real-time reverse transcription polymerase chain reaction (QRT-PCR)

Total RNA from each sample was extracted using Trizol (Ambion), and contaminating DNA was eliminated using the Turbo DNA-free Kit (Ambion, Austin, TX). The concentration of RNA was measured in a NanoDrop ND-1000 spectrophotometer (NanoDrop Technologies, Wilmington, DE). Total RNA was diluted as 5 ng/μL^{-1}. QRT-PCR was carried out in the Agilent Mx3005P System (Agilent Technology) using a Brilliant III Ultra-Fast SYBR Green QRT-PCR Master Mix (Agilent Technology). Glyceraldehyde 3-phosphate dehydrogenase (GAPDH) was used as a reference gene to provide relative quantification for the target gene valencene synthase (*Cstps1*). Primer sequences of *Cstps1* were used according to Sharon-Asa *et al.* [15]

(Additional file 2: Table S4). The results represent normalized mean values and standard error of mean analyzed by using the program in the Agilent Mx3005P System.

Availability of supporting data

The data supporting the results of this article are included within the article.

Additional files

Additional file 1: The excel spread sheet contains 3 tables describing identified volatiles, proteins and results of metabolite pathway analyses in Temple and Murcott. Table S1. Volatiles identified in Temple and Murcott mandarin hybrid fruit. **Table S2.** Total proteins identified in fruit flesh of Temple and Murcott mandarin hybrid fruits, and ratio of Temple versus Murcott. **Table S3.** Metabolite pathways containing differentially expressed proteins between Temple and Murcott mandarin hybrid fruits.

Additional file 2: Detailed information on primers used for amplifying valencene synthase, gene ontology assignment, valencene and carotenoid content during fruit ripening in Temple and Murcott. Table S4. Primers used for amplifying valencene synthase and control genes for real-time PCR. **Figure S1.** Gene Ontology (GO) assignment (2nd level GO terms) of differential proteins between Murcott and Temple. The differential proteins were categorized based on GO annotation and the proportion of each category was displayed according to: Biological process (A), Cellular component (B) and Molecular function (C). Because a gene could be assigned to more than one GO term, the sum of genes in a category would be above the total number 92. X axis indicates number of different expressed proteins. **Figure S2.** (A) Valencene production during fruit ripening in Temple and Murcott; (B) Carotenoid content in Temple and Murcott during ripening.

Abbreviation

ATP: Adenosine triphosphate; Cstps1: Valencene synthase; CaEDTA: Ethylenediaminetetraacetic acid disodium calcium salt; DMPP: Isomer dimethylallyl diphosphate; GC-MS: Gas chromatography–mass spectrometry; GO: Gene ontology; HPLC: High performance liquid chromatography; IPP: Isopentenyl diphosphate; iTRAQ: Isobaric tags for relative and absolute quantification; KEGG: Kyoto encyclopedia of genes and genomes; MEP: Methylerythritol 4-phosphate; MVA: Mevalonate; NAD +: Nicotinamide adenine dinucleotide (oxidized form); NAPDH: Nicotinamide adenine dinucleotide (reduced form); QRT-PCR: Quantitative real-time Reverse transcription polymerase chain reaction; RIs: Retention indices; SOD: Superoxide dismutase; SPME: Solid-phase microextraction; SSC: Soluble solids content; TA: Titratable acidity; TCA: The tricarboxylic acid cycle.

Competing interests

The authors declare that they have no competing interests.

Authors' contributions

QY, AP, and FGG conceived and designed the study; QY, AP, EAB, JB and YY collected and analyzed the volatile and non-volatile data; QY, MH and HSD collected and analyzed the proteomic data; QY wrote the manuscript. All authors read and approved the final manuscript.

Authors' information

Qibin Yu, submitting author.

Acknowledgements

The authors thank Mrs. Misty Holt for collecting fruit samples. This work was partly supported by grants from the New Varieties Development and Management Corporation (NVDMC), and the Citrus Research and Development Foundation Inc. (CRDF), on behalf of the Florida citrus industry.

Author details

[1]University of Florida, Institute of Food and Agricultural Sciences, Citrus Research and Education Center, Lake Alfred, FL 33850, USA. [2]USDA-ARS Horticultural Research Laboratory, Fort Pierce, FL 34945, USA. [3]College of Agriculture, Punjab Agricultural University, Ludhiana, Punjab 141004, India.

References

1. Baldwin IT. Plant volatiles. Curr Biol. 2010;20(9):392–7.
2. Tietel Z, Plotto A, Fallik E, Lewinsohn E, Porat R. Taste and aroma of fresh and stored mandarins. J Sci Food Agric. 2011;91(1):14–23.
3. Miyazaki T, Plotto A, Baldwin EA, Reyes-De-Corcuera JI, Gmitter Jr FG. Aroma characterization of tangerine hybrids by gas-chromatography-olfactometry and sensory evaluation. J Sci Food Agric. 2012;92(4):727–35.
4. Miyazaki T, Plotto A, Goodner K, Gmitter Jr FG. Distribution of aroma volatile compounds in tangerine hybrids and proposed inheritance. J Sci Food Agric. 2011;91(3):449–60.
5. Dudareva N, Klempien A, Muhlemann JK, Kaplan I. Biosynthesis, function and metabolic engineering of plant volatile organic compounds. New Phytol. 2013;198(1):16–32.
6. Goff SA, Klee HJ. Plant volatile compounds: sensory cues for health and nutritional value? Science. 2006;311(5762):815–9.
7. Tieman D, Bliss P, McIntyre LM, Blandon-Ubeda A, Bies D, Odabasi AZ, et al. The chemical interactions underlying tomato flavor preferences. Curr Biol. 2012;22(11):1035–9.
8. Baldwin EA, Goodner K, Plotto A. Interaction of volatiles, sugars, and acids on perception of tomato aroma and flavor descriptors. J Food Sci. 2008;73(6):S294–307.
9. Beekwilder J, Alvarez-Huerta M, Neef E, Verstappen FW, Bouwmeester HJ, Aharoni A. Functional characterization of enzymes forming volatile esters from strawberry and banana. Plant Physiol. 2004;135(4):1865–78.
10. Aharoni A, Keizer LCP, Bouwmeester HJ, Sun ZK, Alvarez-Huerta M, Verhoeven HA, et al. Identification of the SAAT gene involved in strawberry flavor biogenesis by use of DNA microarrays. Plant Cell. 2000;12(5):647–61.
11. Davidovich-Rikanati R, Sitrit Y, Tadmor Y, Iijima Y, Bilenko N, Bar E, et al. Enrichment of tomato flavor by diversion of the early plastidial terpenoid pathway. Nat Biotechnol. 2007;25(8):899–901.
12. Gonzalez Aguero M, Troncoso S, Gudenschwager O, Campos Vargas R, Moya Leon MA, Defilippi BG. Differential expression levels of aroma-related genes during ripening of apricot (*Prunus armeniaca* L.). Plant Physiol Bioch. 2009;47(5):435–40.
13. Zhang B, Shen JY, Wei WW, Xi WP, Xu CJ, Ferguson I, et al. Expression of genes associated with aroma formation derived from the fatty acid pathway during peach fruit ripening. J Agric Food Chem. 2010;58(10):6157–65.
14. Lucker J, Bowen P, Bohlmann J. Vitis vinifera terpenoid cyclases: functional identification of two sesquiterpene synthase cDNAs encoding (+)-valencene synthase and (–)-germacrene D synthase and expression of mono- and sesquiterpene synthases in grapevine flowers and berries. Phytochemistry. 2004;65(19):2649–59.
15. Sharon-Asa L, Shalit M, Frydman A, Bar E, Holland D, Or E, et al. Citrus fruit flavor and aroma biosynthesis: isolation, functional characterization, and developmental regulation of Cstps1, a key gene in the production of the sesquiterpene aroma compound valencene. Plant J. 2003;36(5):664–74.
16. Tietel Z, Feldmesser E, Lewinsohn E, Fallik E, Porat R. Changes in the transcriptome of 'Mor' mandarin flesh during storage: emphasis on molecular regulation of fruit flavor deterioration. J Agric Food Chem. 2011;59(8):3819–27.
17. Muhlemann JK, Klempien A, Dudareva N. Floral volatiles: from biosynthesis to function. Plant Cell Environ. 2014;37(8):1936–49.
18. Klee HJ. Improving the flavor of fresh fruits: genomics, biochemistry, and biotechnology. New Phytol. 2010;187(1):44–56.
19. Pan Z, Zeng Y, An J, Ye J, Xu Q, Deng X. An integrative analysis of transcriptome and proteome provides new insights into carotenoid biosynthesis and regulation in sweet orange fruits. J Proteomics. 2012;75(9):2670–84.
20. Katz E, Fon M, Eigenheer RA, Phinney BS, Fass JN, Lin D, et al. A label-free differential quantitative mass spectrometry method for the characterization and identification of protein changes during citrus fruit development. Proteome Sci. 2010;8:68.

21. Katz E, Fon M, Lee YJ, Phinney BS, Sadka A, Blumwald E. The citrus fruit proteome: insights into citrus fruit metabolism. Planta. 2007;226(4):989–1005.
22. Guillaumie S, Fouquet R, Kappel C, Camps C, Terrier N, Moncomble D et al. Transcriptional analysis of late ripening stages of grapevine berry. BMC Plant Biol. 2011;11:165.
23. Saunt J. Citrus Varieties of the World: an Illustrated Guide. Norwich, England: Sinclair international Limited; 2000.
24. Chen S, Harmon AC. Advances in plant proteomics. Proteomics. 2006;6(20):5504–16.
25. Gan CS, Chong PK, Pham TK, Wright PC. Technical, experimental, and biological variations in isobaric tags for relative and absolute quantitation (iTRAQ). J Proteome Res. 2007;6(2):821–7.
26. Pierce A, Unwin RD, Evans CA, Griffiths S, Carney L, Zhang L, et al. Eight-channel iTRAQ enables comparison of the activity of six leukemogenic tyrosine kinases. Mol Cell Proteomics. 2008;7(5):853–63.
27. Schwab W, Davidovich-Rikanati R, Lewinsohn E. Biosynthesis of plant-derived flavor compounds. Plant J. 2008;54(4):712–32.
28. Feng C, Chen M, Xu CJ, Bai L, Yin XR, Li X, et al. Transcriptomic analysis of Chinese bayberry (*Myrica rubra*) fruit development and ripening using RNA-Seq. BMC Genomics. 2012;13:19.
29. Yun Z, Li WY, Pan ZY, Xu J, Cheng YJ, Deng XX. Comparative proteomics analysis of differentially accumulated proteins in juice sacs of ponkan (*Citrus reticulata*) fruit during postharvest cold storage. Postharvest Biol Tec. 2010;56(3):189–201.
30. Pan ZY, Liu Q, Yun Z, Guan R, Zeng WF, Xu Q, et al. Comparative proteomics of a lycopene-accumulating mutant reveals the important role of oxidative stress on carotenogenesis in sweet orange (*Citrus sinensis* [L.] osbeck). Proteomics. 2009;9(24):5455–70.
31. Trindade H. Molecular biology of aromatic plants and spices. A Rev Flavour Frag J. 2010;25(5):272–81.
32. Liu Q, Zhu A, Chai L, Zhou W, Yu K, Ding J, et al. Transcriptome analysis of a spontaneous mutant in sweet orange [*Citrus sinensis* (L.) Osbeck] during fruit development. J Exp Bot. 2009;60(3):801–13.
33. Aharoni A, Jongsma MA, Bouwmeester HJ. Volatile science? Metabolic engineering of terpenoids in plants. Trends Plant Sci. 2005;10(12):594–602.
34. Yonekura-Sakakibara K, Hanada K. An evolutionary view of functional diversity in family 1 glycosyltransferases. Plant J. 2011;66(1):182–93.
35. Fan J, Chen C, Yu Q, Li ZG, Gmitter Jr FG. Characterization of three terpenoid glycosyltransferase genes in 'Valencia' sweet orange (*Citrus sinensis* L. Osbeck). Genome. 2010;53(10):816–23.
36. Oke M, Pinhero RG, Paliyath G. The effects of genetic transformation of tomato with antisense phospholipase D cDNA on the quality characteristics of fruits and their processed products. Food Biotechnol. 2003;17(3):163–82.
37. Pirrello J, Regad F, Latche A, Pech JC, Bouzayen M. Regulation of tomato fruit ripening. CAB Reviews. 2009;4(051):1–14.
38. Song J, Forney CF. Flavour volatile production and regulation in fruit. Canadian J Plant Sci. 2008;88(3):537–50.
39. Dudareva N, Negre F, Nagegowda DA, Orlova I. Plant volatiles: recent advances and future perspectives. Crit Rev Plant Sci. 2006;25(5):417–40.
40. Lewinsohn E, Sitrit Y, Bar E, Azulay Y, Ibdah M, Meir A, et al. Not just colors – carotenoid degradation as a link between pigmentation and aroma in tomato and watermelon fruit. Trends Food Sci Tech. 2005;16(9):407–15.
41. Lewinsohn E, Sitrit Y, Bar E, Azulay Y, Meir A, Zamir D, et al. Carotenoid pigmentation affects the volatile composition of tomato and watermelon fruits, as revealed by comparative genetic analyses. J Agric Food Chem. 2005;53(8):3142–8.
42. Baldwin E, Plotto A, Manthey J, McCollum G, Bai J, Irey M, et al. Effect of Liberibacter infection (huanglongbing disease) of citrus on orange fruit physiology and fruit/fruit juice quality: chemical and physical analyses. J Agric Food Chem. 2010;58(2):1247–62.
43. Baldwin EA, Nisperos-Carriedo MO, Baker R, Scott JW. Quantitative analysis of flavor parameters in six Florida tomato cultivars (*Lycopersicon esculentum* Mill). J Agric Food Chem. 1991;39(6):1135–40.
44. Baldwin EA, Scott JW, Einstein MA, Malundo TMM, Carr BT, Shewfelt RL, et al. Relationship between sensory and instrumental analysis for tomato flavor. J Am Soc Hortic Sci. 1998;123(5):906–15.
45. Baldwin EA, Goodner K, Plotto A, Pritchett K, Einstein M. Effect of volatiles and their concentration on perception of tomato descriptors. J Food Sci. 2004;69(8):S310–8.
46. Baldwin EA, Bai J, Plotto A, Cameron R, Luzio G, Narciso J, et al. Effect of extraction method on quality of orange juice: hand-squeezed, commercial-fresh squeezed and processed. J Sci Food Agric. 2012;92(10):2029–42.
47. Zhu M, Dai S, McClung S, Yan X, Chen S. Functional differentiation of Brassica napus guard cells and mesophyll cells revealed by comparative proteomics. Mol Cell Proteomics. 2009;8(4):752–66.
48. Conesa A, Gotz S, Garcia-Gomez JM, Terol J, Talon M, Robles M. Blast2GO: a universal tool for annotation, visualization and analysis in functional genomics research. Bioinformatics. 2005;21(18):3674–6.

AtEAF1 is a potential platform protein for Arabidopsis NuA4 acetyltransferase complex

Tomasz Bieluszewski[1], Lukasz Galganski[1], Weronika Sura[1], Anna Bieluszewska[2], Mateusz Abram[1], Agnieszka Ludwikow[1], Piotr Andrzej Ziolkowski[1,3*] and Jan Sadowski[1*]

Abstract

Background: Histone acetyltransferase complex NuA4 and histone variant exchanging complex SWR1 are two chromatin modifying complexes which act cooperatively in yeast and share some intriguing structural similarities. Protein subunits of NuA4 and SWR1-C are highly conserved across eukaryotes, but form different multiprotein arrangements. For example, the human TIP60-p400 complex consists of homologues of both yeast NuA4 and SWR1-C subunits, combining subunits necessary for histone acetylation and histone variant exchange. It is currently not known what protein complexes are formed by the plant homologues of NuA4 and SWR1-C subunits.

Results: We report on the identification and molecular characterization of AtEAF1, a new subunit of Arabidopsis NuA4 complex which shows many similarities to the platform protein of the yeast NuA4 complex. AtEAF1 copurifies with Arabidopsis homologues of NuA4 and SWR1-C subunits ARP4 and SWC4 and interacts physically with AtYAF9A and AtYAF9B, homologues of the YAF9 subunit. Plants carrying a T-DNA insertion in one of the genes encoding AtEAF1 showed decreased *FLC* expression and early flowering, similarly to *Atyaf9* mutants. Chromatin immunoprecipitation analyses of the single mutant *Ateaf1b-2* and artificial miRNA knock-down *Ateaf1* lines showed decreased levels of H4K5 acetylation in the promoter regions of major flowering regulator genes, further supporting the role of AtEAF1 as a subunit of the plant NuA4 complex.

Conclusions: Growing evidence suggests that the molecular functions of the NuA4 and SWR1 complexes are conserved in plants and contribute significantly to plant development and physiology. Our work provides evidence for the existence of a yeast-like EAF1 platform protein in *A. thaliana*, filling an important gap in the knowledge about the subunit organization of the plant NuA4 complex.

Keywords: NuA4, EAF1, YAF9, *Arabidopsis thaliana*, histone acetylation, PIE1

Background

Eukaryotic chromatin has evolved for seemingly contradictory functions. It ensures compaction and protection of genetic material, but also controls diverse processes including transcription, replication and DNA repair that require a relatively open and dynamic chromatin structure. This mixture of robustness and flexibility is achieved by a number of specialized enzymes that remodel chromatin or modify nucleosomes by covalent histone modifications. Different chromatin modifications can have a combinatorial effect on chromatin properties and influence each other by guiding, stimulating or inhibiting chromatin modifying enzymes.

One of the best studied examples of an interplay between different types of chromatin modifications is the strong functional relationship between histone acetylation and chromatin remodeling. Protein domains specialized in specific recognition of acetylated histone tails are often found in chromatin remodeling complexes. Acetylation of nucleosomal histones has a strong influence on the action of chromatin remodeling complexes such as RSC [1], SWI/SNF [2], INO80 [3] or SWR1-C [4].

In the case of SWR1-C, this link seems to go much further. In yeast, the histone variant exchange reaction, catalyzed by SWR1-C is stimulated by the NuA4 complex through acetylation of nucleosomal histones [4]. These two protein complexes not only cooperate, but also share

* Correspondence: paz22@cam.ac.uk; jsad@amu.edu.pl
[1]Department of Biotechnology, Institute of Molecular Biology and Biotechnology, Adam Mickiewicz University, Umultowska 89, 61-614 Poznań, Poland
Full list of author information is available at the end of the article

interesting structural similarities. Each of the two complexes is formed by more than ten different protein subunits organized into modules [5]. One of the modules, composed of proteins ARP4, SWC4, YAF9 and monomeric Actin, is common to both complexes [6]. Furthermore, a crucial role in the integrity of NuA4 and SWR1-C is played by proteins EAF1 and SWR1, respectively, which organize different modules into a functional complex. Although EAF1 lacks the ATPase domain, central to the chromatin remodeling activity of SWR1-C, it shares with SWR1 the HSA domain which is necessary for binding of the common module ARP4-SWC4-YAF9-Actin [5,7].

In human, homologues of yeast NuA4 and SWR1-C subunits form a hybrid complex called TIP60-p400. How far the functions of the fused complexes are conserved is yet to be determined. Whereas the ATPase activity of p400 enables H2A.Z deposition through histone exchange [8], the HAT activity of the TIP60 subunit seems to be blocked by the association with p400 [9]. Nevertheless, subunit conservation between SWR1, NuA4 and TIP60-p400 implies a strong evolutionary link. One attractive explanation of the relationship between SWR1-C, NuA4 and TIP60-p400 points to the fact that the integrity of the three complexes depends on their platform subunits SWR1, EAF1 and p400, respectively. The complex is formed when the remaining subunits bind, directly or as a part of a protein module, to several conserved features of the platform protein. Strikingly, p400 combines the features of SWR1 and EAF1. A synthetic p400-like construct, obtained by insertion of the ATPase domain of SWR1 between the HSA (helicase/SANT–associated) and SANT (Swi3, Ada2, N-Cor, and TFIIIB) domains of EAF1, reconstituted a TIP60-p400-like complex when expressed in yeast [5]. The authors of the study concluded that a similar rearrangement could have given rise to the p400-like architecture in higher eukaryotes [5].

Outside Metazoa, the best characterized example of a domain architecture similar to that of p400 is PIE1, a protein necessary for incorporation of H2A.Z into nucleosomes in *Arabidopsis thaliana* [10,11]. HSA, ATPase and SANT domains are all present in PIE1 and show a high degree of sequence similarity to the corresponding domains of p400. Although most of the subunits of NuA4 and SWR1-C have clear homologues in Arabidopsis, no study has addressed the question of whether PIE1 can organize these proteins into a hybrid complex similar to TIP60-p400. Importantly, a plant homologue of EAF1 has not been identified so far, which calls into question the existence of an independent NuA4 complex in plants.

Up to now, studies in plants have embraced homologues of only three subunits of the yeast NuA4 complex, ESA1, YAF9 and EAF3. ESA1, the only essential histone acetyltransferase in *S. cerevisiae* and the catalytic subunit of the NuA4 complex [12], has two homologues in *A. thaliana*, HAM1 (Histone Acetyltransferase of the MYST family 1) and HAM2. Both proteins specifically acetylate lysine 5 of the histone H4 [13] and are functionally redundant, as the single mutants display no developmental phenotype, while a double mutant is nonviable [14]. Reduction of HAM1 and HAM2 transcript levels results in decreased expression of the negative flowering regulator *FLOWERING LOCUS C* (*FLC*) and premature transition to flowering. This change is accompanied by a decrease in H4 acetylation in the chromatin of *FLC* [15]. Similar effects were observed in plants deficient in AtYAF9A, one of the two Arabidopsis homologues of yeast YAF9, a subunit shared by NuA4 and SWR1-C [16]. Interestingly, a recent study shows that simultaneous loss of function of two *A. thaliana* homologues of EAF3 results in late flowering, also mediated by reduced histone acetylation which, in this case, disrupts the expression of flowering inducer *FLOWERING LOCUS T (FT)* [17].

The aim of this study was to determine whether Arabidopsis homologues of NuA4 subunits are organized into a big protein complex similar to yeast NuA4 or human TIP60-p400. Our initial hypothesis was that PIE1 serves as a platform protein for a plant analog of the human TIP60-p400 complex. Through analysis of proteins that bind to Arabidopsis homologues of ARP4 and SWC4, common subunits of yeast NuA4 and SWR1-C, we confirmed their interaction with PIE1 and other Arabidopsis homologues of NuA4 and SWR1-C subunits. In addition, we revealed their association with an uncharacterized EAF1-like protein, not previously considered a subunit of plant NuA4 or SWR1-C. Subsequently, we focused on a possible role of AtEAF1 as a platform of the plant NuA4 complex. We demonstrated that one of the two isoforms of AtEAF1 interacts with AtYAF9A and AtYAF9B through a conserved HSA domain. Phenotypical analyses of mutants indicated that AtEAF1 and AtYAF9 are necessary for proper timing of transition to flowering, which can be explained by their influence on the H4K5 acetylation in the chromatin of major flowering regulator genes and their transcriptional activity, supporting the role of AtEAF1 as a platform subunit in the plant NuA4 complex.

Results

An uncharacterized plant-specific domain-relative of the yeast EAF1 protein is physically associated with AtARP4 and AtSWC4

We used affinity purification followed by tandem mass spectrometry (AP-MS/MS) to test which Arabidopsis homologues of yeast NuA4 and SWR1-C are associated with the common protein module of the two complexes. For this purpose we chose Arabidopsis homologues of ARP4 and SWC4 as protein baits, because these subunits are encoded by single genes in *A. thaliana*. We overexpressed AtARP4 and AtSWC4 fused to Strep-tag, in an Arabidopsis cell suspension culture. Following purification of proteins bound to the baits, we identified them by mass spectrometry. Proteins detected in control purifications, with no bait or with Strep-GFP as bait, were eliminated as nonspecific hits (Additional file 1). Next, we looked for proteins showing sequence- or domain architecture-similarity to subunits of NuA4 and SWR1-C. We also looked for homologues of INO80 complex subunits, because the yeast and human versions of this complex also contain ARP4 [18,19]. As expected, we found conserved subunits of all three complexes in association with AtARP4, while AtSWC4 copurified only with subunits of NuA4 and SWR1-C (Table 1). In agreement with a recently published study [20], we found multiple subunits of the SWI/SNF complex among proteins copurified with AtARP4, suggesting that besides NuA4, SWR1-C and INO80, AtARP4 interacts with SWI/SNF in plants. We identified no subunits characteristic of INO80 or SWI/SNF among the proteins copurified with AtSWC4, which additionally confirms the specificity of the detected interactions.

According to published data for other species, the presence of ARP4 and SWC4 subunits is restricted to protein complexes closely related to NuA4, SWR1-C and TIP60-p400. Therefore, we considered proteins copurified with both baits as potential subunits of a hypothetical plant TIP60-p400-like complex (Table 1). One uncharacterized protein in this category had a domain architecture similar to that of yeast EAF1 (Figure 1a). We therefore named it AtEAF1.

We aligned the amino acid sequences of AtEAF1 homologues identified in distantly related plant species with the sequences of *S. cerevisiae* EAF1 and its *Schizosaccharomyces pombe* homologue VID21, as well as with human p400 and several plant homologues of PIE1 (Figure 1a). Strikingly, all proteins contained highly conserved HSA and SANT domains (Figure 1b), despite little overall sequence similarity and different domain arrangements (Figure 1c).

AtEAFf1 is encoded by two nearly identical genes that are both transcriptionally active in *A. thaliana*

AtEAF1 is encoded by two genes in all of the sequenced *A. thaliana* ecotypes, except Mt-0 (1001genomes.org). The two genes occupy adjacent loci At3g24870 and At3g24880 on the reverse strand of chromosome 3 and share 98.5% identity in their coding regions. As no full-length cDNA clone was available for AtEAF1, we designed primers using existing annotations (Additional file 2), and cloned the full-length coding sequence (CDS) (Additional file 3). Our *AtEAF1B* CDS clone corresponds to the At3g24870.1 gene model (Additional file 4).

To determine whether both genes were transcriptionally active, we designed three pairs of primers complementary to both coding sequences. Each amplicon contained a restriction site specific for one gene (i.e. not found in the other). The results of restriction digestion of the RT-PCR products were consistent for all three amplicons and indicated that both transcripts are equally abundant in mature rosette leaves (Additional file 4).

Plant homologues of the NuA4 complex subunit YAF9 interact with AtSWC4 and AtEAF1

As shown above, we found evidence for physical association of AtARP4 and AtSWC4 with AtEAF1, a domain relative of yeast NuA4 subunit EAF1. We assumed that AtARP4 and AtSWC4 associated with AtEAF1 through interaction with the latter's HSA domain. Cooperative binding of ARP4 and SWC4 to the N-terminal region of SWR1, containing the HSA domain, requires YAF9 in yeast [7]. One of the two Arabidopsis homologues of YAF9 was shown recently to be required for histone H4 acetylation in the *FLC* locus, in line with its role as a plant NuA4 subunit [16]. To see whether the HSA domain of AtEAF1 can recruit *A. thaliana* YAF9, we transiently coexpressed nEYFP-tagged AtYAF9A or AtYAF9B with the HSA-containing fragment of AtEAF1 fused to the Flag-tag in Arabidopsis mesophyll protoplasts. Coimmunoprecipitation showed that AtYAF9A and AtYAF9B do indeed interact with this fragment of AtEAF1 (Figure 2a, Additional file 5). We also tested HSA-containing fragments of PIE1 and AtINO80 fused to Flag-tag, but no interaction was observed.

AtYAF9A and AtYAF9B interact with AtSWC4 in the nucleus

Our AP-MS/MS results revealed a physical association of AtARP4 and AtSWC4 with AtYAF9A and AtYAF9B. To test whether AtARP4, AtSWC4, AtYAF9A and AtYAF9B associate closely to form a functional module, we screened these proteins for protein-protein interactions, using Bimolecular Fluorescence Complementation (BiFC) in Arabidopsis mesophyll protoplasts. Coexpression of AtYAF9A or AtYAF9B fused to the N- or C-terminal fragment of Enhanced Yellow Fluorescent Protein (EYFP) with AtSWC4 fused to the complementary EYFP fragment produced strong fluorescence localized in the nucleus (Figure 2b,c). No other combination of complementary fusions produced detectable EYFP fluorescence (Additional file 6).

Although pairwise sequence alignment between AtYAF9A and AtYAF9B amino acid sequences shows a high degree of similarity along their whole length, only a shorter splice variant of AtYAF9B has been studied previously [16,21]. We obtained CDS clones of both splice variants (Additional file 3, Additional file 4) and found that the shorter variant lacks the conserved C-terminal region,

Table 1 ***A thaliana*** **homologues of yeast SWR1-C and NuA4 subunits copurify with AtARP4 and AtSWC4**

Purified proteins		Homologues		Baits		Protein complexes			Protein domains		
Locus ID	Name	*H. sapiens*	*S. cerevisiae*	At ARP4	At SWC4	SWR1/NuA4	INO80	SWI/SNF-type	HSA	ATPase	SANT
AT1G18450	AtARP4[3]	BAF53A	ARP4	+	+	+	+	+			
AT5G22330	RVB1	RUVBL1	RVB1	+	+	+	+				
AT5G67630	RVB21	RUVBL2	RVB2	+	+	+	+				
AT3G49830	RVB22	RUVBL2	RVB2	+	+	+	+				
AT3G24880	**AtEAF1A**	-	**EAF1**	**+**	**+**	**+**			+		+
AT3G24870	**AtEAF1B**	-	**EAF1**	**+**	**+**	**+**			+		+
AT4G14385	EAF6	MEAF6	EAF6	+	+	+					
AT1G79020	EPL1B	EPC1	EPL1	+	+	+					
AT5G64610	HAM1[1]	TIP60	ESA1	+	+	+					
AT1G54390	ING2	ING3	YNG2	+	+	+					
AT3G12810	PIE1[1]	p400	SWR1	+	+	+			+	+	+
AT2G47210	AtSWC4[1]	DMAP1	SWC4	+	+	+					+
AT5G45600	AtYAF9A[1]	GAS41	YAF9	+	+	+					
AT3G33520	ARP6[1]	ARP6	ARP6		+	+					
AT4G37280	MRG1[1]	MRG15	EAF3		+	+					
AT1G26470	EAF7	MRGBP	EAF7		+	+					
AT1G16690	EPL1A	EPC1	EPL1		+	+					
AT2G36740	SWC2[1]	YL-1	SWC2		+	+					
AT5G37055	SWC6[1]	ZNHIT1	SWC6		+	+					
AT4G36080	TRA1	TRRAP1	TRA1		+	+					
AT2G18000	AtYAF9B[1]	GAS41	YAF9		+	+					
AT3G12380	ARP5[2]	ARP5	ARP5	+			+				
AT5G43500	ARP9	ARP8	ARP8	+			+				
AT3G57300	INO80[2]	INO80	INO80	+			+		+	+	
AT5G13950	NFRKB	NFRKB	-	+			+				
AT1G65650	UCH2	UCH37	-	+			+				
AT5G16310	UCH2	UCH37	-	+			+				
AT4G06634	YY1	YY1	-	+			+				
AT3G60830	ARP7[3]	-	ARP7	+				+			
AT3G17590	BSH[3]	INI1	SNF5	+				+			
AT2G46020	BRM[3]	BRG1/HBRM	SNF2/STH1	+				+	+	+	
AT2G28290	SYD[3]	BRG1/HBRM	SNF2/STH1	+				+	+	+	
AT3G06010	MINU1	BRG1/HBRM	SNF2/STH1	+				+	+	+	
AT5G19310	MINU2	BRG1/HBRM	SNF2/STH1	+				+	+	+	
AT2G47620	SWI3A[3]	BAF155/BAF170	SWI3	+				+			+
AT2G33610	SWI3B[3]	BAF155/BAF170	SWI3	+				+			+
AT1G21700	SWI3C[3]	BAF155/BAF170	SWI3	+				+			+
AT4G34430	SWI3D[3]	BAF155/BAF170	SWI3	+				+			+
AT3G01890	SWP73A	BAF60A	SNF12	+				+			
AT5G14170	SWP73B	BAF60A	SNF12	+				+			

Proteins copurified with AtARP4 or AtSWC4, fused to Strep-tag, were identified by MS/MS and manually annotated on the basis of sequence- and domain architecture-similarity to yeast and human proteins. The table is limited to proteins that were identified as subunits of chromatin remodeling or histone modifying complexes (multiple subunits were identified). Columns on the right show the distribution of selected protein domains.

[1,2,3]Proteins either directly or indirectly linked to *Arabidopsis* SWR1 and/or NuA4 (1), INO80 (2) or SWI/SNF-type (3) complexes, according to published experimental data.

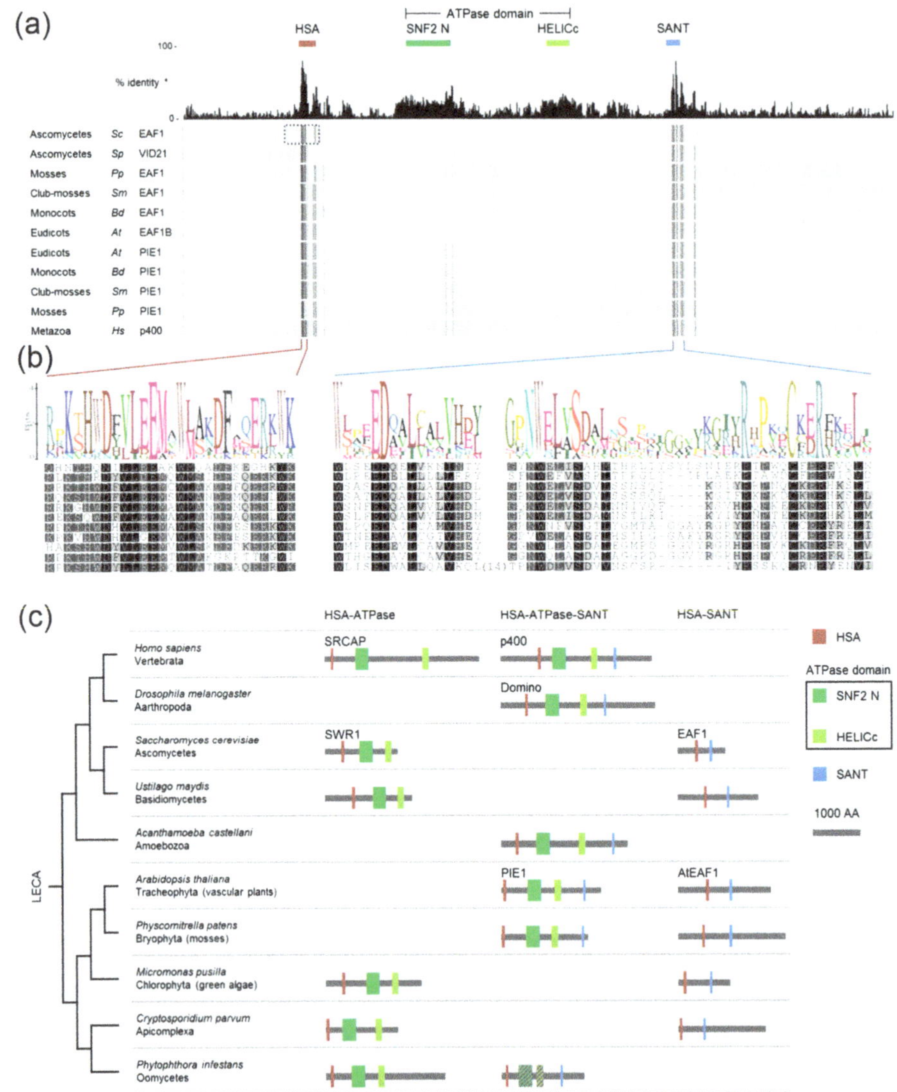

Figure 1 ***Arabidopsis thaliana*** **AtEAF1 and PIE1 represent two distinct protein families that share highly similar HSA and SANT domains and do not coexist outside plants. (a)** Protein alignment of domain relatives of *Arabidopsis thaliana* (*At*) AtEAF1 and PIE1, found in *Saccharomyces cerevisiae* (*Sc*), *Schizosaccharomyces pombe* (*Sp*), *Brachypodium distachyon* (*Bd*), *Selaginella moellendorffii* (*Sm*), *Physcomitrella patens* (*Pp*) and *Homo sapiens* (*Hs*). A dotted rectangle depicts the region of yeast EAF1 that recruits ARP4 and actin, according to Szerlong et al. [28]. Columns containing over 75% gaps were removed from the alignment for clarity. Similarity shading of the alignments in **(a)** and **(b)** was done using the Blosum62 matrix. White indicates < 60% similarity, light grey 60 to 80%, dark grey 80 to 100% and black 100%. **(b)** A close-up of the conserved fragments of the HSA domain and the SANT domain. **(c)** Protein domains HSA, ATPase and SANT form three different domain architectures, which occur in various combinations across eukaryotes. All sequences and alignments used in this figure can be found in the Additional file 3. * Mean pairwise identity over all pairs in the column. A sliding window of 5 was used in the histogram.

responsible for SWC4 binding in yeast [22]. In agreement with this observation, the shorter splice variant of AtYAF9B did not interact with AtSWC4 when tested by BiFC (Figure 2b). Therefore, we used the longer splice variant in all protein-protein interaction assays described herein.

AtEAF1B, AtYAF9A and AtYAF9B are necessary for normal *FLC* levels and timing of flowering

AtYAF9A-deficient plants express *FLC* at reduced levels [16]. Physical interaction between AtYAF9A and AtEAF1, and their similarity to functional counterparts in yeast, YAF9 and EAF1, respectively, suggested that AtEAF1 might also influence *FLC* transcription. To test this prediction we used plants with a T-DNA insertion near the 5′ end of the last exon of the *AtEAF1B* gene (Figure 3a, Additional file 4). We will refer to this line as *Ateaf1b-2*. Compared to wild-type seedlings, *Ateaf1b-2* mutants expressed *FLC* at significantly reduced levels under both long day (LD) and short day (SD) conditions (Figure 3b).

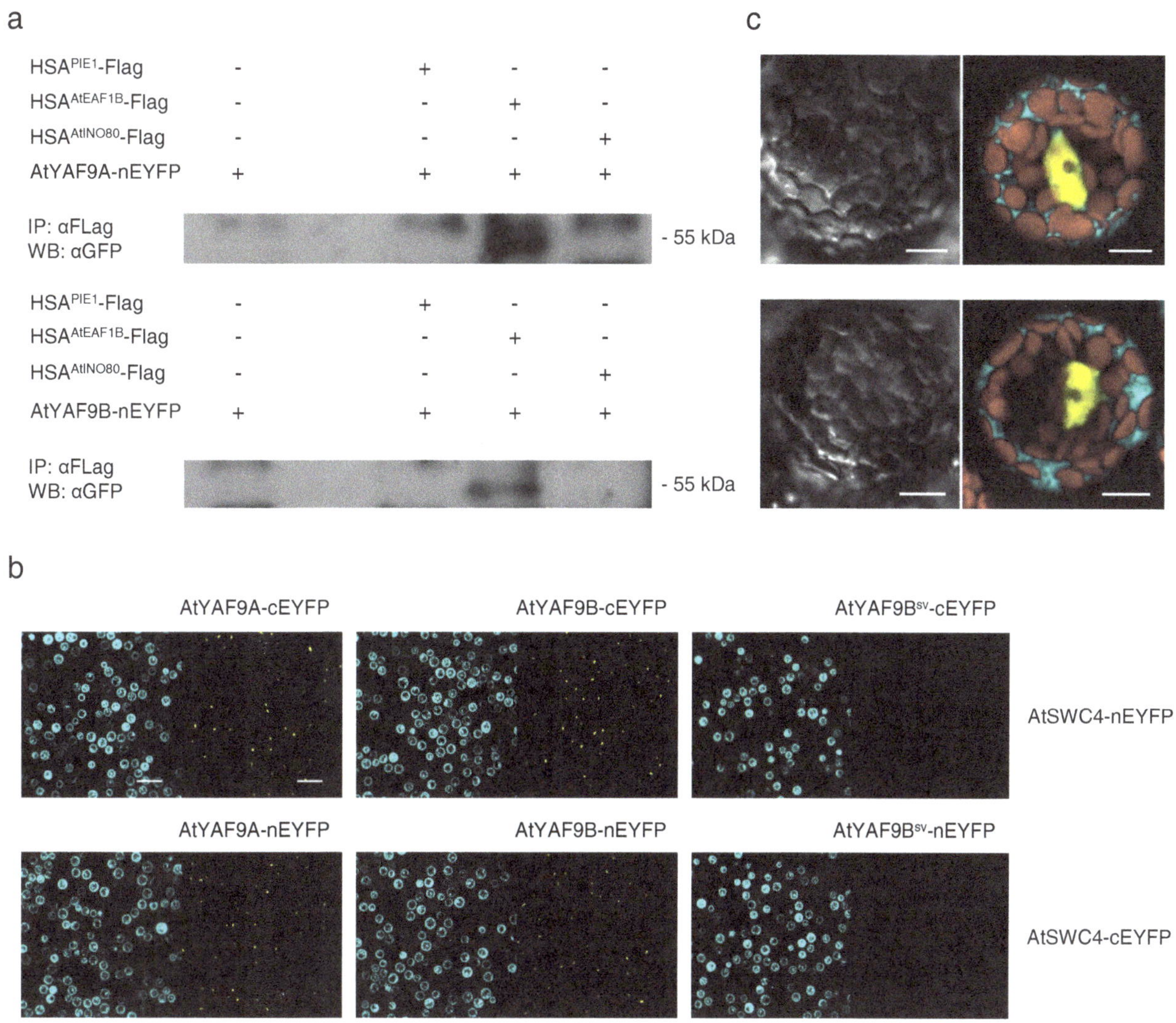

Figure 2 AtYAF9A and AtYAF9B interact with AtSWC4 and the HSA-containing fragment of AtEAF1. (a) Coimmunoprecipitation test of interaction between AtYAF9A and AtYAF9B with the HSA-containing fragments of PIE1, AtEAF1 and AtINO80. Yaf9:nEYFP fusion proteins were detected with antibodies against GFP. Bands above the 55 kDa bars are nonspecific signals from immunoglobulin heavy chains. For additional experimental controls, see Additional file 5. **(b)** BiFC assay in Arabidopsis mesophyll protoplasts. Each co-transfection consisted of a pair of complementary BiFC constructs (nEYFP and cEYFP fusion, right panels, yellow dots indicate fluorescence complementation in the nuclei) and an ECFP construct as an internal transfection control (left panels, cyan). AtYAF9B^{SV} is the short splice variant of AtYAF9B, lacking the putative C-terminal SWC4-binding domain. **(c)** Enlarged images of single protoplasts showing nuclear localization of the EYFP fluorescence, indicating interaction between AtSWC4 and AtYAF9A (upper image) or AtYAF9B (lower image). Chloroplast autofluorescence is shown in red and ECFP fluorescence in cyan. Scale bar: 10 μm. Contrast and brightness were enhanced in all micrographs in **(b)** and **(c)** to improve clarity.

In our protein-protein interaction assays both AtYAF9A and AtYAF9B interacted with AtSWC4 and AtEAF1 (Figure 2). To investigate whether AtYAF9B also contributes to *FLC* expression and whether there is a redundancy of AtYAF9A and AtYAF9B function, we compared the relative *FLC* expression levels of *Atyaf9a-1*, *Atyaf9b-2* and *Atyaf9a-1 Atyaf9b-2* mutants with wild-type plants using the same experimental setup as described for *Ateaf1b-2* above. Interestingly, the *FLC* expression in the double mutant was not significantly reduced when compared to *Atyaf9a-1* (LD $p = 0.18$, SD $p = 0.33$) (Figure 3b).

A decrease in *FLC* expression leads to earlier transition from vegetative to reproductive phase in the *Atyaf9a-1* mutant [16]. We tested the effect of reduced *FLC* expression on the timing of flowering in AtYAF9- and AtEAF1-deficient plants. We grew *Atyaf9a-1*, *Atyaf9b-2*, *Atyaf9a-1 Atyaf9b-2* and *Ateaf1b-2* mutants, as well as wild type plants, under LD and SD conditions. The *Atyaf9a-1 Atyaf9b-2* double mutant flowered significantly

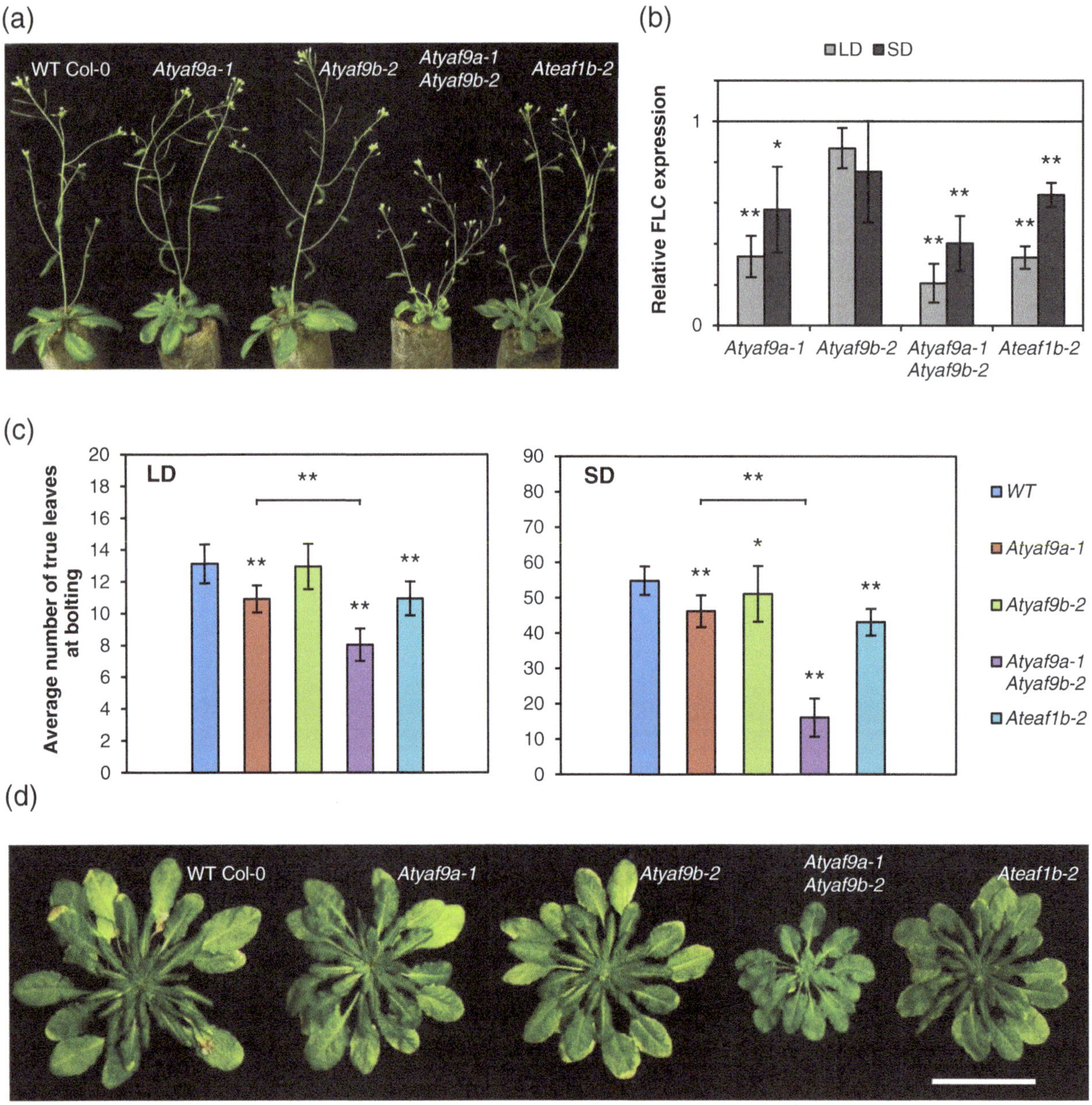

Figure 3 Mutations in *AtEAF1* and *AtYAF9* genes affect *FLC* expression and flowering time. (a) Comparison of plants grown under long day conditions (LD). **(b)** Relative expression levels of *FLC* transcript compared to WT control. Seedlings were collected in the middle of the light photoperiod. **(c)** Comparison of flowering time represented by an average number of true rosette leaves at the stage where the flower stem is 1 cm long. **(d)** Representative rosettes of plants grown under short day conditions (SD) at the stage when the leaves were counted. The bar length is 5 cm. In all graphs, asterisks indicate statistical significance of the difference between each mutant and the WT control. A single asterisk indicates a p-value < 0.05, double asterisk – p-value < 0.01 (t-test).

earlier than single mutants under both LD and SD conditions (Figure 3c,d). Importantly, both *Ateaf1b-2* and *Atyaf9a-1* mutants flowered at a similar growth stage, which was intermediate between that of the *yaf9* double mutant and wild-type plants (WT), while the early flowering phenotype of the *Atyaf9b-2* mutant was least pronounced of all the mutant lines (Figure 3c).

AtEAF1 and AtYAF9 are necessary for normal levels of H4K5 acetylation

Published experimental data support involvement of Arabidopsis homologues of NuA4 subunits YAF9 and ESA1 (HAM1 and HAM2) in the acetylation of lysine 5 of histone H4 [13]. Therefore, if AtEAF1 is a functional subunit of the Arabidopsis NuA4 complex, partial loss of its function should lead to changes in H4K5 acetylation levels. As an initial test of the influence of AtEAF1 on H4K5 acetylation, we grew *Ateaf1b-2* and *Atyaf9a-1 Atyaf9b-2* seedlings for 12 days on MS medium supplemented with 1% sucrose or 1% sucrose and 12.5 μM Trichostatin A (TSA) (Figure 4). TSA is a specific inhibitor of histone deacetylases and has a strong negative effect on plant growth, coinciding with dramatic accumulation of acetylated histones [23,24]. We reasoned that impaired function of an important histone acetylatransferase such as NuA4 could prevent abnormal accumulation of acetylated histones and give mutant plants an advantage over WT plants under TSA challenge. As expected, average fresh weight of *Atyaf9a-1 Atyaf9b-2* and *Ateaf1b-2* mutant plants grown on plates containing TSA was significantly larger than those of WT plants grown in the same conditions, whereas no significant differences where observed under control conditions (Figure 4b). In order to verify if the observed effect can be attributed to differences in global H4K5 acetylation levels, we tested the abundance of histone H4 acetylated on lysine 5 by Western Blot (Additional file 7). Only the double mutant *Atyaf9a-1 Atyaf9b-2* displayed decreased levels of acetylated H4K5 relative to WT which indicates that the increased resistance of *Ateaf1b-2* plants to TSA cannot be due to a global loss of H4K5 acetylation. This observation could be explained if AtEAF1 had a specialized function in the Arabidopsis NuA4 complex. In fact, in yeast *eaf1* mutant a strong decrease in histone H4 acetylation was observed in the promoter region of the PHO5 gene, but no decrease in bulk histone H4 acetylation was reported [5]. Therefore we decided to test if specific genomic targets of histone acetyltraferases are affected in plants with impaired function of AtEAF1. We focused on major regulators of flowering transition *FLC, FT, CONSTANS* (*CO*) and *SUPPRESSOR OF OVEREXPRESSION OF CONSTANS 1* (*SOC1*) as histone acetylation in these genes was found to be deregulated by various H4 acetylation mutants in previous studies [15-17]. Our observations of flowering timing in *Ateaf1b-2* mutant, presented above, further justified that choice.

In order to verify if the flowering phenotype of the *Ateaf1b-2* mutant is related to the function of AtEAF1, we generated transgenic Arabidopsis plants in which transcript levels of *AtEAF1A* and *AtEAF1B* genes were reduced simultaneously through artificial micro RNA (amiRNA) (Figure 5a) [25]. Although independent transgenic lines expressing amiRNA construct displayed a moderate early flowering phenotype (Figure 5b, c), it did not correlate with the decrease in *AtEAF1* transcript (Figure 5), In the line 2.39 which showed the earliest flowering we detected only slightly decreased levels of *AtEAF1* transcript while line 2.29, with stronger *AtEAF1* silencing, showed less obvious early flowering phenotype.

Our next step was to characterize differences in histone H4K5 acetylation levels over *FLC* and *FT* genes between WT, *Ateaf1b-2*, 2.29, 2.39 and *Atyaf9a-1 Atyaf9b-2* plants by chromatin immunoprecipitation (ChIP). We carried out the experiments on 10-day old seedlings grown under long day conditions, collected at the end of the day. For the amplification of the DNA fragments obtained from ChIP we used five pairs of PCR primers for each gene, corresponding to various functional elements of the gene (Figure 6a, b, Additional file 2). We observed a moderate but consistent decrease in H4K5 acetylation over both genes. Interestingly, we observed a stronger reduction in acetylation levels near the 5′ end of the genes, especially in the *FLC* locus. Following this observation, we tested the acetylation levels in the promoter regions of two other major flowering regulators, *CO* and *SOC1*. We observed a significant and consistent drop of H4K5 acetylation in the promoter of *CO* but little change in SOC1, except for the *Atyaf9a-1 Atyaf9b-2* line, which showed significantly lower acetylation at both loci (Figure 7a).

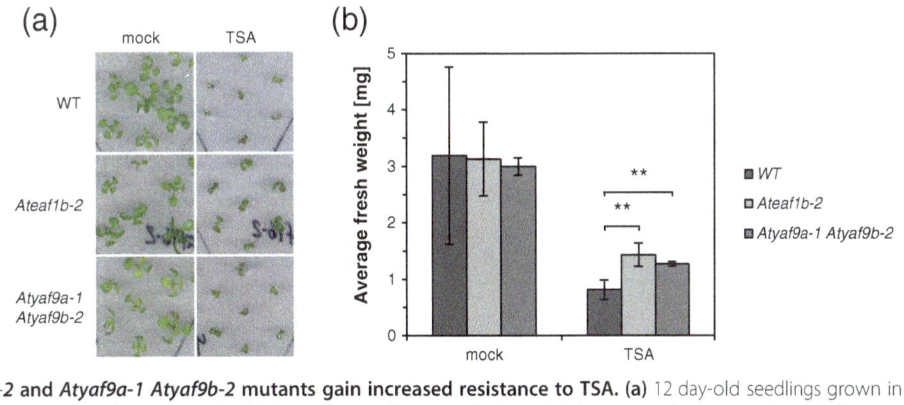

Figure 4 ***Ateaf1b-2*** **and** ***Atyaf9a-1 Atyaf9b-2*** **mutants gain increased resistance to TSA.** **(a)** 12 day-old seedlings grown in the presence of TSA or mock. All images are in the same scale. **(b)** Comparison of average fresh weight between plants treated with TSA or mock. Error bars represent standard deviation of 4 biological replicates. Double asterisks indicate a p-value < 0.01 (t-test).

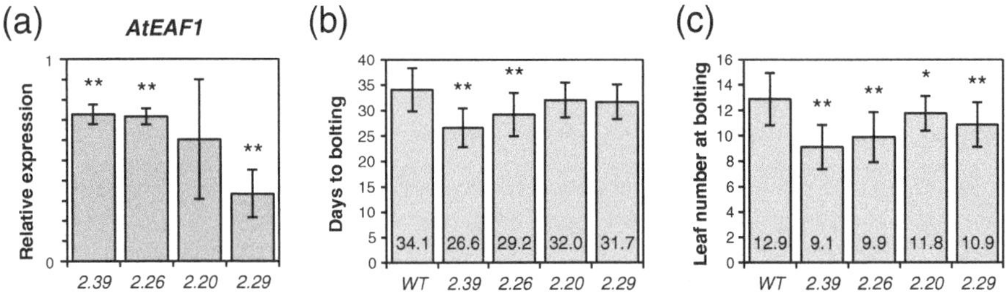

Figure 5 Plants with transcript levels of *AtEAF1* decreased by artificial miRNA show deregulation of flowering time. (a) Expression levels of *AtEAF1* in four independent amiRNA lines relative to WT Col-0. **(b)** Comparison of flowering time represented by an average number of days until the stage where the flower stem is 1 cm long. **(c)** Comparison of flowering time represented by an average number of true rosette leaves at the stage where the flower stem is 1 cm long. Asterisks in **(a, b, c)** indicate a p-value < 0.05 or p-value < 0.01 (double asterisk) (t-test).

Examination of the transcript levels of *FLC*, *FT*, *CO* and *SOC1* genes by RT-qPCR showed their transcriptional deregulation. As expected, in many cases reduction in H4K5 acetylation was accompanied by a decreased level of a given transcript (Figure 6c, Figure 7b). In several cases, however, the observed decrease in acetylation did not result in lower transcript levels. In fact, we observed higher relative levels of *FT* transcript in the *Atyaf9a-1 Atyaf9b-2* line, despite a significant decrease of H4K5 acetylation in the promoter region of *FT* (Figure 6b,c). The lack of clear correlation between H4K5 acetylation and transcriptional activity of tested genes may be at least partially explained by their involvement in a network of functional interactions involving other mechanisms of transcriptional regulation.

Discussion

In this study, we have investigated the role of a previously uncharacterized protein AtEAF1 as a potential subunit of the plant NuA4 histone acetylatransferase complex. We found that AtEAF1 and other Arabidopsis homologues of NuA4 subunits copurify with Arabidopsis homologues of ARP4 and SWC4, common subunits of the yeast NuA4 and SWR1 complexes. Our AtARP4 and AtSWC4 purifications resulted also in detection of peptides belonging to the subunits of the Arabidopsis SWR1 complex including PIE1, which recruits subunits responsible for H2A.Z deposition in Arabidopsis [26,27]. So far, PIE1 has been the only known plant protein with a potential to physically link AtARP4 and AtSWC4 with homologues of other SWR1-C and NuA4 subunits. We argue that the identification of AtEAF1 opens a possibility for a PIE1-independent NuA4 complex formation in plants.

The observed physical association of AtEAF1 with AtAPR4 and AtSWC4 is best explained by the presence of the HSA domain in AtEAF1. HSA domain is a common feature of the platform subunits of SWR1-C, NuA4 and the hybrid complex TIP60-p400. Several published studies suggest that HSA domain may provide the assembly surface for a submodule consisting of ARP4, SWC4, YAF9 and monomeric actin in NuA4 and SWR1 complexes [5,28,29]. Indeed, by coimmunoprecipitation we demonstrated a physical interaction between both Arabidopsis YAF9 homologues and a fragment of AtEAF1 containing the HSA domain (Figure 2a). Formation of a protein module by Arabidopsis homologues of ARP4, SWC4 and YAF9 is further supported by the interaction of AtSWC4 with AtYAF9A and AtYAF9B, as revealed by BiFC (Figure 2b,c), and by reciprocal copurification of AtARP4 and AtSWC4, as shown by AP-MS/MS (Table 1). Interestingly, we were not able to detect the expected interaction between YAF9 homologues and the fragment of PIE1 containing the HSA domain in our CoIP assay (Figure 2a). As YAF9 is necessary for H2A.Z deposition in yeast [5,30], it is assumed to be a subunit of the Arabidopsis SWR1-C. Although our CoIP result alone is not sufficient to question this view, it seems to agree with a relatively mild phenotype of the *Atyaf9a-1 Atyaf9b-2* mutant (Figure 3) as compared to the *pie1-5* or *arp6-1* mutant phenotypes [11].

The other conserved sequence feature of AtEAF1 is the SANT domain, located C-terminal of the HSA domain. We found that the combination of HSA and SANT domains is usually represented by no more than two genes per eukaryotic genome (Figure 1c). For example, in both *S. cerevisiae* and human there is just single gene that encodes an HSA-SANT domain protein, i.e. EAF1 and p400, respectively. Importantly, AtEAF1, unlike p400 or PIE1, does not contain an ATPase domain between the HSA and SANT domains. This characteristic leaves EAF1 as the most probable functional analogue of AtEAF1.

Physical interaction of AtEAF1 with AtYAF9A and AtYAF9B is consistent with our observation that plants carrying a T-DNA insertion in one of the *AtEAF1* genes are phenotypically similar to *Atyaf9a-1* mutants. It has been shown recently that *Atyaf9a-1* mutants display reduced levels of H4 acetylation in *FLC* gene chromatin [16], which leads to a decrease in *FLC* transcript levels and, consequently, partial loss of flowering inhibition. Although the double mutant *Ateaf1a Ateaf1b* is not currently available,

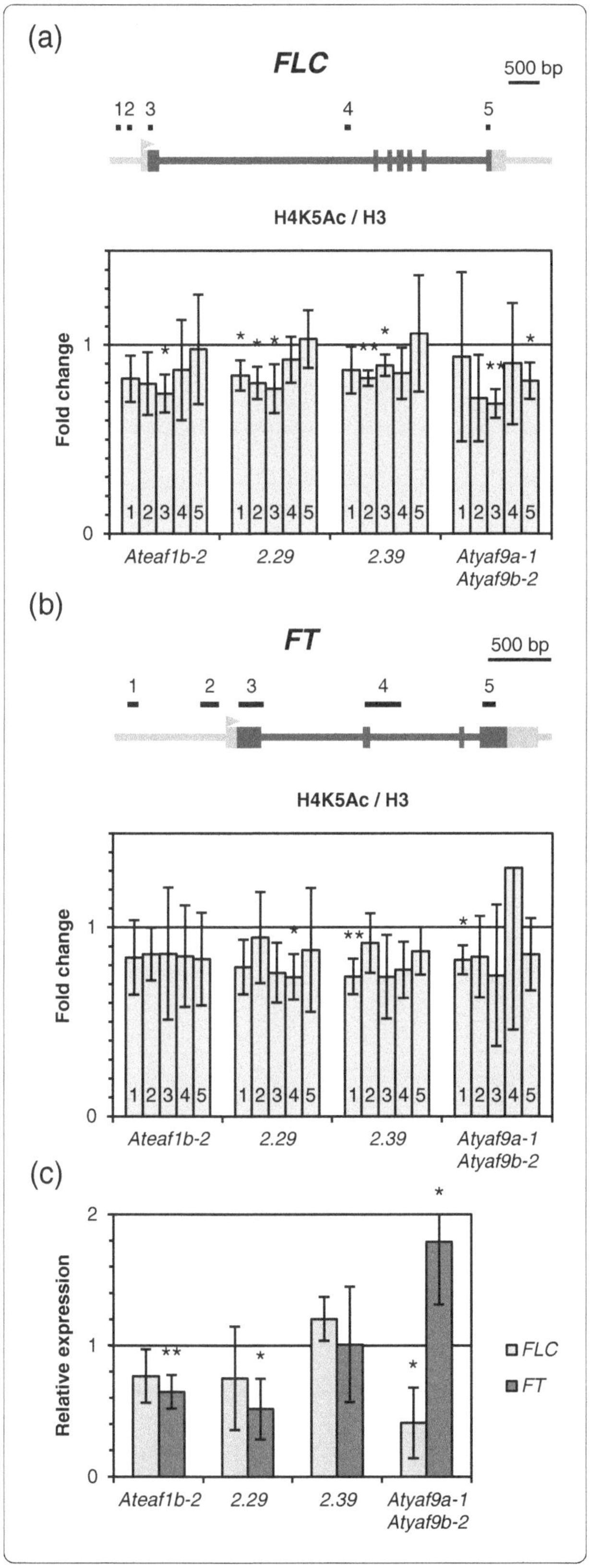

Figure 6 ***Ateaf1b-2*****,** ***Atyaf9a-1 Atyaf9b-2*** **and** ***AtEAF1*****-amiRNA lines display reduced H4K5 acetylation but different activity of** ***FLC*** **and** ***FT*****.** **(a, b)** Acetylation levels in different parts of the *FLC* and *FT* genes normalized to H3 presented as fold change over WT Col-0. **(c)** Expression levels of *FLC* and *FT* relative to WT Col-0. Asterisks in **(a, b, c)** indicate a p-value < 0.05 or p-value < 0.01 (double asterisk) (t-test).

our results show that the single mutant *Ateaf1b-2* is affected in *FLC* expression and flowering time under LD and SD conditions (Figure 3), phenocopying *Atyaf9a-1*.

We demonstrated that AtEAF1 and SWC4 interact with both Arabidopsis homologues of YAF9. If the influence of AtYAF9A on *FLC* expression results from its interaction with AtEAF1, the same could be true for AtYAF9B. Functional redundancy of AtYAF9A and AtYAF9B has been suggested previously on the basis of the phenotype of *Atyaf9a-1 Atyaf9b-kd* plants, which differs from the phenotype of either *Atyaf9a-1* or *Atyaf9b-kd* plants [21]. Our own data show that a double mutant, carrying T-DNA insertions in both YAF9 genes, displays stronger deregulation of flowering time control than either of the single mutants (Figure 3). This result suggests some level of functional redundancy between the two genes in the Arabidopsis NuA4 complex and is in agreement with our finding that both proteins interact with AtEAF1B.

Our analyses of the influence of the histone deacetylase inhibitor TSA revealed an increased resistance to this hyperacetylation-inducing agent in *Atyaf9a-1 Atyaf9b-2* and *Ateaf1b-2* mutant seedlings. This result supports a role for AtYAF9A, AtYA9B and AtEAF1B in histone acetylation. Under TSA treatment *Atyaf9a-1 Atyaf9b-2* accumulated less acetylated H4K5 than WT plants (Additional file 7) which may explain its increased resistance to the drug. No such effect was observed for *Ateafb-2* mutant. Several explanations of this result are possible. We chose to follow a hypothesis that AtEAF1 is only required for H4K5 acetylation in specific genomic targets, similar to yeast EAF1 which is mainly required for the NuA4 activity in the promoter regions [31].

As we were not able to determine which genes may be relevant to the negative effect of TSA on plant growth, we used genes encoding main flowering regulators as models to study the role of AtEAF1 in Arabidopsis NuA4. ChIP experiment, in which we employed artificial miRNA knock-down lines for *AtEAF1* gene showed that decreased levels of *AtEAF1* transcript have similar effect on histone H4K5 acetylation as the disruption of one of the AtEAF1 isoforms in the *Ateaf1b-2* mutant (Figures 6 and 7). As expected, we observed a general decrease in H4K5 acetylation levels over most of the tested regions of *FLC* and *FT* loci with a slight bias towards their 5′ end (Figure 6a,b). We could also detect significant decrease in acetylation in the promoter of the *CO* gene and a weaker effect in the *SOC1* gene (Figure 7a).

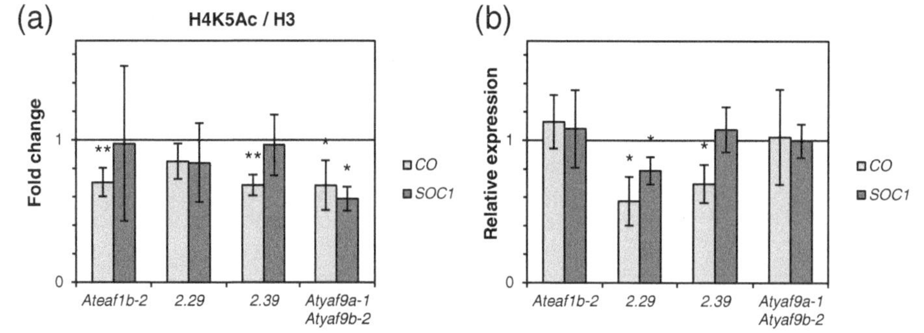

Figure 7 *CO* is more affected than *SOC1* in *Ateaf1b-2* and *AtEAF1*-amiRNA lines. (a) H4K5 acetylation levels in the promoters of *CO* and *SOC1* normalized to H3 presented as fold change over WT Col-0. **(b)** Expression levels of *CO* and *SOC1* relative to WT Col-0. Asterisks in **(a, b)** indicate a p-value < 0.05 or p-value < 0.01 (double asterisk) (t-test).

The gene-specific effects which we observed at the level of histone acetylation were not evenly reflected in the transcript levels of the studied genes. This can be at least in part attributed to the functional interactions between *FLC, FT, CO, SOC1* and multiple other genes involved in the control of flowering transition. Partial loss of NuA4 activity likely affects many genes with diverse functions. Put to such a stress, the system of positive and negative flowering regulators may seek new balance with outcomes that are difficult to predict. Another possible explanation of these results comes from a recent study on plant homologues of the NuA4 subunit EAF3, i.e. MRG1 and MRG2. In contrast to previous reports showing that Arabidopsis homologues of NuA4 subunits YAF9 and ESA1 contribute to flowering time control mainly by ensuring proper levels of the negative flowering regulator FLC, Xu et al. demonstrated that transcriptional activity of the positive flowering regulator *FT* also depends on H4K5 acetylation [17]. According to the report, MRG proteins act as H3K36me3 readers and guide HAM1 and HAM2 acetyltraferases through a direct protein-protein interaction. This does not exclude a role for other NuA4 subunits as in the case of the yeast NuA4 complex or human TIP60-p400 which incorporate MRG homologues as stable subunits [12,32]. In fact, MRG1 was among the NuA4 subunits identified as AtSWC4 interaction partners in our AP-MS/MS analyses (Table 1, Additional file 1). Conservation of the AtEAF1 role as a bridge between functional submodules of NuA4 containing YAF9 and EAF3 homologues would explain the ambivalent effects of its partial loss that we observe.

Conclusions

Our work introduces AtEAF1 as a new subunit of the Arabidopsis NuA4 complex. Characteristic sequence features, interaction with the AtYAF9-AtSWC4-AtARP4 submodule and influence on H4K5 acetylation suggest that AtEAF1 may be a functional analog of the yeast EAF1 protein. Judging from previous studies on human and yeast models, elucidation of the exact role of AtEAF1 subunit in the structural integrity and function of the Arabidopsis NuA4 complex will require significant effort. Our data do not exclude the existence of another large HAT complex in Arabidopsis, organized around PIE1, with additional H2A.Z exchange activity. Addressing this problem may also be required to better understand the regulation of histone H4 acetylation in plants.

Methods

Plant material

The *Ateaf1b-2* (SALK_067053), *Atyaf9a-1* (SALK_106430) and *Atyaf9b-2* (SALK_046223) mutants were obtained from the T-DNA mutant collection at Salk Institute. The seeds were provided by Nottingham Arabidopsis Stock Centre (NASC). Positions of T-DNA insertions were confirmed by Sanger sequencing of cloned PCR amplification products (for primer sequences see Additional file 2). The double mutant *Atyaf9a-1 Atyaf9b-2* was obtained by crossing the homozygous *Atyaf9a-1* and *Atyaf9b-2* mutants. WT Columbia-0 ecotype (Col-0) plants were used as a control in all experiments.

Protoplast preparation and transfection

Protoplasts were prepared from 30-day old plants grown under LD conditions. The Tape-Arabidopsis Sandwich method was used for protoplast preparation [33]. Transfection of protoplasts was carried out in U96 Microwell plates (Thermo Scientific Nunc) according to a published protocol [34].

Coimmunoprecipitation (CoIP)

For each CoIP, 12 wells in the U96 Microwell plate were used to cotransfect protoplasts with plasmids carrying HSA-Flag and YAF9-nEYFP constructs. Isolation of protoplast proteins and CoIP was performed as previously described [35], except that the immunoprecipitation (IP) lysis buffer was modified by replacing HEPES with 10 mM Tris-HCl, pH 7.5. The same buffer was used to wash beads after IP. Immunodetection after western blot was done with anti-GFP antibodies (sc-8334, Santa Cruz, www.scbt.com) and anti-Flag antibodies (OctA-Probe Antibody D-8, sc-807, Santa Cruz). All CoIPs were carried out at least twice.

Affinity purification followed by tandem mass spectrometry (AP-MS/MS)

Expression cassettes containing the coding sequences of the One-STrEP-tag:ATARP4 or One-STrEP-tag:ATSWC4 fusion proteins under the control of the Arabidopsis *UBQ10* promoter and *NOS* terminator were ligated into the binary vector pART27 [36]. *Agrobacterium tumefaciens* (strain GV3101) mediated transformations were carried out to stably express recombinant proteins in the Arabidopsis suspension-cultured cells (T87 line).

To purify AtARP4 or AtSWC4 complexes, 320 ml (~24 g of fresh weight) of a 10-day-old T87 culture were vacuum filtered, ground in liquid nitrogen and resuspended in 5 ml of extraction buffer H (200 mM NaCl, 200 mM Tris-HCl pH 8.0, 200 mM NaF, 15% glycerol, 0.3 mM EDTA, 0.5% TritonX-100, 0.8 mM PMSF, 4 mM DTT, 3.2 mM NaV_2O_3,) or L (150 mM NaCl, 100 mM Tris-HCl pH 8.0, 100 mM NaF, 10% glycerol, 2.5 mM EDTA, 2.5 mM EGTA, 0.4% Triton X-100, 0.8 μM PMSF, 4 mM DTT, 3.2 mM NaV_2O_3) with one tablet of cOmplete EDTA-free Protease Inhibitor Cocktail (Roche, www.roche.com). After centrifugation (15000 *g*, 15 min, 4°C and then 180 000 *g*, 90 min, 4°C) supernatant was loaded onto a column containing 600 μl of Strep-Tactin Superflow High Capacity resin (IBA). After seven washing steps (100 mM Tris-HCl pH 8.0, 150 mM NaCl), bound proteins were eluted with 1.8 ml of 2.5 mM desthiobiotin (IBA) and finally concentrated and desalted using Amicon Ultra 10 K (Millipore, www.millipore.com).

Trypsin–digested peptides were analyzed using Orbitrap Velos (Thermo Scientific, www.thermo.com) coupled with nanoAcquity UPLC (Waters, www.waters.com) according to standard protocols at the Mass Spectrometry Laboratory, Institute of Biochemistry and Biophysics, Polish Academy of Sciences, Warsaw. To identify proteins, Mascot software (Matrix Science, www.matrixscience.com) was employed to search mass spectra against the TAIR10 database using the following parameters: no missed cleavages allowed, 20 ppm for peptide mass tolerance and 0.6 Da for fragment ion mass tolerance, fixed modification – cysteine carbamidomethylation, variable modification – methionine oxidation. Results were filtered using significance threshold p-value < 0.05 and expect cut-off 0.05 (http://www.matrixscience.com/help/scoring_help.html).

A. thaliana T87 suspension-cultured cells were grown at 22°C under continuous light (50 μmol m^{-2} s^{-1}) and shaking at 120 rpm in Gamborg's B5 medium [37] with 1.5% sucrose and 0.1 μg/l 2,4-dichlorophenoxyacetic acid (Sigma-Aldrich, www.sigmaaldrich.com).

Bimolecular fluorescence complementation (BiFC)

For BiFC, single wells in the U96 Microwell plate were used to transfect protoplasts with plasmids encoding proteins of interest fused to the N-terminus of nEYFP and cEYFP. Empty vector (3 μg) encoding Enhanced Cyan Fluorescent Protein (ECFP) was added to each well as a transfection control. After transfection, protoplasts resuspended in 200 μl of W5 solution with 5 mM glucose [33] were transferred to black 96-well black glass-bottom plates (SensoPlate, Greiner Bio-One, www.greinerbioone.com), sealed with transparent sealing tape (Thermo Scientific Nunc, cat. no. 236366), and returned to the growth chamber for overnight incubation. After incubation, the sealing tape was removed and the plate was screened with a Nikon A1Rsi confocal system (www.nikon.com). See Additional file 8 for original files.

Gene expression analyses

RNA was extracted using TRI Reagent (Sigma-Aldrich). For the analysis of *FLC* expression, RNA was extracted from 10- and 16-day old seedlings, grown in plates under LD (16 hours of light) or SD (8 hours of light) conditions, respectively. For the analysis of relative expression levels of AtEAF1A and AtEAF1B, RNA was extracted from mature rosette leaves of Col-0 WT plants grown under LD conditions. The gene expression data presented in Figures 6 and 7 was obtained from plants grown and harvested in the same way as plants grown for ChIP analyses. Reverse transcription was done with the RevertAid kit (Thermo Scientific). Relative expression levels of *FLC* were measured by quantitative PCR. *UBQ10* (Figures 3 and 5) and *UBC21* (Figures 6 and 7) gene transcripts were used as a reference. The $2^{-\Delta\Delta Ct}$ method [38] was used to quantify the relative transcript levels in all experiments. Sequences of all primers can be found in Additional file 2.

Preparation of genetic constructs

Coding sequences of AtYAF9B, AtEAF1, PIE1 and INO80 were cloned from cDNA obtained from WT Col-0 plants. The remaining coding sequences were subcloned from pUNI51 constructs from the FL-cDNA collection at the Salk Institute, provided by Arabidopsis Biological Resource Center (ABRC). All transient expression vectors used in this study are based on pSAT vector series [39]. The Flag vectors were constructed by replacing the nEYFP sequence with Flag sequence. All vectors were modified by inserting two *Sfi*I recognition sites into the multiple cloning site to facilitate transfer of coding sequences between vectors. Sequences of all primers can be found in Additional file 2.

Sequence analyses

Conserved protein domains were found using InterProScan [40]. Domain relatives were identified using CDART [41] and by sequence similarity search using BLAST (blast.ncbi.nlm.nih.gov). All multiple sequence alignments were performed using the T-Coffee program [42] accessed through EMBL-EBI web services (www.ebi.ac.uk). For visualization of sequence alignment data, Geneious version 6.0 was used (Biomatters, www.geneious.com).

Trichostatin A treatment

Plants were grown in plates containing MS medium supplemented with 1% sucrose with pH stabilized by MES at 5.7. DMSO solution of Trichostatin A (Sigma-Aldrich) was added to the medium before pouring the plates to the final concentration of 12.5 μM. Equal amount of DMSO was added as a mock. Seeds were surface-sterilized before sawing and kept in 4°C for four days. After stratification, plates were transferred to a growth chamber with long day conditions (16 h of light). The measurements were done on 12-day old seedlings.

Artificial micro RNA lines

Artificial micro RNA transgene was designed with the WMD3 tool (wmd3.weigelworld.org) and constructed according to the instructions available at the same website. Briefly, the target-specific sites of the Arabidopsis miRNA precursor MIR319a cloned into pRS300 vector were mutated through overlap extension PCR to target a sequence identical in both *AtEAF1* genes. pRS300 was a gift from Detlef Weigel (Addgene plasmid # 22846). After confirmation of the mutations by DNA sequencing, the precursor was placed downstream of the *UBQ10* promoter in a binary vector using standard molecular biology procedures. The plasmid was then transferred into *Agrobacterium tumefaciens* strain GV3101. *A. thaliana* Col-0 plants were transformed using floral deep method [43]. Transgenic plants were selected on soil by spraying with diluted BASTA (Bayer). Presence of the amiRNA construct was confirmed by PCR. Progeny of the BASTA resistant plants of the T1 and T2 plants was used in subsequent experiments. Sequences of all primers can be found in Additional file 2.

Chromatin immunoprecipitation (ChIP)

About 350 mg of 10-day-old seedlings grown in LD conditions were harvested at the end of the light photoperiod, ground in liquid nitrogen and then fixed for 15 min at room temperature in extraction buffer containing 1% formaldehyde (60 mM Hepes pH 8.0, 1 M sucrose, 5 mM KCl, 5 mM $MgCL_2$, 5 mM EDTA, 0.6% Triton X-100, 1% formaldehyde, 1 mM PMSF, 1% Protease Inhibitor Cocktail, Sigma). The reaction was quenched by adding glycine to a final concentration of 100 mM. The homogenate was filtered through Miracloth, centrifuged and the pellet was washed with buffer containing 0.25 M sucrose, 10 mM Tris-HCl pH 8.0, 10 mM MgCl2, 1% Triton X-100, 1 mM EDTA, 5 mM beta-mercaptoethanol, 1 mM PMSF and Protease Inhibitor Cocktail. Then, the pellet was resuspended in the sonication buffer (50 mM Tris-HCL pH 8.0, 10 mM EDTA, 0.5% SDS, 1 mM PMSF and Protease Inhibitor Cocktail) and sonicated using Bioruptor (Diagenode) for 25 cycles of 30 sec on and 30 sec off. Sonicated chromatin was diluted 10-fold and incubated with antibodies (anti-H3, Abcam, ab1791; anti-H4K5ac, Millipore 07-327) overnight at 4°C with gentle agitation. After pre-clearing, magnetic protein A-beads (Dynabeads protein A, Life Sciences) were incubated with the antibodies-chromatin mix for 1 hours. The slurry was then washed and DNA was extracted with 10% Chelex (Biorad) as described previously [44].

All ChIP experiments were carried out with three independent biological replicates and quantified by qPCR (Maxima SybrGreen/ROX, Thermo Scientific). Primer sequences used for the ChIP-qPCR are listed in Additional file 2. The ChIP data are presented as the ratio of percent of input for H4K5ac to percent of input for total H3 normalized to WT levels. Error bars correspond to SD of the mean of three biological replicates. Statistical evaluations were performed using a Student's t-test.

Additional files

Additional file 1: AP-MS/MS data.

Additional file 2: Oligonucleotides used in this study.

Additional file 3: DNA and protein sequences, alignments.

Additional file 4: Gene structures of *AtYAF9A*, *AtYAF9B* and *AtEAF1B* with positions of T-DNA insertions in the mutant lines marked. Details of the amiRNA design and phenotypes of the silenced lines. Relative expression levels of the *AtEAF1A* and *AtEAF1B* genes.

Additional file 5: Uncropped scans of the Western blot films.

Additional file 6: BiFC screening for protein-protein interactions between AtARP4, AtSWC4, AtYAF9A and AtYAF9B.

Additional file 7: Western Blot analyses of bulk H4K5 acetylation levels in mutants and WT plants treated or untreated with TSA.

Additional file 8: Original confocal images (.nd2 format requires free viewer available at www.nikoninstruments.com/Products/Software/NIS-Elements-Viewer).

Competing interests

The authors declare that they have no competing interests.

Authors' contributions

TB drafted the manuscript, developed all main concepts of the work, designed or participated in designing and data analysis of all experiments, performed all sequence analyses, prepared all genetic constructs except those used in the AP-MS/MS, performed BiFC and TSA assays, participated in CoIP, RT-qPCR and phenotypic analyses. LG designed and performed the AP-MS/MS experiments and provided crucial ideas for the modification of the pSAT vectors. AB participated in designing the CoIP experiment and carried out all work on the proteins overexpressed in protoplasts. MA obtained the double mutant *Atyaf9a-1 Atyaf9b-2*, performed major part of the phenotypic analyses and participated in the RT-qPCR analyses.

WS participated in the phenotypic analyses, designed and carried out the ChIP experiments. AL participated in the data analysis and discussions. PAZ and JS provided the conceptual background and supported TB in all experimental designs, interpretation of the results and preparation of the manuscript. All authors read and approved the final manuscript.

Acknowledgements

This work was supported by the Polish National Science Center NCN grants (NCN UMO-2011/01/B/NZ2/01691 to J.S. and N N303 313437 to P.A.Z.) and KNOW RNA Research Centre in Poznan, 01/KNOW2/2014. The PhD fellowship of T.B. is part of the International PhD Programme 'From genome to phenotype: A multidisciplinary approach to functional genomics' (MPD/2010/3) of the Foundation for Polish Science (FNP), cofinanced from European Union, Regional Development Fund (Innovative Economy Operational Programme 2007-2013). T.B. would like to thank Jerzy Paszkowski and Marco Catoni for useful suggestions and Daniel Kierzkowski for the introduction to the protoplast expression system and BiFC.

Author details

[1]Department of Biotechnology, Institute of Molecular Biology and Biotechnology, Adam Mickiewicz University, Umultowska 89, 61-614 Poznań, Poland. [2]Department of Molecular Virology, Institute of Experimental Biology, Adam Mickiewicz University, Umultowska 89, 61-614 Poznań, Poland. [3]Department of Plant Sciences, University of Cambridge, Downing Street, CB2 3EA, Cambridge, UK.

References

1. Cairns BR, Schlichter A, Erdjument-Bromage H, Tempst P, Kornberg RD, Winston F. Two functionally distinct forms of the RSC nucleosome-remodeling complex, containing essential at hook, BAH, and bromodomains. Mol Cell. 1999;4:715–23.
2. Chandy M, Gutiérrez JL, Prochasson P, Workman JL. SWI/SNF Displaces SAGA-Acetylated Nucleosomes. Eukaryot Cell. 2006;5:1738–47.
3. Papamichos-Chronakis M, Watanabe S, Rando OJ, Peterson CL. Global Regulation of H2A.Z Localization by the INO80 Chromatin-Remodeling Enzyme Is Essential for Genome Integrity. Cell. 2011;144:200–13.
4. Altaf M, Auger A, Monnet-Saksouk J, Brodeur J, Piquet S, Cramet M, et al. NuA4-dependent Acetylation of Nucleosomal Histones H4 and H2A Directly Stimulates Incorporation of H2A.Z by the SWR1 Complex. J Biol Chem. 2010;285:15966–77.
5. Auger A, Galarneau L, Altaf M, Nourani A, Doyon Y, Utley RT, et al. Eaf1 is the platform for NuA4 molecular assembly that evolutionarily links chromatin acetylation to ATP-dependent exchange of histone H2A variants. Mol Cell Biol. 2008;28:2257–70.
6. Mizuguchi G, Shen X, Landry J, Wu W-H, Sen S, Wu C. ATP-driven exchange of histone H2AZ variant catalyzed by SWR1 chromatin remodeling complex. Science. 2004;303:343–8.
7. Wu W-H, Wu C-H, Ladurner A, Mizuguchi G, Wei D, Xiao H, et al. N Terminus of Swr1 binds to Histone H2AZ and provides a platform for subunit assembly in the chromatin remodeling complex. J Biol Chem. 2009;284:6200–7.
8. Gévry N, Chan HM, Laflamme L, Livingston DM, Gaudreau L. p21 transcription is regulated by differential localization of histone H2A.Z. Genes Dev. 2007;21:1869–81.
9. Park JH, Sun X-J, Roeder RG. The SANT Domain of p400 ATPase Represses Acetyltransferase Activity and Coactivator Function of TIP60 in Basal p21 Gene Expression. Mol Cell Biol. 2010;30:2750–61.
10. Noh Y-S, Amasino RM. PIE1, an ISWI Family Gene, Is Required for FLC Activation and Floral Repression in Arabidopsis. Plant Cell Online. 2003;15:1671–82.
11. Deal RB, Topp CN, McKinney EC, Meagher RB. Repression of flowering in Arabidopsis requires activation of FLOWERING LOCUS C expression by the histone variant H2A.Z. Plant Cell. 2007;19:74–83.
12. Doyon Y, Selleck W, Lane WS, Tan S, Côté J. Structural and functional conservation of the NuA4 histone acetyltransferase complex from yeast to humans. Mol Cell Biol. 2004;24:1884–96.
13. Earley KW, Shook MS, Brower-Toland B, Hicks L, Pikaard CS. In vitro specificities of Arabidopsis co-activator histone acetyltransferases: implications for histone hyperacetylation in gene activation. Plant J. 2007;52:615–26.
14. Latrasse D, Benhamed M, Henry Y, Domenichini S, Kim W, Zhou D-X, et al. The MYST histone acetyltransferases are essential for gametophyte development in Arabidopsis. BMC Plant Biol. 2008;8:121.
15. Xiao J, Zhang H, Xing L, Xu S, Liu H, Chong K, et al. Requirement of histone acetyltransferases HAM1 and HAM2 for epigenetic modification of FLC in regulating flowering in Arabidopsis. J Plant Physiol. 2013;170:444–51.
16. Zacharaki V, Benhamed M, Poulios S, Latrasse D, Papoutsoglou P, Delarue M, et al. The Arabidopsis ortholog of the YEATS domain containing protein YAF9a regulates flowering by controlling H4 acetylation levels at the FLC locus. Plant Sci. 2012;196:44–52.
17. Xu Y, Gan E-S, Zhou J, Wee W-Y, Zhang X, Ito T. Arabidopsis MRG domain proteins bridge two histone modifications to elevate expression of flowering genes. Nucleic Acids Res. 2014;42:10960–74.
18. Shen X, Mizuguchi G, Hamiche A, Wu C. A chromatin remodelling complex involved in transcription and DNA processing. Nature. 2000;406:541–4.
19. Jin J, Cai Y, Yao T, Gottschalk AJ, Florens L, Swanson SK, et al. A mammalian chromatin remodeling complex with similarities to the yeast INO80 complex. J Biol Chem. 2005;280:41207–12.
20. Vercruyssen L, Verkest A, Gonzalez N, Heyndrickx KS, Eeckhout D, Han S-K, et al. ANGUSTIFOLIA3 binds to SWI/SNF chromatin remodeling complexes to regulate transcription during Arabidopsis leaf development. Plant Cell. 2014;26:210–29.
21. Choi K, Kim J, Hwang H-J, Kim S, Park C, Kim SY, et al. The FRIGIDA Complex Activates Transcription of FLC, a Strong Flowering Repressor in Arabidopsis, by Recruiting Chromatin Modification Factors. Plant Cell Online. 2011;23:289–303.
22. Bittner CB, Zeisig DT, Zeisig BB, Slany RK. Direct Physical and Functional Interaction of the NuA4 Complex Components Yaf9p and Swc4p. Eukaryot Cell. 2004;3:976–83.
23. Yang F, Zhang L, Li J, Huang J, Wen R, Ma L, et al. Trichostatin A and 5-azacytidine both cause an increase in global histone H4 acetylation and a decrease in global DNA and H3K9 methylation during mitosis in maize. BMC Plant Biol. 2010;10:178.
24. Tanaka M, Kikuchi A, Kamada H. The Arabidopsis Histone Deacetylases HDA6 and HDA19 Contribute to the Repression of Embryonic Properties after Germination. Plant Physiol. 2008;146:149–61.
25. Schwab R, Ossowski S, Riester M, Warthmann N, Weigel D. Highly Specific Gene Silencing by Artificial MicroRNAs in Arabidopsis. Plant Cell Online. 2006;18:1121–33.
26. Choi K, Park C, Lee J, Oh M, Noh B, Lee I. Arabidopsis homologues of components of the SWR1 complex regulate flowering and plant. Development. 2007;134:1931–41.
27. March-Díaz R, García-Domínguez M, Florencio FJ, Reyes JC. SEF, a New Protein Required for Flowering Repression in Arabidopsis, Interacts with PIE1 and ARP6. Plant Physiol. 2007;143:893–901.
28. Szerlong H, Hinata K, Viswanathan R, Erdjument-Bromage H, Tempst P, Cairns BR. The HSA domain binds nuclear actin-related proteins to regulate chromatin-remodeling ATPases. Nat Struct Mol Biol. 2008;15:469–76.
29. Nguyen VQ, Ranjan A, Stengel F, Wei D, Aebersold R, Wu C, et al. Molecular architecture of the ATP-dependent chromatin-remodeling complex SWR1. Cell. 2013;154:1220–31.
30. Zhang H, Richardson DO, Roberts DN, Utley R, Erdjument-Bromage H, Tempst P, et al. The Yaf9 component of the SWR1 and NuA4 complexes is required for proper gene expression, histone H4 acetylation, and Htz1 replacement near telomeres. Mol Cell Biol. 2004;24:9424–36.
31. Ginsburg DS, Govind CK, Hinnebusch AG. NuA4 lysine acetyltransferase esa1 is targeted to coding regions and stimulates transcription elongation with Gcn5. Mol Cell Biol. 2009;29:6473–87.
32. Eisen A, Utley RT, Nourani A, Allard S, Schmidt P, Lane WS, et al. The yeast NuA4 and drosophila MSL complexes contain homologous subunits important for transcription regulation. J Biol Chem. 2001;276:3484–91.
33. Wu F-H, Shen S-C, Lee L-Y, Lee S-H, Chan M-T, Lin C-S. Tape-Arabidopsis sandwich - a simpler Arabidopsis protoplast isolation method. Plant Methods. 2009;5:16.
34. Fujikawa Y, Kato N. TECHNICAL ADVANCE: split luciferase complementation assay to study protein–protein interactions in Arabidopsis protoplasts. Plant J. 2007;52:185–95.
35. Li J-F, Bush J, Xiong Y, Li L, McCormack M. Large-Scale Protein-Protein Interaction Analysis in Arabidopsis Mesophyll Protoplasts by Split Firefly Luciferase Complementation. PLoS ONE. 2011;6
36. Gleave AP. A versatile binary vector system with a T-DNA organisational structure conducive to efficient integration of cloned DNA into the plant genome. Plant Mol Biol. 1992;20:1203–7.

37. Yamada H, Koizumi N, Nakamichi N, Kiba T, Yamashino T, Mizuno T. Rapid response of Arabidopsis T87 cultured cells to cytokinin through his-to-Asp phosphorelay signal transduction. Biosci Biotechnol Biochem. 2004;68:1966–76.
38. Livak KJ, Schmittgen TD. Analysis of relative gene expression data using real-time quantitative PCR and the 2 − ΔΔCT method. Methods. 2001;25:402–8.
39. Tzfira T, Tian G-W, Lacroix B, Vyas S, Li J, Leitner-Dagan Y, et al. pSAT vectors: a modular series of plasmids for autofluorescent protein tagging and expression of multiple genes in plants. Plant Mol Biol. 2005;57:503–16.
40. Quevillon E, Silventoinen V, Pillai S, Harte N, Mulder N, Apweiler R, et al. InterProScan: protein domains identifier. Nucleic Acids Res. 2005;33 suppl 2:W116–20.
41. Geer LY, Domrachev M, Lipman DJ, Bryant SH. CDART: protein homology by domain architecture. Genome Res. 2002;12:1619–23.
42. Notredame C, Higgins DG, Heringa J. T-coffee: a novel method for fast and accurate multiple sequence alignment. J Mol Biol. 2000;302:205–17.
43. Clough SJ, Bent AF. Floral dip: a simplified method forAgrobacterium-mediated transformation ofArabidopsis thaliana. Plant J. 1998;16:735–43.
44. Wierzbicki AT, Haag JR, Pikaard CS. Noncoding transcription by RNA Polymerase Pol IVb/Pol V mediates transcriptional silencing of overlapping and adjacent genes. Cell. 2008;135:635–48.

Comparative proteome analysis of embryo and endosperm reveals central differential expression proteins involved in wheat seed germination

Miao He[1†], Chong Zhu[1†], Kun Dong[1†], Ting Zhang[1†], Zhiwei Cheng[1], Jiarui Li[2] and Yueming Yan[1,3*]

Abstract

Background: Wheat seeds provide a staple food and an important protein source for the world's population. Seed germination is vital to wheat growth and development and directly affects grain yield and quality. In this study, we performed the first comparative proteomic analysis of wheat embryo and endosperm during seed germination.

Results: The proteomic changes in embryo and endosperm during the four different seed germination stages of elite Chinese bread wheat cultivar Zhengmai 9023 were first investigated. In total, 74 and 34 differentially expressed protein (DEP) spots representing 63 and 26 unique proteins were identified in embryo and endosperm, respectively. Eight common DEP were present in both tissues, and 55 and 18 DEP were specific to embryo and endosperm, respectively. These identified DEP spots could be sorted into 13 functional groups, in which the main group was involved in different metabolism pathways, particularly in the reserves necessary for mobilization in preparation for seed germination. The DEPs from the embryo were mainly related to carbohydrate metabolism, proteometabolism, amino acid metabolism, nucleic acid metabolism, and stress-related proteins, whereas those from the endosperm were mainly involved in protein storage, carbohydrate metabolism, inhibitors, stress response, and protein synthesis. During seed germination, both embryo and endosperm had a basic pattern of oxygen consumption, so the proteins related to respiration and energy metabolism were up-regulated or down-regulated along with respiration of wheat seeds. When germination was complete, most storage proteins from the endosperm began to be mobilized, but only a small amount was degraded during germination. Transcription expression of six representative DEP genes at the mRNA level was consistent with their protein expression changes.

Conclusion: Wheat seed germination is a complex process with imbibition, stirring, and germination stages, which involve a series of physiological, morphological, and proteomic changes. The first process is a rapid water uptake, in which the seed coat becomes softer and the physical state of storage materials change gradually. Then the germinated seed enters the second process (a plateau phase) and the third process (the embryonic axes elongation). Seed embryo and endosperm display distinct differentially expressed proteins, and their synergistic expression mechanisms provide a basis for the normal germination of wheat seeds.

Keywords: Wheat seeds, Germination, Embryo, Endosperm, Proteome, qRT-PCR

* Correspondence: yanym@cnu.edu.cn
[†]Equal contributors
[1]College of Life Science, Capital Normal University, Beijing 100048, China
[3]Hubei Collaborative Innovation Center for Grain Industry, 434025 Jingzhou, China
Full list of author information is available at the end of the article

Background

Wheat (*Triticum aestivum* L.) is one of the three most important grain crops and is widely cultivated worldwide due to its value as a staple food and protein source. Total wheat production is more than 600 million tons, which accounts for more than 20% of the world's food supply [1]. Wheat grains are primarily composed of the embryo and the endosperm, both of which play important roles in seed germination and subsequent plant growth and development. The embryo forms a radicel, a plumule, and the new plant, whereas the endosperm contains reserve substances to supply nutriments for subsequent plant growth, the basis of wheat yield and quality.

The embryo and endosperm play different roles in wheat seed germination. The embryo contains most of the genetic information that controls germination. Upon imbibition, the substrate and energy starvation activate the embryo to produce phytohormones (mainly gibberellic acid, GA). The GAs can diffuse to aleurone and initiate a signaling cascade that leads to synthesis of α-amylases and other hydrolytic enzymes. These enzymes then secrete into the endosperm to drive the degradation of storage compounds, including starch, lipid, and protein, for seedling establishment [2,3]. Studies of endosperm function have demonstrated that the endosperm can secrete signals to control embryo growth [4]; thus, germination is a systemic response that involves bidirectional interactions between the embryo and endosperm.

Seed germination generally proceeds from heterotrophic to autotrophic and involves a series of physiological, biochemical, and morphological changes. It is highly related to seedling survival rate and subsequent vegetative growth and therefore directly affects wheat yield and quality. Storage materials of starch and protein are mainly deposited in the endosperm. Dry seeds' water content is usually low (5-15%), and their metabolic activity almost ceases. When the dry seeds imbibe water, germination begins, and when the radicle breaks through the episperm, germination is finished [5]. In general, the seed germination process can be divided into three phases: fast water uptake (phase I), metabolism reactivation (phase II), and radicle emergence (phase III) [6]. Phase II is the most critical stage, because all necessary metabolic pathways and physiological processes are reactivated and germination is initiated in this phase. Germination involves many events, such as proteolysis, synthesis of macromolecules, respiration, changes in subcellular structures, and cell elongation [6,7].

The metabolic pathway of wheat germination is highly complex. Early work reported the activity of some key enzymes in glycolysis, pentose phosphate pathway (PPP), the tricarboxylicacid cycle (TCA cycle), and amino acid metabolism during germination [8]. Aoki et al. (2006) detailed the pathway of starch hydrolysis and sucrose transport during germination [9]. In the embryo, some germination-specific gene products were found, including catabolic enzymes associated with the mobilization of reserves from endosperm [10] and germin (oxalate oxidase), which is related to cell wall restructuring [11]. Caliskan et al. (2003) identified four new members of the wheat "germin" gene family [12]. In the past several decades, most of our knowledge about initiation of germination has come from the *Arabidopsis thaliana* model system [13-15]. Several large-scale -omics methods, including transcriptomic, proteomic, and metabolomic methods, recently have been established to investigate the mechanisms of seed germination [13]. The great achievements in genomics have led to application of large-scale gene expression analysis at both mRNA and protein levels to uncover the features of seed germination. For instance, 51,411 transcripts were identified during the five seed germination stages of elite Chinese bread wheat cultivar Jimai 20 using the Affymetrix Wheat Genome Array [16]. Microarrays containing 2600 genes expressed in developing *Arabidopsis* seeds were reported [17]. At the same time, considerable proteomics work on seed germination in different plants has been performed, such as cress [18], sugar beet [19], *Medicago truncatula* [20], barley [21], maize [22], and rice [23,24]. The development of proteomic techniques has provided a unique opportunity to study the proteome changes during wheat seed germination [25-27].

Although different proteomic studies on seed germination in some plants have been conducted, these investigations were focused only on a single organ or tissue such as whole seeds, endosperm, or embryo. This approach does not provide overall understanding of the metabolic and regulatory networks occurring in both embryo and endosperm tissues. In the present study, we performed the first integrative proteome analysis of embryo and endosperm during wheat seed germination by a comparative proteomic approach. Our results revealed that distinct differentially expressed proteins in seed embryo and endosperm are cooperatively involved in seed germination, which provides new insights into the proteomic mechanisms of wheat seed germination.

Results

Seed morphological and structural changes during germination

Seed morphological and structural characteristics in the four germination periods are displayed in Figure 1, including phase I (5 min), phase II (5 hours), phase III (15 hours), and phase IV (35 hours). Along with water imbibitions, seed size increased gradually during the seed germination process. As shown in Figure 1a, at

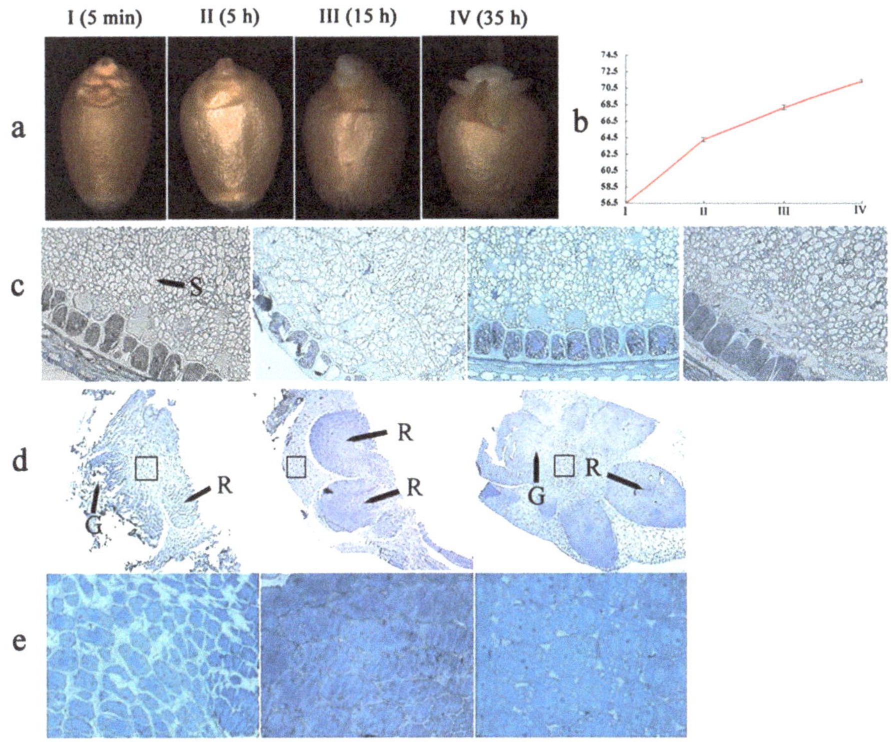

Figure 1 Morphological and structural characteristics of wheat seeds during the four germination stages. a. Seed morphological changes during the four germination stages (phase I: seeds absorb water for 5 min; phase II: seeds absorb water for 5 hours; phase III: seeds absorb water for 15 hours; phase IV: seeds absorb water for 35 hours). **b**. Changes in 1,000-seed weight during the germination process. **c**. Endosperm slice from the four germination periods observed by light microscope magnified 20 times (S: starch granules). **d**. Embryo slice from the first three periods observed by light microscope magnified 5 times (R: radicle; G: germ). **e**. Embryo slice from the first three periods observed by light microscope magnified 40 times.

phase III, radicels broke through the episperm, indicating that seeds would soon sprout. At phase IV, green germs and three radicles were visible. During the germination process, seed weight increased with imbibition (Figure 1b). In early germination, water uptake is rapid, and after 5 hours, it increased slowly. Endosperm slices from four germination periods were observed by light microscope (Figure 1c), and results showed the endosperm cell structure had changed little from germination stages I to III, but stage IV brought significant change: starch granule sizes were enlarged and starch granule numbers clearly decreased in some areas due to the mobilization and degradation of reserved starch. Observation of embryo slices by light microscope showed that embryo structures displayed significant changes during the germination process (Figure 1d) and gradually disappeared; at phase IV in particular, embryo structures could not be observed due to the emergence of germ and radicles. Results from light microscope observation showed that the embryo cells in the dry seed (phase I) were hydropenic, and thus a larger cell interval was present. Cells expanded when they absorbed water (phase II-III), and cell interval became significantly smaller (Figure 1e).

Protein expression profiles of embryo and endosperm during seed germination

The embryo and endosperm proteins from four germination periods were separated by two-dimensional electrophoresis (2-DE), and both tissues displayed distinct protein expression profiles on the gel during seed germination (Figure 2). The molecular weight (*Mr*) of protein spots from embryos in phase I, II, and III ranged from 15 to 98 kDa, and most were present in pH 4-10. In the last phase, the protein *Mr* ranged from 16 to 85 kDa, and most were present in pH 4-7. The protein spots were roughly evenly distributed on the gels (Figure 2a). The protein *Mr* from endosperm ranged from 15 to 85 kDa, and most were present in pH 4-10. The protein spots mainly concentrated in a particular area on gels and showed an uneven distribution compared with those in embryo (Figure 2b).

The protein spots on 2-DE gels from four germination periods of embryo and endosperm were compared, and DEP spots with at least two-fold differences in abundance were determined. After image analysis, 97 and 49 DEP spots from embryo (M) and endosperm (N), respectively, were recognized and selected for MALDI-TOF/TOF-MS analysis. In total, 74 and 34 DEP spots

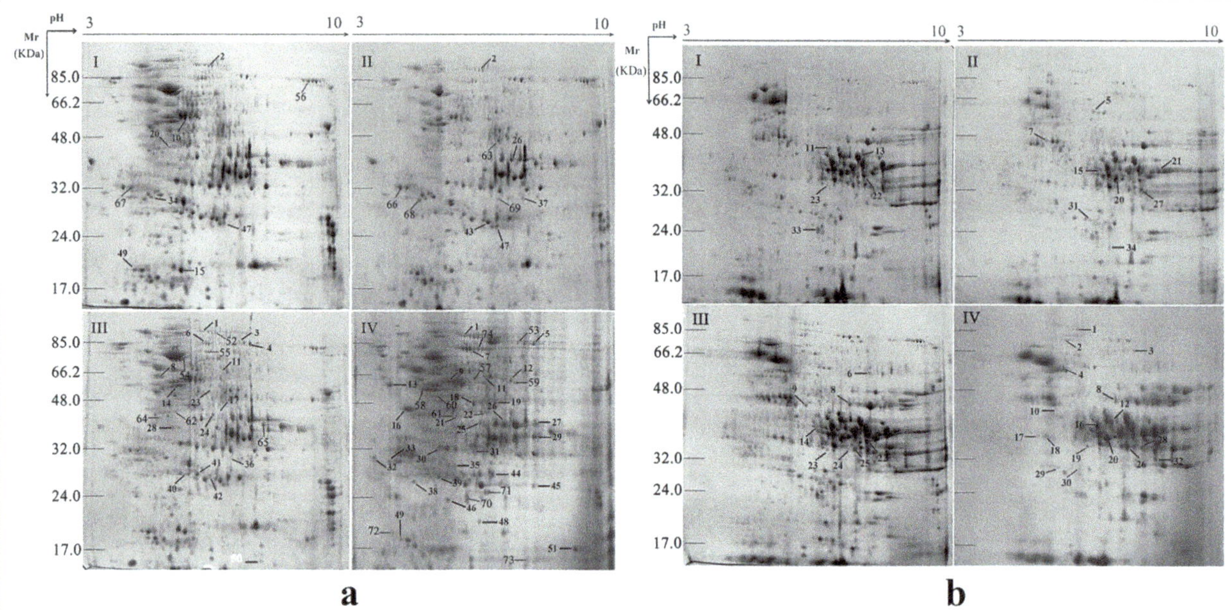

Figure 2 Proteome maps of wheat embryo (a) and endosperm (b) during four germination periods by linear 2-DE. The DEP spots with at least 2-fold changes during seed germination are indicated on 2-DE gels.

representing 63 and 26 unique DEPs from embryo and endosperm were identified, respectively. Eight common DEPs were present in both tissues, and 66 and 26 DEPs were specific to embryo and endosperm, respectively (Additional file 1: Figure S1). Detailed is in Additional file 2: Table S1, and peptide sequences are shown in Additional file 3: Table S3.

The identified DEP spots from the embryos were grouped into 13 different functional categories, mainly including stress-induced proteins (18%), carbohydrate metabolism (19%), proteometabolism (12%), nucleic acid metabolism (5%), inhibitors (5%), sucrose biosynthetic process (4%), and other metabolism processes (19%), as shown in Figure 3. The DEPs from endosperm were classified into seven functional categories: storage proteins (38%), carbohydrate metabolism (23%), stress-related proteins (21%), inhibitors (9%), proteometabolism (3%), amino acid metabolism (3%), and other metabolisms (3%) (Figure 3). In order to obtain more information about the predicted proteins, their protein sequences were used as queries to search for homologues by the BLASTP (http://www.ncbi.nlm.nih.gov/BLAST/) tool. The identified corresponding homologues classified into different functional groups are listed in Additional file 2: Table S2. The majority of spots shared above 90% Positive %, Identity %, or Query Coverage % with homologues at the amino acid level, suggesting similar functions. Following this process 15 DEPs remained as

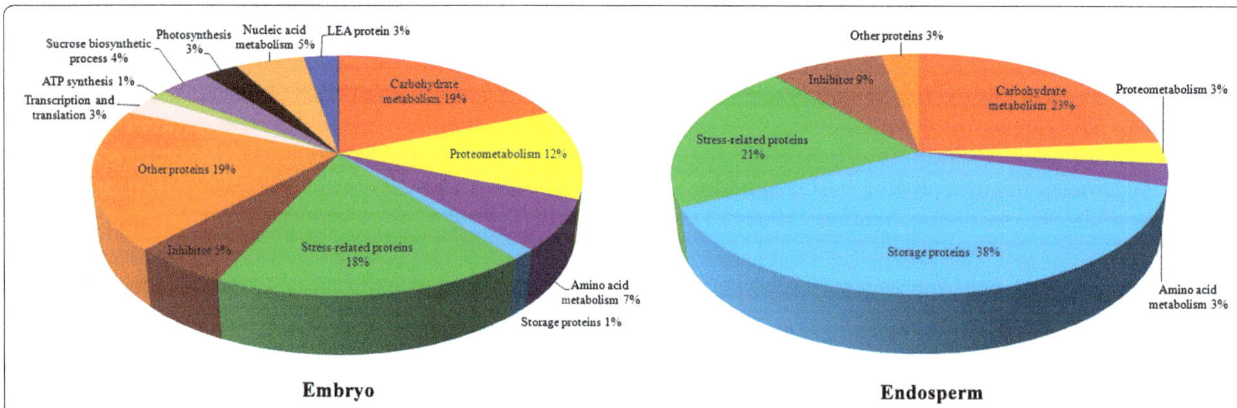

Figure 3 Functional classification and the proportion of DEP spots identified in wheat embryo and endosperm.

having unpredicted functions, namely other proteins metabolism in Additional file 2: Table S1.

The transformed Log (protein relative abundance) was used for cluster analysis. The expression pattern analysis of DEPs was performed using the Euclidean distance method over a complete linkage dissimilarity matrix (Figure 4), which can represent quantitative differences in DEPs during the four germination periods. As shown in Figure 4, the embryo had five distinct expression patterns at protein level, designated as A-E. Pattern A displayed a down- to up-regulated expression trend represented by 5 protein spots (5 unique DEPs). Pattern B, the largest group of identified DEPs, had 30 protein spots representing 23 DEPs and showed an up-regulated expression, with particularly significant up-regulation during the last germination stage (phase IV). This group was mainly involved in carbohydrate metabolism and proteometabolism. In pattern C, 12 protein spots representing 10 DEPs displayed an up-down-up expression trend. These proteins were mainly related to basic metabolic activity such as

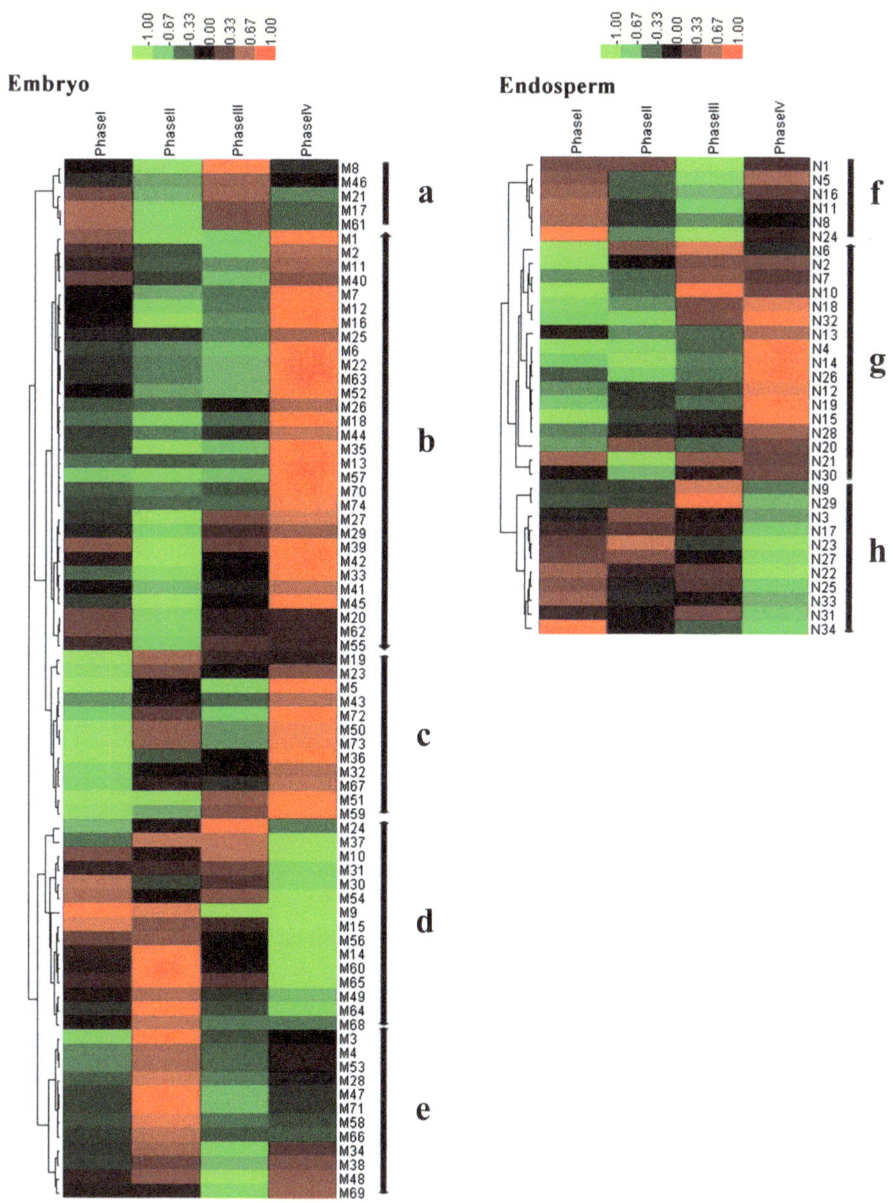

Figure 4 Hierarchical clustering analysis of the expression patterns of DEP spots in wheat embryo and endosperm during seed germination. Clustering was based on the protein expression levels across phase I, II, III, and IV of seed germination. Red color indicates a positive abundance in protein spots; green indicates a negative abundance in protein spots. Different expression trends of DEP spots identified in embryo and endosperm were classified. Embryo: **a**. down-up regulation; **b**. up-regulation; **c**. up-down-up regulation; **d**. down regulation; **e**. up-down regulation. Endosperm: **f**. down-up regulation; **g**. up-regulation; **h**. down-regulation.

carbohydrate metabolism and proteometabolism. Pattern D contained 15 protein spots (15 unique DEPs) that exhibited a down-regulated expression and decreased especially rapidly in phase IV; these proteins were mainly involved in carbohydrate metabolism and stress-related proteins. Pattern E displayed an up- to down-regulated expression trend containing 12 protein spots (10 unique DEPs).

Endosperm exhibited three main expression patterns (F-H) at protein level, as shown in Figure 4. Pattern F, with 6 protein spots representing 6 unique DEPs, displayed a down- to up-regulated expression trend, and proteins were mainly involved in carbohydrate metabolism and protein storage. In late germination, the DEPs in pattern F increased during mobilization of reserved substances. Pattern G showed an up-regulated expression trend, including 17 protein spots and 11 unique DEPs. On the contrary, pattern H included 11 protein spots and 9 unique DEPs and displayed a down-regulated expression trend, and proteins were mainly involved in storage and stress-related roles.

Comparative proteome analysis of embryo and endosperm during different seed germination stages

The results listed in Additional file 4: Table S4 showed greater proteome differences during seed germination between embryo and endosperm. We compared the protein spots between both organs which were different spots only identified as the same protein name. Only eight common DEPs were present in both embryo and endosperm, and they were mainly responsible for carbohydrate metabolism, proteometabolism, amino acid metabolism, stress responsive proteins, and protease inhibitors. However, six of the proteins displayed different expression patterns, and only two DEPs (class II chitinase and proteasome subunit alpha type-3) showed the same expression trend during seed germination. Glyceraldehyde-3-phosphate dehydrogenase (GPD) (M26 and N11) was the first enzyme for which sequence data of the Calvin cycle/glycolysis homologues became available [28]. It was up-regulated (pattern B) in the seed embryo (Figure 5a-b) but had a down- to up-regulated expression (pattern F) in endosperm (Figure 5c-d). As an important physiologically active molecule involved in amino acid metabolism, methionine synthase (Met) (M6 and N2) showed down- to up-regulation (pattern B) and up-regulation (pattern G), respectively (Figures 4 and 5a-b). M54 exhibited pattern D in the embryo, N4 showed pattern G in the endosperm, and both were identified as beta-amylase, a major enzyme that hydrolyzes starch for glucose provision during late seed germination. In addition, M11 and N1 (dihydrolipoyl dehydrogenase 1), M66 and N17 (lactoylglutathione lyase), and M9 and N7 (Serpin-Z2B) also displayed different expression patterns in embryo and endosperm (Figure 4).

Figure 5a-b shows the expression patterns of 12 key DEP spots in the embryo that generally displayed up-

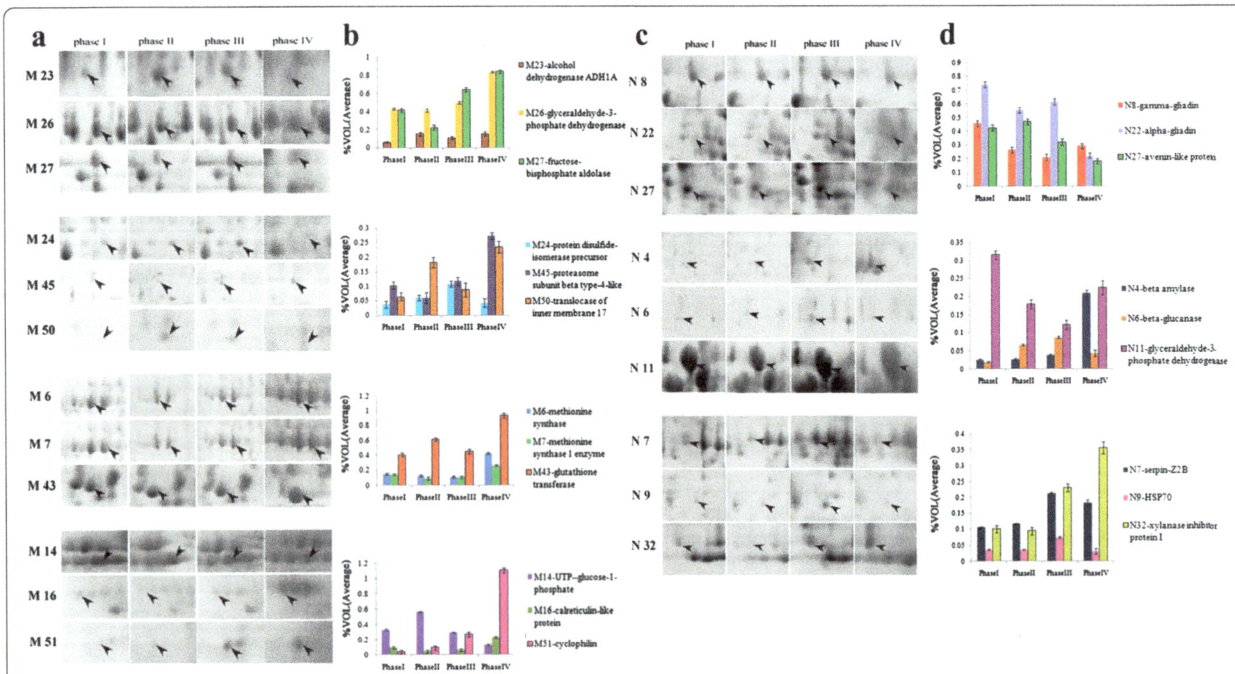

Figure 5 Expression patterns of 21 important DEP spots from embryo (a, b) and endosperm (c, d) during seed germination. a, c: The close-up view of the expression levels for 21 protein spots on 2-DE gels during four germination periods. **b, d**: The expression trends of the abundance (Vol %) of the 21 DEP spots.

regulated expression during seed germination and were mainly involved in basic metabolisms and stress response. Six DEP spots showed particularly significant up-regulation at late germination stages, especially at phase IV, including GPD (M26) and fructose-bisphosphate aldolase (M27), which were involved in carbohydrate metabolism, glutathione transferase (M43) related to amino acid metabolism, proteasome subunit beta type-4-like (M45), translocase of inner membrane 17 (TIM 17) (M50), and cyclophilin (M51) involved in proteometabolism. Five DEP spots exhibited a tender up-regulation during seed germination: alcohol dehydrogenase (ADH) (M23), protein disulfide-isomerase precursor (PDI) (M24), methionine synthase (M6), methionine synthase 1 enzyme (M7), and calreticulin-like protein (M16); however, UTP-glucose-1-phosphate (M14) showed an up-down expression pattern and had a higher expression level at stages I and II.

The expression patterns of nine key DEP spots in endosperm shown in Figure 5c-d were mainly related to storage proteins, carbohydrate metabolism, inhibitors, and stress response proteins. The storage proteins γ-gliadins (N8 and N25), α-gliadins (N14, N20, N22 and N24), gliadin/avenin-like seed proteins (N33 and N34), and avenin-like proteins (N12, 23and N27) were generally down-regulated during germination. Among carbohydrate metabolism, β-glucanase (β-Glu) (N3, N6) participated in the degradation of cell wall related to the mobilization of reserved substances and transportation and showed an up-down expression pattern (Figures 4 and 5c-d). Cytoplasmic aldolase (N13) as an important enzyme in glycolysis and dihydrolipoamide dehydrogenase (N5) as part of the multienzyme complex, which catalyzed pyruvic acid to acetyl-CoA, exhibited a down-up-regulation pattern during germination, as shown in Figure 4. Both xylanase inhibitor protein I (XIP-I) (N32) and serpin-Z2B (N7) displayed an up-regulated expression, particularly with a much higher expression level at stages III and IV (Figure 5c-d). Heat shock protein 70 (Hsp 70) (N9) was related to protein folding and participated in basic metabolic activity, and its expression abundance increased until phase III, then decreased (Figure 5c-d).

Protein-protein interaction network analysis of the key DEPs

Our results showed that the endosperm had limited DEPs, mainly storage proteins and carbohydrate metabolism-related proteins (Figure 3) that could provide nutrition for germination and subsequent seedling growth, during the seed germination process (Figure 4). But in the embryo, considerable DEPs were identified during seed germination, including those involved in multiple metabolic processes such as stress response, carbohydrate metabolism, and proteometabolism (Figure 3). These DEPs could interact each other or with other proteins and function in seed germination. Thus, an interaction network of some key DEPs identified in the embryo was constructed using online analysis software STRING (version 9.1, http://string-db.org/). Proteins with different GI numbers may have the same KOG number, such as spots M6 and M7 sharing KOG2263 and M56 and M57 sharing KOG1370. All 47 KOGs listed in Additional file 5: Table S5, including 57 DEPs in the embryo, were displayed through Cytoscape (version 3.0, http://cytoscape.org/). To improve the reliability of this analysis, the confidence score was set to 0.700. As shown in Figure 6, six main interacting proteins were frequently located at the central part of the interaction network. KOG2854, which represents ATP synthase beta subunit (M67), was used to extract the potential interacting proteins from the whole protein-protein interaction network, and an ATP synthase beta subunit-centered sub-network was constructed (Figure 6) and showed that the ATP synthase beta subunit interact with 47 proteins involved in carbonmetabolism, amino acid metabolism, nucleic acid metabolism, proteometabolism, sucrose biosynthesis, and defense. The ATP synthase beta subunit aids ATP biosynthesis (energy metabolism) and is the universal enzyme that synthesizes ATP from ADP and phosphate using the energy stored in a transmembrane ion gradient [29-31].

Transcriptional expression patterns of six key protein genes during seed germination

The transcriptional expression changes of the genes encoding six representative DEPs identified from wheat embryo and endosperm during seed germination are shown in Figure 7, including *ADH1A*, *GPD*, and *Met* from the embryo and *β-Glu*, *HSP70*, and *XIP-I* from the endosperm. Three types of expression patterns (A, B, and C) were recognized after careful analysis. In pattern A, the gene expression level was first up-regulated then down-regulated. Proteins encoded by genes corresponding to this pattern included *Met*, *β-Glu*, and *HSP 70*. Pattern B showed an up-regulated trend and included *XIP-I* and *GPD*. In pattern C, the *ADH1A* gene was first up-regulated, then down-regulated, and up-regulated in the last stage. A possible explanation is that when germination was complete, oxygen consumption in the embryo increased continuously but decreased in storage tissues because of the exhaustion of reserve substances or decreased dependence of seedlings on reserve substances. The expression patterns of all six genes were consistent with their protein expression trends (Figures 4 and 5).

Discussion

Energy requirements for seed germination

Respiration and energy production play key roles in whole seed germination. Seeds in the activation process

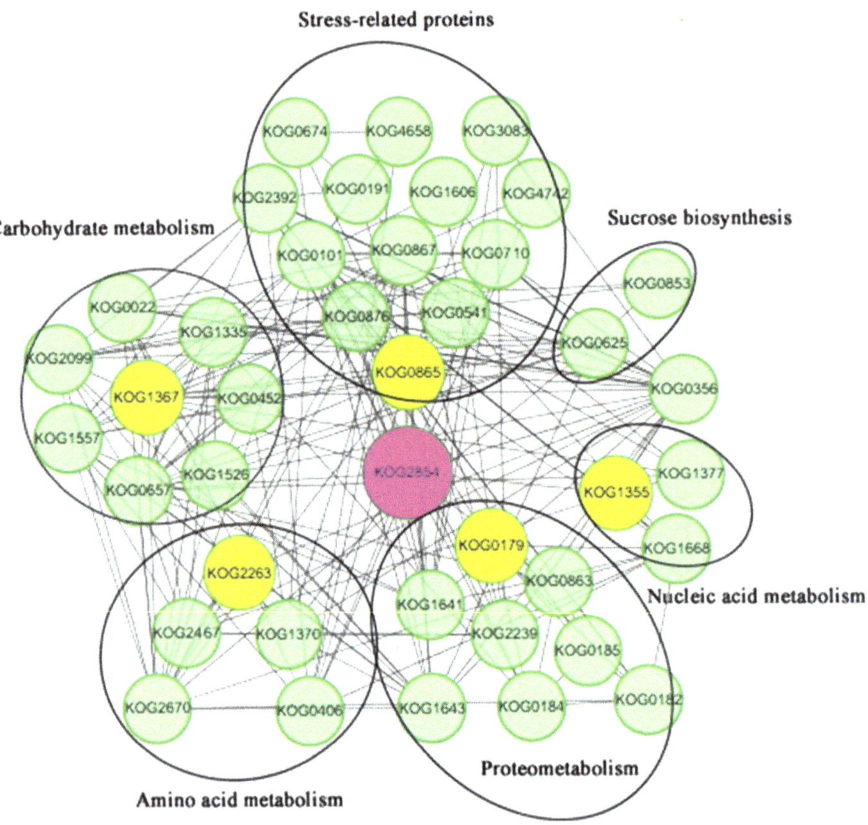

Figure 6 Protein-protein interaction network of ATP synthase and related proteins. Nodes with pink background color represent the KOGs of ATP synthase; nodes with yellow background represent the KOGs of main DEPs in each functional group; nodes with green background represent the KOGs of related DEPs in each functional group.

begin soon after absorbing plenty of water. The activation sequence is amino acid metabolism, glycolysis, and the TCA cycle. Due to limited oxygen content, TCA cycle efficiency is not high in the early germination stages, and energy is supplied mainly by glycolysis and alcohol fermentation [23]. The energy provided by anaerobic respiration cannot satisfy the needs of germinated seeds at this point, but the TCA cycle also provides a large amount of energy in oxygen-rich conditions. The energy for seed germination is provided mostly by anaerobic respiration in the early stages, then respiratory activity increases as oxygen uptake and carbon dioxide release accelerate during imbibition [32]. The embryo and storage tissues of most seeds during germination exhibit a basic pattern of oxygen consumption [5]: they imbibe plenty of oxygen at the beginning of germination, after which they enter a lag period when the need for oxygen in the embryo is little. Oxygen consumption then increases sharply when the radicles break through the episperm, and germination is finished.

In the embryo, ADH (M23) is believed to participate in the glycometabolism pathway from pyranic acid to alcohol. The energy demands of germinating cereal seeds seem to be met mostly by glycolysis in the absence of oxygen [23]. It has been reported that ADH is involved in coleoptile growth under oxygen-limiting conditions, and rice coleoptile growth is depressed by deficiencies of ADH [33-36]. ADH increased from phase I to II and decreased in phase III in this study (Figure 5a-b); the variation trends correspond with the seeds' pattern of oxygen consumption [5], an expression trend that was also found in dark-germinated pea seeds [37]. In addition, the gene expression of ADH at the transcription level also first increased, then decreased (Figure 7), consistent with the expression of ADH at the translation level.

GPD (M26) was a key enzyme during glycolysis, and its expression trend was up-regulated during all four seed germination stages (Figure 5a-b). The transcription expression of this enzyme corresponds well with its protein expression pattern (Figure 7). However, in the endosperm, GPD (N11) decreased quickly in the first three periods and increased when germination was complete (Figure 5c-d), so we can draw the conclusion that the speed of the glycolytic pathway in the seed embryo is slower than in the endosperm. In the process of seed germination, all kinds of stored substance (mainly in the

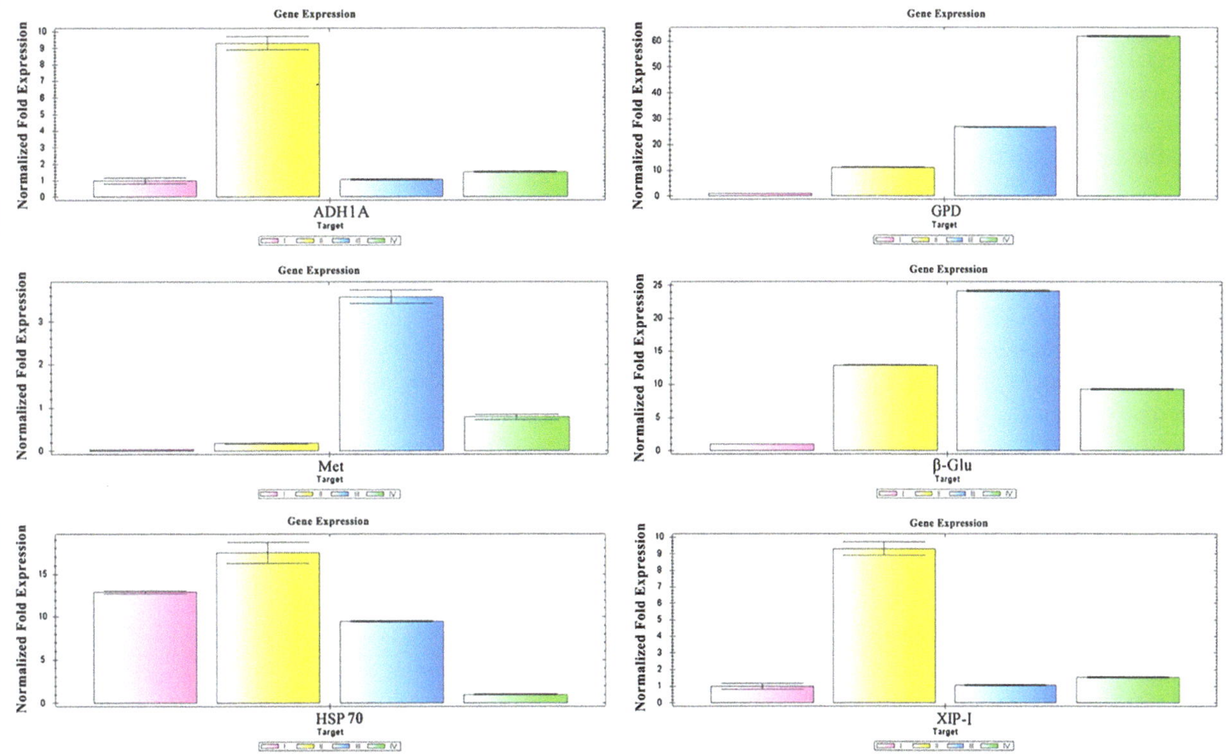

Figure 7 The dynamic transcriptional expression changes of the genes encoded six key DEP spots during seed germination as revealed by qRT-PCR. Three genes encoding alcohol dehydrogenase (ADH1A), glyceraldehyde-3-phosphate dehydrogenase (GPD), and methionine synthase (Met) were from embryo, and three genes encoding β-glucanase (β-Glu) and HSP 70 and xylanase Inhibitor protein-I (XIP-I) were from endosperm.

endosperm) are broken down into a variety of small molecules, thus providing the respiration substrates as well as substance transformation and synthesis and a number of raw materials to form the new cells. Triosephosphatisomerase (M40, M41) also participates in glycolysis. When germination begins, all embryo cells need more energy to form new organs such as the radicel and germ, so glycolysis increases and all enzymes related to glycolysis also increase. When germination is finished, consumption of reserve substances slows down and leads to slower glycolysis. Glycolysis is the dominant pathway for formation of pyruvic acid and ATP during germination. In our study, GPD always increased in protein expression and kept the same pattern at the transcriptional expression level during the seed germination process (Figures 5a-b and 7).

Aconitate hydratase (M1) exists in cells, not only in cytoplasm but also in mitochondria, and participates in the TCA cycle in mitochondria [38,39]. In dry seeds and seeds that imbibed early, mitochondrial structure and function were found to be defective [40]. Our observations showed that the structures of embryo cells in early germination were also defective (Figure 1d); therefore, the requirement of enzymes involved in the TCA cycle was low, and some of them were degraded during early germination such as dihydrolipoyl dehydrogenase 1 (M11) and alpha-glucan phosphorylase (M5). These enzymes displayed lower expression quantity in early germination stages (Figure 4). M5 is an enzyme that catalyzes the reversible phosphorylation reaction, converting malt sugar, starch, or glycogen into glucose-1-phosphate; therefore, it plays an important role in carbohydrate metabolism [41,42]. Its expression was an up-down-up trend, increasing especially rapidly in the last period (Figure 4), which corresponded with the pattern of oxygen consumption by the seeds [5]. In endosperm, dihydrolipoamide dehydrogenase (N5) is a component of the plastidial pyruvate dehydrogenase complex, which participated in the preparation for pyruvic acid entering TCA cycle. Its expression trend first decreased then increased (Figure 4), suggesting that the pattern of oxygen consumption by seed endosperm may be in accord with the basic oxygen consumption pattern by seed. Correspondingly, due to the continuous synthesis of pyruvate, which was turned into alcohol and entered the fermentation pathway, pyruvic acid could be further degraded through the TCA cycle. Thus, most TCA cycle-related proteins identified in this work, such as aconitate hydratase (M1), dihy drolipoyl dehydrogenase 1 (M11), and alpha-glucan phosphorylase (M5) (Figure 4), were generally up-regulated.

When the germination is complete, and the one promoting factor (GA) produced by the embryo induced certain endosperm cells to express a number of enzymes involved in the mobilization of most storage proteins, starch, and triacylglycerol (the main lipid of seed) from the endosperm [43,44]. In fact, most of these enzymes existed in the endosperm in the first place, but they were inactive until they were activated or re-synthesized prior to mobilization [5,45]. At this time, most reserve substances from endosperm began to be mobilized to supply nutriments for seedling growth, so only a small amount was degraded during the germination process [46].

In the late germination stages, the hydrolysis of some storage proteins produces amino acids and peptides that may either remain in the endosperm or be transformed into Asn or Gln, then transported to the seedlings [47]. Most storage proteins, such as γ-gliadin (N25), α-gliadin (N22 and N24), gliadin/avenin-like seed protein (N33 and N34), and avenin-like protein (N23 and N27), are degraded when germination is complete (Figure 4). This phenomenon demonstrated that the main storage proteins were mobilized after germination at the molecular level.

Starch and sucrose metabolism

Starch is the major reserve in mature cereal seed. Wheat seeds contain about 70% starch and 10% storage proteins. The degradation of these materials provides energy for seed germination and seedling growth. When the seed endosperm imbibes a certain amount of water, starch mobilization occurrs after deposited sucrose degradation. The mobilization of starch is a complex process including the lysis of endosperm cells wall, starch hydrolysis, and the synthesis or transport of sucrose. β-glucanase (N3 and N6), one kind of hemicellulase, which participates in the degradation of cell wall, first increased then decreased during seed germination in this study (Figure 5c-d), which corresponded well with its transcriptional expression pattern (Figure 7). The consistency of β-glucanase gene expression at transcriptional and translational levels confirmed that the mobilization of starch happened during germination at the molecular level. The decomposition products of triacylglycerol were glycerol and fatty acid (FA). FA could be oxidized and used for respiratory metabolism, and glycerol could be not only oxidized by TCA cycle but also transformed into glucose by a suite of metabolic reactions [48,49]. Then sucrose was synthesized from glucoses and transported to seedlings in the form of sucrose. Beta-amylase (N4) increased gradually in the first three germination periods and increased dramatically to about four times during the fourth period (Figure 5c-d), suggesting that starch degradation metabolism is vigorous at the end of seed germination. Similar results were also found in the germination of barley seeds [50]. These results imply that large-scale starch mobilization might happen during the late germination stage and after germination to provide nutriments for subsequent seedling growth. Beta-amylase is a major enzyme that hydrolyzes starch for glucose provision during late germination. We can conclude that starch hydrolysis becomes more and more intense during the germination process. Starch is converted to glucose after a series of catabolism, and sucrose is further synthesized, then transported to the seedlings for growth and development. Previous work found that the majority of sucrose resulted from storage lipid degradation, not from other soluble sugars within *Arabidopsis* seeds [15]; however, our results revealed that the majority of sucrose comes from the mobilization of storage starch and lipids, similar to germinating barley seeds [44]. Sucrose may be resolved into glucose and fructose by sucrose synthase, then enter glycolysis. We found that sucrose synthase type 2 (M52) increased in the latter germination stages for seedling growth (Additional file 4: Table S4).

Protein metabolism

Besides starch, the other most important energy reserves are seed proteins, especially storage proteins. Dry seeds of wheat have various kinds of enzymes or proteins, but most of them have no activity. When seeds begin to germinate, they could participate in a series of basic metabolic activities immediately, such as energy metabolism and synthesis of proteins. Dry seeds contain not only the above proteins but also numerous components related to the synthesis of proteins [51]. Once the dry grains contact water, mRNAs and ribosomes assemble rapidly into polysomes, then immediately begin to synthesize proteins with other components [52]. During early germination, the RNA required for protein synthesis exists in dry seeds and supports the synthesis of early germination proteins, including proteins concerned with basic metabolic activity and germination [53]. In addition to early proteins, other proteins are degraded into amino acids for early protein synthesis. As germination progresses, protein synthesis becomes increasingly dependent on the synthesis of mRNA, the compositions of which are changing, and then the types of proteins expressed in the embryo change [54].

Proteasome (M35, M42, M45, and N30), a large protein aggregate with several proteolytic activities, has 7 different α and β subunits, respectively, in eucaryon [55] and participated in proteolysis. Proteins are marked for degradation by the proteasome when they are modified by the addition of ubiquitin [56]. PDI (M24) and TIM 17 (M50) are associated with the transport and processing of protein, and their expression is mainly up-regulated (Figure 5a-b). PDI (M24) is an essential protein that facilitates the isomerization of disulfide bonds [57], especially in assisting the folding of synthesized proteins [58]. TIM 17 was concerned with protein transport. These two enzymes from the embryo participate in basic metabolic activity, so they increase as germination begins. In endosperm, proteins are catabolized upon the initiation of germination by hydrolytic enzymes secreted from the aleurone layer. The hydrolyzed storage proteins act as an energy source, providing carbon and nitrogen for seed germination and subsequent seedling establishment. Most storage proteins, such as γ-gliadins (N8 and N25), α-gliadins (N14, 20, 22 and N24), gliadin/avenin-like seed proteins (N33 and N34), and avenin-like proteins (N12, 23 and N27), are down-regulated after germination (Figure 4).

Amino acid and nucleic acid metabolism

Met (M6,M7) is an important amino acid metabolism enzyme. It catalyzes the transfer of a methyl group from methyl-cobalamin to homocysteine [59]. The expression of this enzyme decreased in early germination then increased, as shown in Figure 5a-b. The expression of M6 quantity increased to its maximum during the fourth period, an expression trend implying that these were not early germination proteins. The expression of the Met gene at the transcriptional level was generally up-regulated and achieved its peak expression in the third period (Figure 7). Protein translation is known to lag behind the transcriptional expression of protein, so gene expression at the transcription and protein level was also consistent. In the endosperm, Met (N2) content increased throughout germination, suggesting that it is the early germination protein in endosperm and undergoes protein degradation. Hence, the embryo is in the process of protein synthesis at this moment, but the endosperm undergoes protein degradation. The metabolism of nucleotides was not active at early stages of germination because few enzymes required for nucleotide metabolism were detected. Both UTP–glucose-1-phosphate (M14) and adenylosuccinate synthetase (M21) were nucleic acid metabolism proteins, which play a central role as glucosyl donors in cellular metabolic pathways. Adenylosuccinate synthetase (AdSS) plays an important role in the de novo synthesis of adenylmonophosphate and in the salvage pathway of purine nucleotide biosynthesis [60]; its express trend is a down-up-down pattern (Figure 4), and a possible reason is injured RNA (especially mRNA) molecules in dry seeds (phase I) inside the cell membrane system. As the water absorps, cells begin repair activities, and new RNA is constantly formed to synthesize various enzymes. When germination is complete (phase IV), RNA synthesis decreases, so Adss content dropped significantly.

Stress-related proteins mobilization

Due to changes in external environments, wheat seeds are able to activate a series of mechanisms to respond to many biotic and abiotic stresses during germination [61]. Calreticulin-like protein (M16 and M13) is an abundant Ca^{2+}-binding protein predominantly located in the endoplasmic reticulum (ER) of higher plants. It is a key player in plant growth and development, a powerful regulator of plant responses to various stresses, and can help assemble other proteins correctly [62]. It also serves as a calcium-binding protein in the endoplasmic reticulum with functions like lectin combined with molecular calcium [63-65]. Its expression trend was mainly up-regulated, especially in the last period of germination (Figure 5a-b), which hints that wheat seed may encounter stress in late germination [66]. Cyclophilin (M51) has multiple biological activities involved in signal transduction, cell apoptosis, and protein folding. It is highly conservative and possesses peptidyl-proly cis-trans isomerase activity [67]. In higher plants, cyclophilin genes from *Zea mays* and *Brassica napus* were separated and expressed [68]. Our results show that M51 increased slowly in first three germination stages but increased sharply in the last stage (Figure 5a-b). Because the cellular structure is not complete in early germination, metabolic activities such as protein synthesis are low, so cyclophilin expression is very low. In the following three phases, however, it increased significantly along with the perfection of the cellular structure and the increase of metabolic activities in germinating seeds.

Reactive oxygen species (ROS) are produced in all living organisms, and their overaccumulation of ROS can result in oxidative stress. Reducing oxidized proteins is another critical way for plants to cope with ROS. Upon imbibition, ROS contents increased gradually. ROS can be scavenged efficiently by the antioxidant enzymes such as superoxide dismutases (SODs) (M47 and M71), glutathione S-transferase (GST) (M69), and peroxidases (M72,N19 and N21). Many of these redox regulation proteins were identified during germination [69]. In this study, these proteins showed gradually up-regulated expression when germination began (Figure 4). All of them can reduce the target proteins' disulfide and keep these proteins functional [70-73]. This result indicates that wheat may prefer to protect functional proteins from ROS attack during seed germination.

Stress responsive proteins, including HSPs and late embryogenesis abundant (LEA) proteins (M56), accumulated during seed maturation [23,24]. HSPs participate in diverse cellular processes by acting as molecular chaperones. They belong to a huge glycoprotein gene family, of which HSP 70 (N9) is the most conservative [74]. They also are described as being developmentally regulated, abundant in dry mature seeds, and disappearing during germination [75]. In agreement with these findings, the expression of Hsp70 (N9) at both transcriptional and translational levels increased at phase III and decreased by the end of the germination process

(Figures 5c-d and 7) in accord with the synthesis of plentiful enzymes that participated in the mobilization of reserve substances. Its down-regulation after germination suggests that the metabolic activity of endosperm weakened when accompanied by the large consumption of reserve substances.

Other protein metabolisms

According to our results, the photosynthesis process was mobilized slightly after initiation of reserve mobilization during seed germination, indicating that subsequent (after 35 hours of imbibition) growth depends not only on mobilization of energy reserves but also on photosynthesis. Ribulose-1,5-bisphosphate carboxylase/oxygenase (RuBisCO, M74) increased rapidly in the late germination stages (Additional file 4: Table S4). It methylates 'Lys-14' of the large subunit of RuBisCO and also can use chloroplastic fructose-bisphosphate aldolases with lower efficiency [76]. So we can conclude that the photosynthetic apparatus was gradually integrated into the seed in late germination.

The ATP synthase beta subunit (M67) contributes to catalytic sites that bind ATP and ADP. It plays a pivotal role in energy transduction in living cells [77]. Its expression trend showed an up-regulated pattern (Figure 4) and indicated that energy metabolism continuously strengthened the process of germination. Translation elongation factor 1-beta (M33) stimulates the exchange of GDP bound to elongation factor-1-alpha to GTP and has actual GDP/GTP exchange activity [78]. It displayed a generally up-regulated trend (Figure 4), especially in the last period. We speculate that new protein synthesis is more active when germination is complete, and this may be related to the emergence of germs and radicles. Pyridoxine biosynthesis protein (M31) is a limited velocity enzyme in the process of vitamin B6 synthesis. Studies have shown that it relates to the plasma membrane and inner membrane systems. Vitamin B6 is related to membrane system repairs in early germination and may play an important role in the function of the membrane [79]. Its expression pattern in this study (M31) was up-down-regulated. Its expression decreased significantly when germination was finished (Figure 4). So we can know that seed undergoes membrane system repairs in early germination and related enzymes content will increase accordingly. In addition, 27 K protein (N31) as a glycoprotein with mannose and fucose residues may belong to the family of γ-interferon-inducible thiol reductases. It is also a kind of serine/arginine-rich (SR) protein and could modulate pre-mRNA splicing reaction [80]. Its expression showed a down-regulated pattern (Figure 4), suggesting that the synthesis of mRNA decreased greatly in the late germination stages.

Overview of central metabolic pathways during wheat seed germination

To better understand the central proteomic changes of embryo and endosperm during wheat seed germination, a putative metabolic pathway was proposed, as shown in Figure 8. In the embryo, the activation process begins soon after seeds absorb plenty of water. The activation sequence is amino acid metabolism, glycolysis, and the TCA cycle. The energy demands of germinating cereal seeds seem to be met mainly by glycolysis [23]. The TCA cycle also provides a large amount of energy in oxygen-rich conditions. Many important related proteins are involved in various metabolic pathways. The embryo synthesizes and secretes GA that is transferred to the aleurone layer of the endosperm. GA then induces the synthesis of β-amylase in the aleurone layer, which secretes hydrolysis enzymes to the starchy endosperm [2]. The embryo utilizes sugars released by starch degradation for its growth. In the endosperm, all storage proteins showed a down-regulated expression trend and were broken into small molecules and used for the development of the embryo and seedling growth. Some stress-related proteins and ROS scavenging proteins were activated and displayed an up-regulated pattern. Upon imbibition, the contents of ROS increased gradually, resulting in further oxidative stress. ROS can be scavenged efficiently by antioxidant enzymes such as SODs (M47), GST (M69), and peroxidases (M72). The cooperative functions of these important proteins provide a sound basis for regular germination of seed and subsequent seedling growth.

Conclusion

In this study, 74 and 34 DEP spots representing 63 and 26 unique proteins were identified from embryo and endosperm, respectively, during seed germination. Seed embryo and endosperm display distinct differentially expressed proteins involved in seed germination; in particular, DEPs from the embryo were mainly related to carbohydrate metabolism, proteometabolism, amino acid metabolism, nucleic acid metabolism, and stress response, whereas those from the endosperm were mainly involved in storage proteins, carbohydrate metabolism, stress response, and protein synthesis. During germination, the embryo and endosperm of wheat seeds possibly have a basic pattern of oxygen consumption. They consume plenty of oxygen at the beginning of germination, enter lag period, then oxygen consumption increases sharply when the radicles have broken through the episperm. When germination is complete, oxygen consumption increases continuously in the embryo but decreases in the endosperm. Reserve substances from the endosperm begin to be mobilized when the germination is finished, and more enzymes involved in these

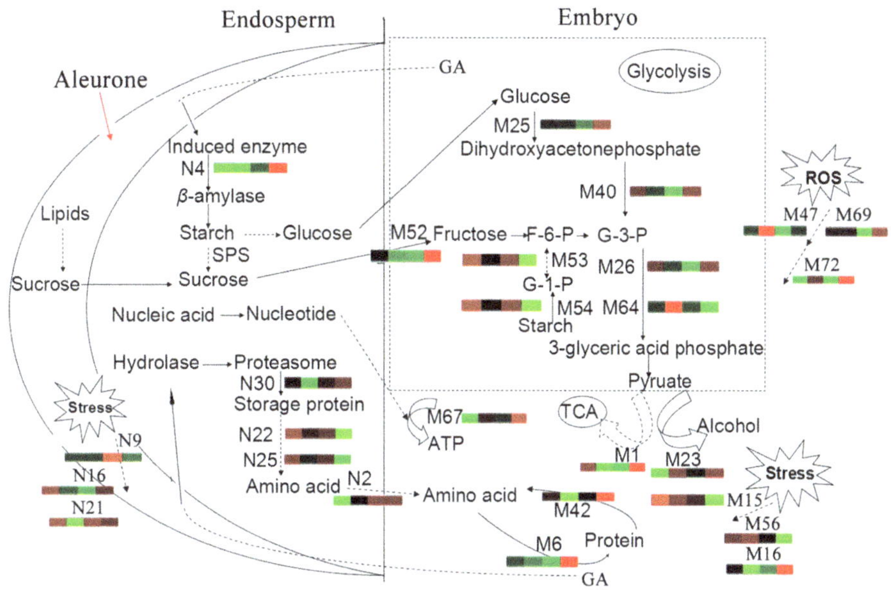

Figure 8 A putative metabolic pathway of wheat embryo and endosperm during the germination process. M1: putative aconitate hydratase; M6: methionine synthase; M15: Em protein H5; M16: calreticulin-like protein; M21: adenylosuccinate synthetase; M23, alcohol dehydrogenase ADH1A; M25: sucrose synthase type 2; M26: glyceraldehyde-3-phosphate dehydrogenase; M40: triosephosphat-isomerase; M42: proteasome; M47: manganese superoxide dismutase; M52: sucrose synthase type 2; M53: phosphoglucomutase; M54: Beta-amylase; M56: putative late embryogenesis abundant protein; M64: cytosolic 3-phosphoglycerate kinase; M67: adenosine kinase 2; M69: glutathione S-transferase; M72: peroxiredoxin-2E-1; N2: methionine synthase; N4: beta amylase; N9: HSP70; N16: class II chitinase; N21: class II chitinase; N22: alpha gliadin; N25: gamma- gliadin; N27: avenin-like protein; N30: proteasome subunit alpha type-3; SPS: Suc phosphate synthase; F-6-P: fructose-6-P; G-6-P: glucose-6-P; G-1-P: glucose-1-P; G-3-P: glyceraldehyde-3-phosphate.

processes are synthesized in large quantities. The mobilization also possibly concerns the activation of enzymes and inactivation of inhibitors. Storage proteins produce amino acids, peptides, and their derivatives, which are used to synthesize new proteins or transformed into Asn or Gln, then transported to the seedling. A putative metabolic pathway was proposed, which displays the central proteomic changes of embryo and endosperm during wheat seed germination.

Methods

Plant material

The elite Chinese bread wheat (*Triticum aestivum* L.) cultivar Zhengmai 9023, which has high yield, superior gluten quality, and good biotic resistance and has been widely cultivated in the main Chinese wheat production areas during the past ten years, was used in this study [81].

Seed germination and sampling

Seeds with similar sizes and weights were selected as the experimental materials and washed with distilled water three times. Seeds were germinated on wet filter paper in Petri dishes and incubated at 25°C in a growth chamber in the dark. Imbibed seeds can progressively lose dormancy in a matter of weeks if they are kept in constant darkness [82]. Water was added as needed to ensure complete imbibitions and normal germination. To understand the protein trends in the wheat embryo and endosperm, our study investigated four periods of the seeds germination, including phase I (seeds absorbing water for 5 min), phase II (seeds absorbing water for 5 hours), phase III (radicles broke through the episperm and seeds will sprout soon), and phase IV (green germ and three radicles visible). The embryo and endosperm collected from four germination periods with three biological replicates were separated and stored at −80°C prior to use.

Light microscopy observation

Transverse slices approximately 1 mm thick were cut from the collected wheat embryo and endosperms during each of the four periods and were fixed, rinsed, dehydrated, infiltrated, and polymerized by the series of steps outlined in Arcalis et al. [83]. For light microscopy observation, sections approximately 800 nm thick were cut, collected on mesh nickel grids, and stained with toluidine blue. Because the seed embryo from phase IV exhibited a large change compared with the first three phases, its cell structures were not observed by light microscope.

Protein extraction

About 0.5 g of wheat embryo and endosperm were ground into a fine powder in liquid nitrogen [84]. After 1 ml of extraction buffer (0.25 M sucrose, 1 M pH7.5 Tris-HCl, 0.1 M EDTA) was added into each sample, 10 μl 0.1 M PMSF and 0.1 M DTT were added immediately. Then the samples were ground for 10 min and 1 ml extraction buffer (0.25 M sucrose, 1 M pH7.5 Tris-HCl, 0.1 M EDTA, 4% (v/v) Triton-100), 10 μl 0.1 M PMSF, and 0.1 M DTT were added, and samples were ground again for 3 min. Then samples were transferred into tubes and vortexed for 10 min at room temperature. After centrifugation for 10 min at 13,000 rpm and 4°C, the supernatant was transferred into new tubes. This step was repeated twice, and the proteins in the supernatant were precipitated by adding 1/4 volumes 50% TCA buffer at -20°C for 2 to 3 hours, followed by centrifugation for 5 min at 13000 rpm and 4°C. The precipitate was washed three times with 1 ml chilled (-20°C) acetone containing 0.002 g DTT and centrifuged at 8,000 rpm and 4°C for 10 min between rinses. The fluid was removed and the pellet was dried under room temperature. Then the pellet was dissolved in 400 μl Lysis buffer (7 M urea, 2 M thiourea, 4% CHAPS) overnight. Finally, the protein concentration was determined with a 2-D Quant Kit (Amersham Bioscience, USA) using BSA (2 mg/ml) as the standard.

2-DE and image analysis

Each protein sample extracted from wheat embryo and endosperm (each about 600 μg), rehydration buffer (8 M urea, 2% w/v CHAPS), 0.5% (v/v) IPG buffer pH 3–10, 1% (w/v) DTT, and 1 μl 1% bromphenol blue were mixed to a final volume of 360 μl. The final solution was loaded onto the IEF linear IPG strips (pH 3-10, 18 cm, GE Healthcare) and actively rehydrated at 30 V for 12 h at 20°C in a Multiphor II apparatus (GE Healthcare). Focusing was performed on the IPGphor apparatus under the following conditions: 300 V for 1 hour, 500 V for 1 hour, 1000 V for 1 hour, 3000 V for 1 hour, and finally run at 8000 V for 80 kV h. After the first-dimension IEF, the IPG strips were equilibrated according to Gao et al. [85], then separated by 12% SDS-PAGE. After electrophoresis, gels were stained by colloidal Coomassie Brilliant blue (CBB) according to R-250/G-250 = 4:1. The images were visualized using a BIO-RAD GS-800 scanner. Protein spots that showed statistically significant changes by Student's *t*-test (abundant difference at least 2-fold, $P < 0.05$) between samples from different germination stages and organs were determined as DEP spots by an ImageMasterTM 2D 5.0 software. After image analysis, corresponding DEP spots of interest were selected for tandem mass spectrometry analysis.

MALDI-TOF/TOF-MS and protein identification

Protein spots were excised from the preparative gels and destained with 100 mM NH_4HCO_3 in 30% ACN. After removing the destaining buffer, the gel pieces were lyophilized and rehydrated in 30 μL of 50 mM NH_4HCO_3 containing 50 ng trypsin (sequencing grade; Promega, Madison, WI, USA). After overnight digestion at 37°C, the peptides were extracted three times with 0.1% TFA in 60% ACN. Extracts were pooled together and lyophilized. The resulting lyophilized tryptic peptides were kept at -80°C prior to mass spectrometric analysis. A protein-free gel piece was treated as above and used as a control to identify autoproteolysis products derived from trypsin.

MS/MS spectra were obtained using the ABI 4800 Proteomics Analyzer MALDI-TOF/TOF (Applied Biosystems, Foster City, CA, USA) operating in a result-dependent acquisition mode. Peptide mass maps were acquired in positive ion reflector mode (20 kV accelerating voltage) with 1000 laser shots per spectrum. Monoisotopic peak masses were automatically determined within the mass range 800–4000 Da with a signal to noise ratio minimum set to 10 and a local noise window width of m/z 250. Up to ten of the most intense ion peaks (minimum signal noise ratio: 50) were selected as precursors for MS/MS acquisition, excluding common trypsin autolysis peaks and matrix ion signals. In MS/MS-positive ion mode, spectra were averaged, collision energy was 2 kV, and default calibration was set. Monoisotopic peak masses were automatically determined with a signal to noise ratio minimum set to 5 and a local noise window width of m/z 250. The MS/MS spectra were searched against the NCBI nonredundant green plant and Triticum_ncbi database using GPS Explorer software, version 3.6 (Applied Biosystems) and MASCOT version 2.1 (Matrix Science) with the following parameter settings: trypsin cleavage, one missed cleavage allowed, carbamidomethylation set as fixed modification, oxidation of methionines allowed as variable modification, peptide mass tolerance set to 100 ppm, fragment tolerance set to ± 0.3 Da, total ion score confidence interval percentage and protein score confidence interval percentage were both set above 95%, and the significance threshold was $p < 0.05$ for the MS/MS.

Quantitative real-time polymerase chain reaction (qRT-PCR)

The wheat embryo and endosperm samples from four germination periods were ground into fine powder in liquid nitrogen. Then, total RNA was extracted from each sample and reverse-transcribed into cDNA according to the procedures in Yang et al. [86]. Gene-specific primers were designed using online Primer3Plus (http://primer3.ut.ee/) according to Untergasser et al. [87]. The sequence of each protein was first used as a BLASTP search term against *Triticum aestivum* genome. The best aligned gene was used for primer design. The specificity of the primers was checked by observing the melting curve of the RT-PCR products and the specific band on agarose gel. The primer sequences for qRT-PCR assays are given in Additional file 6: Table S6. ADP was chosen as an internal control for data normalization. qRT-PCR was performed in a 20-μl volume containing 10 μl 2 × SYBR Premix Ex Taq™ (TaKaRa), 2 μl 50-fold diluted cDNA, 0.15 μl of each gene-specific primer, and 8 μl ddH_2O. The PCR conditions were as follows: 95°C for 3 min, 40 cycles of 15 s at 95°C, 61°C for 30 s. Three replicates were used for each sample. Reaction was conducted on a CFX96 Real-Time PCR Detection System (Bio-Rad). All data were analyzed using the CFX Manager Software (Bio-Rad) and the 2 (-Delta Delta C(T)) method [88]. QRT-PCR efficiency was determined by five serial five-fold dilutions of cDNA, and the standard curve confirmed them at high RT-PCR efficiency rates (Additional file 7: Figure S2).

Additional files

Additional file 1: Figure S1. The DEP spots and unique proteins identified in wheat embryo and endosperm during seed germination.

Additional file 2: Table S1. The DEP spots and their functions from embryo and endosperm during seed germination identified by linear 2-DE (pH 3-10 and 18 cm strips) and MALDI-TOF/TOF mass spectrometry. **Table S2.** The unnamed and predicted proteins from embryo and endosperm in Table S1 grouped by BLASTP (www.ncbi.nlm.nih.gov/BLAST/) in searching for closest homologue.

Additional file 3: Table S3. The peptide sequences of identified DEP spots in embryo and endosperm identified by MALDI-TOF/TOF-MS.

Additional file 4: Table S4. Functions and expression features of some important DEPs in wheat embryo and endosperm.

Additional file 5: Table S5. Eukaryotic orthologous groups (KOGs) of the key DEPs identified in the wheat embryo.

Additional file 6: Table S6. The specific primer sequences for qRT-PCR assays.

Additional file 7: Figure S2. qRT-PCR optimization design: double standard curve (a) and dissolution curve (b) of six genes in embryo and endosperm. The dissolution curves of different genes are indicated.

Abbreviations

ADH: Alcohol dehydrogenase; AdSS: Adenylosuccinate synthetase; β-Glu: β- glucanase; 2-DE: Two-dimensional electrophoresis; CBB: Colloidal Coomassie Brilliant blue; CRT: Calreticulin-like protein; DEP: Different expressed protein; DEPs: Differentially expressed protein spots; FA: Fatty acid; GPD: Glyceraldehyde-3-phosphate dehydrogenase; GA: Gibberellic acid; GST: Glutathione S-transferase; HSP 70: Heat shock protein 70; IEF: Isoelectric focusing; IPG: Immobiline™ DryStrip gel; Met: Methionine synthase; MALDI-TOF/TOF-MS: Matrix assisted laser desorption ionization/ time of flight/tandem mass spectrometry; M: Embryo; N: Endosperm; PDI: Protein disulfide-isomerase; qRT-PCR: Quantitative real-time polymerase chain reaction; ROS: Reactive oxygen species; SODs: Superoxide dismutases; TCA cycle: Tricarboxylicacid cycle; TIM 17: Translocase of inner membrane 17; XIP-I: Xylanase inhibitor protein I.

Competing interests

The authors declare that they have no competing interests.

Authors' contributions

MH, CZ, KD and TZ carried out all experiments and data analysis. CZ and ZC performed the preparation of RNA, cDNA, qRT-PCR and bioinformatics analyses. JL made modifications for the manuscript. YY conceived the study, planned experiments, and helped draft the manuscript. All authors read and approved the final manuscript.

Acknowledgements

This research was financially supported by grants from the National Natural Science Foundation of China (31471485), the International Cooperation Project from Chinese Ministry of Science and Technology (2013DFG30530), the National Key Project for Transgenic Crops of China (2014ZX08009-003) and Natural Science Foundation of Beijing City and the Key Developmental Project of Science Technology, Beijing Municipal Commission of Education (KZ201410028031). This is contribution number 15-180-J from the Kansas Agricultural Experiment Station.

Author details

[1]College of Life Science, Capital Normal University, Beijing 100048, China. [2]Department of Plant Pathology, Kansas State University, Manhattan, KS 66506, USA. [3]Hubei Collaborative Innovation Center for Grain Industry, 434025 Jingzhou, China.

References

1. International Grains Council: Grain Maket Report. 2010, www.lgc.lnt/downloads/gmrsummary/gmrsumme.pdf.
2. Jacobsen JV, Gubler F, Chandler PM. Gibberellin action in germinated cereal grains. In: Davies PJ, editor. Plant hormones: Physiology, biochemistry and molecular biology. Dordrecht: Academic; 1995. p. 246–71.
3. Bethke PC, Schuurink R, Jones RL. Hormonal signaling in cereal aleurone. J Exp Bot. 1997;48:1337–56.
4. Lee KP, Piskurewicz U, Tureckova V, Strnad M, Lopez-Molina L. A seed coat bedding assay shows that RGL2-dependent release of abscisic acid by the endosperm controls embryo growth in Arabidopsis dormant seeds. Proc Natl Acad Sci U S A. 2010;107:19108–13.
5. Bewley JD, Black M. Seeds: Physiology of development and germination. New York, USA: Plenum Press; 1994. p. 199–389.
6. Bewley JD. Seed germination and dormancy. Plant Cell. 1997;9:1055–66.
7. Salon C, Raymond P, Pradet A. Quantification of carbon fluxes through the tricarboxylic acid cycle in early germinating lettuce embryos. J Biol Chem. 1988;263:12278–87.
8. Firenzuoli AM, Vanni P, Ramponi G, Baccari V. Changes in enzyme levels during germination of seeds of *Triticum durum*. Plant Physiol. 1968;43:260–4.
9. Aoki N, Scofield GN, Wang XD, Offler CE, Patrick JW, Furbank RT. Pathway of sugar transport in germinating wheat seeds. Plant Physiol. 2006;141:1255–63.
10. Fincher GB. Molecular and cellular biology associated with endosperm mobilization in germinating cereal grains. Annu Rev Plant Physiol Plant Mol Biol. 1989;40:305–46.
11. Caliskan M. Germin, an oxalate oxidase, has a function in many aspects of plant life. Turk J Biol. 2000;24:717–24.
12. Caliskan M, Bashiardes S, Ozcan B, Cuming AC. Isolation and localization of new germination-related sequences from wheat embryos. J Biochem Mol Biol. 2003;36:580–5.
13. Holdsworth MJ, Bentsink L, Soppe WJ. Molecular networks regulating *Arabidopsis* seed maturation, after-ripening, dormancy and germination. New Phytol. 2008;179:33–54.
14. Finch-Savage WE, Leubner-Metzger G. Seed dormancy and the control of germination. New Phytol. 2006;171:501–23.
15. Pritchard SL, Charlton WL, Baker A, Graham IA. Germination and storage reserve mobilization are regulated independently in Arabidopsis. Plant J. 2002;31(5):639–47.
16. Yu Y, Guo G, Lv D, Hu Y, Li J, Li X, et al. Transcriptome analysis during seed germination of elite Chinese bread wheat cultivar Jimai 20. BMC Plant Biol. 2014;14:20.
17. Girke T, Todd J, Ruuska S, White J, Benning C, Olrogge J. Microarray analysis of developing *Arabidopsis* seeds. Plant Physiol. 2000;124:1570–81.
18. Müller K, Job C, Belghazi M, Job D, Leubner-Metzger G. Proteomics reveals tissue-specific features of the cress (*Lepidium sativum* L.) endosperm cap

proteome and its hormone-induced changes during seed germination. Proteomics. 2010;10:406–16.
19. Catusse J, Strub JM, Job C, Van DA, Job D. Proteome-wide characterization of sugarbeet seed vigor and its tissue specific expression. Proc Natl Acad Sci. 2008;105:10262–7.
20. Boudet J, Buitink J, Hoekstra FA, Rogniaux H, Larré C, Satour P, et al. Comparative analysis of the heat stable proteome of radicles of *Medicago truncatula* seeds during germination identifies late embryogenesis abundant proteins associated with desiccation tolerance. Plant Physiol. 2006;140:1418–36.
21. Finnie C, Maeda K, Østergaard O, Bak-Jensen KS, Larsen J, Svensson B. Aspects of the barley seed proteome during development and germination. Biochem Soc Trans. 2004;32:517–9.
22. Fu Z, Jin X, Ding D, Li Y, Fu Z, Tang J. Proteomic analysis of heterosis during maize seed germination. Proteomics. 2011;11:1462–72.
23. Yang P, Li X, Wang X, Chen H, Chen F, Shen S. Proteomic analysis of rice (*Oryza sativa*) seeds during germination. Proteomics. 2007;7:3358–68.
24. He DL, Han C, Yao JL, Shen SH, Yang PF. Constructing the metabolic and regulatory pathways in germinating rice seeds through proteomic approach. Proteomics. 2011;11:2693–713.
25. Mak Y, Skylas DJ, Willows R, Connolly A, Cordwell SJ, Wrigley CW, et al. A proteomic approach to the identification and characterization of protein composition in wheat germ. Funct Integr Genomics. 2006;6:322–37.
26. Williams KL. Genomes and proteomes: towards a multidimensional view of biology. Electrophoresis. 1999;20:678–88.
27. Irara S, Brinib F, Godaya A, Masmoudib K, Pagès M. Proteomic analysis of wheat embryos with 2-DE and liquid-phase chromatography (ProteomeLab PF-2D) — A wider perspective of the proteome. J Proteomics. 2010;73:1707–21.
28. Weeden NF. Genetic and biochemical implications of the endosymbiotic origin of the chloroplast. J Mol Evol. 1981;17:133–13929.
29. Boyer PD. The ATP synthase—A splendid molecular machine. Ann Rev Biochem. 1997;66:717–49.
30. Weber J, Senior AE. ATP synthase: what we know about ATP hydrolysis and what we do not know about ATP synthesis. Biochim Biophys Acta. 2000;1458:300–9.
31. Senior AE, Nadanaciva S, Weber J. The molecular mechanism of ATP synthesis by F1F0-ATP synthase. Biochim Biophys Acta. 2002;1553:188–211.
32. Botha FC, Potgieter GP, Botha AM. Respiratory metabolism and gene expression during seed germination. Plant Growth Regul. 1992;11:211–24.
33. Matsumura H, Takano T, Yoshida KT, Takeda G. Arice mutant lacking alcohol dehydorogenase. Breed Sci. 1995;45:365–7.
34. Matsumura H, Takano T, Takeda G, Uchimiya H. Adh1 is transcriptionally active but its translational product is reduced in a rad mutant of rice (*Oryza sativa* L.), which is vulnerable to submergence stress. Theor Appl Genet. 1998;97:1197–203.
35. Rahman M, Grover A, Peacock WJ, Dennis ES, Ellis MH. Effects of manipulation of pyruvate decarboxylase and alcohol dehydrogenase levels on the submergence tolerance of rice. Aust J Plant Physiol. 2001;28:1231–41.
36. Edwards JM, Roberts TH, Atwell BJ. Quantifying ATP turnover in anoxic coleoptiles of rice (*Oryza sativa*) demonstrates preferential allocation of energy to protein synthesis. J Exp Bot. 2012;63:4389–402.
37. Kollöffel C. ADH in peas. Acta Bot Neerl. 1968;17:70–7.
38. Pierre P, Pascual P, Monique A. Structure, genomic organization, and expression of the *Arabidopsis thaliana* aconitase gene. J Biol Chem Mol Biol. 1995;14:8131–7.
39. Hayashi M, De BL, Alpi A, Nishimura M. Cytosolic aconitase participates in the glyoxylate cycle in etiolated pumpkin cotyledons. Plant Cell Physiol. 1995;36(4):669–80.
40. Morohashi Y, Bewley JD. Development of mitochondrial activities in pea cotyledons during and following germination of the axis. Plant Physiol. 1980;68:318–23.
41. Schinzel R, Nidetzky B. Bacterial alpha-glucan phosphorylases. Fems Microbiol Lett. 1999;171(2):73–9.
42. Schwarz A, Pierfederici FM, Nidetzky B. Catalytic mechanism of alpha-retaining glucosyl transfer by Corynebacterium callunae starch phosphorylase: the role of histidine-334 examined through kinetic characterization of site-directed mutants. Biochem J. 2005;387(2):437–45.
43. Jones RL, Jacobsen JV. Regulation of synthesis and transport of secreted proteins in cereal aleurone. Int Rev Cytol. 1991;126:49–88.
44. Sreenivasulu N, Usadel B, Winter A, Radchuk V, Scholz U, Stein U. Barley grain maturation and germination: metabolic pathway and regulatory network commonalities and differences highlighted by new MapMan/PageMan profiling tools. Plant Physiol. 2008;146:1738–58.
45. Guo HX, Yin J, Ren JP, Wang ZY, Chen HL. Changes in proteins within germinating seeds of transgenic wheat with an antisense construct directed against the thioredoxin. J Plant Physiol Mol Biol. 2007;33:18–24.
46. Chrispeels MJ, Tenner AJ, Johnson KD. Synthesis and release of sucrose by the aleurone layer of barley: regulation by gibberellic acid. Planta. 1973;113:35–46.
47. Onomo PE, Niemenak N, Ndoumou DO, Lieberei R. Change in amino acids content during germination and seedling growth of Cola sp. African J Biotechnol. 2010;9:5632–42.
48. Cardwell VB. Seed germination and crop production. In: Tesar MB, editor. Physiological basis of crop growth and development. Madison: Academic; 1984. p. 53–92.
49. Penfield S, Rylott EL, Gilday AD, Graham S, Larson TR, Graham IA. Reserve mobilization in the *Arabidopsis* endosperm fuels hypocotyl elongation in the dark, is independent of abscisic acid, and requires phosphoenolpyruvate carboxykinase1. Plant Cell. 2004;16:2705–18.
50. Ostergaard O, Finnie C, Laugesen S, Roepstorff P, Svensson B. Proteome analysis of barley seeds: Identification of major proteins from two-dimensional gels (*pI* 4-7). Proteomics. 2004;4:2437–47.
51. Mory YY, Chen D, Sarid S. Onset of deoxyribonucleic acid synthesis in germinating wheat embryos. Plant Physiol. 1972;49:20–3.
52. Marcus A, Feeley J, Volcani T. Protein synthesis in imbibed seeds III. Kinetics of amino acid incorporation ribosome activation, and polysome formation. Plant Physiol. 1966;41:1167–72.
53. Speigel S, Marcus A. Polyribosome formation in early wheat embryo germination independent of either transcription or polyadenylation. Nature. 1975;256:228–30.
54. Lalonde L, Bewley JD. Desiccation during germination of pea axes and its effects on the messenger RNA population. J Exp Bot. 1986;37:754–64.
55. Groll M, Ditzel L, Löwe J, Stock D, Bochtler M, Bartunik HD, et al. Structure of 20S proteasome from yeast at 24 Å resolution. Nature. 1997;386:463–71.
56. Eytan E, Ganoth D, Armon T, Hershko A. ATP-dependent incorporation of 20S protease into the 26S complex that degrades proteins conjugated to ubiquitin. Proc Nat Acad Sci USA. 1989;86:7751–5.
57. Chivers PT, Raines RT. General acid/base catalysis in the active site of *Escherichia coli* thioredoxin. Biochem. 1997;36:15810–6.
58. Freedman RB, Klappa P, Ruddock LW. Protein disulfide isomerases exploit synergy between catalytic and specific binding domains. EMBO Rep. 2002;3:136–40.
59. Bandarian V, Pattridge KA, Lennon BW, Huddler DP, Matthews RG. Domain alternation switches B(12)-dependent methionine synthase to the activation conformation. Nat Struct Biol. 2002;9:53–6.
60. Prade L, Cowan-Jacob SW, Chemla P, Potter S, Ward E, Fonne-Pfister R. Structures of adenylosuccinate synthetase from *Triticum aestivum* and *Arabidopsis thaliana*. J Mol Biol. 2000;296:569–77.
61. Finnie C, Melchior S, Roepstorff P, Svensson B. Proteome analysis of grain filling and seed maturation in barley. Plant Physiol. 2002;129:1308–19.
62. Helenius AES. Trombetta: Calnexin, calreticulin and the folding of glycoproteins. Trends Cell Biol. 1997;7(5):193–200.
63. Li YY, Ru ZG, Su Q, Fu QY. Identification and analysis of differentially expressed proteins of BNS male sterile line and its conversion line of wheat. Acta Agron Sin. 2011;37(9):1540–50.
64. Menegazzi P, Guzzo F, Baldan B, Mariani P, Treves S. Purification of calreticulin-like protein(s) from spinach leaves. Biochem Biophys Res Commun. 1993;190(3):1130–5.
65. Chen F, Hayes PM. Identification and characterization of cDNA clones encoding plant calreticulin in barley. Plant Cell. 1994;6:835–43.
66. Komatsu S, Yang G. Over-expression of calcium-dependent protein kinase 13 and calreticulin interacting protein 1 confers cold tolerance on rice plants. Mol Genet Genomics. 2007;277:713–23.
67. Galat A. Variations of sequences and amino acid composition of proteins that sustain their biological function: An analysis of the cyclophilin family of proteins. Arch Biochem Biophys. 1999;371(2):149–62.
68. Gasser CS, Gunning DA, Budelier KA. Structure and expression of cytosolic cyclophilin/peptidyl-prolyl cis-thans isomerase of higher plants and production of active tomato cyclophilin in *Escherichia coli*. Proc Natl Acad Sci U S A. 1990;87:9519–23.

69. He D, Han C, Yang P. Gene expression profile changes in germinating rice. J Integr Plant Biol. 2011;53:835–44.
70. Buchanan BB, Balmer Y. Redox regulation: a broadening horizon. Ann Rev Plant Biol. 2005;56:187–220.
71. Meyer Y, Buchanan BB, Vignols F, Reichheld JP. Thioredoxins and glutaredoxins: unifying elements in redox biology. Ann Rev Genet. 2009;43:335–67.
72. Raychaudhuri SS, Deng XW. The role of superoxide dismutase in combating oxidative stress in higher plants. Bot Rev. 2000;66:89–98.
73. Wood ZA, Schröder E, Robin HJ, Poole LB. Structure, mechanism and regulation of peroxiredoxins. Trends Biochem Sci. 2003;28:32–40.
74. Pol Merkur L. Heat shock proteins and their characteristics. J Article. 2005;19((110):215–9.
75. Wehmeyer N, Hernandez LD, Finkelstein RR, Vierling E. Synthesis of small heat-shock proteins is part of the development program of late seed maturation. Plant Physiol. 1996;112:747–57.
76. Elizabeth A, Kellogg E, Nickolas DJ. The structure and function of Rubisco and their implication for systematic studies. Am J Bot. 1997;84(3):413–28.
77. John EW, Matti S, Michael JR, Nicholas JG. Distantly related sequences in the a- and β-subunits of ATP synthase, myosin, kinases and other ATP-requiring enzymes and a common nucleotide binding fold. EMBO J. 1982;8:945–51.
78. Bec G, Kerjan P, Jean-Pierre Waller S. Reconstitution *in vitro* of the valyl-tRNA synthetase-elongation factor (EF) Iβγδ complex. J Biol Chem. 1994;269:2086–92.
79. Chen H, Xiong L. Pyridoxine is required for post embryonic root development and tolerance to osmotic and oxidative stresses. Plant J. 2005;44(3):396–408.
80. Fetzer S, Lauber J, Cindy LW, Luhrmann R. The [U4/U6.U5] tri-snRNP-specific 27 K protein is a novel SR protein that can be phosphorylated by the snRNP-associated protein kinase. RNA-Pub RNA Soc. 1997;3:344–55.
81. Gao L, Yan X, Li X, Guo G, Hu Y, Ma W, et al. Proteome analysis of wheat leaf under salt stress by two-dimensional difference gel electrophoresis (2D-DIGE). Phytochem. 2011;72:1180–91.
82. Steadman KJ. Dormancy release during hydrated storage in *Lolium rigidum* seeds is dependent on temperature, light quality, and hydration status. J Exp Bot. 2004;55:929–37.
83. Arcalis E, Sylvain M, Altmann F, Kolarich D, Drakakaki G, Fisher R, et al. Unexpected deposition patterns of recombinant proteins in post-endoplasmic reticulum compartments of wheat endosperm. Plant Physiol. 2004;136:3457–66.
84. Yang P, Shen S, Kuang TJ. Proteomic analysis of the response of *Liangyoupeijiu* (super high-yield hybrid rice) seeding to cold stress. Integra Plant Biol. 2006;48:1023–33.
85. Gao L, Wang A, Li X, Dong K, Wang K, Appels R, et al. Wheat quality related differential expressions of albumins and globulins revealed by two-dimensional difference gel electrophoresis (2-D DIGE). J Protemics. 2009;73:279–96.
86. Yang X, Xu H, Li W, Li L, Sun J, Li Y, et al. Screening and identification of seed-specific genes using digital differential display tools combined with microarray data from common wheat. BMC Genomics. 2011;12:513–24.
87. Untergasser A, Nijveen H, Rao X, Bisseling T. Primer3Plus, an enhanced web interface to Primer3. Nucleic acid res. 2007;35:71–4.
88. Livak KJ, Schmittgen TD. Analysis of relative gene expression data using real-time quantitative PCR and the 2 – ΔΔCTmethod. Methods. 2001;25:402–8.

Specific pools of endogenous peptides are present in gametophore, protonema, and protoplast cells of the moss *Physcomitrella patens*

Igor A Fesenko[1*], Georgij P Arapidi[1,3], Alexander Yu Skripnikov[1,6], Dmitry G Alexeev[2,3], Elena S Kostryukova[2], Alexander I Manolov[2,3], Ilya A Altukhov[2,3], Regina A Khazigaleeva[1], Anna V Seredina[1], Sergey I Kovalchuk[1,2], Rustam H Ziganshin[1], Viktor G Zgoda[4], Svetlana E Novikova[4], Tatiana A Semashko[2], Darya K Slizhikova[1], Vasilij V Ptushenko[5], Alexey Y Gorbachev[2], Vadim M Govorun[1,2,3] and Vadim T Ivanov[1]

Abstract

Background: Protein degradation is a basic cell process that operates in general protein turnover or to produce bioactive peptides. However, very little is known about the qualitative and quantitative composition of a plant cell peptidome, the actual result of this degradation. In this study we comprehensively analyzed a plant cell peptidome and systematically analyzed the peptide generation process.

Results: We thoroughly analyzed native peptide pools of *Physcomitrella patens* moss in two developmental stages as well as in protoplasts. Peptidomic analysis was supplemented by transcriptional profiling and quantitative analysis of precursor proteins. In total, over 20,000 unique endogenous peptides, ranging in size from 5 to 78 amino acid residues, were identified. We showed that in both the protonema and protoplast states, plastid proteins served as the main source of peptides and that their major fraction formed outside of chloroplasts. However, in general, the composition of peptide pools was very different between these cell types. In gametophores, stress-related proteins, e.g., late embryogenesis abundant proteins, were among the most productive precursors. The Driselase-mediated protonema conversion to protoplasts led to a peptide generation "burst", with a several-fold increase in the number of components in the latter. Degradation of plastid proteins in protoplasts was accompanied by suppression of photosynthetic activity.

Conclusion: We suggest that peptide pools in plant cells are not merely a product of waste protein degradation, but may serve as important functional components for plant metabolism. We assume that the peptide "burst" is a form of biotic stress response that might produce peptides with antimicrobial activity from originally functional proteins. Potential functions of peptides in different developmental stages are discussed.

Keywords: Endogenous peptides, LC-MS/MS, Physcomitrella patens, Proteome, Transcriptome profiling

Background

Peptides are well known to be key regulators of many animal physiological processes, including defense reactions and hormonal, neurohumoral, and signaling functions. In recent years, a number of small peptides with similar activities have been also discovered in land plants [1-4]. As in animals, peptide signals regulating plant growth and development act as ligands of receptor-like kinases [5]. Over 400 homologs of receptor-like kinases and more than 1000 genes predicted to encode precursors of secreted peptides are found in the genome of *Arabidopsis thaliana* [6,7]. Thus, the currently known regulatory plant peptides very likely constitute just a tiny portion of the total number of secreted peptides really involved in the control of physiological processes [7]. Bioactive peptides are assumed to be primarily translated as inactive precursor proteins that are cleaved by various proteases to produce matured bioactive factors. In recent years, a new source of bioactive peptides has been found. Small open

* Correspondence: fesigor@gmail.com
[1]Department of Proteomics, Shemyakin-Ovchinnikov Institute of Bioorganic Chemistry of the Russian Academy of Sciences, 16/10, Miklukho-Maklaya, GSP-7, Moscow 117997, Russian Federation
Full list of author information is available at the end of the article

reading frames (ORFs) can be directly translated into peptides that play essential roles in eukaryotes [8-11]. In addition, degradation of originally functional proteins can also contribute to functional peptidomes in eukaryotic organisms [2,12-17]. Examples of such peptides in plants are inseptin, which is a fragment of chloroplast ATP synthase from cowpea (*Vigna unguiculata*) [18], and the GmSubPep and GmPep914 peptides produced from soy (*Glycine max*) subtilisin-like protease [19,20]. Still, little is known about the generation of the proteolytic degradome in plant cells and tissues or its physiological role.

The moss *Physcomitrella patens* is a promising model organism in plant biology [21-23]. Mosses are descendants of early divergent embryophyte lines and therefore occupy an ideal phylogenetic position for reconstructing the evolutionary history of terrestrial plants and understanding the changes that accompanied the emergence of land plants. Furthermore, *P. patens* exhibits the highest rate of homologous recombination among land plants, giving it the unique ability to be genetically manipulated using targeted gene replacements. In addition to its nuclear genome [24], numerous studies of the proteome [25-31], transcriptome [32-36], and metabolome [37] of *P. patens* have been published.

The gametophyte, the haploid generation that prevails in the moss life cycle, goes through two stages of development. In the first stage, called the protonema, the gametophyte is a net of filaments that develops in a wet environment. Protonema cells differentiate into buds that give rise to the leafy adult stage, termed the gametophore. Gametophores grow as three-dimensional leafy shoots on which the reproductive organs, antheridia and archegonia, form under suitable environmental conditions. Protoplasts prepared from protonema filaments are of particular interest because, during the first hours of regeneration, they are reprogrammed into protonemal apical stem cells without forming a callus. Protoplasts are useful for studies of stress because the isolation of protoplasts from cell walls appears to have similar effects to plasmolysis induced by drought or salinity stress [35]. We previously described the significant change in the proteome of *P. patens* protonema cells that occurs during protoplast isolation [38].

Peptides formed by degradation of functionally active proteins can represent a significant fraction of the cell peptidome, but this fraction is poorly understood in plant cells. The aim of this work was to identify the pools of native peptides, elucidate their patterns of formation, and evaluate the effects of stress factors on the peptidome. We comprehensively analyzed the peptidomes of protoplasts, protonemata, and gametophores of moss *P. patens* cells and performed transcriptional profiling and quantitative proteomic analysis of precursor proteins. Significant differences between the peptidomes of the three cell types were found. We did not observe direct proportionality between intact protein concentrations and their corresponding native peptide fragments; the intensity of degradation and proteolysis patterns depended, rather, on the moss cell form. This fact suggests that differentially regulated mechanisms of protein degradation are involved at different growth stages and that different peptides may be important for different cell forms. Under stress conditions, we found significant differences in the peptidome of moss protoplasts compared with protonemata and gametophores. An increase in the number of chloroplast protein peptides was accompanied by suppression of photosynthetic activity. We suggested that peptide pools generated by protein turnover and degradation have a significant potential for biological activity. We identified 81 peptides in protoplasts with probable antimicrobial activity. Finally, we suggested a scheme of processes leading to and affecting peptidome formation in protoplast cells.

Methods

Physcomitrella patens protonema and gametophore growth conditions

The protonemata of the moss *P. patens* subsp. *patens* Gransden 2004 were grown on Knop medium with 500 mg/L ammonium tartrate with 1.5%agar (Helicon, Moscow, Russian Federation) in a Sanyo Plant Growth Incubator MLR-352H (Panasonic, Osaka, Japan) with a photon flux of 61 μmol/m^2•s during a 16-hour photoperiod at 24°C. For transcriptomic and peptidomic analyses, we used 5-day-old protonema tissue. The moss gametophores were grown on Knop medium in 9-cm Petri dishes in the same incubator with a 16-hour photoperiod at 24°C and 61 μmol/m^2•s. We used 8-week-old gametophores for analyses.

Protoplast preparation and driselase treatment

Five-day-old protonema filaments were harvested with a spatula from the agar surface, and 1 g well-drained protonema tissue was placed in 14 mL 0.5% (w/v) Driselase (Sigma-Aldrich, St. Louis, MO, USA) solution in 0.48 M mannitol and incubated for 60 min with constant shaking in darkness. Then, the suspension was filtered through 100 μm steel mesh (Sigma-Aldrich), and the protoplasts obtained were incubated in Driselase solution for 15 minutes more. The protoplasts were then precipitated by centrifugation in 50-mL plastic tubes using a swinging bucket rotor at 100 × *g* for 5 min. Next, protoplasts were washed twice with 0.48 M mannitol with centrifugation under the same conditions and sedimented again. The supernatant was removed and the protoplast pellet was frozen in liquid nitrogen for peptide extraction or RNA isolation. The number of protoplasts was measured with a hematocytometer.

The treatment of protonemata with 0.025% w/v and 0.0025% w/v Driselase solution followed a similar protocol. As a control, we also incubated protonema tissue in 0.48 M mannitol. After 1-h incubation, the protonema tissue was washed and peptides extracted.

Isolation of chloroplasts from moss protoplasts

Chloroplasts were isolated from protoplasts as previously described [27]. In short, protoplasts were resuspended in buffer A (50 mM HEPES-KOH, pH 7.5, 330 mM sorbitol, 2 mM EDTA, and 0.4 mM phenylmethylsulfonyl fluoride) and filtered through a double layer of Miracloth (Calbiochem Behring, La Jolla, CA, USA). Protoplast disintegration was evaluated with a light microscope. The filtrate was then centrifuged at 1200 × *g* for 3 min in 50-mL plastic tubes using a bucket rotor. The pellet was resuspended in a small volume of buffer A and fractionated by centrifugation in a bucket rotor at 3800 × *g* for 10 min in a 10%-40%-85% Percoll (Sigma-Aldrich) stepwise gradient in 15-mL plastic tubes. Intact chloroplasts between the 40% and 85% Percoll layers were gathered, washed with buffer A, and centrifuged at 1200 × *g* for 3 min in 15-mL plastic tubes (Falcon) in a bucket rotor. The resulting chloroplast pellet was used for native peptide extraction.

Peptide extraction

Endogenous peptides were extracted from moss tissue, protoplasts, and intact chloroplasts as previously described with some modification [38]. To minimize artifacts during peptide extraction, we used an acid extraction buffer with a mixture of plant protease inhibitors, and all steps were performed on ice. For peptide extraction from moss tissues and protoplasts, the extraction buffer was 1 M acetic acid in 10% acetonitrile and 10 μL/mL of Protease Inhibitor Cocktail (Sigma-Aldrich). Protoplasts and intact chloroplasts were disrupted directly in the extraction buffer with a Ultra-Turrax T10 basic homogenizer (IKA, Staufen, Germany) using a S10N10G nozzle at a rotation speed of 3000 rpm at 4°C. For peptide extraction, protonemata were harvested from the surface of the agar medium and gametophores were excised 1 mm above the agar surface. The tissue was then placed into a porcelain mortar pre-cooled to −70°C, where it was immediately frozen with liquid nitrogen and ground to fine dust with a pestle pre-cooled to −70°C. The ground material was placed into cooled extraction buffer containing proteinase inhibitors and homogenized using a Dismembrator S ball mill (Sartorius, Göttingen, Germany) at 2600 rpm for 2 min with a mix of glass balls of 0.1, 0.3, and 1 mm diameter (Sartorius). The suspension was centrifuged at 11,000 × *g* for 10 min at 4°C. The supernatant was then transferred to a clean test tube and centrifuged again at 11,000 × *g* for 10 min at 4°C, after which the pellet was discarded.

Samples were immediately placed into a gel filtration column to extract and fractionate the peptides. Gel filtration was carried out on a 2.5 cm × 30 cm column filled with Sephadex G-25 superfine in 0.1 M acetic acid. The elution was with 0.1 M acetic acid at a flow rate of 1 mL/min. Proteins and peptides were detected on an LKB Bromma 2518 Uvicord SD device (LKB, Vienna, Austria) at a wavelength of 280 nm. The fractions containing peptides were lyophilized and resuspended in 5% acetonitrile-0.1% trifluoroacetic acid. Before recording the mass spectra, samples were desalted on reversed-phase C18 microcolumns, which were prepared in 200 μL tips for an automatic pipette with two layers of Empore™ extraction disk reversed-phase C18 membrane (Supelco, Bellefonte, PA, USA) 1.6 mm in diameter, as previously described [39]. The desalted peptide preparations were concentrated on a SpeedVac Concentrator vacuum centrifugal evaporator (Savant, Waltham, MA, USA) to a volume of 5 μL and diluted with 3% acetonitrile in 0.1% trifluoroacetic acid to 20 μL.

Protein extraction

Proteins were extracted using a modified phenol extraction procedure [40]. Plant tissue was ground to fine powder in liquid nitrogen, and three volumes of ice-cold extraction buffer (500 m Tris–HCl, pH 8.0, 50 mM EDTA, 700 mM sucrose, 100 mM KCl, 1 mM phenylmethylsulfonyl fluoride, 2% 2-mercaptoethanol, 1% Triton X-100) were added, followed by 10 min incubation on ice. An equal volume of ice-cold Tris–HCl (pH 8.0)-saturated phenol was added, and the mixture was vortexed and incubated for 10 min with shaking. After centrifugation (10 min, 5500 × *g*, 4°C), the phenol phase was collected and re-extracted twice with extraction buffer. Proteins were precipitated from the final phenol phase with three volumes of ice-cold 0.1 M ammonium acetate in methanol overnight at −20°C. The pellets were rinsed with ice-cold 0.1 M ammonium acetate in methanol three times and with ice-cold acetone containing 13 mM dithiothreitol once and then dried. Pellets were solubilized in a sample buffer (8 M urea, 2 M thio urea, 17% solution of 30% CHAPS (3-[(3-cholamidopropyl) dimethylammonio]-1-propanesulfonate) and 10% NP40 (octylphenoxypolyethoxyethanol)). Protein concentration in the samples was determined according to Bradford procedure using the Quick Start Bradford protein assay (Bio-Rad, Hercules, CA USA); bovine serum albumin was used to prepare standard solutions.

Mass-spectrometry analysis

Analysis was performed on a TripleTOF 5600+ mass-spectrometer with NanoSpray III ion source (ABSciex, Framingham, MA 01701, USA) coupled with a NanoLC

Ultra 2D+ nano-HPLC system (Eksigent, Dublin, CA, USA). The HPLC system was configured in a trap–elute mode. For sample loading buffer and buffer A, a mix of 98.9% water, 1% methanol (v/v), and 0.1% formic acid (v/v) was used. Buffer B was 99.9% acetonitrile and 0.1% formic acid (v/v). Samples were loaded on a trap column Chrom XP C18 (3 μm, 120 Å, 350 μm × 0.5 mm; Eksigent) at a flow rate of 3 μL/min for 10 min and eluted through the separation column 3C18-CL-120 (3 μm, 120 Å, 75 μm × 150 mm; Eksigent) at a flow rate of 300 nL/min. The gradient was from 5% to 40% buffer B over 120 min. The column and precolumn were regenerated between runs by a wash with 95% buffer B for 7 min and equilibrated with 5% buffer B for 25 min. To thoroughly wash the column and precolumn between different samples and to prevent possible crosstalk, a 45-min blank run consisting of 5 × 5 min waves (5%, 95%, 95%, and 5% B) was performed, followed by column equilibration for 10 min with 5% B.

An information-dependent mass-spectrometer (MS) experiment included one survey MS1 scan followed by 50 dependent MS2 scans. MS1 acquisition parameters were 300–1250 m/z mass range for analysis and subsequent ion selection for MS2 analysis and 250 ms signal accumulation time. Ions for MS2 analysis were selected on the basis of intensity with a threshold of 400 cps and a charge state from 2 to 5. MS2 acquisition parameters were: resolution of quadrupole set to UNIT (0.7 Da), measurement mass range 200–1800 m/z, optimization of ion beam focus to obtain maximal sensitivity, and signal accumulation time of 50 ms for each parent ion. Collision activated dissociation was performed with nitrogen gas with collision energy ramping from 25 to 55 V within the 50-ms signal accumulation time. Analyzed parent ions were sent to a dynamic exclusion list for 15 sec to get an MS2 spectrum at the chromatographic peak apex (minimum peak width throughout the gradient was about 30 s).

An LTQ Orbitrap Velos system was equipped with an Agilent HPLC System 1100 Series (Agilent Technologies, Santa Clara, CA, USA) and a nanoelectrospray ion source (Thermo Scientific, Waltham, MA, USA). The peptide separation was carried out on an RP-HPLC column Zorbax 300SB-C18 (Agilent Technologies, Santa Clara, CA 95051, USA) (3.5 μm × 75 μm × 150 mm) using a linear gradient from 95% solvent A (100% water, 0.1% formic acid) and 5% solvent B (20% water, 80% acetonitrile, 0.1% formic acid) to 40% solvent A and 60% solvent B over 85 minutes at a flow rate of 300 nL/min.

Mass spectra were acquired in positive ion mode. Data were acquired in the Orbitrap analyzer with a resolution of 30,000 (m/z 400) for MS and 7,500 (m/z 400) for MS/MS scans. A survey MS scan was followed by acquisition of MS/MS spectra of the five most abundant precursors. For peptide fragmentation, high-energy collisional dissociation (HCD) was used; the signal threshold was set to 5,000 for an isolation window of 2 Th and the first mass of an HCD spectrum was set to 100 m/z. The collision energy was set to 35 eV. Fragmented precursors were dynamically excluded from targeting for 60 s. Singly charged ions and ions with a non-defined charge state were excluded from triggering MS/MS scans.

Relative protein quantification

Protein concentrations were evaluated by label-free MS1 intensity-based quantification with the use of the Progenesis LC-MS (Nonlinear Dynamics, Durham, NC, USA) software package, which estimated correlation of tryptic peptidogenicity and protein expression levels. Raw data files (.wiff format) were converted into .mzML files using AB SCIEX MS Data Converter (version 1.3, ABSciex) and loaded into the Progenesis LC-MS. Progenesis LC-MS generated mascot generic files (.mgf format) that were searched using Mascot version 2.4.1 (Matrix Science, Boston, MA 02110, USA) against the UniProt sequence database (UniProt Consortium, (ftp.uniprot.org/pub/databases/uniprot/current_release/knowledgebase/complete, downloaded April 19, 2010) filtered by *P. patens* proteins (35,414 amino acid sequences). The Mascot search was performed with the following parameters: tryptic-specific peptides; maximum of one missed cleavage; peptide charge state limited to 1+, 2+, and 3+; precursor mass tolerance 20 ppm; MS/MS mass tolerance 50 ppm; variable modifications caused by oxidation (M) and carbamidomethylation (C). Using decoy (reversed) databases, false discovery rates (FDRs) were calculated, and the ion score cut-off was set to an FDR less than 5%.

Two-sided unpaired Student's t-test (R version 3.0.2; R Foundation, Vienna, Austria) was conducted to evaluate the validity of the quantification results. Log (base 2) fold changes between different conditions were calculated for median values.

Peptide analysis by mass spectrometry and data integration

Moss native peptide identifications were performed on the basis of a single LC-MS run for each sample. The .wiff data files were analyzed with the ProteinPilot software 4.5 revision 1656 (ABSciex) using the search algorithm Paragon 4.5.0.0 revision 1654 and the default parameter set for protein identification with the following adjustments: uniref100_Physco_35213 protein sequence database no Cys alkylation, no digestion, TripleTOF5600 equipment, organism type not specified, search effort – thorough ID, detection protein threshold – unused protein score 0.05. Spectrum grouping was performed with default parameters using the ProGroup algorithm embedded in ProteinPilot. Peptide identification FDR statistical analysis was performed using the ProteomicS Performance Evaluation Pipeline Software (PSPEP) algorithm also embedded in the ProteinPilot software. Peptides with probability over 95% were selected for analysis. Additionally, spectra acquired with TripleTOF 5600+ and LTQ Orbitrap Velos were searched with Mascot Version: 2.2.07 (Matrix Science), using the following parameters: precursor mass tolerance 20 ppm, MS/MS mass tolerance 50 ppm, no fixed modifications. Peptides with Mascot scores above the threshold were selected for analysis.

Peptide identification data was integrated in an *ad hoc* SQL database based on protein accessions. The number of peptides per protein was calculated as a sum of unique peptides found by both search algorithms. The functional analysis of precursor proteins was performed with the Database for Annotation, Visualization and Integrated Discovery (DAVID) (http://david.abcc.ncifcrf.gov/). The following parameters were used: count threshold 2, EASE threshold 0.01. Parameters of cluster analysis were as follows: similarity term overlap 10, similarity threshold 0.50, initial and final group membership 3, multiple linkage threshold 0.5, and enrichment thresholds 0.01.

Antimicrobial peptide potential

The native peptide theoretical antimicrobial potential was assessed on the basis of sequence with special AMPA software (http://tcoffee.crg.cat/apps/ampa/do) [41]. All identified native peptides longer than six amino acids were searched for any internal part with high antimicrobial potential. For the AMPA analysis, the recommended parameters were used: threshold value of 0.225 and window size of seven amino acids. Only antimicrobial peptides with probability of misclassification less than 5% were used.

RNA extraction and cDNA library preparation

To analyze the transcriptomes of protonemata, gametophores, and protoplasts and to validate the RNA-seq results, we extracted RNA as previously described [42]. The quality and quantity of the extracted total RNA was initially evaluated by electrophoresis in agarose gels with ethidium bromide staining. Quantification of the total RNA in the sample was carried out with the Quant-iT™ RNA Assay Kit (5–100 ng; Life Technologies, Carlsbad, CA, USA) in a Qubit fluorometer (Invitrogen, Carlsbad, CA, USA). The quality of the total RNA samples was evaluated using an Agilent RNA 6000 Nano kit and a 2100 Bioanalyzer (Agilent Technologies). The RNA was evaluated on the basis of peaks for 28S and 18S ribosomal RNA. The mRNA fraction was isolated using a MicroPoly(A)Purist™ Kit (Ambion, Carlsbad, CA, USA) according to the manufacturer's recommendations. To achieve maximum removal of ribosomal and noncoding RNA from the sample, the procedure was repeated twice. The mRNA was quantified and the quality evaluated as described above. To generate a fragment library, about 500 ng mRNA of each sample was used. The mRNA fragment library was prepared with the SOLiD™ Total RNA-Seq Kit (Ambion) according to the manufacturer's recommendations.

SOLiD sequencing and sequence assembly

The sequencing of the mRNA fragment library was performed with a SOLiD 4 genetic analyzer (Applied Biosystems, Foster City, CA, USA) according to the manufacturer's recommendations with both biological and technical repeats (gametophores: 2 biological and 2 technical repeats, protonema and protoplast: 3 biological and 2 technical repeats). We obtained 173, 197, and 204 million reads for the gametophore, protonema, and protoplast samples, respectively. The length of each read was 50 bp. The number of uniquely mapped filtered reads was 31, 36, and 38 million for gametophore, protonema, and protoplast samples, respectively. The reads were filtered with the SOLiD_preprocess_meanFilter_v2.pl utility using default parameters [reads with unread positions were rejected (hole filtering), as were those with average quality below 20].

As a reference, we used the *P. patens* genome v.1.6 (http://cosmoss.org) [43]. Reference genome mapping was performed with TopHat v2.0.7 software [44] using default parameters. To evaluate the gene expression level in RPKM, the produced .bam file was processed with the Cufflinks utility [45], and we used HTSeq to count the number of mapped reads for each gene. The number of uniquely mapped filtered reads was 31, 36, and 38 million for gametophore, protonema, and protoplast samples, respectively. We found evidence of expression of 18,412 coding sequences (CDS; at the level more than one read per million). To validate the accuracy and to evaluate the distortion that occurred during library preparation, the transcriptional levels of 17 genes were analyzed by quantitative real-time PCR (qRT-PCR). The Spearman correlation values of gene expression obtained by qRT-PCR and RNA-seq methods were 0.7, 0.7, and 0.8 for gametophore, protonema, and protoplast samples, respectively (see Additional file 1). For analysis of differential expression, the edgeR [46] package was used, and the analysis was performed according to the recommendations in the edgeR vignette. We used read count per gene data as input for edgeR. The genes up-regulated in protoplasts were identified using the following criteria for differential expression: a FDR level less than 0.05 and expression level difference between samples of at least four fold.

Quantitative real-time PCR

Real-time PCR was performed using iQ SYBR Green Supermix (Bio-Rad) and the CFX96™ Real-Time PCR Detection System (Bio-Rad). Droplet digital PCR allows direct quantification of DNA molecules in a sample [47]. It was performed using ddPCR™ Supermix for Probes (Bio-Rad) and the QX100 system (droplet generator and droplet reader) along with a DNA Engine Tetrad 2 PCR machine (Bio-Rad). Real-time and ddPCR data were analyzed with CFX Manager and QuantaSoft (Bio-Rad) software, respectively. Primers and probes are listed in Additional file 2. PCR experiments were carried out using three biological and two technical replicates. We used the bootstrap method to determine the Pearson correlation coefficient.

Analysis of photosynthetic activity of *P. patens* protonemata and protoplasts

To analyze changes in the photosynthetic activity of *P. patens* cells, we monitored the induction of chlorophyll fluorescence (Maxwell and Johnson, 2000; Adams and Demmig-Adams, 2004; Baker, 2008). The measurements were carried out with a FluorPen FP100 PAM-fluorometer (Photon Systems Instruments, Brno, Czech Republic). Fluorescence was measured in response to short (<50 μs) flashes of measuring light with average intensity not exceeding 0.1 $\mu mol{\cdot}m^{-2}{\cdot}s^{-1}$. Flashes of saturating light (3000 $\mu mol{\cdot}m^{-2}{\cdot}s^{-1}$) were 1 s long. The actinic light intensity ranged from 10 to 1000 $\mu mol{\cdot}m^{-2}{\cdot}s^{-1}$. The wavelength of measuring, saturating, and actinic light was 475 nm. Fluorescence was monitored in the range of 697–750 nm. To estimate the maximum quantum efficiency of photosystem (PSII) photochemistry, Φ^{PSII}_{max}, the sample was first adapted to darkness for 15 min. The operating quantum efficiency of PSII at light intensity X $\mu mol{\cdot}m^{-2}{\cdot}s^{-1}$, $\Phi^{PSII}_{X\mu E}$, depending on closing a part of PSII centers, was evaluated after cell adaptation to the given light intensity for 1 min after darkness or short-term treatment with light of lower intensity. The coefficient of non-photochemical quenching, q_{NPQ}, was also evaluated while lighting cells after adaptation to darkness for 15 min.

Results

The *P. patens* gametophyte peptidome

Extracts of gametophores and protonemata were analyzed by tandem mass spectrometry (LC-MS/MS) against a uniref100 protein database for *P. patens*. Peptides ranging in size from 5 to 78 aa were identified (Additional file 3). A total of 4,361 peptide fragments of 761 precursor proteins were identified in gametophore extracts (Figure 1; Additional file 4). Since the majority of the identified precursors were annotated as predicted proteins, to assign their possible functions and localizations, we used BLAST homology analysis against the protein database of green plants (Viridiplantae). However, the most peptidogenic gametophore precursor proteins, A9U4I0, A9RXW5, and A9RXR3, with 144, 115, and 111 unique peptides respectively, were uncharacterized (see Additional file 4). Also, among the most peptidogenic precursor proteins were such major chloroplast proteins as A9SPD7 (photosystem I reaction center subunit IV, 52 native peptides) and A9U1R6 (outer envelope pore protein 16, 42 native peptides), as well as chaperone-like proteins, such as the A9SU24 (putative late embryogenesis abundant protein, group 3, 41 native peptides).

In protonema cells, mainly represented by chloronema cells, we identified 4,333 peptides that are fragments of 855 precursor proteins (see Additional file 5). Precursor proteins represented by a large number of peptides included major proteins like the large subunit of ribulose-1, 5-bisphosphate carboxylase/oxygenase (RuBisCO), P34915 (giving rise to 111 native peptides), and elongation factor 1-alpha, A9RGA5 (80 native peptides), A9RX76 (chloroplastic fructose-bisphosphate aldolase, 48 native peptides), and A9SPD7 (photosystem I reaction center subunit IV, 47 native peptides), among others.

We analyzed the peptide abundance in peptidomes for several groups of precursor proteins. Gametophore and protonema peptidomes contain a prominent fraction of peptides (approximately 25% to the total peptidome) from the chloroplast precursor proteins (see Additional file 6). The share of fragments of mitochondrial (3% for protonema and 10% for gametophores) and nuclear (4% for protonema and 5% for gametophores) precursor proteins, as well as of those proteins involved in translation (ribosome proteins, elongation factor alpha, etc.; 8% for protonema and 3% for gametophores), was also rather high. Since LEA proteins were among the most peptidogenic precursor proteins in gametophores, we evaluated their contribution to the peptidome as well. LEA protein content was higher in gametophores than in protonema (164 native peptides in gametophores vs 38 peptides in protonema), while peptides of other chaperon-like proteins, such as the heat shock proteins, were more abundant in protonema (125 peptides in protonema vs 66, in gametophores) (see Additional file 6).

Dramatic changes in the *P. patens* protoplast peptidome

Physcomitrella patens protoplasts are released after treating young protonema tissue with Driselase, a natural enzyme mixture containing laminarinase, xylanase, and cellulase activities [48]. This process severely stresses plant cells through the loss of the cell wall, as well as through the effects of the compounds in the crude Driselase preparation. We also considered that, as protoplasts are a good model for studying reprogramming of somatic plant cells [35], peptidomic information must be essential to describe the metabolic landscape.

We found that *P. patens* protoplast peptidome comprised 20,427 peptides, ranging in size from 6 to 78 aa (Additional file 3), that were fragments of 1,572

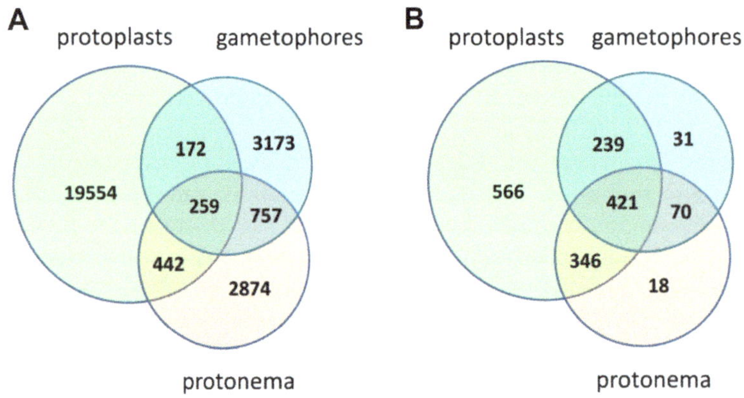

Figure 1 Venn diagrams of peptide and precursor distributions in protonemata, gametophores, and protoplasts. A, Peptide distributions. **B**, Precursor protein distributions.

precursor proteins (see Additional file 7). Notably, a considerable part of the peptides in protoplasts differed from each other by deletion of either the C- or N-terminal amino acid. Apparently, the "peptide ladders" result from degradation of the peptides by amino- and carboxypeptidases (Additional file 3).

In the protoplast peptidome, we identified a large number of peptides derived from chloroplast proteins (see Additional file 6). The large subunit of RuBisCO underwent the most severe degradation. Among the other highly represented precursor proteins were photosystem I reaction center subunit II-2, RuBisCO activase, carbonic anhydrase, elongation translation factor 1-alpha, lipoxygenase, and plastocyanin. All these data point to intensive protein degradation in protoplasts. There is evidence that chloroplast proteins like RuBisCO and, particularly, RuBisCO activase are the main targets for cysteine protease in vacuoles of plant cells [49]. We tested whether the observed peptides were generated in chloroplasts. In chloroplasts isolated from protoplasts, we identified 82 unique peptides that were fragments of 21 precursor proteins (see Additional file 8); only three of the peptides were fragments of the large subunit of RuBisCO, one of the most abundant proteins in cell. These data confirm that the major chloroplast proteins are degraded outside the intact chloroplasts.

To examine the factors involved in this marked difference in the amount/diversity of peptides, we tested the effects of Driselase at a concentration lower than that used to isolate the protoplasts. When protonema tissue was treated with 0.025% (w/v) Driselase solution, protoplasts did not form. However, treatment of protonemata with Driselase at this concentration resulted in 2.5 times more native peptides (Additional file 9). This finding is indicative of the "biotic" stress that protonema cells undergo when being treated with Driselase.

To evaluate the fraction of peptides that could be isolated from dead cells, we assessed the viability of protoplasts using the Trypan blue dye. In the analysis, we were not able to identify a substantial number of stained cells (data not shown). Such results indicate that dead or dying protoplasts did not significantly affect of our results. Neither did we detect a significant number of peptides in the protoplast wash solution (data not shown).

Earlier, functionally active proteins were shown to potentially have encrypted sequences possessing antimicrobial activity [16]. To evaluate the antimicrobial potential of the identified peptides, we used the Antimicrobial Sequence Scanning System (AMPA) [41]. We identified 117 endogenous peptides that might have antimicrobial activity, 81 of which were unique to protoplasts, 11 to protonemata, and 27 to gametophores (see Additional file 10).

Functional analysis of precursor proteins

We used the identified precursor proteins of peptides to carry out functional analysis using the Database for Annotation, Visualization and Integrated Discovery (DAVID) (http://david.abcc.ncifcrf.gov/). After 1,691 of the precursor proteins were correlated with proteins in the DAVID database, we performed clustering with a high threshold to decrease the number of clusters. We obtained 15 clusters comprising 1,030 proteins (Additional file 11). Precursor proteins localized in chloroplasts and ribosome structural components were the two most represented clusters in all three life forms (Figure 2). Many other precursor proteins were involved in the processes of carbohydrate metabolism, transmembrane transporter activity, and biosynthesis of plant hormones, terpenoids, and steroids. The number of precursor proteins referred to the ribosomal protein cluster was highest in protoplasts. However, in the protoplast peptidome, more peptides were derived from chloroplast precursor proteins than from ribosomal precursor proteins. The total amount of endogenous peptides identified in protoplasts differed

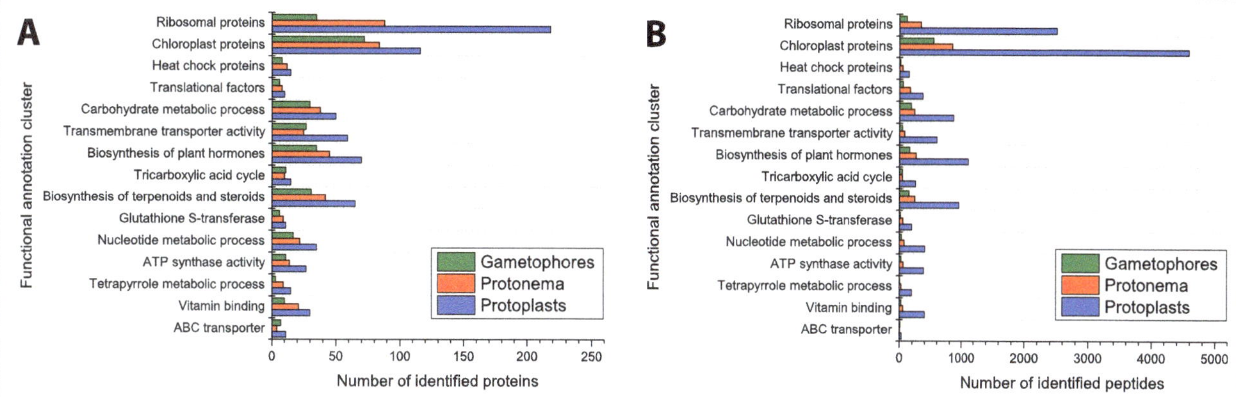

Figure 2 Distribution of the clusters of precursor proteins. A, Distribution of the clusters by the number of peptides identified in the gametophores, protonemata, and protoplasts. **B**, Distribution of the clusters by the number of identified precursor proteins in the gametophores, protonemata, and protoplasts. Statistically significant ($p < 0.05$) clusters of precursor proteins were grouped by the DAVID web service based on the default functional annotation database set.

considerably compared to gametophores or protonemata, indicating that peptide generation depends on cell type. For example, in protoplasts, we identified 4,580 peptides derived from chloroplast precursor proteins, more than five times the number in gametophores (543 peptides) or protonemata (848). The number of identified precursor proteins in protoplasts related to chloroplasts (116 proteins) was only about 1.5 times greater than those in gametophores (72) or protonemata (84).

The change of gene expression profiles in *P. patens* protoplasts

To evaluate the transcription levels of the identified peptide precursor proteins and to examine what cellular mechanisms were activated to cause the significant differences observed in the protoplast peptidome, we performed transcriptome analysis of gametophores, protonemata, and protoplasts. We identified 1936 genes upregulated in protoplasts (Additional file 12). Analysis of Gene Ontology (GO) terms showed that, in protoplasts, the transcriptional levels increased of genes involved in responses to different stress factors, e.g., abiotic stimulus (GO:0009628), cold (GO:0009409), temperature (GO:0009266), and oxidative stress (GO:0006979; Additional file 13; Additional file 14). We identified several genes that participate in jasmonic acid (JA) biosynthesis. In addition, the transcription levels of eight WRKY transcriptional factors involved in gene regulation in a range of processes, including biotic and abiotic stresses, senescence, and different developmental processes, increased [50].

We also examined the differentially expressed genes that could contribute to the protoplast peptide pool. There was a significant increase in the transcription of Pp1s166_98V6, encoding putative proteasome activating protein 200 (PA200). Also, a number of genes encoding proteases, for example Pp1s78_186V6 (*subtilisin-like serine protease 2*), Pp1s112_240V6 (*subtilisin-like serine protease 3*), Pp1s39_149V6 (a homolog of mitochondrial protease *FtsH3*), and Pp1s5_15V6 (a homolog of protease *Lon1*), showed higher transcription levels in protoplasts. In addition, we observed increased transcription of genes responsible for ubiquitin-mediated protein breakdown, e.g. RING/U-box superfamily protein and E3 ubiquitin ligase family protein (see Additional file 15 online). Interestingly, we identified significant increases in the transcription levels of the peptide transporter genes, such as Pp1s114_7V6 (*peptide transporter 1*, $\log_2 = 6.3$) and Pp1s72_96V6 (*peptide transporter 5*, $\log_2 = 2.3$). We also noted an increase in the transcription of Pp1s1_60V6 (*AtTAP2*), a plant analogue of *TAP* (*Transporter associated with Antigenic Processing*), which transports peptides generated by the proteasome complex into the endoplasmic reticulum where they are loaded onto a newly synthesized MHC class I complex in humans [51]. This fact could be related to an increase in the pool of free amino acids and/or oligopeptides or to enhanced direct transport of peptides from the cell.

Comparative quantitative analysis of precursor proteins

We tried to determine whether there was any correlation between the quantity of precursor proteins in a cell and the amount of endogenous peptides in its peptidome. We roughly estimated the quantitative relation of precursor proteins in protonemata vs gametophores and protonemata vs protoplasts. We correlated the protein abundance with the transcription level and the number of endogenous peptides of corresponding precursor proteins. In the protonemata vs gametophores case, we can see a week correlation between precursor protein abundance and their transcriptional levels as well as with the number of endogenous peptides (Additional file 16). Yet, when comparing protonemata and protoplasts, we found no correlation between transcription levels and protein abundances nor between the number of endogenous peptides and the abundances of the corresponding proteins (Additional file 17).

The detailed analyses of certain precursor proteins showed that even when cells of two developmental states had equal abundances of a protein, the number of endogenous peptides could differ significantly (Additional files 18 and 19). As an example, a range of stromal chloroplast proteins, such as P34915 (ribulose bisphosphate carboxylase large chain), A9TC11 (phosphoglycerate kinase, chloroplastic), A9T3W5 (small chain of ribulose bisphosphate carboxylase, chloroplastic), and A9TBP0 (ribulose bisphosphate carboxylase/oxygenase activase 1, chloroplastic), showed very different peptide contents in the two developmental stages (Additional files 18 and 19). Precursor proteins such as A9U1H4 (probable linoleate 9S-lipoxygenase 4), A9RD61 (major allergen Mald 1), and A9U188 (glycerate dehydrogenase) produced more endogenous peptides in protonemata than in gametophores although they were less abundant in the former. For such precursor proteins as A9SCV0 (glutathione S-transferase F9), A9RT52 (5-methyltetrahydropteroyltriglutamate-homocysteine methyltransferase), A9RWS2 (5-methyltetrahydropteroyltriglutamate-homocysteine methyltransferase), A9TIY2 (glutamine synthetase, chloroplastic), and a range of others, we observed not only increased levels of proteins and peptides, but also higher transcription of the corresponding genes in protonemata. Only a few precursor proteins, such as A9RPL4 (ribosomal protein S25 family protein) or A9RWN4 (small nuclear ribonucleoprotein family protein) showed a decrease of protein abundance in protoplasts with an increase of the number of endogenous peptides from corresponding proteins in the peptidome, which may indicate degradation of these proteins (Additional file 18). In other cases, we observed an increase of endogenous peptide number along with an increase in protein abundance. Examples include the precursor proteins A9RAS2 (chlorophyll a-b binding protein, chloroplastic), A9SL09 (photosystem I reaction center subunit VI, chloroplastic), A9SX31 (superoxide dismutase [Cu-Zn], chloroplastic), A9T3W5 (ribulose bisphosphate carboxylase small chain, chloroplastic), and others (Additional file 18). Also, we did not observe any significant changes in the transcriptional levels of corresponding genes in protonemata and protoplasts.

Thus, our results showed that the degradation pathway of a protein can be different in gametophores, protonemata, and protoplasts. The peptide alignments of several most represented proteins supported this assumption (for example, see Figure 3).

Photosynthetic activity changes during protoplast isolation

As stated earlier, most precursor proteins identified in protoplasts were chloroplast proteins. Therefore, we analyzed the changes in protoplast photosynthetic activity during their isolation from protonema cells.

In cells exposed to light, the operating quantum efficiency of PSII photochemical activity $\left(\Phi^{PSII}_{light}\right)$ in protonema cells started to diminish at the beginning of maceration and, by the time protoplasts were released from protonema tissue, both the maximal Φ^{PSII} value Φ^{PSII}_{max} and, especially, the operating value at moderate light intensity (100 μmol·m^{-2}·s^{-1}), $\Phi^{PSII}_{100\mu E}$ (Additional file 20) were significantly lower than in intact protonema cells. The decrease in Φ^{PSII}_{light} indicates disrupted electron outflow from PSII, which can be due to suppressed uptake of photosynthetic light-phase outputs (ATP and NADPH). Under normal conditions, the main consumer of ATP and NADPH in photosynthetic cells is the carbon dioxide fixation system; thus, damage to Calvin cycle enzymes can reduce Φ^{PSII}_{light} (both Φ^{PSII}_{max} and, especially, $\Phi^{PSII}_{100\mu E}$). This result is in agreement with the observed degradation of the large subunit of RuBisCO. Notably, Φ^{PSII} of protonemata also diminishes upon treatment with low Driselase concentrations (which do not lead to protoplast isolation, but nevertheless degrade some cell proteins). Thus, the decrease in Φ^{PSII} was not related to mechanical injury of cells or chloroplasts.

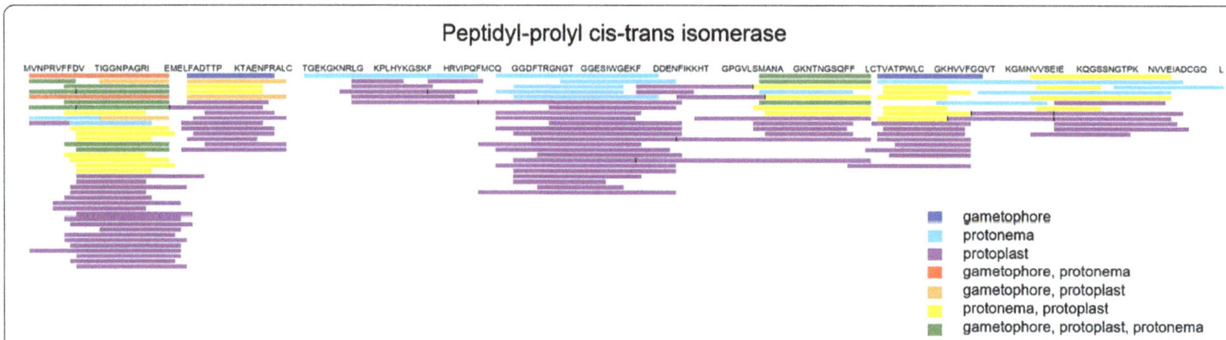

Figure 3 Peptide alignment of a precursor protein A9TK88 (peptidyl-prolyl *cis*-trans isomerase).

Discussion

Little is known about the qualitative and quantitative compositions of the peptide pools that result from protein degradation in cells. The processes of protein synthesis and degradation are constantly ongoing in cells. The rate of protein turnover, as well as the composition of proteins subjected to degradation, changes in response to many factors [52,53]. We found that the two developmental stages of *P. patens*, namely protonemata and gametophores, as well as protoplasts, contain thousands of endogenous peptides resulting from degradation of functionally active proteins. We observed differences in the peptide pools of the three moss cell types, which could reflect both their different protein contents and differences in the regulatory mechanisms of degradation. Our attempt to elucidate the correlation between protein abundance and the amount of endogenous peptides revealed that degradation pathways of a given protein could be different for gametophores, protonemata, and protoplasts. This result may indicate that the identified endogenous peptide products of protein degradation are not a mere "noise" but may play some role in the cell. Thus, this area offers a vast field for further research.

Peptide pools in moss gametophores and protonemata differ

In the peptidome of young and growing protonema tissue, we identified a large number of peptides that are derived from chloroplast proteins. The most abundant precursor proteins in protonemata were major cell proteins such as elongation factor 1a and the large subunit of RuBisCO, as well as fructose-bisphosphate aldolase. The predominance of peptides derived from chloroplast proteins might be due to the rapid growth of protonema cells, resulting in intensive metabolism by proteins related to cellular energy processes. For example, Nelson et al. demonstrated that some major proteins involved in photosynthesis turn over at an above-average rate. Also, chloroplast proteins can serve as a source of amino acids, and in growing tissues RuBisCO acts as a nitrogen source [52]. We found that proteins involved in carbohydrate metabolism, transmembrane transporter activity, and biosynthesis of plant hormones, terpenoids, and steroids are also among the most peptidogenic precursor proteins. We assume this fact correlates with the rate of turnover of these proteins in cells. Besides, as shown by Nelson et al., proteins involved in tetrapyrrole metabolism also degrade rapidly. We identified a separate cluster of such precursor proteins that generated the highest numbers of peptides in protoplasts (see Figure 2).

In addition to chloroplast protein fragments, the peptidome of the mature gametophyte stage, the gametophore, also contains large amounts of peptides from chaperone- and stress-related proteins, such as LEA proteins, aquaporins, AWPM-19-like, and formate dehydrogenases (Additional file 6). A possible explanation could be the fact that gametophores, unlike protonemata, grow in the air environment, and water balance regulation is therefore important in this tissue. Notably, Widiez et al. detected activation of genes responsible for the response to lack of water as early as at the stage of gametophore development from protonemata and hypothesized that such changes play protective roles [54]. Our data on the elevated transcription levels of these genes, as well as the greater numbers of endogenous peptides derived from these proteins, support that hypothesis. One may suppose that an increase in a protein's representation in the proteome leads to more endogenous peptides—products of hydrolysis of that protein—in the peptidome. However, until we know the rate of the protein's turn over, we cannot reliably conclude whether the elevated peptide levels are due to rapid turnover of the protein or to its targeted degradation.

LEA proteins were first discovered in plant seeds [55,56]. Plants biosynthesize LEA proteins in response to drought or abscisic acid [57]. These proteins protect other cellular proteins from denaturation and aggregation under low water content [58]. In *P. patens*, LEA proteins are expressed at a basal level in gametophores, which could be related to protection from water stress [30]. We do not know whether endogenous peptides of chaperone-like proteins like LEA play protective roles in the moss cells, although previous research shows that short LEA protein fragments can independently serve as chaperones [59]. Peptides of the LEA protein family were also identified in both protonemata and protoplasts, but the number of unique peptides for these proteins was higher in gametophores. According to Lienard et al. [60], the *P. patens* aquaporin genes *PIP2-1* and *PIP2-2* are expressed only in gametophore cells and not in protonemata, consistent with our data that native peptides derived from *PIP2-1* and *PIP1-4* are present only in gametophores. Such agreement between mass spectrometry and transcriptomic data holds for many precursor proteins, peptides of which can be identified only in gametophores or protonemata. Quantitative analysis of peptides in protonema and gametophore cells will be the subject of our further research.

Peptidome of *P. patens* protoplasts

We suggest several hypotheses to explain the dramatic increase in the numbers of precursor proteins, as well as of protoplast-specific peptides, that occur in the protoplast peptidome: 1) an immune response leading to specific degradation of cell proteins; 2) an increase in protein degradation rate induced by stress; and 3) protein degradation due to programmed cell death (Figure 4). We analyzed the differentially expressed genes of protonemata and

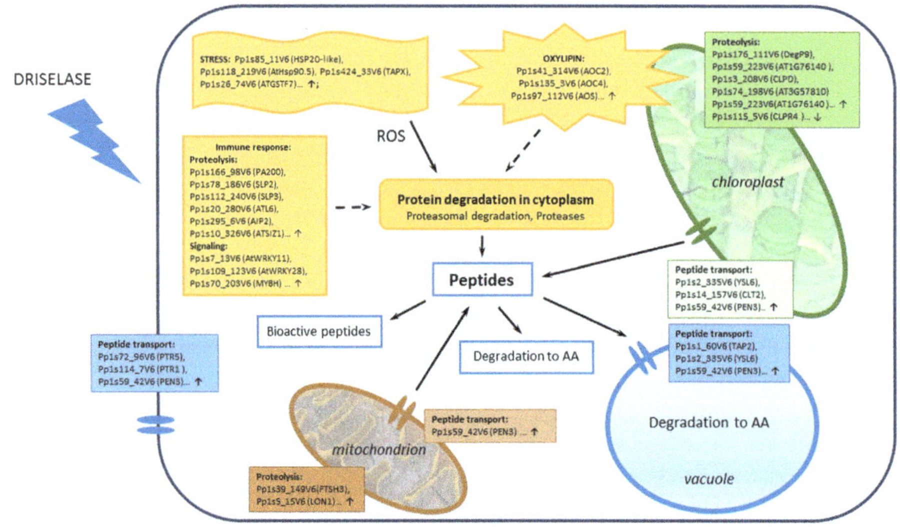

Figure 4 Peptide generation process in protoplasts. We used transcription profiling data to construct a model describing the peptidogenesis in protoplasts. Up-regulated genes associated with factors affecting the protein degradation, such as stress response leading to increase in ROS, immune reactions, and synthesis of stress hormones, are indicated. We inferred that some peptides are degraded to amino acids, while another fraction can be biologically active. The scheme also indicates up-regulated genes encoding proteases and peptide transporters, which also can contribute to the cell peptidome. Arrows show some genes that were up-regulated 2-fold. A complete list of genes and quantitative data for this experiment are given in Additional file 12.

protoplasts to elucidate what molecular mechanisms could be responsible for the significant changes observed in the protoplast peptidome. In addition to the reactions to abiotic stress, which may be associated with the change in cellular form, we also observed increased expression of genes that respond to biotic stress. This result points to the complex nature of stress experienced by cells during protoplast isolation. Previously, Xiao et al. analyzed the protoplast transcriptome during regeneration [35]. According to their findings, the expression of some stress genes as well as of genes responsible for biosynthesis of a range of phytohormones, increases in protoplasts, while photosynthetic pathways are inhibited. However, comparative analysis of protonema and protoplast transcriptomes was not performed in that work.

Cell wall degradation, active elicitors, and the contents of the disrupted cell could induce immune reactions during protoplast isolation [61]. According to our findings, protoplasts have significantly increased expressions of allene oxide cyclase 2 (*AOC2*, Pp1s41_314V6), allene oxide cyclase 4 (*AOC4*, Pp1s135_3V6), and allene oxide synthase (*AOS*, Pp1s97_112V6) (see Additional file 12), which participate in JA biosynthesis [62,63]. *Physcomitrella patens* protonemata and gametophores do not contain JA, but instead use its precursor 12-oxo-phytodienoic acid [63]. However, *P. patens* responds to methyl jasmonate by reducing moss colony growth and rhizoid length, suggesting that jasmonate perception is present in mosses [64]. Because the synthesis of oxylipins in plants is related to biotic stress, findings like these constitute additional evidence favoring the inference that peptidome rearrangement in protoplasts can be related to a severe "biotic" stress due to the effects of Driselase on cells. JA causes degradation of specific proteins and synthesis of specific peptides and promotes senescence [65,66]. Plant defense response is related to the decreased expression genes involved in photosynthesis and is also accompanied by a slowdown in plant growth [67]. However, the physiological importance and mechanisms of the phenomenon have been poorly studied thus far. Our data on photosynthetic activity of moss cells during protoplast isolation show that the decrease in the photosynthesis activity of protonemata treated with Driselase depends on the concentration of the enzyme applied to protonemata and correlates with the increase in the amount of peptides—the products of chloroplast protein degradation. In other words, the observed significant increase in the number of native peptides from chloroplast proteins in protoplasts might be linked to important physiological effects.

Growth of the peptide number in plants that underwent biotic stress can also be due to the protective or signaling role of such peptides [4,17]. Whether the observed effect of the dramatic peptidome change in protoplasts is the result of non specific degradation or whether it serves a protective function is not yet clear. Peptides, such as Systemin, HypSys, GmSubPep, GmPep914, AtPeps, and inceptins induce plant defense mechanisms under biotic stress [20,68]. Three of these, GmSubPep, GmPep914, and inceptin, are produced during the breakdown of cellular proteins that have their own metabolic functions unrelated to stress reactions. Furthermore, biotic stress leads to proteolysis of specific proteins, yielding endogenous peptide elicitors [66]. We evaluated the antimicrobial potential of the identified peptides and revealed 117 endogenous peptides with high potential antimicrobial activity. Of these, 81 were identified in protoplasts. Induction of antimicrobial peptides under the stress conditions used here sheds light on the probable function of peptide pools in *P. patens*. However, checking the antimicrobial activity of such peptides demands further research. This hypothesis agrees with the increase in transcription levels of peptide transporters, which can play an important role in bidirectional transport of peptides. Interestingly, a range of peptide transporters, such as PEN3 (Pp1s59_42V6) and CLT2 (Pp1s14_157V6), can function in transportation of such peptides and are induced under biotic stress.

Another mechanism influencing protein degradation is the effect of stress factors. Stress conditions affect degradation of cell proteins and, consequently, can significantly change the cell peptidome. Severe stress in protoplasts could be detected by the increased expression of genes participating in protection from reactive oxygen species (ROS), such as Pp1s424_33V6, *thylakoidal ascorbate peroxidase*; Pp1s26_74V6, *glutathione S-transferase PHI 9*, and others. An increase in ROS level can lead to oxidation and subsequent protein degradation. Degradation of RuBisCO might be triggered by oxidation of critical cysteine residues in the protein [69]. Our data indicate that, in protoplasts, peptide generation from stromal and membrane thylakoid proteins occurs outside of chloroplasts. In protoplasts, none of the genes encoding chloroplast proteases exhibited elevated transcription levels, except for the gene encoding Pp1s3_208V6 (ATP-dependent Clp protease regulatory subunit), which regulates the activity of the Clp protease complex in chloroplasts (Additional file 15). At the same time, the level of transcription of the Pp1s115_5V6 gene encoding the R4 subunit of Clp protease complex decreased. The existence of membrane compartments containing RuBisCO (called RuBisCO vesicular bodies, RVB) in the cytoplasm also suggests that there is a relationship between protein degradation systems in the chloroplasts, cytosol, and vacuoles [49] [70]. Today, little is known about the mechanisms of degradation of other chloroplast proteins, especially as part of the peptide generation process. The available data on stress-related degradation of chloroplast membranes with subsequent release of proteins to the cytoplasm suggests the hypothesis that they may be degraded outside chloroplasts under stress conditions [71].

For some chloroplast proteins, such as A9S1E1 (photosystem I reaction center subunit XI), A9S6S7 (LHCB2, light harvesting complex of photosystem II), and A9TIY2 (GLN2, glutamine synthetase) we observed a decrease in their abundance and an increase in the number of their endogenous peptides. However, for the majority of chloroplast proteins, we observed a positive correlation between protein and endogenous peptide abundances. This finding can be explained by an elevated rate of protein turnover, leading to enhanced degradation and, consequently, more peptides from these proteins.

Although little is known about the generation of peptides *en masse* in plant cells, biotic stress triggers a number of cellular destruction mechanisms. For example, the hypersensitive reaction involves programmed cell death and can lead to massive protein degradation [72]. Along with that, a number of proteases are induced by biotic stress [73], which also could lead to peptide generation and a significant change in the peptide pool of protoplasts. Also, we detected an increase in transcription levels of the genes of heat shock proteins (HSPs), such as Pp1s97_106V6 (HSP 21) and Pp1s3_114V6 (HSP17.6II), which can indicate the initialization of cell death mechanisms [74], similarly to apoptosis in animal cells [75,76].

Mechanisms of peptide generation in plant cells

The identified peptide pools apparently result from protein degradation processes in the cell. What is the nature of the observed proteolytic degradome of plant cells? A number of studies have shown that plant cells, like animal cells, have a complex protein degradation system. Plants have two main ways to degrade intracellular proteins: the proteasome-mediated pathway and autophagy [77].

The proteasome is a proteolytic complex that degrades short- and long-lived intracellular proteins. Plants have two types of proteasomes, 20S and 26S, that are responsible for ubiquitin-dependent and ubiquitin-independent proteolysis, respectively [78]. In proteasomes, proteins are degraded to peptides of 2–25 residues. The generally accepted view is that peptides generated by proteasomes are labile in the cytoplasm and hydrolyzed rapidly to amino acids [79]. However, there is some evidence that, in plants, these peptides can be transferred to the vacuole for further degradation [51]. In mammals, a set of proteases degrades proteasome-generated peptides [79,80]. In plants, a similar set of proteases with the same function was also discovered [81-83]. In a previous report, oligopeptides generated by proteasomes were not found in cell extracts [84] possibly because of rapid degradation *in vivo* [80]. However, the tools available at that time did not allow accurate identification of peptide pools. Still, accumulation of peptide fragments in a cell was assumed to be undesirable, because it may prevent protein-to-protein interactions and the peptides could possibly aggregate and become toxic [85,86]. However, we identified several thousand native peptides in gametophore and protonema cells and about 20,000 peptides in protoplasts. The results of our study provide ground for a more detailed description of peptide generation in *P. patens*. Examining peptide families derived from individual proteins allows several common principles of protein degradation to be outlined. The original amino acid sequence of the protein is clearly divided into segments 15–25 amino acid long that give rise to sets of structurally related peptides. We believe that primary proteolysis occurs at sites located between these segments, presumably within proteasomes. After the primary segments are released from the proteasome, they are further degraded to smaller peptides by amino- and carboxypeptidases [81-83]. Degradation products of peptidyl-prolyl cis-trans isomerase shown in Figure 3 serve as an example.

Proteins or peptides also can be degraded by a large number of other unspecified proteases. Plant genomes encode hundreds of proteases that belong to a dozen unrelated families [87]. The biological roles of the majority of proteases are unknown, but the very mechanism of their effects on their substrates suggests that they may also contribute to peptide pool formation in the cell [87].

Stress conditions might modify the above-described picture. Stimuli such as infection, drought, cold, and ultraviolet light result in increased ROS in the cell [52]. The elevated ROS levels result in oxidized protein, which might be cytotoxic [88]. In animal cells, proteasomes recognize and degrade oxidized proteins in the cytosol, nucleus, and endoplasmic reticulum, thus protecting the cell from their cytotoxicity [89]. *In vitro*, the 20S proteasome recognizes and degrades the oxidized proteins while components of the ubiquitin-dependent 26S proteasome pathway become inactive [90]. In plants, severe stress results in excess amounts of proteolytic substrate, and oxidation stress directly inhibits the 26S proteasome degradation pathway [78,91].

We assume that the mechanisms described above can be responsible for the formation of peptide pools in cells. We suppose, under stress conditions, proteins undergo active proteolysis, leading to rapid and extensive changes in the peptidome. We found that expression of some genes encoding proteases and proteins involved in ubiquitin-dependent degradation of proteins increased in protoplasts. For example, transcription of the Pp1s166_98V6 gene encoding PA200 increased in protoplasts. PA200 is a non-ATPase activator of 20S particles of proteasomes, thus it could be responsible for the increase in hydrolysis of small peptides [92]. Notably, the level of transcription of eleven ubiquitin ligase E3 genes belonging to RING/U-box class and involved in proteasome-dependent protein degradation was also higher in protoplasts (Additional file 15). These proteins are known to participate in responses to abiotic and biotic stresses in plants [93,94]. Interestingly, expression of PA200, as well as subtilisin-like serine protease 2 (Pp1s78_186V6) and subtilisin-like serine protease 3 (Pp1s112_240V6), is induced by JA in *Arabidopsis* [95,96]. This fact agrees with our data on increased transcription of genes involved in oxylipin synthesis. Besides, the transcription levels of mitochondrial proteases, such as Pp1s39_149V6 (a homolog of mitochondrial protease FtsH 3) and Pp1s5_15V6 (a homolog of mitochondrial protease Lon1), is increased in protoplasts, perhaps indicating active proteolysis in mitochondria. At the same time, transcription of a number of genes encoding proteases, such as Pp1s90_35V6 and Pp1s24_95V6 (aspartyl protease family proteins), Pp1s8_260V6 and Pp1s283_60V6 (prolyl oligopeptidase family proteins), Pp1s233_64V6 (Peptidase C78), Pp1s58_185V6 (serine carboxypeptidase-like 50), Pp1s113_206V6 (carboxypeptidase D), is lower in protoplasts. Some proteases with higher expression in protoplasts localize in chloroplasts, for example, Pp1s8_260V6 and Pp1s283_60V6 (prolyl oligopeptidase family proteins) and Pp1s233_64V6 (Peptidase C78). This finding probably indicates the fine regulation of the processes of protein degradation and peptidome generation in plant cells.

Conclusions

In conclusion, our results show that moss cells contain extended peptide pools that are hydrolysis products of cell proteins. The peptide pool composition depends on the type of tissue yet always contains peptides derived from the major chloroplast proteins. We observe no correlation between protein abundance, its transcription level, and the amount of endogenous peptides. Active peptidogenesis in protoplasts is probably due to a range of mechanisms, with stress during isolation and immune reaction to Driselase treatment being the key ones. Eighty-nine peptides of protoplasts possess high antimicrobial potential. Genes involved in JA synthesis, as well as those associated with biotic stress, had increased transcription levels in protoplasts. Changes in the peptidome in protoplasts are accompanied by suppression of photosynthetic activity. In our future research, we aim to study which mechanisms of degradation are responsible for the formation of endogenous peptide pools in cells, evaluate biological activity of the peptides, and study the effects of hormones on peptidome formation.

Availability of supporting data

The mass spectrometry proteomics data have been deposited in the ProteomeXchange Consortium (http://proteomecentral.proteomexchange.org) via the PRIDE partner repository [97] with the dataset identifier PXD000227 and doi:10.6019/PXD000227. Sequence data described in this article can be found in the ArrayExpress (http://www.ebi.ac.uk/arrayexpress/) database under accession number E-MTAB-1637.

Additional files

Additional file 1: Correlation between RNA-seq data and quantitative real-time PCR analysis.

Additional file 2: List of primers and probes used in quantitative real-time PCR.

Additional file 3: List of peptides identified in gametophores, protonemata, and protoplasts.

Additional file 4: List of precursor proteins of endogenous peptides identified in gametophores.

Additional file 5: List of precursor proteins of endogenous peptides identified in protonemata.

Additional file 6: Ratios of endogenous peptides derived from eight groups of precursor proteins in peptidomes of gametophores, protonemata, and protoplasts.

Additional file 7: List of precursor proteins of endogenous peptides identified in protoplasts.

Additional file 8: List of precursor proteins of endogenous peptides identified in intact chloroplasts.

Additional file 9: Venn diagram showing the amount of peptides identified in protonemata treated with 0.025% (w/v) Driselase compared with untreated protonemata (control).

Additional file 10: List of endogenous peptides with high antimicrobial potential (AMPA results).

Additional file 11: Functional annotation of precursor proteins using the DAVID service.

Additional file 12: Differentially expressed (DE) genes in protoplasts compared to protonemata.

Additional file 13: GO terms of up-regulated genes in protoplasts compared to protonemata.

Additional file 14: Distribution of differentially expressed (DE) genes over the clusters obtained with the DAVID tools.

Additional file 15: Differentially expressed (DE) genes related to protein degradation in protoplasts compared to protonemata.

Additional file 16: Change in protein abundance across gametophores and protonemata.

Additional file 17: Change in protein abundance across protoplasts and protonemata.

Additional file 18: Quantitative relationship of precursor proteins in protonemata versus gametophores.

Additional file 19: Quantitative relationship of precursor proteins in protonemata versus protoplasts.

Additional file 20: Time-courses of maximal (Φ^{PSII}_{max}) and operating value at moderate light intensity (100 µmol·m^{-2}·s^{-1}; $\Phi^{PSII}_{100\mu E}$) of quantum efficiency of photosystem II (PSII) photochemical activity in *Physcomitrella patens* protonema cells under treatment with 0.5%, 0.025%, and 0.0025% Driselase solution.

Abbreviations

qRT-PCR: Quantitative real-time PCR; RuBisCO: Ribulose-bisphosphate carboxylase; LEA: Late embryogenesis abundant protein; CDS: Coding sequences; ROS: Reactive oxygen species; NADPH: Nicotinamide adenine dinucleotide phosphate; ATP: Adenosine-triphosphate; JA: Jasmonic acid; PSII: Photosystem II.

Competing interests

The authors declare that they have no competing interests.

Authors' contributions

IAF, AYS, and VMG designed and performed research, analyzed data, and co-wrote the article. GPA and DGA contributed new analytic/computational tools, analyzed data, and co-wrote the article. ESK, IAA, SEN, SIK, and RHZ performed research and contributed new analytic/computational tools. VVP, VGZ, TAS performed research and analyzed data. AIM contributed new analytic/computational tools and analyzed data. DKS, RAC, AVS, and AYG performed research. VTI designed research, analyzed data, and co-wrote the article. All authors revised the manuscript critically and approved the final version for publication.

Acknowledgment

The reported study was partially supported by the Russian Foundation for Basic Research (RFBR), research project No. 13-04-40102-H comfi and the Russian Science Foundation (project No. 14-50-00131).

Author details

[1]Department of Proteomics, Shemyakin-Ovchinnikov Institute of Bioorganic Chemistry of the Russian Academy of Sciences, 16/10, Miklukho-Maklaya, GSP-7, Moscow 117997, Russian Federation. [2]Research Institute of Physical-Chemical Medicine, Federal Medical & Biological Agency, 1a, Malaya Pirogovskaya, Moscow 119992, Russian Federation. [3]Moscow Institute of Physics and Technology, 9 Institutskiy per., Dolgoprudny, Moscow Region 141700, Russian Federation. [4]Institute of Biomedical Chemistry RAMS im. V.N. Orehovicha, 10, Pogodinskaya Street, Moscow 119121, Russian Federation. [5]A. N. Belozersky Institute of Physico-Chemical Biology, M.V. Lomonosov Moscow State University, Leninskye Gory, House 1, Building 40, Moscow 119992, Russian Federation. [6]Biology Department, Lomonosov Moscow State University, Moscow 199234, Russian Federation.

References

1. Matsubayashi Y, Sakagami Y. Peptide hormones in plants. Annu Rev Plant Biol. 2006;57:649–74.
2. Farrokhi N, Whitelegge JP, Brusslan JA. Plant peptides and peptidomics. Plant Biotechnol J. 2008;6(2):105–34.
3. Murphy E, Smith S, De Smet I. Small signaling peptides in Arabidopsis development: how cells communicate over a short distance. Plant Cell. 2012;24(8):3198–217.
4. Marmiroli N, Maestri E. Plant peptides in defense and signaling. Peptides. 2014;56C:30–44.
5. De Smet I, Voss U, Jurgens G, Beeckman T. Receptor-like kinases shape the plant. Nat Cell Biol. 2009;11(10):1166–73.
6. Lease KA, Walker JC. The Arabidopsis unannotated secreted peptide database, a resource for plant peptidomics. Plant Physiol. 2006;142(3):831–8.
7. Butenko MA, Vie AK, Brembu T, Aalen RB, Bones AM. Plant peptides in signalling: looking for new partners. Trends Plant Sci. 2009;14(5):255–63.
8. Kondo T, Plaza S, Zanet J, Benrabah E, Valenti P, Hashimoto Y, et al. Small peptides switch the transcriptional activity of Shavenbaby during Drosophila embryogenesis. Science. 2010;329(5989):336–9.
9. Hanada K, Higuchi-Takeuchi M, Okamoto M, Yoshizumi T, Shimizu M, Nakaminami K, et al. Small open reading frames associated with morphogenesis are hidden in plant genomes. Proc Natl Acad Sci U S A. 2013;110(6):2395–400.
10. Slavoff SA, Mitchell AJ, Schwaid AG, Cabili MN, Ma J, Levin JZ, et al. Peptidomic discovery of short open reading frame-encoded peptides in human cells. Nat Chem Biol. 2013;9(1):59–64.
11. Ladoukakis E, Pereira V, Magny EG, Eyre-Walker A, Couso JP. Hundreds of putatively functional small open reading frames in Drosophila. Genome Biol. 2011;12(11):R118.
12. Ivanov VT, Yatskin ON. Peptidomics: a logical sequel to proteomics. Expert Rev Proteomics. 2005;2(4):463–73.
13. Iatskin ON, Karelin AA, Ivanov VT. Peptidomes of the brain, heart, lung, and spleen of a rat: similarity and differences. Bioorg Khim. 2009;35(4):471–82.
14. Blishchenko EY, Sazonova OV, Yatskin ON, Kalinina OA, Tolmazova AG, Philippova MM, et al. beta-Actin-derived peptides isolated from acidic extract of rat spleen suppress tumor cell growth. J Pept Sci. 2008;14(7):811–8.
15. Sasaki K, Takahashi N, Satoh M, Yamasaki M, Minamino N. A peptidomics strategy for discovering endogenous bioactive peptides. J Proteome Res. 2010;9(10):5047–52.
16. Brand GD, Magalhaes MT, Tinoco ML, Aragao FJ, Nicoli J, Kelly SM, et al. Probing protein sequences as sources for encrypted antimicrobial peptides. PLoS One. 2012;7(9):e45848.
17. Chen YL, Lee CY, Cheng KT, Chang WH, Huang RN, Nam HG, et al. Quantitative Peptidomics Study Reveals That a Wound-Induced Peptide

from PR-1 Regulates Immune Signaling in Tomato. Plant Cell. 2014;26(10):4135–48.

18. Schmelz EA, Carroll MJ, LeClere S, Phipps SM, Meredith J, Chourey PS, et al. Fragments of ATP synthase mediate plant perception of insect attack. Proc Natl Acad Sci U S A. 2006;103(23):8894–9.
19. Pearce G, Munske G, Yamaguchi Y, Ryan CA. Structure-activity studies of GmSubPep, a soybean peptide defense signal derived from an extracellular protease. Peptides. 2010;31(12):2159–64.
20. Yamaguchi Y, Barona G, Ryan CA, Pearce G. GmPep914, an eight-amino acid peptide isolated from soybean leaves, activates defense-related genes. Plant Physiol. 2011;156(2):932–42.
21. Kamisugi Y, Schlink K, Rensing SA, Schween G, Von Stackelberg M, Cuming AC, et al. The mechanism of gene targeting in Physcomitrella patens: homologous recombination, concatenation and multiple integration. Nucleic Acids Res. 2006;34(21):6205–14.
22. Strepp R, Scholz S, Kruse S, Speth V, Reski R. Plant nuclear gene knockout reveals a role in plastid division for the homolog of the bacterial cell division protein FtsZ, an ancestral tubulin. Proc Natl Acad Sci U S A. 1998;95(8):4368–73.
23. Decker EL, Frank W, Sarnighausen E, Reski R. Moss systems biology en route: phytohormones in Physcomitrella development. Plant Biol (Stuttg). 2006;8(3):397–405.
24. Rensing SA, Lang D, Zimmer AD, Terry A, Salamov A, Shapiro H, et al. The Physcomitrella genome reveals evolutionary insights into the conquest of land by plants. Science. 2008;319(5859):64–9.
25. Sarnighausen E, Wurtz V, Heintz D, Van Dorsselaer A, Reski R. Mapping of the Physcomitrella patens proteome. Phytochemistry. 2004;65(11):1589–607.
26. Cho SH, Hoang QT, Kim YY, Shin HY, Ok SH, Bae JM, et al. Proteome analysis of gametophores identified a metallothionein involved in various abiotic stress responses in Physcomitrella patens. Plant Cell Rep. 2006;25(5):475–88.
27. Skripnikov AY, Polyakov NB, Tolcheva EV, Velikodvorskaya VV, Dolgov SV, Demina IA, et al. Proteome analysis of the moss Physcomitrella patens (Hedw.) B.S.G. Biochemistry (Mosc). 2009;74(5):480–90.
28. Wang X, Yang P, Gao Q, Liu X, Kuang T, Shen S, et al. Proteomic analysis of the response to high-salinity stress in Physcomitrella patens. Planta. 2008;228(1):167–77.
29. Wang X, Yang P, Zhang X, Xu Y, Kuang T, Shen S, et al. Proteomic analysis of the cold stress response in the moss. Physcomitrella patens. Proteomics. 2009;9(19):4529–38.
30. Cui S, Hu J, Guo S, Wang J, Cheng Y, Dang X, et al. Proteome analysis of Physcomitrella patens exposed to progressive dehydration and rehydration. J Exp Bot. 2012;63(2):711–26.
31. Lang EG, Mueller SJ, Hoernstein SN, Porankiewicz-Asplund J, Vervliet-Scheebaum M, Reski R. Simultaneous isolation of pure and intact chloroplasts and mitochondria from moss as the basis for sub-cellular proteomics. Plant Cell Rep. 2011;30(2):205–15.
32. Cuming AC, Cho SH, Kamisugi Y, Graham H, Quatrano RS. Microarray analysis of transcriptional responses to abscisic acid and osmotic, salt, and drought stress in the moss. Physcomitrella patens. New Phytol. 2007;176(2):275–87.
33. Richardt S, Timmerhaus G, Lang D, Qudeimat E, Correa LGG, Reski R, et al. Microarray analysis of the moss Physcomitrella patens reveals evolutionarily conserved transcriptional regulation of salt stress and abscisic acid signalling. Plant Mol Biol. 2010;72(1–2):27–45.
34. Xiao L, Wang H, Wan P, Kuang T, He Y. Genome-wide transcriptome analysis of gametophyte development in Physcomitrella patens. BMC Plant Biol. 2011;11:177.
35. Xiao L, Zhang L, Yang G, Zhu H, He Y. Transcriptome of protoplasts reprogrammed into stem cells in Physcomitrella patens. PLoS One. 2012;7(4):e35961.
36. Nishiyama T, Miyawaki K, Ohshima M, Thompson K, Nagashima A, Hasebe M, et al. Digital gene expression profiling by 5'-end sequencing of cDNAs during reprogramming in the moss Physcomitrella patens. PLoS One. 2012;7(5):e36471.
37. Erxleben A, Gessler A, Vervliet-Scheebaum M, Reski R. Metabolite profiling of the moss Physcomitrella patens reveals evolutionary conservation of osmoprotective substances. Plant Cell Rep. 2012;31(2):427–36.
38. Skripnikov A, Anikanov NA, Kazakov VS, Dolgov SV, Ziganshin R, Govorun VM, et al. Exploration and identification of Physcomitrella patens moss peptides. Bioorg Khim. 2011;37(1):108–18.
39. Rappsilber J, Mann M, Ishihama Y. Protocol for micro-purification, enrichment, pre-fractionation and storage of peptides for proteomics using StageTips. Nat Protoc. 2007;2(8):1896–906.
40. Faurobert M, Pelpoir E, Chaib J. Phenol extraction of proteins for proteomic studies of recalcitrant plant tissues. Methods Mol Biol. 2007;355:9–14.
41. Torrent M, Di Tommaso P, Pulido D, Nogues MV, Notredame C, Boix E, et al. AMPA: an automated web server for prediction of protein antimicrobial regions. Bioinformatics. 2012;28(1):130–1.
42. Cove DJ, Perroud PF, Charron AJ, McDaniel SF, Khandelwal A, Quatrano RS. Isolation of DNA, RNA, and protein from the moss Physcomitrella patens gametophytes. Cold Spring Harb Protoc. 2009;2009(2):prot5146.
43. Zimmer AD, Lang D, Buchta K, Rombauts S, Nishiyama T, Hasebe M, et al. Reannotation and extended community resources for the genome of the non-seed plant Physcomitrella patens provide insights into the evolution of plant gene structures and functions. BMC Genomics. 2013;14:498.
44. Trapnell C, Pachter L, Salzberg SL. TopHat: discovering splice junctions with RNA-Seq. Bioinformatics. 2009;25(9):1105–11.
45. Trapnell C, Williams BA, Pertea G, Mortazavi A, Kwan G, Van Baren MJ, et al. Transcript assembly and quantification by RNA-Seq reveals unannotated transcripts and isoform switching during cell differentiation. Nat Biotechnol. 2010;28(5):511–5.
46. Robinson MD, McCarthy DJ, Smyth GK. edgeR: a Bioconductor package for differential expression analysis of digital gene expression data. Bioinformatics. 2010;26(1):139–40.
47. Hindson BJ, Ness KD, Masquelier DA, Belgrader P, Heredia NJ, Makarewicz AJ, et al. High-throughput droplet digital PCR system for absolute quantitation of DNA copy number. Anal Chem. 2011;83(22):8604–10.
48. Blowers DP, Boss WF, Trewavas AJ. Rapid Changes in Plasma Membrane Protein Phosphorylation during Initiation of Cell Wall Digestion. Plant Physiol. 1988;86(2):505–9.
49. Prins A, Van Heerden PD, Olmos E, Kunert KJ, Foyer CH. Cysteine proteinases regulate chloroplast protein content and composition in tobacco leaves: a model for dynamic interactions with ribulose-1,5-bisphosphate carboxylase/oxygenase (Rubisco) vesicular bodies. J Exp Bot. 2008;59(7):1935–50.
50. Rushton PJ, Somssich IE, Ringler P, Shen QJ. WRKY transcription factors. Trends Plant Sci. 2010;15(5):247–58.
51. Ramos MS, Abele R, Nagy R, Grotemeyer MS, Tampe R, Rentsch D, et al. Characterization of a transport activity for long-chain peptides in barley mesophyll vacuoles. J Exp Bot. 2011;62(7):2403–10.
52. Khanna-Chopra R. Leaf senescence and abiotic stresses share reactive oxygen species-mediated chloroplast degradation. Protoplasma. 2012;249(3):469–81.
53. Nelson CJ, Alexova R, Jacoby RP, Millar AH. Proteins with high turnover rate in barley leaves estimated by proteome analysis combined with in planta isotope labeling. Plant Physiol. 2014;166(1):91–108.
54. Widiez T, Symeonidi A, Luo C, Lam E, Lawton M, Rensing SA. The chromatin landscape of the moss Physcomitrella patens and its dynamics during development and drought stress. Plant J. 2014;79(1):67–81.
55. Dure L, Galau GA. Developmental Biochemistry of Cottonseed Embryogenesis and Germination : XIII. Regulation of biosynthesis of principal storage proteins. Plant Physiol. 1981;68(1):187–94.
56. Battaglia M, Olvera-Carrillo Y, Garciarrubio A, Campos F, Covarrubias AA. The enigmatic LEA proteins and other hydrophilins. Plant Physiol. 2008;148(1):6–24.
57. Wise MJ, Tunnacliffe A. POPP the question: what do LEA proteins do? Trends Plant Sci. 2004;9(1):13–7.
58. Chakrabortee S, Meersman F, Kaminski Schierle GS, Bertoncini CW, McGee B, Kaminski CF, et al. Catalytic and chaperone-like functions in an intrinsically disordered protein associated with desiccation tolerance. Proc Natl Acad Sci U S A. 2010;107(37):16084–9.
59. Furuki T, Shimizu T, Chakrabortee S, Yamakawa K, Hatanaka R, Takahashi T, et al. Effects of Group 3 LEA protein model peptides on desiccation-induced protein aggregation. Biochim Biophys Acta. 2012;1824(7):891–7.
60. Lienard D, Durambur G, Kiefer-Meyer MC, Nogue F, Menu-Bouaouiche L, Charlot F, et al. Water transport by aquaporins in the extant plant Physcomitrella patens. Plant Physiol. 2008;146(3):1207–18.
61. Heil M, Ibarra-Laclette E, Adame-Alvarez RM, Martinez O, Ramirez-Chavez E, Molina-Torres J, et al. How plants sense wounds: damaged-self recognition is based on plant-derived elicitors and induces octadecanoid signaling. PLoS One. 2012;7(2):e30537.

62. Feussner I, Wasternack C. The lipoxygenase pathway. Annu Rev Plant Biol. 2002;53:275–97.
63. Stumpe M, Gobel C, Faltin B, Beike AK, Hause B, Himmelsbach K, et al. The moss Physcomitrella patens contains cyclopentenones but no jasmonates: mutations in allene oxide cyclase lead to reduced fertility and altered sporophyte morphology. New Phytol. 2010;188(3):740–9.
64. Ponce De Leon I, Schmelz EA, Gaggero C, Castro A, Alvarez A, Montesano M. Physcomitrella patens activates reinforcement of the cell wall, programmed cell death and accumulation of evolutionary conserved defence signals, such as salicylic acid and 12-oxo-phytodienoic acid, but not jasmonic acid, upon Botrytis cinerea infection. Mol Plant Pathol. 2012;13(8):960–74.
65. Pollier J, Moses T, Gonzalez-Guzman M, De Geyter N, Lippens S, Vanden Bossche R, et al. The protein quality control system manages plant defence compound synthesis. Nature. 2013;504(7478):148–52.
66. Yamaguchi Y, Huffaker A. Endogenous peptide elicitors in higher plants. Curr Opin Plant Biol. 2011;14(4):351–7.
67. Attaran E, Major IT, Cruz JA, Rosa BA, Koo AJ, Chen J, et al. Temporal Dynamics of Growth and Photosynthesis Suppression in Response to Jasmonate Signaling. Plant Physiol. 2014;165(3):1302–14.
68. Pearce G, Yamaguchi Y, Barona G, Ryan CA. A subtilisin-like protein from soybean contains an embedded, cryptic signal that activates defense-related genes. Proc Natl Acad Sci U S A. 2010;107(33):14921–5.
69. Moreno J, Garcia-Murria MJ, Marin-Navarro J. Redox modulation of Rubisco conformation and activity through its cysteine residues. J Exp Bot. 2008;59(7):1605–14.
70. Wada S, Ishida H, Izumi M, Yoshimoto K, Ohsumi Y, Mae T, et al. Autophagy plays a role in chloroplast degradation during senescence in individually darkened leaves. Plant Physiol. 2009;149(2):885–93.
71. Kwon KC, Verma D, Jin S, Singh ND, Daniell H. Release of proteins from intact chloroplasts induced by reactive oxygen species during biotic and abiotic stress. PLoS One. 2013;8(6):e67106.
72. Coll NS, Epple P, Dangl JL. Programmed cell death in the plant immune system. Cell Death Differ. 2011;18(8):1247–56.
73. Vartapetian AB, Tuzhikov AI, Chichkova NV, Taliansky M, Wolpert TJ. A plant alternative to animal caspases: subtilisin-like proteases. Cell Death Differ. 2011;18(8):1289–97.
74. Van Doorn WG. Classes of programmed cell death in plants, compared to those in animals. J Exp Bot. 2011;62(14):4749–61.
75. Garrido C, Gurbuxani S, Ravagnan L, Kroemer G. Heat shock proteins: endogenous modulators of apoptotic cell death. Biochem Biophys Res Commun. 2001;286(3):433–42.
76. Lanneau D, Brunet M, Frisan E, Solary E, Fontenay M, Garrido C. Heat shock proteins: essential proteins for apoptosis regulation. J Cell Mol Med. 2008;12(3):743–61.
77. Araujo WL, Tohge T, Ishizaki K, Leaver CJ, Fernie AR. Protein degradation - an alternative respiratory substrate for stressed plants. Trends Plant Sci. 2011;16(9):489–98.
78. Kurepa J, Wang S, Li Y, Smalle J. Proteasome regulation, plant growth and stress tolerance. Plant Signal Behav. 2009;4(10):924–7.
79. Saric T, Graef CI, Goldberg AL. Pathway for degradation of peptides generated by proteasomes: a key role for thimet oligopeptidase and other metallopeptidases. J Biol Chem. 2004;279(45):46723–32.
80. Reits E, Neijssen J, Herberts C, Benckhuijsen W, Janssen L, Drijfhout JW, et al. A major role for TPPII in trimming proteasomal degradation products for MHC class I antigen presentation. Immunity. 2004;20(4):495–506.
81. Chao WS, Pautot V, Holzer FM, Walling LL. Leucine aminopeptidases: the ubiquity of LAP-N and the specificity of LAP-A. Planta. 2000;210(4):563–73.
82. Book AJ, Yang P, Scalf M, Smith LM, Vierstra RD. Tripeptidyl peptidase II. An oligomeric protease complex from Arabidopsis. Plant Physiol. 2005;138(2):1046–57.
83. Polge C, Jaquinod M, Holzer F, Bourguignon J, Walling L, Brouquisse R. Evidence for the Existence in Arabidopsis thaliana of the Proteasome Proteolytic Pathway: ACTIVATION IN RESPONSE TO CADMIUM. J Biol Chem. 2009;284(51):35412–24.
84. Beninga J, Rock KL, Goldberg AL. Interferon-gamma can stimulate post-proteasomal trimming of the N terminus of an antigenic peptide by inducing leucine aminopeptidase. J Biol Chem. 1998;273(30):18734–42.
85. Hughes E, Burke RM, Doig AJ. Inhibition of toxicity in the beta-amyloid peptide fragment beta -(25–35) using N-methylated derivatives: a general strategy to prevent amyloid formation. J Biol Chem. 2000;275(33):25109–15.
86. Tenidis K, Waldner M, Bernhagen J, Fischle W, Bergmann M, Weber M, et al. Identification of a penta- and hexapeptide of islet amyloid polypeptide (IAPP) with amyloidogenic and cytotoxic properties. J Mol Biol. 2000;295(4):1055–71.
87. van der Hoorn RA. Plant proteases: from phenotypes to molecular mechanisms. Annu Rev Plant Biol. 2008;59:191–223.
88. Xiong Y, Contento AL, Nguyen PQ, Bassham DC. Degradation of oxidized proteins by autophagy during oxidative stress in Arabidopsis. Plant Physiol. 2007;143(1):291–9.
89. Davies KJ. Degradation of oxidized proteins by the 20S proteasome. Biochimie. 2001;83(3–4):301–10.
90. Asher G, Reuven N, Shaul Y. 20S proteasomes and protein degradation "by default". Bioessays. 2006;28(8):844–9.
91. Reinheckel T, Sitte N, Ullrich O, Kuckelkorn U, Davies KJ, Grune T. Comparative resistance of the 20S and 26S proteasome to oxidative stress. Biochem J. 1998;335(Pt 3):637–42.
92. Savulescu AF, Glickman MH. Proteasome activator 200: the heat is on. Mol Cell Proteomics. 2011;10(5):R110–R006890.
93. Stone SL. The role of ubiquitin and the 26S proteasome in plant abiotic stress signaling. Frontiers Plant Sci. 2014;5:135.
94. Marino D, Peeters N, Rivas S. Ubiquitination during plant immune signaling. Plant Physiol. 2012;160(1):15–27.
95. Gfeller A, Baerenfaller K, Loscos J, Chetelat A, Baginsky S, Farmer EE. Jasmonate controls polypeptide patterning in undamaged tissue in wounded Arabidopsis leaves. Plant Physiol. 2011;156(4):1797–807.
96. Golldack D, Vera P, Dietz KJ. Expression of subtilisin-like serine proteases in Arabidopsis thaliana is cell-specific and responds to jasmonic acid and heavy metals with developmental differences. Physiol Plant. 2003;118(1):64–73.
97. Vizcaino JA, Cote RG, Csordas A, Dianes JA, Fabregat A, Foster JM, et al. The PRoteomics IDEntifications (PRIDE) database and associated tools: status in 2013. Nucleic Acids Res. 2013;41(Database issue):D1063–9.

11

Comparative metabolic and transcriptional analysis of a doubled diploid and its diploid citrus rootstock (*C. junos* cv. Ziyang xiangcheng) suggests its potential value for stress resistance improvement

Feng-Quan Tan, Hong Tu, Wu-Jun Liang, Jian-Mei Long, Xiao-Meng Wu, Hong-Yan Zhang and Wen-Wu Guo*

Abstract

Background: Polyploidy has often been considered to confer plants a better adaptation to environmental stresses. Tetraploid citrus rootstocks are expected to have stronger stress tolerance than diploid. Plenty of doubled diploid citrus plants were exploited from diploid species for citrus rootstock improvement. However, limited metabolic and molecular information related to tetraploidization is currently available at a systemic biological level. This study aimed to evaluate the occurrence and extent of metabolic and transcriptional changes induced by tetraploidization in Ziyang xiangcheng (*Citrus junos* Sieb. ex Tanaka), which is a special citrus germplasm native to China and widely used as an iron deficiency tolerant citrus rootstock.

Results: Doubled diploid Ziyang xiangcheng has typical morphological and anatomical features such as shorter plant height, larger and thicker leaves, bigger stomata and lower stomatal density, compared to its diploid parent. GC-MS (Gas chromatography coupled to mass spectrometry) analysis revealed that tetraploidization has an activation effect on the accumulation of primary metabolites in leaves; many stress-related metabolites such as sucrose, proline and γ-aminobutyric acid (GABA) was remarkably up-regulated in doubled diploid. However, LC-QTOF-MS (Liquid chromatography quadrupole time-of-flight mass spectrometry) analysis demonstrated that tetraploidization has an inhibition effect on the accumulation of secondary metabolites in leaves; all the 33 flavones were down-regulated while all the 6 flavanones were up-regulated in 4x. By RNA-seq analysis, only 212 genes (0.8% of detected genes) are found significantly differentially expressed between 2x and 4x leaves. Notably, those genes were highly related to stress-response functions, including responses to salt stress, water and abscisic acid. Interestingly, the transcriptional divergence could not explain the metabolic changes, probably due to post-transcriptional regulation.

Conclusion: Taken together, tetraploidization induced considerable changes in leaf primary and secondary metabolite accumulation in Ziyang xiangcheng. However, the effect of tetraploidization on transcriptome is limited. Compared to diploid, higher expression level of stress related genes and higher content of stress related metabolites in doubled diploid could be beneficial for its stress tolerance.

Keywords: Citrus, Doubled diploid, Stress tolerance, Primary and secondary metabolism, Transcriptome

* Correspondence: guoww@mail.hzau.edu.cn
Key Laboratory of Horticultural Plant Biology (Ministry of Education), Key Laboratory of Horticultural Crop Biology and Genetic Improvement (Central Region) (Ministry of Agriculture), College of Horticulture and Forestry Sciences, Huazhong Agricultural University, Wuhan 430070, China

Background

Polyploidy is a common biological phenomenon and plays an important role in evolutionary history of plants [1-3]. Almost all angiosperms have undergone at least one round of whole-genome duplication in the course of their evolution [4,5]. Polyploids are classified into autopolyploids and allopolyploids. The first comes from doubling a diploid genome. And the latter arises from the combination of two or more sets of divergent genomes [6,7]. Many major crop plants including wheat (allohexaploid), cotton (allotetraploid), oilseed rape (allotetraploid), sweet potato (autotetraplooid), rice and maize (paleopolyploid) are polyploids. Moreover, polyploidy cultivars are prevalent in fruit plants, such as banana (triploid), grape (tetraploid), kiwifruit and persimmon (hexaploid), strawberry (octaploid). Phenotypic variations caused by polyploidization possess the potential to improve agricultural productivity and efficiency, especially in increasing biomass and stress tolerance.

Polyploidy has a significant influence on morphology and physiology of newly formed offspring. Compared with the corresponding diploids, autopolyploids tend to have larger cells, which result in the enlargement of single organs, such as leaves, flowers and seeds [8,9]. Physiological traits such as plant height, growth rate, flowering time, and fertility also can be altered by polyploidization [10-12]. It has been shown that tetraploidization might significantly increase stress tolerance [13,14].

A limited number of studies have investigated metabolic changes caused by autopolyploidization, and those studies focused on only specific metabolites [12]. The production of alkaloids was enhanced in artificial autotetraploids *Hyoscyamus niger* [15]. More artemisinin was produced in hairy roots of autotetraploid *Artemisia annua* [16]. Similarly, essential oils were accumulated much more in autotetraploid aromatic grasses (*Cymbopogon*) [17]. Moreover, the concentration of some metabolites like GAs (glycoalkaloids) were differentially influenced by autotetraploidy, increasing the content of minor GAs and decreasing the content of major GAs in autotetraploid *Solanum commersonii* [18].

Gene expression variations caused by allopolyploidization have been widely reported in many species including *Arabidopsis* [19,20], citrus [21], maize [22], and tobacco [23]. However, the studies on autopolyploidization aimed at identifying the alterations of genome expression patterns are relatively less than those on allopolyploidization. It is probably because autopolyploidy has long been viewed as less frequent and less important. The number of the genes differentially expressed between diploid and autotetraploid potato was about 10% [24]. A much lower rate (less than 2%) was observed when autotetraploid *Arabidopsis* was compared with diploid progenitor [25]. Similarly, study performed in autotetraploid and diploid Rangpur lime (*Citrus limonia*) showed about 1% variation in transcriptome [26]. Notably, the differentially expressed genes induced by autotetraploidization were highly related to stress response [14,25].

Citrus is one most important fruit crop in the world. However, citrus production is influenced by many environmental stresses including drought, salinity and extreme temperature [27]. Citrus rootstock improvement is required to cope with these abiotic stresses. Ziyang xiangcheng is a local citrus rootstock originated from southwest China. It was considered a putative hybrid of *Citrus ichangensis* and *Citrus reticulata* [28]. Because of its excellent performance in biotic and abiotic stresses, it has been widely used as a citrus rootstock in China [28,29]. Citrus rootstocks are propagated through polyembryonic seeds and genetically identical to the maternal plant [30-32]. The majority of citrus genotypes are apomictic, and all the apomictic embryos originate from nucellar cells [30]. Tetraploidization events are frequent in apomictic citrus genotypes [30,33]. Doubled diploid seedlings in apomictic genotypes are considered to arise from somatic chromosome doubling of maternal cells and should be genetically identical to the seed source tree [30,31]. Recent studies demonstrate that genome doubling is often considered to confer plants a better adaptability to various environmental stresses [13,14,33,34]. Therefore, doubled diploid citrus rootstocks were expected to have substantial advantage over diploid in stress tolerance. In our previous citrus breeding program, we obtained plenty of spontaneous doubled diploids from various citrus rootstock varieties, including Ziyang xiangcheng (*Citrus junos* Sieb. ex Tanaka) [35,36].

To test the effects of tetraploidization on Ziyang xiangcheng, we performed comparative metabolic and transcriptional analysis of doubled diploid and its diploid parent. Our results revealed that doubled diploid Ziyang xiangcheng had a distinct metabolic phenotype, compared with diploid. Many stress related metabolites such as sucrose, proline and GABA were enhanced in doubled diploid. However, less than 1% of genes were differentially expressed between doubled diploid and its diploid parent. Interestingly, these differentially expressed genes were highly related to stress response.

Results

Ploidy determination and analysis of genetic constitution

Eight uniform 4× seedlings out of previously identified fifteen doubled diploids were selected and further verified by flow cytometry. These eight 4× seedlings together with thirteen 2× seedlings were then analyzed by the SSR markers. All the SSR makers revealed that the eight 4× and nine 2× plants possessed the same alleles (Additional file 1). This signified that the 4× seedlings derived from genome doubling of the 2× genotype. And three diploids with heterozygous loci (Additional file 1) were excluded for further study.

Morphological changes following tetraploidization

In order to investigate morphological changes caused by tetraploidization, morphological analysis on plant height, stem diameter, leaf area, leaf thickness, stomata size and density was conducted. Compared to 2×, 4× has typical tetraploid morphological features, such as shorter plant height, larger and thicker leaf, larger stomata size and lower stomata density (Figure 1 and Additional file 2). Additionally, enlargement in leaf structure of 4x was observed by anatomical analysis (Additional files 3 and 4).

Changes of primary metabolic profiles following tetraploidization

In order to investigate the effect of tetraploidization on primary metabolism, leaf samples of double diploid and diploid lines were analyzed by using an established

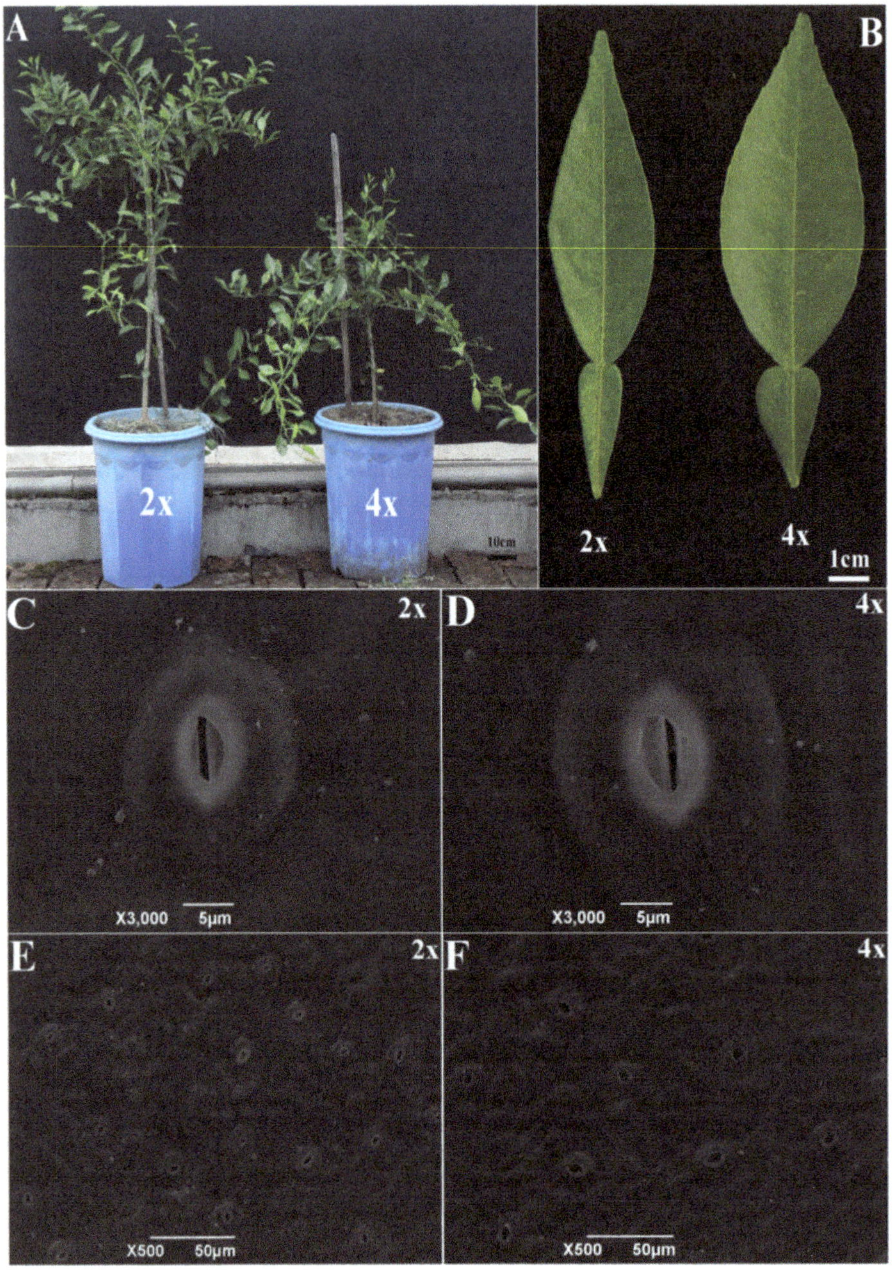

Figure 1 Morphological characterization of 2× and 4× Ziyang xiangcheng. **(A)** 2× and 4× seedlings; **(B)** Leaves of 2× and 4×; **(C)**, **(D)** Stomata size of 2× and 4×; **(E)**, **(F)** Stomata density of 2× and 4×.

GC-MS platform [37]. A total of 30 metabolites were identified by using an available chromatogram library. Utilizing the quantification internal standard, the content of every metabolite was calculated (Table 1).

Principal component analysis (PCA) served as an unsupervised statistical method to study the differences of the major metabolites of 4× and 2× (Figure 2). Parameters of the PCA model based on the primary metabolic data were: two principle components were calculated by cross validation, 58.6% of variables can be explained by first component and 17.2% of variables can be explained by the second component. A clear separation trend could be observed in the score plot (Figure 2), implying that extensive changes in the major metabolites were induced by tetraploidization.

Among the 30 metabolites, the levels of 24 metabolites in 4× leaves were significantly higher than those in 2×.

Table 1 24 of 30 primary metabolites were significantly accumulated in 4× Ziyang xiangcheng

Compound	2× (mean ± SE)[a]	4× (mean ± SE)	Flod change	P-value	Trend[b]
Sugars					
Turanose	7.64 ± 0.37	9.65 ± 0.87	1.3		
Galactose	1.15 ± 0.16	5.97 ± 1.09	5.2	0.01	up
Fructose	41.48 ± 6.07	100.74 ± 4.67	2.4	0.01	up
Glucose	15.16 ± 1.54	55.49 ± 7.59	3.7	0.05	up
Sucrose	4403.79 ± 25.33	9472.04 ± 785.87	2.2	0.01	up
Glucopyranose	93.03 ± 6.78	86.91 ± 9.97	0.9		
Arabinose	36.55 ± 2.03	176.59 ± 29.03	4.8	0.01	up
Mannose	50.12 ± 2.86	185.1 ± 25.59	3.7	0.01	up
Myo-inositol	460.53 ± 12.61	634.93 ± 49.72	1.4	0.01	up
Organic acids					
Ethanedioic acid	21.9 ± 1.29	223.95 ± 16.25	10.2	0.01	up
Succinic acid	21.07 ± 2.24	26.3 ± 2.24	1.2		
citric acid	23.06 ± 1.75	98.79 ± 1.42	4.3	0.01	up
Isocitric acid	1132.63 ± 22.07	2067.81 ± 33.25	1.8	0.01	up
GABA	1.13 ± 0.04	35.55 ± 4.88	31.5	0.01	up
2-Ketoglutaric acid	63.8 ± 6.7	50.41 ± 0.79	0.8		
Malic acid	349.14 ± 42.51	1891.19 ± 90.58	5.4	0.01	up
2,3,4-Trihydroxybutyric acid	50.89 ± 2.74	235.16 ± 10.28	4.6	0.01	up
2-Keto-d-gluconic acid	8.45 ± 0.69	32.15 ± 1.17	3.8	0.01	up
Amino acids					
Glycine	4.23 ± 0.22	16.29 ± 1.45	3.9	0.01	up
Alanine	ND	11.59 ± 3.02			up
Threonine	ND	4.01 ± 0.61			up
Proline	ND	109.17 ± 14.01			up
Serine	ND	8.07 ± 1.75			up
Acetyl-lysine	ND	39.55 ± 3.48			up
Fatty acids					
Octadecanoic acid	77.69 ± 7.32	123.2 ± 6.8	1.6	0.01	up
Octadecanoic acid,2,3-bisoxypropylester	167.7 ± 13.83	352 ± 19.42	2.1	0.01	up
Hexadecanoic acid	19.53 ± 1.46	36.06 ± 2.05	1.8	0.01	up
Hexadecanoic acid,2,3-bisoxypropylester	41.66 ± 3.67	55.75 ± 5.24	1.3		
Alcohols					
Glycerol	204.19 ± 13.38	559.62 ± 63.31	3.3	0.01	up
Rhamnitol	61.76 ± 5.93	73.17 ± 3.98	1.2		

The quantities of metabolites were analyzed using GC-MS, and their levels were normalized to ribitol and calculated as ug per g fresh weight of leaves. The data presented represent mean ± SE of six biological repetitions of leaves collected from eight plants per line. [a]ND represents the metabolite was not detected due to low concentration. [b]Up represents the metabolite is up-regulated in 4× as compared to 2× (Student's *t*-test).

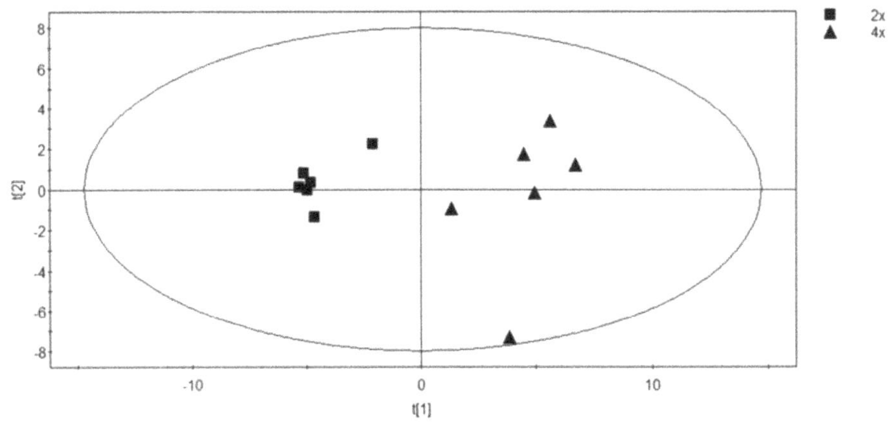

Figure 2 Principal component analysis of GC-MS metabolite profiling data from 4× and 2× leaves. First two components could explain 75.8% of the metabolite variance. Component 1 explained 58.6% of the variance and component 2 explained 17.2%.

But no significant changes in the rest 4 metabolites were observed. This indicated that tetraploidization has an activation effect on the accumulation of primary metabolites in leaves. Seven sugars were significantly accumulated in 4× (Table 1). It should be noted that in 4×, there was a 2.2-fold increase in the content of sucrose, which was the main sugar. Seven of nine identified organic acids exhibited 1.8- and 10.2-fold higher concentrations (Table 1), including γ-aminobutyric acid (GABA). Six amino acids, namely, glycine, alanine, threonine, proline, serine, and lysine, were detected in 4×, while only one amino acid, namely, glycine was detected in 2×. In addition, the content of three fatty acids and one alcohol in 4× increased (Table 1).

Changes of secondary metabolic profiles following tetraploidization

To test whether the alteration of the ploidy has an influence on the level of leaf secondary metabolism, we performed non-targeted metabolite analysis using LC-QTOF-MS metabolomics technologies. In total, 3254 mass signals were detected in positive mode. PCA was performed to promote the classification of the metabolic phenotypes and the identification of the differential metabolites. The PCA effectively clusters biological replicates of the metabolomes of 2× and 4× into two categories, demonstrating extensive changes in the secondary metabolism caused by tetraploidization (Figure 3). Of these mass signals, 898 mass signals were significantly different between 4× and 2× (corrected p-value <0.05). 196 signals were up-regulated, and 702 signals were down-regulated in 4×, reflecting a decreased trend of secondary metabolite accumulation in 4×.

Significantly changed metabolites were analyzed by LC-ESI/MS/MS to obtain structure information. A total of 9 metabolites, namely, narirutin, naringin, hesperidin, neohesperidin, didymin, sinensetin, limonin, nobiletin and nomilin were identified by matching their mass spectra and retention time with known standards. The other 34 metabolites were tentatively identified according to ESI-MS fragmentation patterns (Table 2). These identified metabolites were mainly comprised of phenolic flavonoids, including 6 flavanones and 33 flavones. These flavones were mainly made up of polymethoxyflavones (PMFs), which are widely distributed in citrus. These identified metabolites also included an aromatic amine (octopamine), a cinnamic acid (coumaric acid) and two limonoids (limonin and nomilin). Notably, all the 33 identified flavones were down-regulated in 4×, while all the 6 flavanones were up-regulated.

Global transcriptome analysis

To investigate global transcriptome changes caused by tetraploidization, four cDNA libraries of 2× and 4× mature leaves were constructed. These libraries were sequenced by Illumina Hiseq 2500 platform. And 50 bp single-end reads were then generated. In total, 25,860,712 raw reads were generated from 2× and a total of 24,428,874 raw reads came from 4× (Additional file 5). After we removed reads containing adapter, reads containing poly-N, and low quality reads from raw data, 25,830,902 and 24,402,540 clean reads remained in 2× and 4×, respectively. The GC-contents were 43.30% in 2× and 43.16% in 4× respectively. To assess the sequencing quality, the reads were mapped to the *Citrus sinensis* reference genome. Of the two groups of duplicate data, 11,115,785 (86.06%) and 11,383,064 (88.14%) reads successfully mapped were generated from 2×-1-2×-2 and 11,250,774 (88.87%) and 10,531,271(89.69%) reads from 4×-1-4×-2 (Additional file 6).

More than 50% of the genes were expressed at a low level (<3 RPKM) and less than 8% of genes were expressed at a high level (>15 RPKM) in all samples (Additional file 7). Notably, there were no obvious differences between 2× and 4× in the percentage of genes at

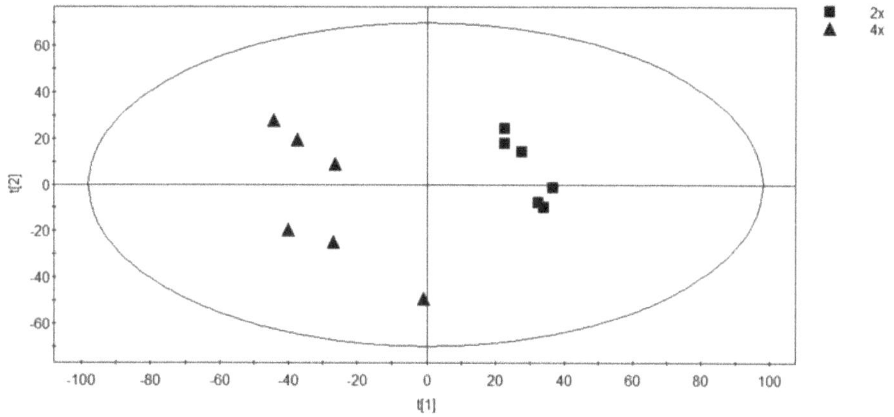

Figure 3 Principal component analysis of LC-QTOF-MS metabolite profiling data from 4× and 2× leaves. First two components could explain 49.3% of metabolite variance. Component 1 explained 32.8% of the variance and component 2 explained 16.5%.

low, medium and high expression levels. This suggested tetraploidization didn't have an effect on the inhibition or activation of gene expression.

Genes with an adjusted p-value <0.05 found by DESeq (R package, version 1.10.1) were assigned as differentially expressed. Totally 24073 genes were detected in all samples, while only 212 genes (0.8% of detected genes) were significantly differentially expressed between 2× and 4× seedling leaves. Of 212 DEGs, 96 genes were up-regulated and 116 genes were down-regulated in 4×, relative to 2×. For up-regulated genes, differences ranged from1.4-fold to 12.5- fold; for down-regulated genes, differences ranged from 1.4-fold and 13.4- fold. These results indicated that the range of gene expression changes between 2× and 4× was very limited.

The functional gene ontology annotation of these DEGs was further performed by using Blast2Go software. 163 out of the 212 DEGs were assigned to at least one term in GO biological process, cellular component, and molecular function categories. Then the DEGs were classified into 38 subcategories in terms of function, almost covering all important categories of biological processes and molecular functions (Figure 4). In the biological process category, metabolic process and cellular process were the two largest groups, suggesting that extensive metabolic activities were taking place in 4× leaves. In the cellular component category, cell and cell part represented two major sub-categories, while catalytic and binding were dominant in molecular function category.

GO enrichment analysis was performed by using BiNGO [38]. In biological process category, DEGs were found to be highly related to stress-response functions, such as response to salt stress, to water, and to abscisic acid (Figure 5). This indicated that some processes related to stress were induced in response to tetraploidization. The other two functions, namely anion transport and polyamine catabolic process, were also significantly enriched (Figure 5). In molecular function category, only two terms were overrepresented, namely, inorganic anion transmembrane transporter activity, inorganic phosphate transmembrane transporter activity (Figure 5). In cellular component category, no terms were overrepresented.

To identify the biological pathways in which the DEGs were involved, we mapped DEGs to the reference canonical pathways in KEGG. In total, 40 out of 212 DEGs were assigned to 46 KEGG pathways. The two largest clusters were metabolic pathways with 19 members and biosynthesis of secondary metabolites with 13 members (Additional file 8). It indicated that many DEGs involved in metabolic process in 4×. However, no KEGG terms was over-represented in DEGs.

To validate the RNA-seq data, the following top 10 up-regulated functionally characterized genes were selected for qPCR assays: Fe(II)/ascorbate oxidase (SRG1, Cs9g09290), UDP-glucoronosyl/UDP-glucosyltransferase family protein (UGT, Cs5g11620), myb family transcription factor (RL6, Cs3g24870), caffeic acid O-methyltransferase (COMT, orange1.1 t02085), aminocyclopropane 1-carboxylic acid oxidase (ACO, Cs9g08990), u-box armadillo repeat protein (PUB19, Cs7g08470), ethylene response factor (ERF4, Cs1g07950), tracheary element vacuolar protein (XCP1, Cs2g27860), glycosyltransferase (GATL9, Cs7g07900), ethylene response factor (ERF9, Cs2g05620) (Additional file 9). As shown in Figure 6, all the 10 genes were verified to be up-regulated by qPCR analysis, although their fold changes differed from the result of RNA-seq. Notably, six of these genes, namely, SRG1 [39], COMT [40], ACO [41], PUB19 [42], ERF4 [43] and ERF9 [44] were involved in abiotic stress response.

Table 2 Identified metabolites showing statistically significant changes between 2× and 4× Ziyang xiangcheng

Peak no.	RT (min)	Component name	$[M + H]^+$	MS/MS fragments	Family	Trend[c]
1	1.0	Octopamine[b]	154	91	Alkaloid	up
2	5.2	Coumaric acid[b]	165	147/120/65/91	Cinnamic Acid	down
3	8.8	Narirutin[a]	581	273/434	Flavanone	up
4	8.9	Neodiosmin[b]	609	301/463	Flavone	down
5	9.0	Naringin[a]	581	273	Flavanone	up
6	9.1	Hesperidin[a]	611	303	Flavanone	up
7	9.3	Neohesperidin[a]	611	303	Flavanone	up
8	9.9	Brutieridin[b]	755	303	Flavanone	up
9	10.5	Didymin[a]	595	287	Flavanone	up
10	10.6	PMFs-1[b]	359	344/329/298	Flavone	down
11	11.2	PMFs-2[b]	359	329/344	Flavone	down
12	11.5	PMFs-3[b]	375	360/345	Flavone	down
13	11.6	PMFs-4[b]	359	298/326/344	Flavone	down
14	11.8	PMFs-5[b]	359	326/344	Flavone	down
15	12.4	PMFs-6[b]	389	359/341/374	Flavone	down
16	12.5	PMFs-7[b]	359	329/344	Flavone	down
17	12.5	Isosinensetin[b]	373	343	Flavone	down
18	12.7	PMFs-8[b]	403	373/388	Flavone	down
19	12.8	PMFs-9[b]	389	359/374	Flavone	down
20	12.8	PMFs-10[b]	375	317/342	Flavone	down
21	12.9	PMFs-11[b]	419	389/371/404	Flavone	down
22	13.0	PMFs-12[b]	403	373/327/388	Flavone	down
23	13.1	PMFs-13[b]	345	330/315	Flavone	down
24	13.3	Sinensetin[a]	373	343/312/329/357	Flavone	down
25	13.3	PMFs-14[b]	343	328/313	Flavone	down
26	13.6	PMFs-15[b]	345	330/284/312	Flavone	down
27	13.6	PMFs-16[b]	405	375/390	Flavone	down
28	13.7	Limonin[a]	471	161/425	Limonoid	up
29	13.7	PMFs-17[b]	375	360/345/317	Flavone	down
30	14.0	PMFs-18[b]	359	343/329	Flavone	down
31	14.2	Nomilin[a]	515	161	Limonoid	down
32	14.2	Nobiletin[a]	403	373	Flavone	down
33	14.3	Tetramethyl-O-scutellarein[b]	343	313/282/299	Flavone	down
34	14.4	PMFs-19[b]	389	331/356/313/374	Flavone	down
35	14.5	PMFs-20[b]	359	329/346	Flavone	down
36	14.8	Heptamethoxyflavone[b]	433	418/403	Flavone	down
37	15.0	PMFs-21[b]	343	313/328	Flavone	down
38	15.0	PMFs-22[b]	359	298/326/343	Flavone	down
39	15.2	PMFs-23[b]	419	389/404	Flavone	down
40	15.8	PMFs-24[b]	405	375/347/357/390	Flavone	down
41	15.9	PMFs-25[b]	389	359/374/341	Flavone	down
42	16.5	PMFs-26[b]	419	389/404	Flavone	down
43	16.9	5-Demethyl tangeretin[b]	359	344/329/301	Flavone	down

$[M + H]^+$, protonated molecular ion. [a]Identified by matching their retention time and mass spectra with known standard. [b]Putatively identified using ESI-MS fragmentation patterns. [c]Relative increased (up) or decreased (down) concentration in 4× as compared to 2×. Student's *t*-test was used and a p-value of less than 0.05 was considered significant. PMFs, polymethoxyflavones.

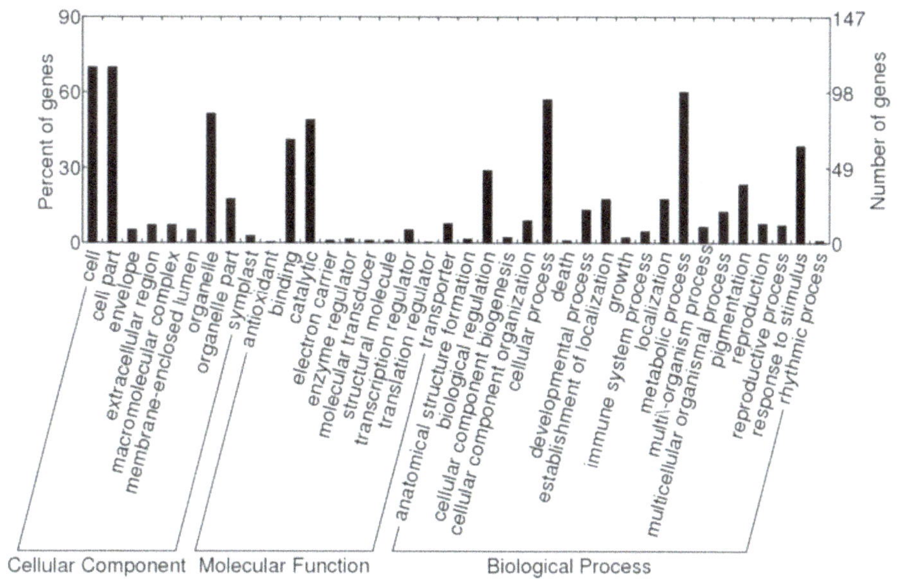

Figure 4 GO categories of the DEGs between 2× and 4× Ziyang xiangcheng. 163 out of the 212 DEGs were assigned to 957 GO annotations, which were divided into three categories: biological processes, cellular components, and molecular functions.

Discussion

Stress related metabolites were significantly up-regulated in doubled diploid Ziyang xiangcheng

Metabolic alterations induced by tetraploidization might confer plant a better adaptation to environmental stresses. Primary metabolites are required for growth, development and interactions of plants with their environment [45]. In this study, most of the detected primary metabolites were up-regulated in 4× Ziyang xiangcheng (Table 1). It indicated that tetraploidization had an activation effect on primary metabolism. These up-regulated metabolites include sugars, amino acids, organic acids, and fatty

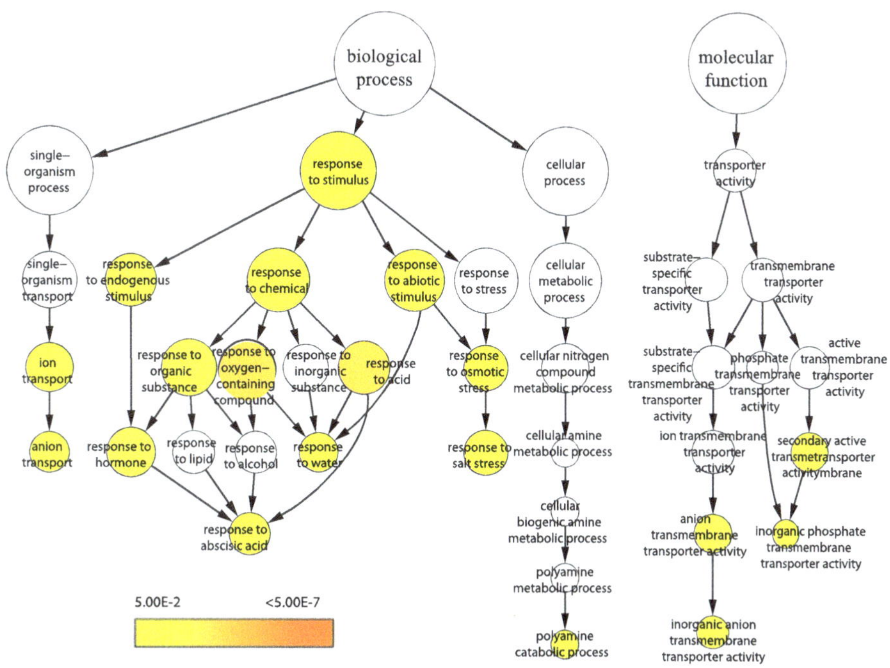

Figure 5 Significantly enriched GO categories in DEGs between 2× and 4× Ziyang xiangcheng. The colored nodes represent the significantly over-represented GO terms. The colored bar shows the significance.

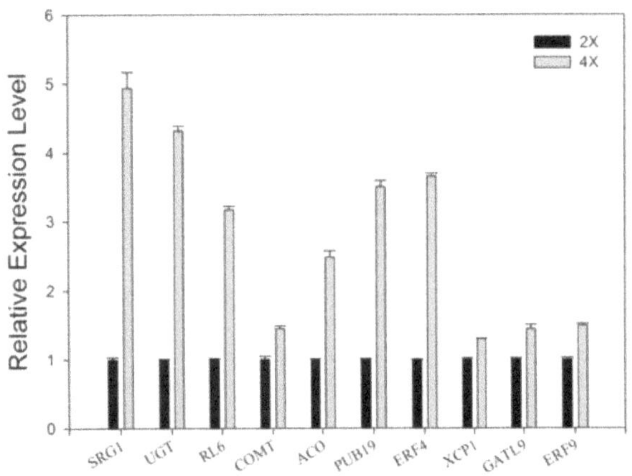

Figure 6 Expression analysis of top 10 up-regulated functionally characterized DEGs in 4× Ziyang xiangcheng by qPCR.

acids. Notably, these metabolites play an important role during plant adaptations to environmental stresses.

Sugars are involved in various abiotic stresses. They have several functions in plants suffering abiotic stresses: acting as osmoprotectants to maintain osmotic balance and stabilize macromolecules or as metabolite signaling molecules to activate specific signal transduction pathway, and providing energy source to recover from water deficit [46,47]. Accumulation of sugars is strongly correlated with improved plant stress tolerance to drought stress [46,48,49]. For example, sucrose accumulates in almost all desiccation-tolerant flowering plants [50] and fern [51]. In this study, seven out of nine detected sugars including sucrose, glucose and fructose were up-regulated in 4×, which implied 4× might have advantages over 2× under drought stress.

A case in point is that increased levels of proline correlate with enhanced stress tolerance [48,52]. Proline was considered to have several functions under stress conditions, including osmotic adjusting, reactive oxygen species (ROS) scavenger and protection of proteins from denaturation [52-54]. Therefore, higher concentration of proline might promote abiotic stress tolerance in 4×. Additionally, Yobi et al. [55] found that desiccation-tolerant species *Selaginella lepidophylla* had significantly higher concentration of sugars, sugar alcohols and amino acids than desiccation-sensitive species *Selaginella moellendorffii*. Compared to 2×, higher concentration of stress metabolites in 4× might be also beneficial for the cultivar grafted on it. A study performed on Rangpur lime (*Citrus limonia*) rootstock demonstrated that tetraploids increase drought tolerance via enhanced constitutive root abscisic acid production [26]. In that study, diploid and tetraploid clones of Rangpur lime rootstocks were grafted with 2× Valencia Delta sweet orange (*Citrus sinensis*) scions, named V/2×RL and V/4×RL, respectively; V/4×RL leaves had greater abscisic acid (ABA) content under normal condition, compared to V/2×RL [26]. Studies of *Arabidopsis* polyploids revealed that the content of leaf potassium and rubidium was evaluated in in diploid leaves on shoots grafted to tetraploid roots, whereas leaves from tetraploid shoots grafted to diploid roots showed the same leaf K as diploid [13]. So we may presume that a distinct metabolic phenotype would be observed between the scion cultivars grafted on 4× and 2× Ziyang xiangcheng respectively. Higher content of stress-related metabolites in 4× might be beneficial for the cultivar grafted on it. In addition, tetraploid rootstock may also have a dwarfing effect on scion cultivar being grafted on it, compared with the diploid rootstock [56].

Gene expression divergence caused by tetraploidization is involved with stress response

A small genome expression change was observed between diploid and autotetraploid according to studies performed on several species. In *Paspalum notatum* and *Isatis indigotica*, about 0.6% and 4% variations in transcript abundance were detected between diploid and autotetraploid by using the *Arabidopsis thaliana* whole genome gene chip [57,58]. In *Arabidopsis thaliana* Col-0 ecotype and Ler-0 ecotype, Yu et al. [25] found about 1% and 0.1% variations between diploid and autotetraploid, respectively. We found less than 1% genes were differentially expressed between diploid and doubled diploid Ziyang xiangcheng. A similar number of genes were also detected between diploid and tetraploid *Citrus limonia* [26]. These studies altogether with our study suggested that the effect of genome doubling on gene expression is relatively limited. Here, we should point out that the 4× Ziyang xiangcheng came from doubling a hybrid (*C. ichangensis* × *C. reticulata*). Theoretically, the doubled diploid should be an allotetraploid rather than an autotetraploid (doubling a homozygous diploid) [6]. Therefore, the expression pattern of doubled diploid Ziyang xiangcheng should consist with the one of an allotetraploid rather than the one of an autotetraploid. Genome expression changes in allotetraploids are considered to be more strongly affected by genome hybridization than by changes in ploidy levels [19,59]. So we presume that a relatively large change in genome expression could be detected between doubled diploid Ziyang xiangcheng and its putative parents (*C. ichangensis* and *C. reticulata*). Herein, we only focused on the effect of genome doubling on gene expression.

Genes involved in the response to abscisic acid and abiotic stimulus, were differentially expressed following genome doubling according to GO enrichment analysis (Figure 5). This indicates that 4× Ziyang xiangcheng might be able to respond to abiotic stresses in a flexible and fast way, to some extent [14]. Interestingly, the phenomenon that tetraploidization influences the expression of genes involved in hormone and abiotic stress responses was also reported in autotetraploid *A. thaliana* [14,25]. We also found that the expression of genes involved in ion transport was also affected by genome doubling. It is known that ion transport is highly related to salt tolerance [60].

Higher potassium accumulation and salinity tolerance has been found in *Arabidopsis* polyploids [13]. The higher potassium accumulation might be partly due to altered expression of genes involved in ion transport.

Moreover, six out of ten top up-regulated genes were involved in ABA- and stress-related process (Additional file 9). The first gene, namely SRG1, was associated with senescence-related processes, encoding a member of the Fe(ll)/ascorbate oxidase superfamily protein, and its expression was induced under drought and heat stress [61,62]. Caffeic acid O-methyltransferases encoded by COMT genes are key enzymes of lignin biosynthesis [63], affecting cell wall structure, and COMT was up-regulated by drought stress in maize [40]. ACO genes encode 1-aminocyclopropane-1-carboxylate (ACC) oxidases which catalyze the reaction from ACC to ethylene [64], and water stress induced ACO gene expression in sunflower leaves was previously reported [65]. PUB19 encodes a U-Box E3 ubiquitin ligase and it was up-regulated by drought, salt, and cold stress and ABA [42]. The last two genes, namely ERF4 and ERF9, which are the members of the ERF/AP2 transcription factor family, are involved in various reactions to abiotic stresses [66]; these two genes bind to the GCC box, DRE/CRT, CE1 elements, and they acted as repressors of gene transcription, enhancing plant tolerance to multiple stresses [67]. Overexpression of ERF4 gene increased tolerance to salt and drought stress in *Arabidopsis* [66]. These reports, together with our results suggest 4× Ziyang xiangcheng may be pre-adapted to abiotic stresses, compared to 2×.

The transcriptome divergence cannot explain the metabolic changes

In order to integrate leaf transcriptome data with the metabolic profiling, attention was focused on the DEGs involved in metabolic pathway. Among these DEGs, 40 were assigned to 46 pathways and no significantly enriched KEEG pathways were found. It implies that the limited DEGs involve in a wide range of pathways, but their functions are dispersive.

To a great extent, the accumulation pattern of the DEGs encoding proteins or enzymes involved in metabolic processes was not consistent with the differences observed in the metabolite profiling (Additional file 10). Most of the detected sugars, amino acids and fatty acids were significantly accumulated in 4×. However, most of the genes involved in these metabolic processes were down-regulated in 4×. For example, the sucrose content of 4x leaves was 2-fold than that of 2×. But the gene encoding sucrose synthase was significantly down-regulated in 4×. In another example, in flavone and flavonol biosynthesis, only one gene, namely, COMT was differentially expressed between 4× and 2×. The gene encoding a caffeic acid O-methyltransferase, positively regulates flavonoid biosynthetic process and may be involved in PMFs (polymethoxyflavones) synthesis [68]. Theoretically, the up-regulation of COMT should promote the accumulation of PMFs in 4×. However, all detected PMFs were down-regulated in 4×. The discordance between transcriptomic and metabolomic data is probably related to several factors. First, it is not easy to find a strict correlation between metabolite accumulation and gene expression because of the complexity in metabolic networks [69,70]. Second, small RNAs, including microRNAs and small interfering RNAs might play an important role in some gene regulation [71]. Third, reactivation of transposable elements (TEs) following polyploidization in synthetic hexaploid wheats (*Triticum*) was considered to participate in regulation of the transcription of neighbouring genes [72]. At last, post-translational modifications may contribute to the discordance between transcriptomic and metabolomic data. The transcriptome divergence might not reflect the protein divergence between 4× and 2× Ziyang xiangcheng, leading to the discordance. In support of this hypothesis, percentage of differentially accumulated proteins between autotetraploid and diploid *Arabidopsis thaliana* that matched the differentially expressed genes was relatively low, due to post-transcriptional regulation and translational modifications of proteins during polyploidization [73]. Similarly, transcriptional changes do not explain differential protein regulation in resynthesized *Brassica napus* allotetraploids [74].

Conclusions

Our results suggest that tetraploidization has multi-level effects on Ziyang xiangcheng. Morphological and anatomical traits like leaf thickness, stoma number, stomatal density and vessel size were altered as a consequence of tetraploidization. The metabolic phenotype was also significantly altered following tetraploidization and many stress-related metabolites, such as sucrose, proline and GABA were significantly up-regulated in 4×. However, relatively small transcriptome alterations were induced by tetraploidization. Notably, the transcriptome alterations were highly related to hormone and stress responses, and many top up-regulated genes in 4× were associated with stress response. Interestingly, the transcriptional divergence could not adequately explain the metabolic changes, probably due to post-transcriptional regulation. Compared to diploid, higher expression level of stress related genes and higher content of stress related metabolites in doubled diploid could be beneficial for its stress tolerance. Our data will help better understanding the consequences caused by polyploidization and be valuable for citrus rootstock breeding in the future.

Methods

Plant materials

Seeds of diploid (2n = 2× = 18) Ziyang xiangcheng (*Citrus junos* Sieb. ex Tanaka) were collected and provided by Professor Keling Chen, Institute of Horticulture Research, Sichuan Provincial Academy of Agricultural Science, China. Seeds were grown in pots filled with commercial soil in the greenhouse. Doubled diploid and diploid Ziyang xiangcheng were prepared by Liang et al. [30]. Thirteen uniform 1-year-old 2× and eight uniform 1-year-old 4× seedlings were respectively selected and transplanted in commercial soil. Seedlings were grown in the greenhouse under natural photoperiod conditions and they were irrigated twice a week. The ploidy levels of these seedlings were determined by flow cytometry (CyFlow Space, Partec, Germany). The genetic constitution of these seedlings was further analyzed by the published SSR markers [75,76], namely, Ci01C09, Ci06A05b, Ci07C07, Ci07E06, mCrCIR08A03, CCSM40, CCSM46, CCSM69.

Sampling

Fully expended leaves (from fourth or fifth leaf from the top, 3–5 month old, from the spring flush of the current season) were collected in the morning. The leaves being used for metabolic and transcriptional profiling were immediately frozen in liquid nitrogen and stored at −80°C.

Morphological and anatomic analysis

Leaf thickness and area was determined using a micrometer and a portable area meter (Yaxin-1241, Beijing), respectively. Three leaves of each individual seedling were measured. SEM analysis was performed using one leaf of each individual plant according to the method described by Yi et al. [77]. Photographs were taken to measure stomatal size and density. About 60 stomata were measured for each genotype.

Samples were fixed, dehydrated and embedded according to Hu et al. [78]. Transverse sections about 1–5 μm thick were cut using a Leica Ultracut R ultramicrotome (Leica, Bensheim, Germany). The sections were stained with Toluidine Blue O (Aldrich, Milwaukee) and photographed with a BX61 fluorescence microscope (Olympus, Tokyo). The morphometrical analysis was performed using Image-Pro Plus 6.0 software (Media Cybernetics, USA).

The primary metabolic profiling

Six independent plants were used as biological replicates, and about five leaves were sampled from each plant in primary metabolic profiling. Non-targeted metabolite profiling was carried out by GC-MS using a modified method described by Yun et al. [37]. A total of 200 mg ground leaf samples were extracted in 2,700 μl methanol and ribitol solution (300 μl, 0.2 mg ml^{-1}) was added as an internal standard. The samples were centrifuged, dried and derivatized. GC-MS analysis was performed by using a Thermo Trace GC Ultra, coupled with Thermo Fisher a DSQ II mass spectrometer (Thermo Fisher Scientific, Waltham, MA, USA). Metabolites were identified by using an available chromatogram library and PCA analysis was performed by using the software Simca-P (Ver 11, Umetrics, Umea, Sweden).

The secondary metabolic profiling

Six independent plants were used as biological replicates, and about five leaves were sampled from each plant in secondary metabolic profiling. The secondary metabolic profiling was performed by LC-QTOF-MS using a modified method according to Yun et al. [37]. 100 mg freeze-dried powder was extracted with 80% methanol over night at 4°C. The mixture was centrifuged and filtered. Then, the metabolic profiling were performed using a QTOF 6520 mass spectrometer (Agilent Technologies, Palo Alto, CA, USA) coupled to a 1200 series Rapid Resolution HPLC system.

The raw data was processed by Agilent Mass Hunter Qualitative Analysis (version B. 04.00, Aglient Technologies) and Mass Profiler Software (version B.02.02, Aglient Technologies). Then PCA analysis was performed using the filtered and normalized data. Metabolite identification was carried out by comparing mass spectra and retention time with those of authentic standards. Nine authentic standards, namely, narirutin, naringin, hesperidin, neohesperidin, didymin, sinensetin, limonin, nomilin, and nobiletin were obtained from Sigma Chemical Co. (St. Louis, MO).

RNA sequencing and data analysis

Leaves from four plants were pooled as an independent biological replicate and leaves from other four trees were pooled as the other independent biological replicate in transcriptomic analysis. Total RNA extraction and a quality assessment were performed according to the protocol described by Zheng et al. [79]. RNA Samples were sent to Novogene Bioinformatics Technology Co. Ltd (Beijing), where the libraries were constructed. Sequencing libraries were generated from 3 μg total RNA using NEBNext Ultra RNA Library Prep Kit (NEB, USA) and sequenced on an IlluminaHiseq 2500 platform and 50 bp single-end reads were generated.

Clean reads were obtained by removing reads containing adapter, reads containing ploy-N and low quality reads from raw data and were aligned to the *Citrus sinensis* genome [80] (http://211.69.128.148/

orange/index.php) using TopHat (2.0.9) software [81]. To estimate gene expression level, RPKM of each gene was calculated based on the length of the gene and reads count mapped to this gene. Genes RPKM values were calculated based on all the uniquely mapped reads. The genes with RPKMs ranging from 0 to 3 were considered at a low expression level; the genes with RPKMs ranging from 3 to 15 at a medium expression level; and the genes with RPKMs above15 at high expression level.

Differential expression analysis was implemented using the DESeq R package (1.10.1) [82]. Genes with an adjusted P-value <0.05 found by DESeq were assigned as differentially expressed. GO (Gene Ontology) annotation was performed by using Blast2GO software (GO association done by a BLASTX against the NCBI NR database). Then GO enrichment analysis of differentially expressed genes was performed by the BiNGO plugin for Cytoscape [38]. Over-presented GO terms were identified by using a hypergeometric test with a significance threshold of 0.05 after a Benjamini and Hochberg FDR correction [83]. KEGG enrichment analysis of differentially expressed genes was performed by KOBAS (2.0) software [84].

Verification of RNA-seq by q-PCR

To test the reliability of RNA-seq, a set of top ten up-regulated genes in 4× were selected for qRT-PCR. Specific primers were designed with the Primer Express software (Applied Biosystems) and synthesized by Sangon (Shanghai, China). The cDNA was synthesized from 1 μg of total RNA using PrimeScript RT reagent Kit (Takara, Dalian, China). Real-time RT–PCR was performed on the ABI 7500 Real-Time PCR System (Applied Biosystems) using the 2× SYBR green PCRmaster mix (Applied Biosystems). Three independent biological replicates were analyzed for each sample and data were indicated as mean ± SE (n = 3).

Availability of supporting data

Raw sequencing data is available through the Gene Expression Omnibus (GEO) under accession NO. GSE65416 at website: http://www.ncbi.nlm.nih.gov/geo/query/acc.cgi?acc=GSE65416.

Additional files

Additional file 1: SSR analysis of eight 4× and thirteen 2× Ziyang xiangcheng seedlings. Black arrows represent seedlings possessed heterozygous loci.

Additional file 2: Comparison between 2× and 4× Ziyang xiangcheng seedlings with respect to height, stem diameter, leaf area, stoma size and density. [a]An asterisk (*) indicates significant differences (Student's *t*-test; P < 0.01).

Additional file 3: Transversal sections of mature leaves (fourth or fifth leaf from the top) in 2× (A and C) and 4× (B and D) Ziyang xiangcheng. Anatomy of leaf blades (A and B) and leaf central vein (C and D) were shown. Abe, abaxial epidermis; Ade, adaxial epidermis; Cut, cuticle; Is, intercellular space; Pa, parenchyma; Pi, pith; Ph, phloem; Pp, palisade parenchyma; Sc, sclerenchyma; Sp, spongy parenchyma; X, xylem.

Additional file 4: Anatomical characteristics of leaves from 2× and 4× Ziyang xiangcheng seedlings. [a]An asterisk (*) indicates significant differences (Student's *t*-test; P < 0.01).

Additional file 5: Quality of illumina sequencing.

Additional file 6: Summary of read mapping statistics.

Additional file 7: Statistics of genes in different expression-level intervals.

Additional file 8: KEGG classification of the DEGs between 2× and 4× Ziyang xiangcheng. 44 out of the 212 DEGs were assigned to 46 KEGG pathways. The top 10 most abundant KEGG pathways are shown.

Additional file 9: Primer sequences for Real-time PCR analysis.

Additional file 10: DEGs involved in major metabolic pathways.

Abbreviations

2x: Diploid Ziyang xiangcheng; 4x: Doubled diploid Ziyang xiangcheng; GAs: Glycoalkaloids; GC-MS: Gas chromatography coupled to mass spectrometry; LC- QTOF- MS: Liquid chromatography quadrupole time-of-flight mass spectrometry; PCA: Principal component analysis; GABA: γ-aminobutyric acid; PMFs: Polymethoxyflavones; DEGs: Differentially expressed genes; ROS: Reactive oxygen species.

Competing interests

The authors declare that they have no competing interests.

Authors' contributions

WWG conceived the study. FQT designed the experiments, performed RNA-seq analysis and wrote the manuscript. HT and HYZ carried out primary and secondary metabolite detection and statistical analysis. WJL and JML performed morphological and anatomical analysis. WWG and XMW interpreted the experimental data and revised the manuscript. All the authors read and approved the final manuscript.

Acknowledgments

This research was financially supported by the National NSF of China (nos. 31125024, 31221062), the Ministry of Education of China (no. IRT13065), and the Ministry of Agriculture of China (no. 201203075). The authors thank Prof. Ping Liu in Foreign Language College (HZAU) for her effort to polish the language.

References

1. Otto SP, Whitton J. Polyploid incidence and evolution. Annu Rev Genet. 2000;34:401–37.
2. Comai L. The advantages and disadvantages of being polyploid. Nat Rev Genet. 2005;6:836–46.
3. Doyle JJ, Flagel LE, Paterson AH, Rapp RA, Soltis DE, Soltis PS, et al. Evolutionary genetics of genome merger and doubling in plants. Annu Rev Genet. 2008;42:443–61.
4. Wendel JF. Genome evolution in polyploids. Plant Molecular Evolution. 2000;42:225–49.
5. Adams KL, Wendel JF. Polyploidy and genome evolution in plants. Curr Opin Plant Biol. 2005;8:135–41.
6. Chen ZJ. Genetic and epigenetic mechanisms for gene expression and phenotypic variation in plant polyploids. Annu Rev Plant Biol. 2007;58:377.
7. Doyle JJ, Egan AN. Dating the origins of polyploidy events. New Phytol. 2010;186:73–85.
8. Abel S, Becker H. The effect of autopolyploidy on biomass production in homozygous lines of *Brassica rapa* and *Brassica oleracea*. Plant Breed. 2007;126:642–3.

9. Li X, Yu E, Fan C, Zhang C, Fu T, Zhou Y. Developmental, cytological and transcriptional analysis of autotetraploid *Arabidopsis*. Planta. 2012;236:579–96.
10. Yao H, Kato A, Mooney B, Birchler JA. Phenotypic and gene expression analyses of a ploidy series of maize inbred Oh43. Plant Mol Biol. 2011;75:237–51.
11. Martin SL, Husband BC. Whole genome duplication affects evolvability of flowering time in an autotetraploid plant. PLoS One. 2012;7:e44784.
12. Cohen H, Tel-Zur N. Morphological changes and self-incompatibility breakdown associated with autopolyploidization in *Hylocereus* species (Cactaceae). Euphytica. 2012;184:345–54.
13. Chao DY, Dilkes B, Luo H, Douglas A, Yakubova E, Lahner B, et al. Polyploids exhibit higher potassium uptake and salinity tolerance in *Arabidopsis*. Science. 2013;341:658–9.
14. del Pozo JC, Ramirez-Parra E. Deciphering the molecular bases for drought tolerance in Arabidopsis autotetraploids. Plant Cell Environ. 2014. doi:10.1111/pce.12344.
15. Lavania U, Srivastava S. Enhanced productivity of tropane alkaloids and fertility in artificial autotetraploids of *Hyoscyamus niger* L. Euphytica. 1991;52:73–7.
16. De Jesus-Gonzalez L, Weathers P. Tetraploid *Artemisia annua* hairy roots produce more artemisinin than diploids. Plant Cell Rep. 2003;21:809–13.
17. Lavania UC, Srivastava S, Lavania S, Basu S, Misra NK, Mukai Y. Autopolyploidy differentially influences body size in plants, but facilitates enhanced accumulation of secondary metabolites, causing increased cytosine methylation. Plant J. 2012;71:539–49.
18. Caruso I, Lepore L, De Tommasi N, Dal Piaz F, Frusciante L, Aversano R, et al. Secondary metabolite profile in induced tetraploids of wild *Solanum commersonii* Dun. Chem Biodivers. 2011;8:2226–37.
19. Wang J, Tian L, Lee HS, Wei NE, Jiang H, Watson B, et al. Genomewide nonadditive gene regulation in *Arabidopsis* allotetraploids. Genetics. 2006;172:507–17.
20. Ni Z, Kim ED, Ha M, Lackey E, Liu J, Zhang Y, et al. Altered circadian rhythms regulate growth vigour in hybrids and allopolyploids. Nature. 2008;457:327–31.
21. Bassene J, Froelicher Y, Dubois C, Ferrer R, Navarro L, Ollitrault P, et al. Non-additive gene regulation in a citrus allotetraploid somatic hybrid between *C. reticulata* Blanco and *C. limon* (L.) Burm. Heredity. 2009;105:299–308.
22. Riddle NC, Jiang H, An L, Doerge R, Birchler JA. Gene expression analysis at the intersection of ploidy and hybridity in maize. Theor Appl Genet. 2010;120:341–53.
23. Bombarely A, Edwards KD, Sanchez-Tamburrino J, Mueller LA. Deciphering the complex leaf transcriptome of the allotetraploid species *Nicotiana tabacum*: a phylogenomic perspective. BMC Genomics. 2012;13:406.
24. Stupar RM, Bhaskar PB, Yandell BS, Rensink WA, Hart AL, Ouyang S, et al. Phenotypic and transcriptomic changes associated with potato autopolyploidization. Genetics. 2007;176:2055–67.
25. Yu Z, Haberer G, Matthes M, Rattei T, Mayer KF, Gierl A, et al. Impact of natural genetic variation on the transcriptome of autotetraploid *Arabidopsis thaliana*. Proc Natl Acad Sci U S A. 2010;107:17809–14.
26. Allario T, Brumos J, Colmenero-Flores JM, Tadeo F, Froelicher Y, Talon M, et al. Large changes in anatomy and physiology between diploid Rangpur lime (*Citrus limonia*) and its autotetraploid are not associated with large changes in leaf gene expression. J Exp Bot. 2011;62:2507–19.
27. Gong XQ, Liu JH. Genetic transformation and genes for resistance to abiotic and biotic stresses in Citrus and its related genera. Plant Cell Tissue Organ Cult. 2013;113:137–47.
28. Liu JJ, Chen KL, Hu Q, Yang M, Zhou QM, Li HW, et al. Preliminary study on Ziyang xiangcheng (*Citrus junos* Sieb ex Tanaka), a special local citrus germplasm. Southwest China J Agric Sci. 2008;21:1658–60 (in Chinese with English abstract).
29. Zhou GF, Peng SA, Liu YZ, Wei QJ, Han J, Islam MZ. The physiological and nutritional responses of seven different citrus rootstock seedlings to boron deficiency. Trees. 2014;28:295–307.
30. Aleza P, Froelicher Y, Schwarz S, Agustí M, Hernández M, Juárez J, et al. Tetraploidization events by chromosome doubling of nucellar cells are frequent in apomictic citrus and are dependent on genotype and environment. Ann Bot. 2011;108:37–50.
31. Rao MN, Soneji JR, Chen C, Huang S, Gmitter Jr FG. Characterization of zygotic and nucellar seedlings from sour orange-like citrus rootstock candidates using RAPD and EST-SSR markers. Tree Genet Genom. 2008;4:113–24.
32. Ruiz C, Breto MP, Asins M. A quick methodology to identify sexual seedlings in citrus breeding programs using SSR markers. Euphytica. 2000;112:89–94.
33. Saleh B, Allario T, Dambier D, Ollitrault P, Morillon R. Tetraploid citrus rootstocks are more tolerant to salt stress than diploid. C R Biol. 2008;331:703–10.
34. Mouhaya W, Allario T, Brumos J, Andrés F, Froelicher Y, Luro F, et al. Sensitivity to high salinity in tetraploid citrus seedlings increases with water availability and correlates with expression of candidate genes. Funct Plant Biol. 2010;37:674–85.
35. Liang WJ, Xie KD, Guo DY, Xie ZZ, Yi HL, Guo WW. Spontaneous generation and SSR molecular characterization of autotetraploids in ten citrus rootstocks. J Fruit Sci. 2014;31:1–6 (in Chinese with English abstract).
36. Liang WJ, Xie KD, Guo DY, Xie ZZ, Xu Q, Yi HL, et al. Spontaneous generation and SSR characterization of polyploids from ten citrus cultivars. Acta Hort Sin. 2014;41:409–16 (in Chinese with English abstract).
37. Yun Z, Gao H, Liu P, Liu S, Luo T, Jin S, et al. Comparative proteomic and metabolomic profiling of citrus fruit with enhancement of disease resistance by postharvest heat treatment. BMC Plant Biol. 2013;13:44.
38. Maere S, Heymans K, Kuiper M. BiNGO: a Cytoscape plugin to assess overrepresentation of gene ontology categories in biological networks. Bioinformatics. 2005;21:3448–9.
39. Zhang X, Ju HW, Chung MS, Huang P, Ahn SJ, Kim CS. The RR-type MYB-like transcription factor, AtMYBL, is involved in promoting leaf senescence and modulates an abiotic stress response in *Arabidopsis*. Plant Cell Physiol. 2011;52:138–48.
40. Vincent D, Lapierre C, Pollet B, Cornic G, Negroni L, Zivy M. Water deficits affect caffeate O-methyltransferase, lignification, and related enzymes in maize leaves. A proteomic investigation. Plant Physiol. 2005;137:949–60.
41. Yuan R, Wu Z, Kostenyuk IA, Burns JK. G-protein-coupled α_{2A}-adrenoreceptor agonists differentially alter citrus leaf and fruit abscission by affecting expression of ACC synthase and ACC oxidase. J Exp Bot. 2005;56:1867–75.
42. Liu YC, Wu YR, Huang XH, Sun J, Xie Q. AtPUB19, a U-box E3 ubiquitin ligase, negatively regulates abscisic acid and drought responses in Arabidopsis thaliana. Mol Plant. 2011;4:938–46.
43. Ding Y, Liu N, Virlouvet L, Riethoven JJ, Fromm M, Avramova Z. Four distinct types of dehydration stress memory genes in *Arabidopsis thaliana*. BMC Plant Biol. 2013;13:229.
44. Mizoi J, Shinozaki K, Yamaguchi-Shinozaki K. AP2/ERF family transcription factors in plant abiotic stress responses. Biochim Biophys Acta. 1819;2012:86–96.
45. Neilson EH, Goodger JQ, Woodrow IE, Moller BL. Plant chemical defense: at what cost? Trends Plant Sci. 2013;18:250–8.
46. Couee I, Sulmon C, Gouesbet G, El Amrani A. Involvement of soluble sugars in reactive oxygen species balance and responses to oxidative stress in plants. J Exp Bot. 2006;57:449–59.
47. Patterson JH, Newbigin E, Tester M, Bacic A, Roessner U. Metabolic responses to salt stress of barley (*Hordeum vulgare* L.) cultivars, Sahara and Clipper, which differ in salinity tolerance. J Exp Bot. 2009;60:4089–103.
48. Vinocur B, Altman A. Recent advances in engineering plant tolerance to abiotic stress: achievements and limitations. Curr Opin Biotechnol. 2005;16:123–32.
49. Seki M, Umezawa T, Urano K, Shinozaki K. Regulatory metabolic networks in drought stress responses. Curr Opin Plant Biol. 2007;10:296–302.
50. Ghasempour H, Gaff D, Williams R, Gianello R. Contents of sugars in leaves of drying desiccation tolerant flowering plants, particularly grasses. Plant Growth Regul. 1998;24:185–91.
51. Farrant JM, Lehner A, Cooper K, Wiswedel S. Desiccation tolerance in the vegetative tissues of the fern Mohria caffrorum is seasonally regulated. Plant J. 2009;57:65–79.
52. Krasensky J, Jonak C. Drought, salt, and temperature stress-induced metabolic rearrangements and regulatory networks. J Exp Bot. 2012;63:1593–608.
53. Vendruscolo ECG, Schuster I, Pileggi M, Scapim CA, Molinari HBC, Marur CJ, et al. Stress-induced synthesis of proline confers tolerance to water deficit in transgenic wheat. J Plant Physiol. 2007;164:1367–76.
54. Verbruggen N, Hermans C. Proline accumulation in plants: a review. Amino Acids. 2008;35:753–9.
55. Yobi A, Wone BW, Xu W, Alexander DC, Guo L, Ryals JA, et al. Comparative metabolic profiling between desiccation-sensitive and desiccation-tolerant species of *Selaginella* reveals insights into the resurrection trait. Plant J. 2012;72:983–99.

56. Grosser JW, Gmitter Jr FG. Protoplast fusion for production of tetraploids and triploids: applications for scion and rootstock breeding in citrus. Plant Cell Tiss Org Cult. 2011;104:343–57.
57. Martelotto LG, Ortiz JPA, Stein J, Espinoza F, Quarin CL, Pessino SC. A comprehensive analysis of gene expression alterations in a newly synthesized *Paspalum notatum* autotetraploid. Plant Sci. 2005;169:211–20.
58. Lu B, Pan X, Zhang L, Huang B, Sun L, Li B, et al. A genome-wide comparison of genes responsive to autopolyploidy in *Isatis indigotica* using *Arabidopsis* thaliana Affymetrix genechips. Plant Mol Biol Rep. 2006;24:197–204.
59. Auger DL, Gray AD, Ream TS, Kato A, Coe EH, Birchler JA. Nonadditive gene expression in diploid and triploid hybrids of maize. Genetics. 2005;169:389–97.
60. Zhu JK. Regulation of ion homeostasis under salt stress. Curr Opin Plant Biol. 2003;6:441–5.
61. Rizhsky L, Liang H, Shuman J, Shulaev V, Davletova S, Mittler R. When defense pathways collide. The response of *Arabidopsis* to a combination of drought and heat stress. Plant Physiol. 2004;134:1683–96.
62. Schafleitner R, Gutierrez Rosales RO, Gaudin A, Alvarado Aliaga CA, Martinez GN, Tincopa Marca LR, et al. Capturing candidate drought tolerance traits in two native Andean potato clones by transcription profiling of field grown plants under water stress. Plant Physiol Biochem. 2007;45:673–90.
63. Tu Y, Rochfort S, Liu Z, Ran Y, Griffith M, Badenhorst P, et al. Functional analyses of caffeic acid O-methyltransferase and cinnamoyl-CoA-reductase genes from perennial ryegrass (*Lolium perenne*). Plant Cell. 2010;22:3357–73.
64. Barry CS, Blume B, Bouzayen M, Cooper W, Hamilton AJ, Grierson D. Differential expression of the 1-aminocyclopropane-1-carboxylate oxidase gene family of tomato. Plant J. 1996;9:525–35.
65. Ouvrard O, Cellier F, Ferrare K, Tousch D, Lamaze T, Dupuis JM, et al. Identification and expression of water stress-and abscisic acid-regulated genes in a drought-tolerant sunflower genotype. Plant Mol Biol. 1996;31:819–29.
66. Seo YJ, Park JB, Cho YJ, Jung C, Seo HS, Park SK, et al. Overexpression of the ethylene-responsive factor gene BrERF4 from *Brassica rapa* increases tolerance to salt and drought in *Arabidopsis* plants. Mol Cells. 2010;30:271–7.
67. Miller G, Suzuki N, Ciftci-Yilmaz S, Mittler R. Reactive oxygen species homeostasis and signalling during drought and salinity stresses. Plant Cell Environ. 2010;33:453–67.
68. Ballester A-R, Teresa Lafuente M, González-Candelas L. Citrus phenylpropanoids and defence against pathogens. Part II: Gene expression and metabolite accumulation in the response of fruits to *Penicillium digitatum* infection. Food Chem. 2013;136:285–91.
69. Fontaine J-X, Tercé-Laforgue T, Armengaud P, Clément G, Renou J-P, Pelletier S, et al. Characterization of a NADH-dependent glutamate dehydrogenase mutant of *Arabidopsis* demonstrates the key role of this enzyme in root carbon and nitrogen metabolism. Plant Cell. 2012;24:4044–65.
70. Fernie AR, Stitt M. On the discordance of metabolomics with proteomics and transcriptomics: coping with increasing complexity in logic, chemistry, and network interactions scientific correspondence. Plant Physiol. 2012;158:1139–45.
71. Ha M, Lu J, Tian L, Ramachandran V, Kasschau KD, Chapman EJ, et al. Small RNAs serve as a genetic buffer against genomic shock in *Arabidopsis* interspecific hybrids and allopolyploids. Proc Natl Acad Sci U S A. 2009;106:17835–40.
72. Parisod C, Alix K, Just J, Petit M, Sarilar V, Mhiri C, et al. Impact of transposable elements on the organization and function of allopolyploid genomes. New Phytol. 2010;186:37–45.
73. Ng DW, Zhang C, Miller M, Shen Z, Briggs S, Chen Z. Proteomic divergence in *Arabidopsis* autopolyploids and allopolyploids and their progenitors. Heredity. 2011;108:419–30.
74. Marmagne A, Brabant P, Thiellement H, Alix K. Analysis of gene expression in resynthesized *Brassica napus* allotetraploids: transcriptional changes do not explain differential protein regulation. New Phytol. 2010;186:216–27.
75. Froelicher Y, Dambier D, Bassene J, Costantino G, Lotfy S, Didout C, et al. Characterization of microsatellite markers in mandarin orange (*Citrus reticulata* Blanco). Mol Ecol Resour. 2008;8:119–22.
76. Cristofani-Yaly M, Novelli VM, Bastianel M, Machado MA. Transferability and level of heterozygosity of microsatellite markers in Citrus species. Plant Mol Biol Rep. 2011;29:418–23.
77. Yi B, Zeng F, Lei S, Chen Y, Yao X, Zhu Y, et al. Two duplicate CYP704B1-homologous genes BnMs1 and BnMs2 are required for pollen exine formation and tapetal development in *Brassica napus*. Plant J. 2010;63:925–38.
78. Hu Z, Zhang M, Wen Q, Wei J, Yi H, Deng X, et al. Abnormal microspore development leads to pollen abortion in a seedless mutant of 'Ougan'mandarin (*Citrus suavissima* Hort. ex Tanaka). J Am Soc Hortic Sci. 2007;132:777–82.
79. Zheng BB, Wu XM, Ge XX, Deng XX, Grosser JW, Guo WW. Comparative transcript profiling of a male sterile cybrid pummelo and its fertile type revealed altered gene expression related to flower development. PLoS One. 2012;7:e43758.
80. Xu Q, Chen LL, Ruan X, Chen D, Zhu A, Chen C, et al. The draft genome of sweet orange (*Citrus sinensis*). Nat Genet. 2013;45:59–66.
81. Kim D, Pertea G, Trapnell C, Pimentel H, Kelley R, Salzberg SL. TopHat2: accurate alignment of transcriptomes in the presence of insertions, deletions and gene fusions. Genome Biol. 2013;14:R36.
82. Anders S, Huber W. Differential expression analysis for sequence count data. Genome Biol. 2010;11:R106.
83. Benjamini Y, Hochberg Y. Controlling the false discovery rate: a practical and powerful approach to multiple testing. J R Stat Soc Series B Stat Methodol. 1995;57:289–300.
84. Xie C, Mao X, Huang J, Ding Y, Wu J, Dong S, et al. KOBAS 2.0: a web server for annotation and identification of enriched pathways and diseases. Nucleic Acids Res. 2011;39 Suppl 2:316–22.

AKIN10 delays flowering by inactivating IDD8 transcription factor through protein phosphorylation in *Arabidopsis*

Eun-Young Jeong[1], Pil Joon Seo[1], Je Chang Woo[2] and Chung-Mo Park[1,3*]

Abstract

Background: Sugar plays a central role as a source of carbon metabolism and energy production and a signaling molecule in diverse growth and developmental processes and environmental adaptation in plants. It is known that sugar metabolism and allocation between different physiological functions is intimately associated with flowering transition in many plant species. The INDETERMINATE DOMAIN (IDD)-containing transcription factor IDD8 regulates flowering time by modulating sugar metabolism and transport under sugar-limiting conditions in *Arabidopsis*. Meanwhile, it has been reported that SUCROSE NONFERMENTING-1-RELATED PROTEIN KINASE 1 (SnRK1), which acts as a sensor of cellular energy metabolism, is activated by sugar deprivation. Notably, *SnRK1*-overexpressing plants and *IDD8*-deficient mutants exhibit similar phenotypes, including delayed flowering, suggesting that SnRK1 is involved in the IDD8-mediated metabolic control of flowering.

Results: We examined whether the sugar deprivation-sensing SnRK1 is functionally associated with IDD8 in flowering time control through biochemical and molecular genetic approaches. Overproduction of AKIN10, the catalytic subunit of SnRK1, delayed flowering in *Arabidopsis*, as was observed in *IDD8*-deficient *idd8-3* mutant. We found that AKIN10 interacts with IDD8 in the nucleus. Consequently, AKIN10 phosphorylates IDD8 primarily at two serine (Ser) residues, Ser-178 and Ser-182, which reside in the fourth zinc finger (ZF) domain that mediates DNA binding and protein-protein interactions. AKIN10-mediated phosphorylation did not affect the subcellular localization and DNA-binding property of IDD8. Instead, the transcriptional activation activity of the phosphorylated IDD8 was significantly reduced. Together, these observations indicate that AKIN10 antagonizes the IDD8 function in flowering time control, a notion that is consistent with the delayed flowering phenotypes of *AKIN10*-overexpressing plants and *idd8-3* mutant.

Conclusion: Our data show that SnRK1 and its substrate IDD8 constitute a sugar metabolic pathway that mediates the timing of flowering under sugar deprivation conditions. In this signaling scheme, the SnRK1 signals are directly integrated into the IDD8-mediated gene regulatory network that governs flowering transition in response to fluctuations in sugar metabolism, further supporting the metabolic control of flowering.

Keywords: *Arabidopsis thaliana*, Flowering time, Sugar metabolism, IDD8, SnRK1, AKIN10, Protein phosphorylation

* Correspondence: cmpark@snu.ac.kr
[1]Department of Chemistry, Seoul National University, Seoul 151-742, South Korea
[3]Plant Genomics and Breeding Institute, Seoul National University, Seoul 151-742, South Korea
Full list of author information is available at the end of the article

Background

Appropriate timing of flowering is important for propagation and reproductive success in plants. Therefore, flowering time is precisely regulated through the coordinated actions of endogenous developmental cues, such as plant aging and gibberellic acid (GA), and environmental signals, including changes in the length of day or photoperiod and temperature [1-3]. The floral inductive and repressible signals are transduced through well-established flowering genetic pathways, such as photoperiod, vernalization, GA, autonomous, and thermosensory pathways [1,4], and converge at the floral promoters *FLOWERING LOCUS T* (*FT*) and *SUPPRESSOR OF CONSTANS OVEREXPRESSION 1* (*SOC1*) and the floral repressor *FLOWERING LOCUS C* (*FLC*) [4,5].

Accumulating evidence support that sugar metabolism and distribution is intimately associated with flowering time control in many plant species [1,6]. Plants that are defective in sugar biosynthesis and metabolism exhibit alterations in developmental traits and flowering time [6,7]. It is widely perceived that plants do not flower even under photo-inductive conditions until they accumulate enough sugar reserves for the induction of flowering [6-8], which is consistent with the observations that low-starch-containing mutants, such as *pgm1* and *pgi*, exhibit retarded growth and delayed flowering [9,10]. Endogenous sugar levels are directly linked with photosynthetic carbon assimilation [6], indicating that photosynthetic activity also influences flowering transition [11].

While the effects of sugar on flowering time have been widely documented in many plant species, it is still unclear how sugar regulates the timing of flowering. In some cases, sugar promotes flowering, whereas flowering is inhibited in other cases, depending on different plant genotypes and growth conditions [8,12]. The functional ambiguity of sugar in flowering time control reflects the complexity of sugar homeostasis, which is attributed to the combined regulation of biosynthesis, degradation, and distribution in different plant tissues [6,8,12]. Sugar transport also plays a role in flowering time control. *Arabidopsis* mutants that have mutations in *SUCROSE TRANSPORTER9* (*AtSUC9*) gene exhibits early flowering under short days [13]. It has been suggested that AtSUC9 mediates the directional transport of sugar from the phloem to the sink organs and thus reduces sugar transport to the shoot apical meristem. It is also known that down-regulation of *StSUT4* gene in potato promotes flowering [14], supporting the linkage between sugar transport and flowering induction.

Roles of sucrose-regulated protein kinases and trehalose-6-phosphate (T6P) have been studied in linking sugar metabolism with flowering transition [15-17]. The T6P pathway has been shown to function upstream of the floral integrator FT in the leaves and regulates a flowering pathway that involves microRNA156 and SQUAMOSA PROMOTER-BINDING PROTEIN-LIKE (SPL) proteins in the shoot apical meristem [17], supporting a linkage between sugar and a distinct flowering pathway. In addition, it has been shown that photoperiodic control of sugar metabolism is associated with flowering induction in *Arabidopsis* and soybean [18]. Notably, CONSTANS (CO), which is a central regulator of photoperiodic flowering in *Arabidopsis* [4], plays a key role in the signaling pathway by regulating the expression of genes that are involved in sugar metabolism [19], providing a direct evidence that sugar metabolism is linked with photoperiod flowering.

The INDETERMINATE DOMAIN (IDD)-containing transcription factor IDD8 has been shown to regulate photoperiodic flowering under sugar deprivation conditions [20]. Whereas *IDD8*-defective *idd8* mutants exhibit late flowering, *IDD8*-overexpressing plants exhibit early flowering. The expression of *SUC* and sucrose synthase (*SUS*) genes is altered in the transgenic plants and *idd8* mutants. It has been reported that IDD8 regulates the *SUS* genes by directly binding to the gene promoters [20]. Moreover, the *SUS* genes are regulated by photoperiods, indicating that IDD8 regulation of sucrose metabolism and transport is associated with photoperiodic flowering. However, it is not known how sugar deprivation signals regulate IDD8 activity at the molecular level.

It is notable that T6P inhibits the activity of the Sucrose-non-fermenting1 (Snf1)-related kinase 1 (SnRK1) in sugar metabolic control of flowering [21]. SnRK1 is a serine/threonine protein kinase that is homologous to yeast Snf1 and animal AMP-dependent protein kinase 1 (AMPK1) kinases [22,23]. SnRK1/Snf1/AMPK acts as a metabolic sensor in eukaryotes and is activated under energy deprivation conditions [24,25]. In particular, *snrk1* knockdown plants exhibit early flowering, whereas *SnRK1* overexpression delays flowering [24,26]. In addition, SnRK1 is activated, but IDD8 is inactivated under sugar-limiting conditions, suggesting that SnRK1 and IDD8 are functionally interrelated in the sugar metabolic control of flowering.

In this work, we found that AKIN10, the catalytic α-subunit of SnRK1 kinases [27], phosphorylates IDD8 in the nucleus. While AKIN10-mediated phosphorylation did not affect the nuclear location and DNA-binding property of IDD8, it significantly reduced the transcriptional activation activity of IDD8. These results demonstrate that low-sugar levels trigger the SnRK1-mediated inactivation of IDD8 through protein phosphorylation, leading to delay of flowering. The SnRK1-IDD8 module would also be involved in the timing of flowering under abiotic stress conditions, which limit photosynthetic activity and disturb sugar metabolism in plants [28,29].

Results

idd8-3 and *AKIN10*-overexpresser exhibit delayed flowering under long days

As an initial step to investigate the functional relationship between AKIN10 and IDD8 in flowering time control, we compared the flowering phenotypes of *Arabidopsis* plants that have altered expression of *IDD8* and *AKIN10* genes. T-DNA insertional mutants of *AKIN10* and *AKIN11* genes (*akin10-1* and *akin11-1*, respectively) were obtained from the *Arabidopsis* Biological Resource Center (ABRC, Ohio state university, OH). Gene expression analysis revealed that they are loss-of-function mutants (Additional file 1). We also produced transgenic plants overexpressing either *AKIN10* or *AKIN11* gene under the control of the cauliflower mosaic virus (CaMV) 35S promoter, resulting in *10*-ox or *11*-ox, respectively (Additional file 2). We similarly produced transgenic plants overexpressing *IDD8*, resulting in *8*-ox.

We examined the flowering phenotypes of the plants grown under long days (LDs, 16-h light and 8-h dark) by counting the numbers of rosette leaves at bolting and the days to bolting. The *8*-ox plants and the *akin10-1* and *akin11-1* mutants did not exhibit any discernible flowering phenotypes under our assay conditions (Figures 1A and 1B). In contrast, the *10*-ox and *11*-ox plants exhibited delayed flowering, as observed in *idd8-3* mutant. The delay of flowering time was more prominent in *10-ox* than in *11-ox* (Figure 1B). The similar flowering phenotypes raised a possibility that loss of IDD8 function is related with overproduction of AKIN10 and AKIN11 in regulating flowering time. In support of this hypothesis, the expression of *SUS4* and *SUC* genes was suppressed in the *10*-ox plants but up-regulated in the *akin10-1* mutant (Additional file 3), as observed in the *idd8-3* mutant and the *8*-ox plants, respectively [20].

IDD8 interacts with AKIN10 in the nucleus

On the basis of the similar flowering phenotypes of *idd8-3* mutant and AKIN-overexpressing plants and the biochemical nature of IDD8 transcription factor and SnRK1 kinases, we hypothesized that IDD8 interacts with the SnRK1 kinases.

Yeast two-hybrid assays did not show any positive interactions between IDD8 and AKIN10 (data not shown). We therefore employed *in vitro* pull-down assays using recombinant glutathione S-transferase-AKIN10 (GST-AKIN10) and GST-AKIN11 fusion proteins, which were produced in *E.coli* cells, and ^{35}S-labelled IDD8 polypeptides produced by *in vitro* translation. While IDD8 did not interact with GST alone, it strongly interacted with GST fusions of AKIN10 and AKIN11 (Figure 2A). The lack of IDD8-AKIN interactions in yeast cells might be due to an intrinsic property of AKIN proteins, as has been observed previously [27,30].

We also performed bimolecular fluorescence complementation (BiFC) assays to examine whether the IDD8-

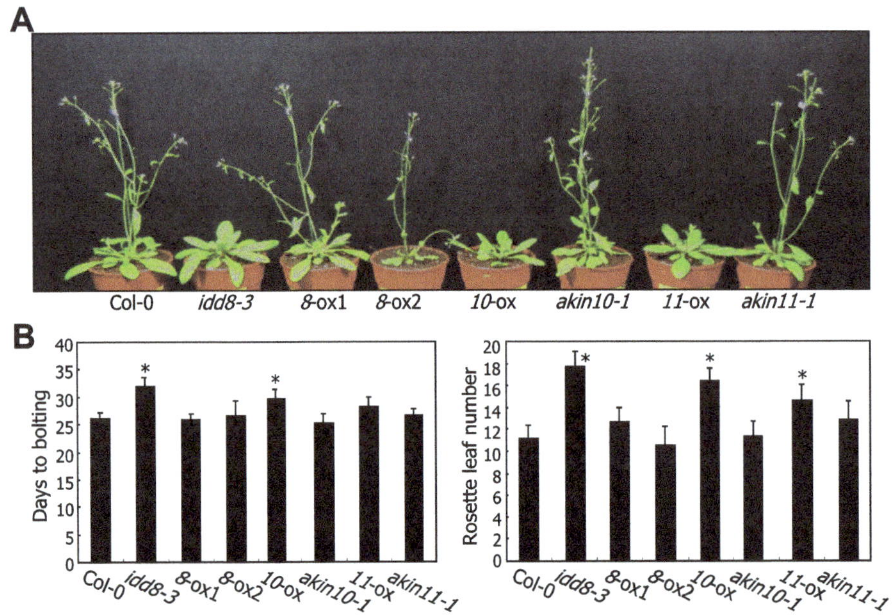

Figure 1 *AKIN10* overexpression delays flowering. Plants were grown in soil under LDs for 6 weeks before taking photographs **(A)**. Flowering times were measured by counting the days to bolting and rosette leaf numbers at bolting (**B**, left and right panels, respectively). Transgenic plants overexpressing *IDD8* (*8*-ox1 and *8*-ox2), *AKIN10* (*10*-ox), and *AKIN11* (*11*-ox) and their gene knockout mutants were analyzed. The countings of approximately 20 plants were averaged and statistically analyzed using Student *t*-test (**P* < 0.01, difference from col-0). Bars indicate standard error of the mean.

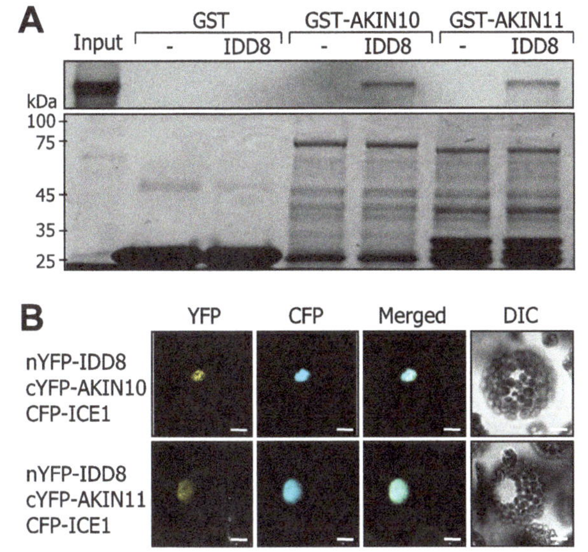

Figure 2 IDD8 interacts with AKIN proteins in the nucleus. **A** *in vitro* pull-down assay. Recombinant GST-AKIN10 and GST-AKIN11 fusion proteins produced in *E. coli* cells and *in vitro* translated, radio-labelled IDD8 polypeptides were used (upper panel). Recombinant GST was used as negative control. The 'Input' represents 20% of the labeling reaction. Part of Coomassie Blue-stained gel was displayed as a loading control (lower panel). kDa, kilodalton. **B** BiFC assay. nYFP-IDD8 and cYFP-AKIN fusions and cyan fluorescent protein (CFP)-ICE1 fusion, which was used as a nuclear marker, were coexpressed transiently in *Arabidopsis* protoplasts. IDD8-AKIN interactions were visualized by differential interference contrast (DIC) and fluorescence microscopy. Scale bars, 10 μm.

AKIN interactions occur in plant cells. Coexpression of the N-terminal half of yellow fluorescent protein (YFP) fused to IDD8 (nYFP-IDD8) and the C-terminal half of YFP fused to AKIN10 (cYFP-AKIN10) or AKIN11 (cYFP-AKIN11) in *Arabidopsis* protoplasts revealed that the IDD8-AKIN interactions occur in the nucleus (Figure 2B, Additional file 4), indicating that IDD8 interacts with AKIN proteins *in planta*.

AKIN10 phosphorylates IDD8

AKIN10 and AKIN11 are the catalytic subunits of SnRK1 kinases [24,27]. Protein phosphorylation is one of the primary biochemical mechanisms that modulate the activities of transcription factors in plants [26,31,32]. We therefore examined whether AKIN proteins phosphorylate IDD8.

We produced recombinant maltose-binding protein-IDD8 (MBP-IDD8) and GST-AKIN fusion proteins in *E.coli* cells, which were purified by affinity chromatography and immunologically quantified (Additional file 5A). The *in vitro* kinase assays showed that AKIN10 possesses an autophosphorylation activity, while AKIN11 does not (Figure 3). It was also evident that AKIN10, but not AKIN11, phosphorylates IDD8. Although both *10*-ox and *11*-ox plants exhibited delayed flowering (Figure 1) and IDD8 interacts with both AKIN10 and AKIN11, IDD8 may not be directly targeted by AKIN11 at least in controlling flowering time.

To identify the Ser and Thr residues of IDD8 targeted by AKIN10, we searched for putative target residues using the NetPhos2 algorithm (http://www.cbs.dtu.dk/services/NetPhos/). The computer-assisted analysis identified 18 Ser and 5 Thr residues that were predicted to be phosphorylated by SnRK1. Among the 23 residues, only the sequence contexts around Thr-98, Ser-178, and Ser-182 partially matched to the consensus sequence established for SnRK1 kinases [26] (Additional file 6). The three residues were mutated to alanine, resulting in T98A, S178A, and S182A (Figure 4A), and the mutated IDD8 proteins were prepared as MBP fusions in *E. coli* cells and immunologically quantified (Additional file 5B). The recombinant MBP-IDD8 proteins were then subjected to *in vitro* phosphorylation assays. It was found that the phosphorylation of S182A was significantly reduced by more than 90% compared to that of wild-type IDD8 protein (Figure 4B). In contrast, T98A and S178A were still phosphorylated with a reduction of approximately 50%. Liquid chromatography-tandem mass spectrometry (LC-MS/MS) also supported the notion that S182 is a major site for AKIN10-mediated phosphorylation (Additional file 7).

AKIN10 does not affect the subcellular localization of IDD8

Protein phosphorylation influences diverse structural and functional aspects of transcription factors, such as protein stability, subcellular localization, and transcriptional activation activity [26,32,33]. It has been reported that AKIN10 regulates the protein stability of the B3-

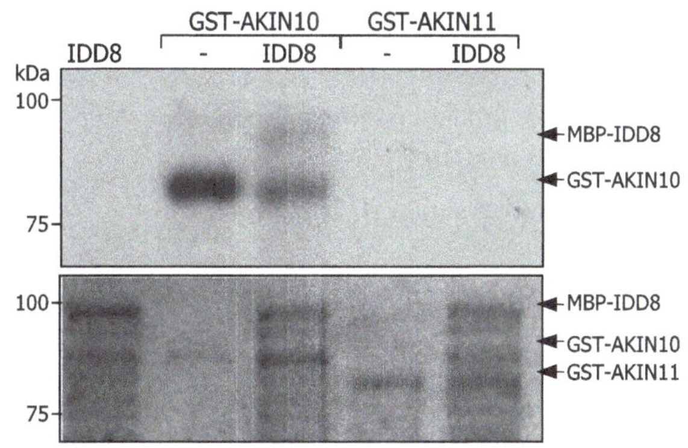

Figure 3 Phosphorylation of IDD8 by AKIN10. The *in vitro* phosphorylation assays were conducted using recombinant GST-AKIN10 and GST-AKIN11 fusion proteins and MBP-IDD8 fusion protein prepared in *E. coli* cells (upper panel). Part of Coomassie Blue-stained gel was displayed as a loading control (lower panel). kDa, kilodalton.

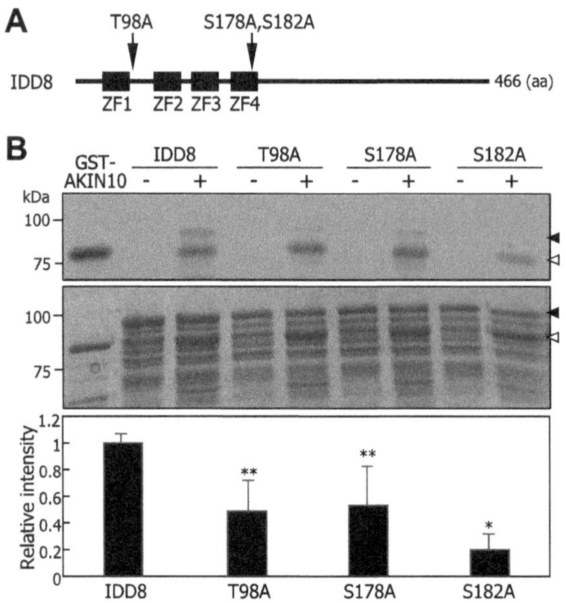

Figure 4 Identification of phosphorylation residues in IDD8. **A** Predicted phosphorylation residues in IDD8. Potential phosphorylation residues were predicted using the NetPhos-based analysis tool (http://www.cbs.dtu.dk/services/NetPhos/). The predicted serine (S) and threonine (T) residues were mutated to alanine (A). ZF, zinc finger. aa, amino acid. **B** *in vitro* phosphorylation assay. The assays were conducted using recombinant GST-AKIN10 and MBP-IDD8 fusion proteins prepared in *E. coli* cells (upper panel). Part of Coomassie Blue-stained gel was displayed as a loading control (middle panel). Black arrowheads indicate IDD8 protein. White arrowheads indicate AKIN10 protein. kDa, kilodalton. The relative intensities of the phosphorylation bands were calculated in comparison to those on Coomassie Blue-stained gel (lower panel). Experimental triplicates were averaged and statistically analyzed using Student *t*-test (**P* < 0.01, ***P* < 0.05, difference from wild-type IDD8). Bars indicate standard error of the mean.

domain-containing transcription factor FUSCA3 (FUS3) during lateral organ development and floral transition [26]. Therefore, a question was how AKIN10-mediated phosphorylation regulates IDD8 function in flowering time control.

We first examined whether protein phosphorylation affects the stability of IDD8 protein using transgenic plants overexpressing *IDD8-MYC* fusion driven by the CaMV 35S promoter in either Col-0 plant or *akin10-1* mutant. The transgenic plants were incubated either in constant light or in complete darkness for 2 days. They were also incubated in the presence of 3-(3,4-dichloro-phenyl)-1,1-dimethylurea (DCMU), which is a specific inhibitor of photosynthesis [24], in constant light. IDD8 proteins were then immunologically detected using an anti-MYC antibody. The results showed that in the Col-0 background, the IDD8 levels were reduced in darkness, and the reduction was more prominent in the presence of DCMU (Additional file 8A, upper panel), which is probably due to dark-induced degradation of IDD8 protein. Alternatively, the reduction would be at least in part attributable to the transcriptional suppression of *IDD8* gene by low sugar levels. Notably, the patterns of IDD8 abundance were similarly observed in *akin10-1* background, although the overall levels were lower than those in Col-0 background. Quantitative real-time RT-PCR (qRT-PCR) showed that the levels of *IDD8* transcripts were lower in *akin10-1* background (Additional file 8A, lower left panel). However, the levels of IDD8 protein relative to those of *IDD8* transcripts were similar in Col-0 and *akin10-1* backgrounds (Additional file 8A, lower right panel). Together, these observations indicate that AKIN10 does not affect the stability of IDD8 protein.

We next examined whether AKIN10-mediated phosphorylation influences the subcellular localization of IDD8 by transient expression of a green fluorescent protein (GFP)-IDD8 fusion in *Arabidopsis* protoplasts prepared from Col-0, *akin10-1*, and *10*-ox plants and using transgenic plants overexpressing a *GFP-IDD8* fusion in Col-0 and *10*-ox backgrounds. The roots of the transgenic plants were visualized by fluorescence microscopy. GFP signals were detected predominantly in the nuclei of root cells of both Col-0 and *10*-ox backgrounds (Additional files 8B and C), indicating that the subcellular distribution of IDD8 is not affected by AKIN10-mediated protein phosphorylation.

AKIN10 inhibits the transcriptional activation activity of IDD8

IDD8 binds directly to *SUS4* gene promoter containing the conserved CTTTTGTCC motif [20]. We therefore asked whether AKIN10 affects the DNA-binding property of IDD8. We performed chromatin immunoprecipitation (ChIP) assays using 35S:*MYC-IDD8* and 35:*MYC-IDD8 akin10-1* plants. IDD8-binding sequence (BS) and non-binding sequence (nBS) within the *SUS4* gene promoter were included in the assays (Additional file 9A). The assays revealed that IDD8 does not bind to nBS sequence (Additional file 9B). In contrast, IDD8 efficiently bound to BS sequence. Notably, IDD8 also bound efficiently to BS sequence in *akin10-1* background, indicating that AKIN10 does not affect the DNA-binding property of IDD8.

A remaining question was whether AKIN10 affects the transcriptional activation activity of IDD8. To address this question, we performed transient β-galactosidase (GUS) expression assays by coexpressing a series of reporter and effecter vectors in *Arabidopsis* protoplasts (Figure 5A). Notably, AKIN10 reduced the transcriptional activation activity of IDD8 by approximately 65% (Figure 5B). In contrast, AKIN11 reduced the IDD8 activity only slightly, further supporting the notion that AKIN11 is not directly related with IDD8.

The transient GUS expression assays also showed that a mutated IDD8 protein (mIDD8) harboring the S178A and S182A substitutions is transcriptionally active comparable to the wild-type IDD8 protein (Figure 5C). It was notable that whereas AKIN10 reduced the IDD8 activity, it did not affect the mIDD8 activity, indicating that IDD8 phosphorylation by AKIN10 is important for the suppression of the IDD8 activity.

It is known that AKIN10 is activated under low-sugar conditions [25]. We therefore examined the effects of sugar deprivation on IDD8 activity by transient GUS expression assays using *Arabidopsis* protoplasts prepared from Col-0 plants and *akin10-1* mutant. *Arabidopsis* protoplasts were treated with DCMU to mimic sugar deprivation conditions before the assays. It was found that whereas DCMU detectably reduced the IDD8 activity in Col-0 plants, it did not affect the IDD8 activity in *akin10-1* mutant (Figure 5D), demonstrating that AKIN10 suppresses IDD8 activity under sugar deprivation conditions.

AKIN10-mediated phosphorylation of IDD8 is relevant for flowering time control

Our data showed that AKIN10 phosphorylates IDD8 to reduce its transcriptional activation activity in response to sugar deprivation. We next examined whether the phosphorylation of IDD8 by sugar deprivation-activated AKIN10 is functionally relevant for flowering time control. We crossed *idd8-3* with *akin10-1*, resulting in *idd8-3 akin10-1* double mutant (Additional file 10). Flowering time measurements showed that the *idd8-3 akin10-1* double mutant exhibited delayed flowering as observed in the *idd8-3* mutant (Figure 6A). What was unexpected was that the delay of flowering was more severe in the double mutant, suggesting that AKIN10 might target additional flowering time modulators other than IDD8 (see below).

qRT-PCR assays on flowering time genes showed that *FT* gene and its downstream targets *SOC1* and *APPETALA 1* (*AP1*) genes were suppressed in the single and double mutants (Figure 6B), consistent with their delay flowering phenotypes. Notably, the floral repressor *FLC* was significantly induced in the *idd8-3 akin10-1* mutants, which might be related with the severity of delayed flowering in the double mutant (Figure 6A).

Altogether, our data demonstrate that SnRK1 inhibits the transcriptional activation activity of IDD8 transcription factor through protein phosphorylation to delay flowering

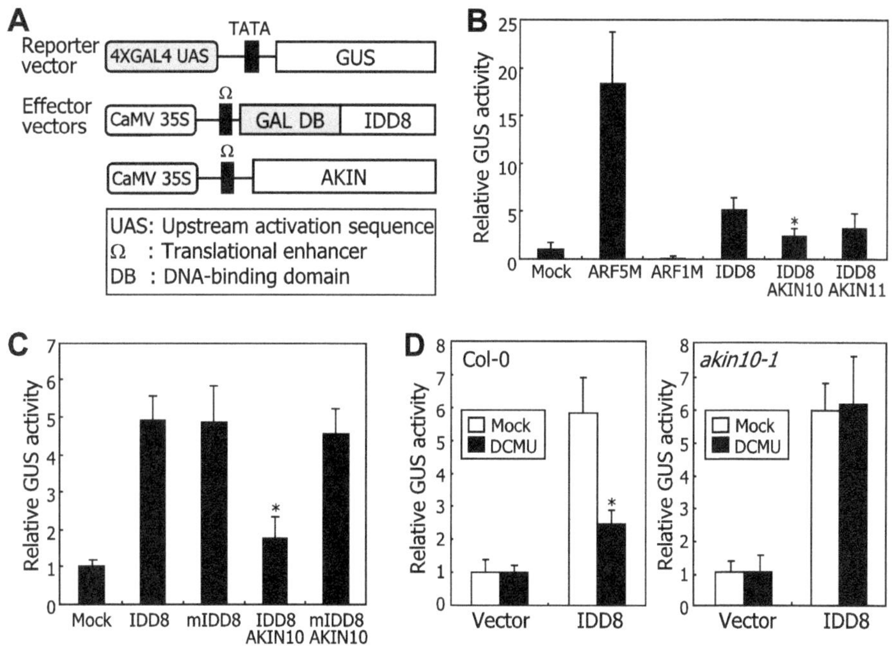

Figure 5 AKIN10 inhibits IDD8 transcription factor activity. **A** Reporter and effector vector constructs. A full-size *IDD8* cDNA was fused in-frame to the 3′ end of GAL4 DNA-binding domain (DB)-coding sequence in the effector vector. **B** SnRK1-mediated inhibition of IDD8 transcriptional activation activity. GAL4 transient expression assays were performed using *Arabidopsis* protoplasts, as described previously [20]. The *Renilla* luciferase gene was used as an internal control to normalize the values in individual assays. ARF5M is a transcriptional activator control. ARF1M is a transcriptional repressor control. Three measurements of GUS activity were averaged and statistically analyzed using Student *t*-test (*$P < 0.01$, difference from IDD8). Bars indicate standard error of the mean. **C** Transcription factor activity of mutated IDD8. The mutated IDD8 (mIDD8) harbors S178A and S182A substitutions. GUS activity measurements were performed as described in **(B)**. Bars indicate standard error of the mean (*t*-test, *$P < 0.01$, difference from IDD8). **D** Effects of sugar deprivation on IDD8 transcription factor activity. The GUS reporter and the IDD8 effector vectors were cotransformed into *Arabidopsis* protoplasts that were prepared from either Col-0 plant or *akin10-1* mutant (left and right panels, respectively). The *Arabidopsis* protoplasts were then treated with 20 μM DCMU for 6 h before GUS activity measurements. Three measurements were averaged and statistically analyzed (*t*-test, *$P < 0.01$, difference from mock). Bars indicate standard error of the mean.

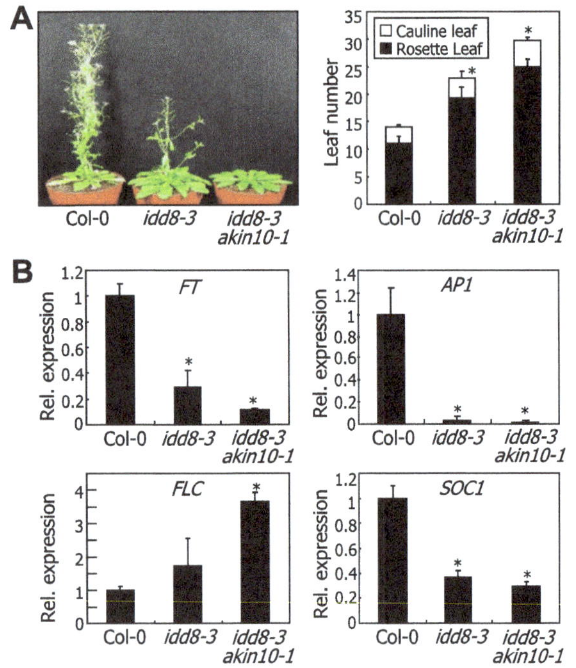

Figure 6 Flowering phenotypes and molecular characterization of *idd8-3 akin10-1* double mutant. The *idd8-3* mutant was crossed with the *akin10-1* mutant, resulting in *idd8-3 akin10-1* double mutant. **A** Flowering phenotypes. Plants were grown in soil under LDs for 6 weeks before taking photographs (left panel). Leaf numbers of 20 plants at bolting were averaged and statistically analyzed using the Student t-test (*$P < 0.01$, difference from Col-0) (right panel). Bars indicated standard error of the mean. **B** Expression of flowering time genes. Aerial parts of two-week-old plants grown in soil were harvested at zeitgeber time 16 for the extraction of total RNA. Transcript levels were examined by qRT-PCR. Biological triplicates were averaged and statistically analyzed using Student *t*-test (*$P < 0.01$, difference from Col-0). Bars indicate standard error of the mean.

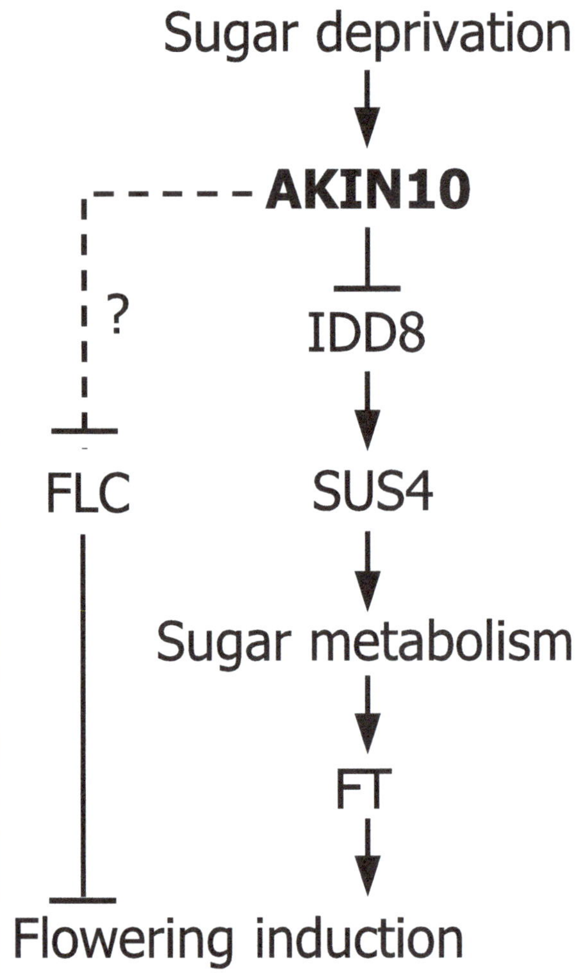

Figure 7 Schematic model of AKIN10 function in flowering time control. Sugar deprivation conditions, which are encountered in early vegetative phase, activate AKIN10 that negatively regulates IDD8 transcription factor. During the reproductive phase transition, increased sugar availability deactivates AKIN10, resulting in flowering transition. It is also likely that AKIN10 negatively regulates FLC function either directly or indirectly via an unidentified regulator of FLC.

under low-sugar conditions (Figure 7). This working scenario explains the suppression of IDD8 function under sugar deprivation conditions [20]. We propose that the SnRK1-IDD8 signaling module provides a molecular clue for the long-lasting interest in the metabolic control of flowering in plants.

Discussion

In this work, we demonstrated that the serine/threonine-specific kinase SnRK1 and its target IDD8 transcription factor constitute a sugar metabolism-mediated flowering pathway. On the basis of molecular characterization of *idd8-3* and *akin10-1* mutants and transgenic plants overexpressing *IDD8* or *AKIN10* genes and biochemical examination of AKIN10-mediated phosphorylation of IDD8, we suggest that the SnRK1 pathway senses fluctuations in sugar metabolism and integrates the metabolic signals into the IDD8-mediated gene regulatory network that regulates flowering time.

There has been a controversy on the molecular nature of *akin10-1* mutant. It has been reported that the *akin10-1* mutant is a null mutant through *AKIN10* gene expression study and immunological detection of AKIN10 proteins using two-dimensional SDS-PAGE [34]. Meanwhile, is has been shown that *AKIN10* gene sequence was amplified and AKIN10 protein was detected in the *akin10-1* mutant [26]. We verified that the *AKIN10* gene is disrupted by the insertion of T-DNA element and it is not expressed in the mutant by PCR-based genotyping and qRT-PCR using different sets of primers. We also found that *SUS4* gene expression is altered in the *akin10-1* mutant that exhibits differential response to DCMU. We believe that *akin10-1* is a loss-of-function mutant. The amplification of *AKIN10* sequence and detection of AKIN10 protein in the previous report would be due to a high sequence similarity among *AKIN* gene members and similar sizes of AKIN proteins in *Arabidopsis*.

SnRK1-IDD8 module in sugar metabolic control of flowering

Floral transition is one of the most energy-consuming developmental processes in plants. Therefore, it is not surprising that the timing of flowering is closely associated with sugar homeostasis. In view of metabolic control of flowering, it is notable that SnRK1 plays a fundamental role in the developmental process in response to carbon availability [35]. SnRK1 members coordinate diverse transcriptional regulatory networks that stimulate catabolism but suppress anabolism to sustain cellular energy homeostasis under stressful conditions [24,35,36]. While the roles of SnRK1 members have been reported in various cellular responses, only a few substrates have been identified so far.

One of the best characterized targets is the FUS3 transcription factor, which regulates seed maturation in *Arabidopsis* [37]. It has been shown that AKIN10-mediated phosphorylation enhances the FUS3 activity by improving its protein stability [26]. Accordingly, FUS3 is involved in the SnRK1-mediated control of developmental phase transitions. Molecular genetic assays have shown that the *fus3-3* mutation partially rescued the delayed flowering of *AKIN10*-overexpressing plants [26]. However, the *FUS3* gene is not detectably induced during the vegetative-to-reproductive phase transition, and the flowering phenotype of the *fus3-3* mutant is similar to that of control plants [26,37]. Together with the partial recovery of the flowering phenotype of *AKIN10*-overexpressing plants by the *fus3-3* mutation, it has been suggested that the SnRK1-mediated metabolic signals are not solely mediated by FUS3 in regulating flowering time control [26].

In this study, we demonstrated that AKIN10, which is the catalytic subunit of SnRK1 kinases [27], negatively regulates the transcriptional activation activity of IDD8 transcription factor through protein phosphorylation. While our data strongly support that IDD8 is phosphorylated by AKIN10, it is still possible that other kinases would also phosphorylate IDD8, assuming the roles of IDD8 in sugar homeostasis and flowering time control [20, see below]. IDD8 induces *SUS4* gene by directly binding to the gene promoter, leading to the promotion of photoperiodic flowering [20]. The *IDD8* gene is suppressed by sugar deprivation [27]. Together with the previous observations, our data show that SnRK1 mediates the inactivation of IDD8 in flowering time control under low-sugar conditions. It is currently unclear whether IDD8 is functionally connected with FUS3 in the process of sensing sugar metabolic status by SnRK1.

SnRK1-mediated inactivation of IDD8 activity

Protein phosphorylation influences the activity of transcription factors through diverse mechanisms, such as modulation of their nucleo-cytoplasmic distributions, DNA-binding properties, and protein stabilities and modification of their interactions with other regulatory proteins [33,38,39]. AKIN10 does not affect the nuclear localization and DNA binding of IDD8. The protein stability of IDD8 is also unaffected by protein phosphorylation. Instead, AKIN10-mediated phosphorylation inhibits the transcriptional activation activity of IDD8.

A critical question is how protein phosphorylation affects the IDD8 activity. We found that AKIN10 phosphorylates IDD8 primarily at Ser-182, which resides in the fourth ZF domain. IDD8 has four copies of ZF domains, which are known to mediate DNA binding and protein-protein interactions [40,41]. It has been reported that a central amino acid sequence region of IDD8, which includes residues 171–320 and thus harbors the fourth ZF domain, contains a potential transcriptional activation domain [20]. It has been suggested that the fourth ZF domain mediates the interactions of IDD transcription factors with other interacting partners in regulating the expression of target genes [20]. We suspect that a similar regulatory scheme is applicable to the inhibition of the IDD8 activity by AKIN10: AKIN10 might inhibit the interaction of IDD8 with other regulatory proteins by phosphorylating the critical residues in the fourth ZF domain. In this regard, it will be interesting to investigate whether FUS3 interacts with IDD8 through the fourth ZF domain.

Additional roles of SnRK1-IDD8 module beyond metabolic control of flowering?

Plant adaptation responses to stressful conditions, such as drought, high salinity, and extreme temperatures, frequently accompany alterations in sugar metabolism and transport [42-44]. It has been known that SnRK1 kinases are associated with plant responses to environmental stress conditions by linking cellular energy status to stress adaptation [24,27]. It is notable that transgenic plants overexpressing *AKIN10* or *FUS3* gene are sensitive to abscisic acid (ABA), a pivotal stress hormone that modulates a broad spectrum of stress responses [45], and exhibit delayed seed germination [26]. SnRK1 kinases have also been implicated in aging process and cell death in eukaryotes [27,28], indicating that SnRK1 is a central regulator of sugar metabolism in linking plant development with environmental adaptation.

The observed role of IDD8 in the SnRK1-mediated control of photoperiodic flowering under sugar starvation conditions suggest that IDD8 function is not limited to flowering time control but might be extended to a range of stress responses. It has been observed that transgenic plants overexpressing *IDD8* gene exhibit a plethora of growth and developmental defects, such as growth retardation and architecturally distorted, pale-green leaves [20]. It will be worthy of examining the responses of *IDD8*-overexpressing plants and *idd8-3* mutant to ABA and abiotic stresses and investigating whether SnRK1 is involved in the IDD8-mediated stress responses.

Conclusions

We aimed to improve our understanding on how IDD8 perceives sugar deprivation signals in regulating photoperiodic flowering. We found that the energy metabolic sensor SnRK1 inhibits the transcriptional activation activity of IDD8 transcription factor, which regulates photoperiodic flowering in response to sugar deprivation. AKIN10, the α-catalytic subunit of SnRK1 kinases, phosphorylates IDD8 predominantly at two serine residues, Ser-178 and Ser-182 that reside in the fourth ZF domain. While protein phosphorylation does not affect the nuclear localization and DNA-binding property of IDD8, it significantly reduces the transcriptional activation activity of IDD8. The reduction of the IDD8 activity was also observed under sugar starvation conditions, which is consistent with the activation of SnRK1 activity by low energy status. Our data show that the SnRK1-IDD8 transcriptional regulatory module serves as a web that integrates sugar metabolic signals into flowering time control in *Arabidopsis*.

Methods

Bioinformatics software

Nucleotide sequences of genes and amino acid sequences of proteins were obtained from the *Arabidopsis* Information Resource (TAIR, http://www.arabidopsis.org/). Protein phosphorylation sites were predicted using the NetPhos 2.0 software (http://www.cbs.dtu.dk/services/NetPhos/).

Plant materials and growth conditions

All *Arabidopsis thaliana* lines used were in the Col-0 background. *Arabidopsis* plants were grown in a controlled culture room set at 23°C with relative humidity of 55% under LDs with white light illumination (120 μM photons $m^{-2}s^{-1}$) provided by fluorescent FLR40D/A tubes (Osram, Seoul, Korea). The *idd8-3*, *akin10-1*, and *akin11-1* mutants have been described previously [20,26,45].

To generate 35S:*MYC-IDD8* transgenic plant, a full-size *IDD8* cDNA (At5g44160) was fused in-frame to the 3′ end of the MYC-coding sequence under the control of the CaMV 35S promoter in the myc-pBA vector [46]. The expression construct was transformed into Col-0 plants. To generate transgenic plants overexpressing *AKIN10* and *AKIN11* genes (At3g01090 and At3g29160, respectively), full-size cDNAs were subcloned under the control of the CaMV 35S promoter into the binary pB2GW7 vector [47]. *Agrobacterium*-mediated transformation was performed according to a modified floral-dip method [48].

For dark treatments, 10-day-old plants grown on $^1/_2$ X Murashige and Skoog-agar plates (MS-agar plates) were covered with aluminum foil and incubated at 23°C for 2 days in complete darkness. For DCMU treatments, 10-day-old plants grown on MS-agar plates were transfer to MS liquid culture containing 50 μM DCMU for 2 days under constant light conditions.

Gene expression analysis

Gene transcript levels were analyzed by qRT-PCR. Reverse transcription and quantitative PCR reaction were performed according to the rules that have been proposed to ensure reproducible and accurate measurements of transcript levels [49]. Total RNA samples were pretreated with RNase-free DNase to get rid of any contaminating genomic DNA before use.

qRT-PCR reactions were performed in 96-well blocks with the Applied Biosystems 7500 Real-Time PCR System (Foster City, CA) using the SYBR Green I master mix in a volume of 20 μl. The PCR primers were designed using the Primer Express Software installed into the system and listed in Additional file 11. The two-step thermal cycling profile used was at 94°C for 15 s and at 68°C for 1 min. An *eIF4A* gene (At3g13920) was included in the reactions as internal control for normalizing the variations in the cDNA amounts used. All qRT-PCR reactions were performed in biological triplicates using RNA samples extracted from three independent plant materials grown under identical conditions. The comparative $\Delta\Delta C_T$ method was employed to evaluate the relative quantity of each amplified product in the samples. The threshold cycle (C_T) was automatically determined for each reaction by the system set with default parameters. The specificity of the PCR reactions was determined by melting curve analysis of the amplified products using the standard method installed in the system.

Flowering time measurement

Plants were grown in soil at 23°C under LDs until flowering. Flowering times were determined by counting the days to bolting and the number of rosette and cauline leaves at bolting. Fifteen to 20 plants were counted and averaged for each measurement.

in vitro pull-down assay

Recombinant AKIN10 and AKIN11 proteins were prepared as GST-AKIN10 and GST-AKIN11 fusions in *Escherichia coli* Rosetta2 (DE3) pLysS strain (Novagen, Madison, WI) and affinity-purified, as described previously [50]. The [^{35}S] methionine-labeled IDD8 polypeptides were prepared by *in vitro* translation using the TNT-coupled reticulocyte lysate system (Promega, Madison, WI).

The *in vitro* pull-down assays were performed as described previously [50]. The bound proteins were eluted with 1X SDS-PAGE loading buffer by boiling for 5 min and subjected to SDS- PAGE and autoradiography.

Subcellular localization assay

A GFP-coding sequence was fused in-frame to the 5′ end of *IDD8* gene, and the gene fusion was subcloned into the p2FGW7 expression vector (Invitrogen, Carlsbad, CA). Protoplasts were prepared from fully expanded leaves of four-week-old plants grown in soil, as described previously [51]. Approximately 2×10^4 protoplasts were mixed with 10 μg of plasmid DNA and 110 μl of polyethylene glycol (PEG)-calcium transfection solution [40% PEG 4000 (w/v), 0.2 M mannitol, 100 mM $CaCl_2$]. After incubation at 22°C for 15 min, the protoplast suspension was centrifuged at 100 × g for 2 min. The protoplasts were resuspended in 1 ml of WI solution (0.5M mannitol, 4 mM Mes, pH 5.7, 20 μM KCl) and incubated in the dark at 22°C for 15 h. The subcellular distribution of green fluorescence was visualized by fluorescence microscopy.

The *GFP-IDD8* gene fusion was overexpressed under the control of the CaMV 35S promoter in Col-0 and *10*-ox plants. The roots of ten-day-old transgenic plants were visualized by differential interference contrast (DIC) and fluorescence microscopy. The root samples were also stained with 4′,6-diamidino-2-phenylindole (DAPI) to visualize the nuclei.

Liquid chromatography-tandem mass spectrometry (LC-MS/MS)

Recombinant MBP-IDD8 and GST-AKIN10 protein fusions, in which the tags were fused to the N termini of the proteins, were prepared in *E. coli* cells. Phosphorylation reactions *in vitro* were induced by incubating with non-radioactive ATP. The MBP-IDD8 protein was excised from 6% SDS-PAGE gel and digested with trypsin. LC-MS/MS was performed in the National Instrumentation Center for Environmental Management (NICEM, Seoul National University, Seoul). Protein Pilot program (Applied Biosystems, Foster City, CA) was used to assign the phosphorylation sites. The serine (S) and threonine (T) phosphorylation sites identified by the Protein Pilot program were calculated with a confidence > 0.95.

***in vitro* protein phosphorylation assay**

The assays were performed in 10 μl kinase reaction buffer (20 mM HEPES, pH 7.4, 10 mM $MgCl_2$, 1 mM Na_3VO_4, 2 mM DTT, 0.5 mM PMSF, 2 mM EDTA), as described previously [50]. Purified recombinant AKIN10 and IDD8 proteins were added to the reaction buffer supplemented with 1 μCi of [γ^{-32}P] ATP. The reaction mixture was incubated at 30°C for 30 min, and the reaction was terminated by adding 4 μl of 6 X SDS-PAGE sample loading buffer. The mixture was boiled for 5 min before loading onto 10% SDS-PAGE gels. The gels were stained with Coomassie Brilliant Blue R250, vacuum-dried onto 3MM paper, and subjected to autoradiography.

BiFC assay

BiFC assays were performed as described previously [51]. A full-size *IDD8* cDNA was fused in-frame to the 3′ end of the gene sequence encoding the N-terminal half of enhanced yellow fluorescent protein (EYFP) in the pSATN-nEYFP-C1 vector (E3081). A full-size *AKIN10* cDNA was fused in-frame to the 5′ end of the gene sequence encoding the C-terminal half of EYFP in the pSATN-cEYFP-C1 vector (E3082). The nYFP-IDD8 and AKIN10-cYFP vectors were cotransfected into *Arabidopsis* mesophyll protoplasts by the PEG-calcium transfection method [51]. The transfected protoplasts were incubated at 23°C for 16 h. The subcellular localization of IDD8-AKIN10 complexes was monitored by DIC microscopy and fluorescence microscopy. Reconstitution of YFP fluorescence was observed using a Zeiss LSM510 confocal microscope (Carl Zeiss, Yena, Germany) with the following YFP filter set up: excitation 515 nm, 458/514 dichroic, and emission 560- to 615-nm band-pass filter.

Chromatin immunoprecipitation (ChIP) assay

ChIP assays were performed using two-week-old plants grown on MS-agar plates, as described previously [52]. Whole plants were vacuum-infiltrated with 1% (v/v) formaldehyde for cross-linking and ground in liquid nitrogen after quenching the cross-linking process. Chromatin preparations were sonicated into 0.4- to 0.7-kb fragments and precleared with salmon sperm DNA/Protein G agarose beads (Roche, Indianapolis, IN), and an anti-MYC antibody (Millipore, Billerica, MA) was added to the mixture. The precipitates were eluted from the beads, and cross-links were reversed. Residual proteins were removed by incubation with proteinase K. DNA was then recovered using a SV minicolumn (Promega). Quantitative PCR was performed to determine the amounts of genomic DNA enriched in the chromatin preparations.

Transcriptional activation activity assay

For transcriptional activation activity assays, a series of reporter and effector vectors was constructed. In the reporter vector, four copies of the GAL4 upstream activation sequence (UAS) were fused to a gene encoding GUS. A full-size *IDD8* cDNA was fused to the GAL4 DNA-binding domain-coding sequence under the control of the CaMV 35S promoter in the effector vector. Full-size *AKIN10* and *AKIN11* cDNAs were subcloned into the expression vector harboring the CaMV 35S promoter. A positive control was ARF5M construct, in which a full-size *ARF5M* cDNA was subcloned into the GAL4 expression vector [53]. The reporter, effector, and expression vectors were cotransformed into *Arabidopsis* mesophyll protoplasts by the PEG-calcium transfection method [51]. The CaMV 35S promoter-luciferase construct was also cotransformed as an internal control. GUS activity was measured by the fluorometric method as described previously [54]. Luciferase activity assays were performed using the Luciferase Assay System kit (Promega).

Accession numbers

The *Arabidopsis* Genome Initiative locus identifiers for the genes mentioned in this article are: At5g44160, *IDD8*; At3g01090, *AKIN10*; At3g29160, *AKIN11*; At3g43190, *SUS4*; At3g13920, *eIF4A*; At1g22710, *SUC2*; At5g43610, *SUC6*; At1g66570, *SUC7*; At2g14670, *SUC8*; and At3g26744, *ICE1*.

Availability of supporting data

The data sets supporting the results of this article are included within the article and its additional files.

Additional files

Additional file 1: Molecular characterization of *akin10-1* and *akin11-1* mutants. A. Mapping of T-DNA insertions. The *AKIN10*-defective *akin10-1* (SALK-127939) and *AKIN11*-defective *akin11-1* (WiscDsLox320B03) mutants were isolated from a pool of T-DNA insertional lines deposited in the *Arabidopsis* Biological Resource Center (ABRC, Ohio State University, OH). Black boxes represent exons, and white boxes represent 5′ and 3′ untranslated regions. F and R, forward and reverse primers, respectively. bp, base pair. **B**. *AKIN10* expression in *akin10-1* mutant. Gene expression was examined by PCR-based genotyping (left panel), in which left and right primers (LP and RP, respectively) that are specific to the flanking sequences of the T-DNA insertion site and a T-DNA-specific LBb1.3 primer were used, quantitative real-time RT-PCR (qRT-PCR, middle panel), and RT-PCR (right panel). The primer sequences were obtained from the Salk Institute Genomic Analysis Laboratory (http://signal.salk.edu/cgi-bin/tdnaexpress). SM, size marker. In RT-PCR, a tubulin gene (*TUB*) was included as control of constitutive expression. In qRT-PCR, biological triplicates were averaged and statistically treated using Student *t*-test (**P* < 0.01, difference from Col-0). Bars indicate standard error of the mean. **C**. *AKIN11* expression in *akin11-1* mutant. Gene expression was examined by PCR-based genotyping (left panel) and qRT-PCR (right panel), as described in (**B**). The T-DNA-specific P745 primer sequence was obtained from the *Arabidopsis* Information Resource (TAIR, http://www.arabidopsis.org/). Bars indicate standard error of the mean (*t*-test,**P* < 0.01, difference from Col-0).

Additional file 2: Expression of transgenes in *10*-ox and *11*-ox transgenic plants. Four independent transgenic plants overexpressing *AKIN10* (*10*-ox) and *AKIN11* (*11*-ox) genes were grown on ½ X Murashige and Skoog-agar plates (hereafter, referred to as MS-agar plates) for 2 weeks under long days (LDs, 16-h light and 8-h dark) before harvesting whole plant materials for total RNA extraction. Transcript levels of *AKIN10* gene (**A**) and *AKIN11* gene (**B**) were determined by qRT-PCR. Biological triplicates were averaged and statistically analyzed using Student *t*-test (**P* < 0.01, difference from Col-0). Bars indicate standard error of the mean.

Additional file 3: Expression of IDD8 downstream genes in *10*-ox plant and *akin10-1* mutant. *AKIN10*-overexpressing (*10*-ox) and -deficient (*akin10-1*) plants were grown on MS-agar plates for 2 weeks under LDs before harvesting whole plant materials for total RNA extraction. Transcript levels of *SUCROSE SYNTHASE 4* (*SUS4*) gene (**A**), which is a target of IDD8, and *SUCROSE-PROTON SYMPORTER* (*SUC*) genes (**B**), which function downstream of IDD8, were determined by qRT-PCR. Biological triplicates were averaged and statistically analyzed using Student *t*-test (**P* < 0.01, difference from Col-0). Bars indicate standard error of the mean.

Additional file 4: Bimolecular fluorescence complementation (BiFC) assay. The nYFP-IDD8 and cYFP-AKIN fusions were coexpressed with cYFP vector and nYFP vector, respectively, in *Arabidopsis* protoplasts. A cyan fluorescent protein (CFP)-tagged ICE1 nuclear marker was also coexpressed in *Arabidopsis* protoplasts. The protoplasts were visualized by differential interference contrast microscopy (DIC) and fluorescence microscopy. Scale bars, 10 μm.

Additional file 5: Immunological detection of IDD8 and AKIN proteins. Recombinant GST-AKIN and MBP-IDD8 proteins used in phosphorylation assays *in vitro* were detected immunologically using anti-GST and anti-MBP antibodies. Wild-type and mutated IDD8 proteins, which were used in Figures 3 and 4, were detected in (**A**) and (**B**), respectively. S, serine. T, threonine. A, alanine. kDa, kilodalton.

Additional file 6: Amino acid sequences surrounding T98, S178, and S182 in IDD8 protein. The consensus sequence for the SnRK1-mediated phosphorylation of serine (S) and threonine (T) residues is shown (upper panel). Amino acid sequences surrounding the putative phosphorylation residues of IDD8, such as T98, S178, and S182, are shown in the lower panel. The phosphorylated S and T residues are marked in bold and underline. Basic residues are marked in black shadow, and hydrophobic residues are marked in grey shadow.

Additional file 7: Analysis of phosphorylated residues in IDD8 protein by mass spectrometry. Recombinant MBP-IDD8 and GST-AKIN10 protein fusions were prepared in *E. coli* cells. Phosphorylation reactions *in vitro* were induced by incubating with non-radioactive ATP. MBP-IDD8 protein was excised from 6% SDS-PAGE gel, digested with trypsin, and analyzed by liquid chromatography-tandem mass spectrometry (LC-MS/MS). Protein Pilot program (Applied Biosystems, Foster City, CA) was used to assign the phosphorylation sites. The serine (S) and threonine (T) phosphorylation sites identified by the Protein Pilot program were calculated with a confidence > 0.95.

Additional file 8: Effects of AKIN10 on protein stability and nuclear localization of IDD8. A. IDD8 protein stability. The *IDD8-MYC* fusion was overexpressed driven by the Cauliflower Mosaic Virus (CaMV) 35S promoter in either Col-0 plant or *akin10-1* mutant. Ten-day-old plants grown on MS-agar plates were incubated for 2 days either in complete darkness or in the light in the presence or absence (mock) of 50 μM DCMU, a specific inhibitor of photosynthesis. Protein extracts were prepared from whole plant materials. IDD8 proteins were detected immunologically using an anti-MYC antibody (upper panel). Part of Coomassie Blue-stained gel was displayed as a loading control (middle panel). Total RNA was extracted from the light-grown plants, and transcript levels of *IDD8* gene were determined by qRT-PCR (lower left panel). Biological triplicates were averaged and statistically analyzed (*t*-test, **P* < 0.01, difference from Col-0 background). Relative protein intensity was calculated by dividing the band intensity with the transcript level (lower right panel). Bars indicate standard error of the mean. **B**. Subcellular localization of IDD8 in *Arabidopsis* protoplasts. A green fluorescent protein (GFP)-coding sequence was fused in-frame to the 5′ end of a full-size *IDD8* cDNA. The *GFP-IDD8* fusion was transiently expressed in *Arabidopsis* protoplasts and visualized by fluorescence microscopy. Chloroplasts appear red because of autofluorescence. Scale bars, 10 μm. **C**. Subcellular localization of IDD8 in the roots of 35S:*GFP-IDD8* transgenic plants. The transgenic were generated from Col-0 and *10*-ox plants. Roots of ten-day-old plants grown on MS-agar plates were visualized by DIC and fluorescence microscopy. The roots were also stained with 4′,6-diamidino-2-phenylindole (DAPI) to visualize the nuclei. Scale bars, 10 μm.

Additional file 9: IDD8 binding to *SUS4* promoter in *akin10-1* mutant. A. IDD8-binding sequence in *SUS4* promoter. The IDD8-binding sequence (IDD8-BS) containing a conserved CTTTTGTCC motif covers residues −2553 to −2348 upstream of the translation start site. A non-binding sequence (IDD8-nBS) covering residues −1363 to −1158 was included as negative control in the assay. Black boxes indicate exons, and white boxes indicate 5′ and 3′ untranslated regions. kbp, kilobase pair. **B**. Chromatin immunoprecipitation (ChIP) assay on IDD8 binding to *SUS4* chromatin. Plants grown on MS-agar plates for 12 days under LDs were used for chromatin preparation. An *eIF4A* DNA fragment was used for normalization. Four measurements were averaged for each plant genotype and statistically analyzed using Student *t*-test (*$P < 0.01$, difference from mock). Bars indicate standard error of the mean. IP, immunoprecipitation.

Additional file 10: Expression of *IDD8* and *AKIN10* genes in *idd8-3 akin10-1* double mutant. Ten-day-old plants grown on MS-agar plates were used for total RNA extraction. Transcript levels were determined by qRT-PCR. Biological triplicates were averaged and statistically analyzed using Student *t*-test (*$P < 0.01$, difference from Col-0). Bars indicate standard error of the mean.

Additional file 11: Primers used. F, forward primer; R, reverse primer.

Abbreviations

IDD: INDETERMINATE DOMAIN; SNF1: Sucrose non-fermenting 1; SnRK1: SNF1-related protein kinase 1; AKIN10: *Arabidopsis* SNF1 kinase homolog 10; ZF: Zinc finger; GA: Gibberellic acid; FT: FLOWERING LOCUS T; SOC1: SUPPRESSOR OF CONSTANS OVEREXPRESSION 1; FLC: FLOWERING LOCUS C; AtSUC9: SUCROSE TRANSPORTER9; T6P: Trehalose-6-phosphate; CO: CONSTANS; SUS: Sucrose synthase; AMPK1: AMP-dependent protein kinase 1; CaMV: Cauliflower mosaic virus; GST: Glutathione S-transferase; BiFC: Bimolecular fluorescence complementation; YFP: Yellow fluorescence protein; MBP: Maltose-binding protein; FUS3: FUSCA3; DCMU: 3-(3,4-dichlorophenyl)-1,1-dimethylurea; qRT-PCR: Quantitative real-time RT-PCR; GFP: Green fluorescence protein; ChIP: Chromatin immunoprecipitation; GUS: β-glucuronidase; AP1: APPETALA 1; LD: Long day; LC-MS/MS: Liquid chromatography-tandem mass spectrometry.

Competing interests

The authors declare that they have no competing interests.

Authors' contributions

CMP conceived and designed the project. EYJ performed the molecular and biochemical assays. EYJ and PJS performed the assays on protein-protein interaction and subcellular localization. CMP, PJS, JCW analyzed the data. CMP and EYJ wrote the manuscript. All authors discussed the results and approved the final form of the manuscript.

Acknowledgements

This work was supported by the Leaping Research (NRF-2010-0014373) and Global Research Lab (NRF-2012K1A1A2055546) Programs provided by the National Research Foundation of Korea and the Next-Generation BioGreen 21 Program (PMBC No. PJ8103012014) provided by the Rural Development Administration of Korea. It was also supported in part by the Human Frontier Science Program (RGP0002/2012) and a grant from the National Research Foundation of Korea (NRF-2012R1A1A2043550 to JCW).

Author details

[1]Department of Chemistry, Seoul National University, Seoul 151-742, South Korea. [2]Department of Biological Science, Mokpo National University, Jeonnam 534-729, South Korea. [3]Plant Genomics and Breeding Institute, Seoul National University, Seoul 151-742, South Korea.

References

1. Matsoukas IG, Massiah AJ, Thomas B. Florigenic and antiflorigenic signaling in plants. Plant Cell Physiol. 2012;53:1828–42.
2. Kim DH, Doyle MR, Sung S, Amasino RM. Vernalization: winter and the timing of flowering in plants. Annu Rev Cell Dev Biol. 2009;25:277–99.
3. Huijser P, Schmid M. The control of developmental phase transitions in plants. Development. 2011;138:4117–29.
4. Srikanth A, Schmid M. Regulation of flowering time: all roads lead to Rome. Cell Mol Life Sci. 2011;68:2013–37.
5. Seo E, Lee H, Jeon J, Park H, Kim J, Noh YS, et al. Crosstalk between cold response and flowering in *Arabidopsis* is mediated through the flowering time gene *SOC1* and its upstream negative regulator *FLC*. Plant Cell. 2009;21:3185–97.
6. Ruan YL. Sucrose metabolism: gateway to diverse carbon use and sugar signaling. Annu Rev Plant Biol. 2014;65:33–67.
7. Streb S, Zeeman SC. Starch metabolism in *Arabidopsis*. Arabidopsis Book. 2012;10, e0160.
8. Bolouri Moghaddam MR, den Ende WV. Sugars, the clock and transition to flowering. Front Plant Sci. 2013;4:22.
9. Yu TS, Lue WL, Wang SM, Chen J. Mutation of *Arabidopsis* plastid phosphoglucose isomerase affects leaf starch synthesis and floral initiation. Plant Physiol. 2000;123:319–26.
10. Matsoukas IG, Massiah AJ, Thomas B. Starch metabolism and antiflorigenic signals modulate the juvenile-to-adult phase transition in *Arabidopsis*. Plant Cell Environ. 2013;36:1802–11.
11. King RW, Hisamatsu T, Goldschmidt EE, Blundell C. The nature of floral signals in *Arabidopsis*. I. Photosynthesis and a far-red photoresponse independently regulate flowering by increasing expression of *FLOWERING LOCUS T* (*FT*). J Exp Bot. 2008;59:3811–20.
12. Gibson SI. Control of plant development and gene expression by sugar signaling. Curr Opin Plant Biol. 2005;8:93–102.
13. Sivitz AB, Reinders A, Johnson ME, Krentz AD, Grof CP, Perroux JM, et al. *Arabidopsis* sucrose transporter AtSUC9. High-affinity transport activity, intragenic control of expression, and early flowering mutant phenotype. Plant Physiol. 2007;143:188–98.
14. Chincinska IA, Liesche J, Krügel U, Michalska J, Geigenberger P, Grimm B, et al. Sucrose transporter StSUT4 from potato affects flowering, tuberization, and shade avoidance response. Plant Physiol. 2008;146:515–28.
15. Paul MJ, Primavesi LF, Jhurreea D, Zhang Y. Trehalose metabolism and signaling. Annu Rev Plant Biol. 2008;59:417–41.
16. Gómez LD, Gilday A, Feil R, Lunn JE, Graham IA. *AtTPS1*-mediated trehalose 6-phosphate synthesis is essential for embryogenic and vegetative growth and responsiveness to ABA in germinating seeds and stomatal guard cells. Plant J. 2010;64:1–13.
17. Wahl V, Ponnu J, Schlereth A, Arrivault S, Langenecker T, Franke A, et al. Regulation of flowering by trehalose-6-phosphate signaling in *Arabidopsis thaliana*. Science. 2013;339:704–7.
18. Lu Q, Zhao L, Li D, Hao D, Zhan Y, Li W. A *GmRAV* ortholog is involved in photoperiod and sucrose control of flowering time in soybean. PLoS One. 2014;9, e89145.
19. Ortiz-Marchena MI, Albi T, Lucas-Reina E, Said FE, Romero-Campero FJ, Cano B, et al. Photoperiodic control of carbon distribution during the floral transition in *Arabidopsis*. Plant Cell. 2014;26:565–84.
20. Seo PJ, Ryu J, Kang SK, Park CM. Modulation of sugar metabolism by an INDETERMINATE DOMAIN transcription factor contributes to photoperiodic flowering in *Arabidopsis*. Plant J. 2011;65:418–29.
21. Zhang Y, Primavesi LF, Jhurreea D, Andralojc PJ, Mitchell RA, Powers SJ, et al. Inhibition of SNF1-related protein kinase1 activity and regulation of metabolic pathways by trehalose-6-phosphate. Plant Physiol. 2009;149:1860–71.
22. Crawford RM, Halford NG, Hardie DG. Cloning of DNA encoding a catalytic subunit of SNF1-related protein kinase-1 (SnRK1-alpha1), and immunological analysis of multiple forms of the kinase, in spinach leaf. Plant Mol Biol. 2001;45:731–41.
23. Halford NG, Hey S, Jhurreea D, Laurie S, McKibbin RS, Paul M, et al. Metabolic signalling and carbon partitioning: role of Snf1-related (SnRK1) protein kinase. J Exp Bot. 2003;54:467–75.
24. Baena-González E, Rolland F, Thevelein JM, Sheen J. A central integrator of transcription networks in plant stress and energy signalling. Nature. 2007;448:938–42.
25. Polge C, Thomas M. SNF1/AMPK/SnRK1 kinases, global regulators at the heart of energy control? Trends Plant Sci. 2007;12:20–8.
26. Tsai AY, Gazzarrini S. AKIN10 and FUSCA3 interact to control lateral organ development and phase transitions in *Arabidopsis*. Plant J. 2012;69:809–21.
27. Halford NG, Hey SJ. Snf1-related protein kinases (SnRKs) act within an intricate network that links metabolic and stress signalling in plants. Biochem J. 2009;419:247–59.
28. Cho YH, Hong JW, Kim EC, Yoo SD. Regulatory functions of SnRK1 in stress-responsive gene expression and in plant growth and development. Plant Physiol. 2012;158:1955–64.

29. Rolland F, Baena-Gonzalez E, Sheen J. Sugar sensing and signaling in plants: conserved and novel mechanisms. Annu Rev Plant Biol. 2006;57:675–709.
30. Brückner A, Polge C, Lentze N, Auerbach D, Schlattner U. Yeast two-hybrid, a powerful tool for systems biology. Int J Mol Sci. 2009;10:2763–88.
31. Meng X, Xu J, He Y, Yang KY, Mordorski B, Liu Y, et al. Phosphorylation of an ERF transcription factor by *Arabidopsis* MPK3/MPK6 regulates plant defense gene induction and fungal resistance. Plant Cell. 2013;25:1126–42.
32. Zhai Q, Yan L, Tan D, Chen R, Sun J, Gao L, et al. Phosphorylation-coupled proteolysis of the transcription factor MYC2 is important for jasmonate-signaled plant immunity. PLoS Genet. 2013;9, e1003422.
33. Bigeard J, Rayapuram N, Pflieger D, Hirt H. Phosphorylation-dependent regulation of plant chromatin and chromatin-associated proteins. Proteomics. 2014;14:2127–40.
34. Fragoso S, Espíndola L, Páez-Valencia J, Gamboa A, Camacho Y, Martínez-Barajas E, et al. SnRK1 isoforms AKIN10 and AKIN11 are differentially regulated in *Arabidopsis* plants under phosphate starvation. Plant Physiol. 2009;149:1906–16.
35. Robaglia C, Thomas M, Meyer C. Sensing nutrient and energy status by SnRK1 and TOR kinases. Curr Opin Plant Biol. 2012;15:301–7.
36. Tsai AY, Gazzarrini S. Trehalose-6-phosphate and SnRK1 kinases in plant development and signaling: the emerging picture. Front Plant Sci. 2014;5:119.
37. Gazzarrini S, Tsuchiya Y, Lumba S, Okamoto M, McCourt P. The transcription factor FUSCA3 controls developmental timing in *Arabidopsis* through the hormones gibberellin and abscisic acid. Dev Cell. 2004;7:373–85.
38. Nietzsche M, Schießl I, Börnke F. The complex becomes more complex: protein-protein interactions of SnRK1 with DUF581 family proteins provide a framework for cell- and stimulus type-specific SnRK1 signaling in plants. Front Plant Sci. 2014;5:54.
39. Mosley AL, Lakshmanan J, Aryal BK, Ozcan S. Glucose-mediated phosphorylation converts the transcription factor Rgt1 from a repressor to an activator. J Biol Chem. 2003;278:10322–7.
40. Akhtar A, Becker PB. The histone H4 acetyltransferase MOF uses a C2HC zinc finger for substrate recognition. EMBO Rep. 2001;2:113–8.
41. Gamsjaeger R, Liew CK, Loughlin FE, Crossley M, Mackay JP. Sticky fingers: zinc-fingers as protein-recognition motifs. Trends Biochem Sci. 2007;32:63–70.
42. Smeekens S, Ma J, Hanson J, Rolland F. Sugar signals and molecular networks controlling plant growth. Curr Opin Plant Biol. 2010;13:274–9.
43. Lastdrager J, Hanson J, Smeekens S. Sugar signals and the control of plant growth and development. J Exp Bot. 2014;65:799–807.
44. Bolouri-Moghaddam MR, Le Roy K, Xiang L, Rolland F, Van den Ende W. Sugar signalling and antioxidant network connections in plant cells. FEBS J. 2010;277:2022–37.
45. Sreenivasulu N, Harshavardhan VT, Govind G, Seiler C, Kohli A. Contrapuntal role of ABA: does it mediate stress tolerance or plant growth retardation under long-term drought stress? Gene. 2012;506:265–73.
46. Seo PJ, Kim MJ, Park JY, Kim SY, Jeon J, Lee YH, et al. Cold activation of a plasma membrane-tethered NAC transcription factor induces a pathogen resistance response in *Arabidopsis*. Plant J. 2010;61:661–71.
47. Karimi M, Inzé D, Depicker A. GATEWAY vectors for *Agrobacterium*-mediated plant transformation. Trends Plant Sci. 2002;7:193–5.
48. Zhang X, Henriques R, Lin SS, Niu QW, Chua NH. *Agrobacterium*-mediated transformation of *Arabidopsis thaliana* using the floral dip method. Nat Protoc. 2006;1:641–6.
49. Udvardi MK, Czechowski T, Scheible WR. Eleven golden rules of quantitative RT-PCR. Plant Cell. 2008;20:1736–7.
50. Kim MJ, Park MJ, Seo PJ, Song JS, Kim HJ, Park CM. Controlled nuclear import of the transcription factor NTL6 reveals a cytoplasmic role of SnRK2.8 in the drought-stress response. Biochem J. 2012;448:353–63.
51. Yoo SD, Cho YH, Sheen J. *Arabidopsis* mesophyll protoplasts: a versatile cell system for transient gene expression analysis. Nat Protoc. 2007;2:1565–72.
52. Lawrence RJ, Earley K, Pontes O, Silva M, Chen ZJ, Neves N, et al. A concerted DNA methylation/histone methylation switch regulates rRNA gene dosage control and nucleolar dominance. Mol Cell. 2004;13:599–609.
53. Tiwari SB, Hagen G, Guilfoyle T. The roles of auxin response factor domains in auxin-responsive transcription. Plant Cell. 2003;15:533–43.
54. Yang SD, Seo PJ, Yoon HK, Park CM. The *Arabidopsis* NAC transcription factor VNI2 integrates abscisic acid signals into leaf senescence via the *COR/RD* Genes. Plant Cell. 2011;23:2155–68.

Analysis of conglutin seed storage proteins across lupin species using transcriptomic, protein and comparative genomic approaches

Rhonda C Foley[1], Jose C Jimenez-Lopez[2], Lars G Kamphuis[1,2], James K Hane[1,3], Su Melser[1] and Karam B Singh[1,2*]

Abstract

Background: The major proteins in lupin seeds are conglutins that have primary roles in supplying carbon, sulphur and nitrogen and energy for the germinating seedling. They fall into four families; α, β, γ and δ. Interest in these conglutins is growing as family members have been shown to have beneficial nutritional and pharmaceutical properties.

Results: An in-depth transcriptome and draft genome from the narrow-leafed lupin (NLL; *Lupinus angustifolius*) variety, Tanjil, were examined and 16 conglutin genes were identified. Using RNAseq data sets, the structure and expression of these 16 conglutin genes were analysed across eight lupin varieties from five lupin species. Phylogenic analysis suggest that the α and γ conglutins diverged prior to lupin speciation while β and δ members diverged both prior and after speciation. A comparison of the expression of the 16 conglutin genes was performed, and in general the conglutin genes showed similar levels of RNA expression among varieties within species, but quite distinct expression patterns between lupin species. Antibodies were generated against the specific conglutin families and immunoblot analyses were used to compare the levels of conglutin proteins in various tissues and during different stages of seed development in NLL, Tanjil, confirming the expression in the seed. This analysis showed that the conglutins were expressed highly at the mature seed stage, in all lupin species, and a range of polypeptide sizes were observed for each conglutin family.

Conclusions: This study has provided substantial information on the complexity of the four conglutin families in a range of lupin species in terms of their gene structure, phylogenetic relationships as well as their relative RNA and protein abundance during seed development. The results demonstrate that the majority of the heterogeneity of conglutin polypeptides is likely to arise from post-translational modification from a limited number of precursor polypeptides rather than a large number of different genes. Overall, the results demonstrate a high degree of plasticity for conglutin expression during seed development in different lupin species.

Keywords: Seed storage, Seed development, Legumes, Seed transcriptome, Lupin varieties

Background

Legumes form a very important group of crop plants for both animal and human consumption [1]. The genus *Lupinus* comprises between 200 and 600 species, the majority of which have not been domesticated. *Lupinus* are part of the genistoid clade which is quite distinct from other legume sister clades and are the least exploited group of legumes. Interest in lupins is growing as they have a number of attractive nutritional attributes relating to their high protein and dietary fibre contents and negligible starch. Recently, studies have also associated lupins with playing positive roles in health in areas such as combating obesity, diabetes and cardiovascular disease [2-4]. For example, studies have demonstrated that γ conglutin has an ability to reduce glycaemia to levels comparable to those obtained with metformin, a widely used hypoglycaemic drug [4]. Lupins are also valuable in sustainable agricultural systems through their

* Correspondence: Karam.Singh@csiro.au
[1]CSIRO Agriculture Flagship, Centre for Environment and Life Sciences, Floreat, WA, Australia
[2]The UWA Institute of Agriculture, University of Western Australia, Crawley, WA, Australia
Full list of author information is available at the end of the article

ability to fix nitrogen, they also contribute high organic matter to the soil and are effective in breaking disease cycles for other crops such as cereals [5]. Cultivation of lupins has occurred for thousands of years, and extensive boiling and steeping were required to rid the grain of their bitter and poisonous alkaloids. Modern lupins have been bred to have reduced alkaloid levels and are referred to as sweet lupins [6,7].

The main seed storage proteins in lupins, referred to as conglutins, have been classified into four families; α, β, γ and δ conglutins. α conglutins have been shown to be degraded proteolytically at germination, confirming their role as a classical seed storage protein [8]. β conglutins are abundant seed proteins as verified by quantitating protein [9] and transcript expression [10]. The bean β conglutin homolog, phaseolin, degrades during germination in a phosphorylation-dependent manner [11]. As well as having a seed storage protein function, metabolism of β conglutins also acts to release lectins during germination [12]. β conglutins have also been shown to be potential allergens [10]. γ conglutins may not fall into the classical seed storage category as the protein is not degraded like traditional seed storage proteins during germination, and γ conglutins may also possess lectin-like activity [13]. Although γ conglutins share structural similarities to xyloglucan-specific endo-beta-1,4-glucanase inhibitors, they lack the functional activity of glucanase inhibitors [14]. δ conglutins have been the least studied conglutin family. They are small proteins that are localised in the vacuole [15], and have low-digestibility properties [16].

In this study we analysed the conglutins in a number of lupin species. *L. angustifolius,* (narrow-leafed lupin, NLL) is currently the most extensively cultivated lupin crop and is grown predominantly in South-Western Australia. Three NLL varieties were used in this study; P27255 is a wild variety available prior to domestication of NLL within Australia; Unicrop was one of the first domesticated Australian NLL varieties (released in 1973) as a sweet, non-shattering pod variety, and Tanjil is a more recent variety (released in 1998). Tanjil has resistance to the fungal pathogen, anthracnose (*Colletotrichum spp.*) and has emerged as the reference NLL/lupin variety for which a number of genomic resources have/are being developed [17-23].

L. albus (white lupin), is the lupin of preference in Europe and much of the nutritional work has been done on this species [3,9,12,14,20]. The two varieties of *L. albus* used in this study were the Kiev mutant, a Ukrainian bred variety, and Andromeda, a variety with intermediate resistance to anthracnose. Similarly to *L. angustifolus* and *L. albus, L. luteus* is an 'Old World'-smooth seeded species. *L. luteus* (yellow lupin) originates from the Mediterranean coastal regions and was domesticated as a grain legume in Europe by selecting for low alkaloid content, non-shattering pods and soft seeds. The *L. luteus* variety, Pootalong, used in this study is well adapted to acidic soils and the species is favoured by the aquaculture industry because of the high protein content. A *L. luteus'* transcriptome has been generated using 454-expressed sequence tag libraries, and comparative studies using model legume species have identified a comprehensive set of molecular markers for yellow lupin [24]. In addition, a comprehensive seed-protein catalog has been developed for *L. luteus* [25]. *L. mutabilis* (Andean or Pearl Lupin) is known as a 'New World' lupin as it originated from South America. *L. mutabilis* has high oil, low alkaloid levels and high protein content. The *L. mutabilis* variety, ID13 was chosen for this study. *L. cosentinii* Guss (sandplain blue lupin) is referred to as an 'Old World' – rough seeded lupin and originated from North Africa. It grows prolifically in coastal South Western Australia. Breeding of this variety to produce a sweeter version developed the variety, Erregulla used in this study. However Erregulla, was highly susceptible to aphids and anthracnose and therefore was not released commercially.

This work expands on the study of NLL lupin conglutins reported by Foley et al. [10], and adds new information on their gene structure and RNA and protein expression patterns both in NLL and other lupin species. Previously we reported the identification of sixteen individual seed storage protein genes and investigation of their expression in the species NLL cv Tanjil using EST sequencing [10]. In this study, analysis of an in-depth Tanjil transcriptomic and draft genome sequence [23] did not reveal any additional conglutins in Tanjil. The 16 Tanjil conglutin genes were used to identify homologous genes in seven other lupin varieties from five different lupin species using RNAseq data, which was also used to compare their respective RNA expression levels in seeds. Antibodies designed specifically against each conglutin family were used to analyse the protein levels of these four conglutin protein families in different tissues and at specific stages of seed development. In addition, the levels of these proteins among lupin varieties were compared.

Results and discussion

Identification of conglutins sequence reads in lupin species

Transcriptomic studies using RNAseq data are valuable in identifying homologous sequences and RNA expression levels. In NLL a total of three α conglutins, seven β conglutins, two γ conglutins and four δ conglutins were previously identified [10]. A draft genome survey assembly and comprehensive transcriptome assembly for Tanjil [23] were searched to confirm the sequences of these conglutins and to potentially identify any additional conglutins with sequence similarity (<1^{e-5}). This analysis confirmed the sequences of all α, γ and δ conglutins identified previously [10] and did not identify any additional conglutins in the NLL cv. Tanjil. In the case of the β conglutins, sequences for BETA1, 2, 5 and 7 were identified, whereas BETA3, 4 and 6 appeared to be mis-assembled in the survey genome assembly and formed contigs with alternate isoforms in the transcriptome assembly, which is most likely due to their high sequence similarity (85-99 % identity) [10]. As the genome survey only comprises 521 Mb of an estimated 924 Mb of genome sequence there remains the possibility that some additional conglutins have not been identified in the genome sequence. However, given that the Tanjil transcriptome, which contained a wide range of different developmental seed stages, and the additional seed transcriptome sequencing for Tanjil presented herein did not show any other transcripts with homology to the conglutins identified previously [10], it would appear unlikely that other conglutins are present in NLL cv. Tanjil.

The conglutins from NLL cv. Tanjil were subsequently used as reference sequences to assemble homologous sequences from transcriptome datasets generated for NLL cv. Unicrop and wild accession P27255, as well as other species in the genus *Lupinus*. This allowed consensus sequences for the eight lupin varieties/species to be retrieved by aligning the RNA seq data for each variety to the 16 Tanjil conglutin sequences. In the case of Tanjil, the extracted 16 conglutin consensus sequences from the RNAseq dataset were compared to the published NLL Tanjil sequences and these aligned very closely (99.6% -100% identity). In some instances stretches of unknown nucleotides were generated from the consensus extraction for other species, likely because of stretches of weaker alignment to Tanjil conglutins.

Phylogenetic relationships among conglutins from different lupin species

Each family of conglutin sequences generated from the eight varieties were aligned using the K-mer program within the CLC Genomics Workbench as shown in Figure 1. With respect to the α family, each lupin variety had three α sequences that were assembled against the Tanjil reference genes. In all instances, the α conglutins were able to be subdivided into three subgroups; ALPHA1, ALPHA2 or ALPHA3 (Figure 1A). This is in agreement with our previous observations which showed that the α conglutin, also known as 11S globulin subgroups, have duplicated and diverged prior to speciation in soybean, pea, peanut, medicago and NLL [10].

In contrast to the α conglutins, the β conglutin sequence relationships were not as clear cut (Figure 1B). While there are some β conglutins that appear to have diverged prior to lupin speciation (for example, BETA1 and BETA2), other examples suggest that the β conglutins are most closely aligned with each other within a species (e.g. *L. mutabilis, L. albus* and *L. luteus*), indicating that they diverged after speciation. Whether these different β sequences are distinct genes or alleles remain to be determined. When comparing β conglutin-like protein (or 7S globulins) relationships in legumes such as soybean, pea, peanut, medicago and NLL, gene duplication in each legume species appeared to have occurred after speciation.

As observed with the α conglutins the γ conglutins fall into two classes among nearly all the lupin varieties examined; GAMMA1 and GAMMA2, with the closest homology found within species (e.g. *L. angustifolius* and *L. albus*) (Figure 1C). Interestingly, *L. mutabilis* GAMMA1 and GAMMA2 did not fall neatly into the two classes with most sequence homology found with each other rather than respective members of the GAMMA1 and GAMMA2 orthologs.

The phylogenic relationships among δ conglutins is more complex compared to α and γ conglutins (Figure 1D). DELTA4 appeared to have arisen prior to speciation, while for *L. cosentinii, L. luteus* and *L. mutabilis* the other three δ conglutins have most homology to the δ conglutins within their own species. For *L. albus* and *L. angustifolius* DELTA2 conglutins show strong sequence similarity between the two species, whereas DELTA1 and DELTA3 conglutins show stronger homology to each other, within the respective species. The phylogenic tree thus illustrates that varieties within the same species are most similar to each other and among the lupin species examined *L. mutabilis* appears most phylogenetically distinct. These results are in agreement with previous findings using ITS sequencing and mass spectrometric fingerprints of seed proteins for defining *Lupinus* spp. relationships [26,27].

RNAseq vs ESTseq: Analysis of conglutin RNA expression in narrow-leafed lupin cv Tanjil

Gene expression has been studied through a variety of methods including northern blots, quantitative reverse transcriptase-PCR, microarray analyses and most recently, RNAseq. RNAseq has the advantage over previous techniques in that it is more sensitive, is not limited to a set of predetermined probes and also detects transcriptional modification such as alternative splicing and gene fusion. A standard method of describing transcript expression in RNAseq datasets is to express the number of sequence reads as *r*eads *p*er *k*ilobase per *m*illion reads (RPKM), [28]. However, there have been concerns that the accuracy of the calculation of RPKM used in RNA-seq may be influenced by gene length, GC content and dinucleotide frequencies [29].

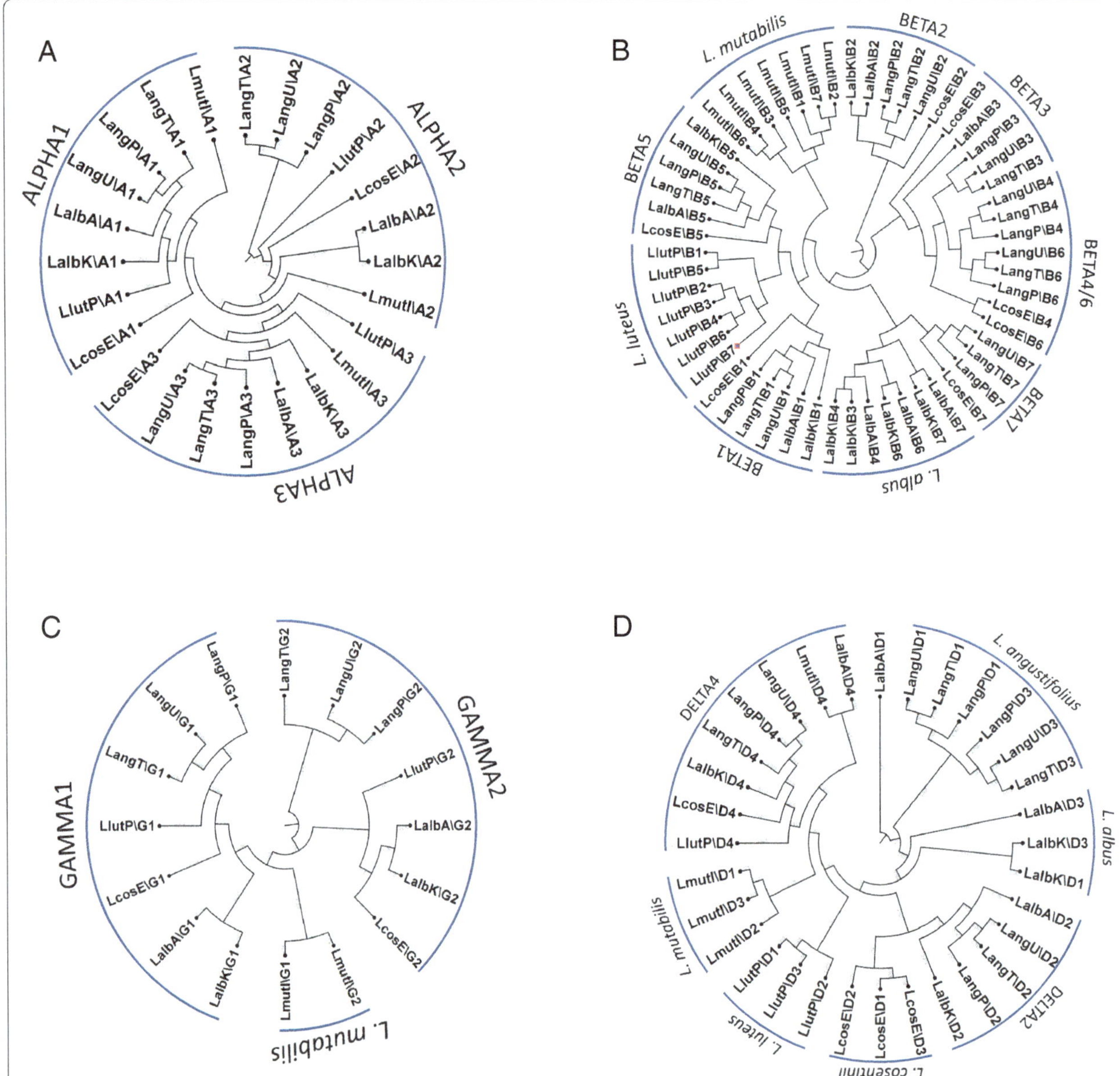

Figure 1 Alignment using the Kmer program of **A)** ALPHA (1-3), **B)** BETA (1-7), **C)** GAMMA (1-2) and **D)** DELTA (1-4) DNA sequences extracted from RNAseq data sets from *L. albus* Andromeda (LalbA), *L. albus* Kiev (LalbK), *L. angustifolius* P27255 (LangP), *L. angustifolius* Tanjil (LangT), *L. angustifolius* Unicrop (LangU), *L. cosentinii* Erregulla (LcosE), *L. luteus* Pootalong (LlutP) and *L. mutabilis* ID13 (Lmutl). The branch numbers and relative distance of the tree represent the amount of evolutionary divergence between two nodes in the tree, which was calculated using a k-mer based method in the CLC Genomics WorkBench v7.5.1.

RNAseq analysis was used to study the expression of the conglutins in NLL cv. Tanjil. The RPKM for the data retrieved for Tanjil was compared to the data obtained from sequencing 3017 clones from a cDNA library, which generated 2381 sequence-readable ESTs [10] from the same stage of seed development and was expressed as *r*eads *p*er *m*illion (RPM) (Figure 2A). The comparison between the two techniques resulted in 32% more reads for α, 7% more reads for β, 36% more reads for γ and over three times more reads for δ using RNAseq compared to ESTseq. This resulted in conglutin genes representing 63% of the total reads using RNAseq compared to 42% using ESTseq [10]. The differences may be due to biological variation, as even though the experiment was replicated as closely as possible, differences may have been introduced unintentionally. For example, plant nutrition and environmental conditions have been shown to greatly affect seed storage protein composition [30,31]. Interestingly, the two data sets have similar expression reads if the EST data is also divided by the gene size (Figure 2B). This suggests

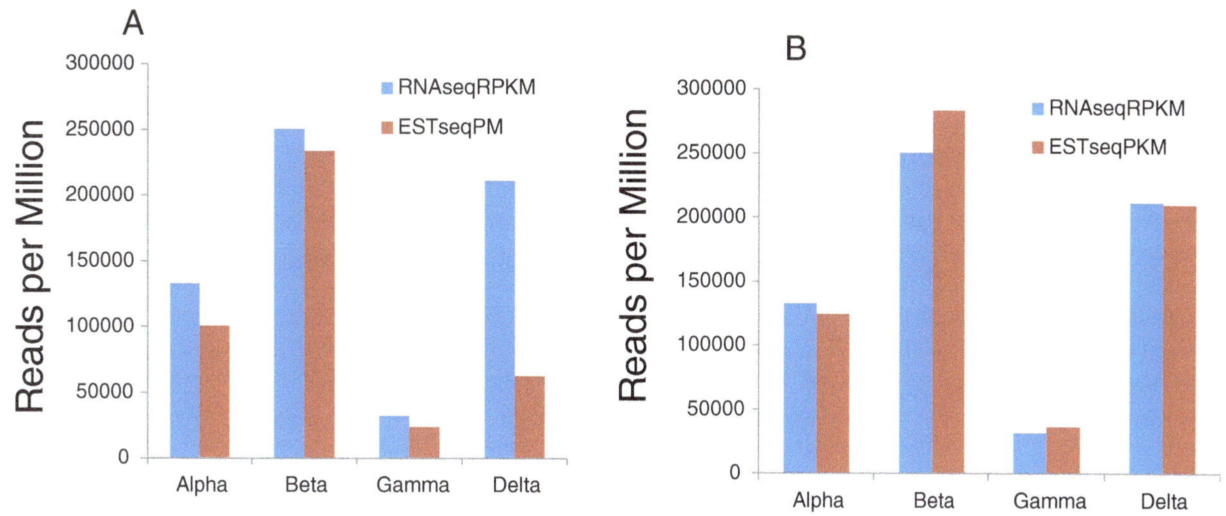

Figure 2 Transcription expression of the four conglutin families from *L. angustifolius*, Tanjil. The RNAseq data was compared to ESTseq data from the same stage of seed development. The EST sequences are represented as **A)** Reads per million (RPM) and **B)** reads per kilobase per million (RPKM).

that the library construction for the EST data set may be proportionally biased towards larger clones and that the RNAseq datasets generated in this study are a more useful means of comparing conglutin gene expression in and across different lupin species.

Comparison of conglutin RNA expression among lupin species

The transcriptomic studies allowed us to directly compare conglutin RNA expression across the five lupin species studied. The RPKM for each family of conglutins in the different lupin species is displayed in Figure 3. As expected, the three varieties of *L. angustifolius* and the two varieties of *L. albus* have similar expression patterns within their respective species. There were some striking differences in conglutin RNA levels between the lupin species. For example, *L. luteus* and *L. cosentinii* have high levels of expression of δ conglutins, with over 60% of conglutin transcripts being δ. In contrast in *L. albus* δ expression was lower and only constituted around 17%

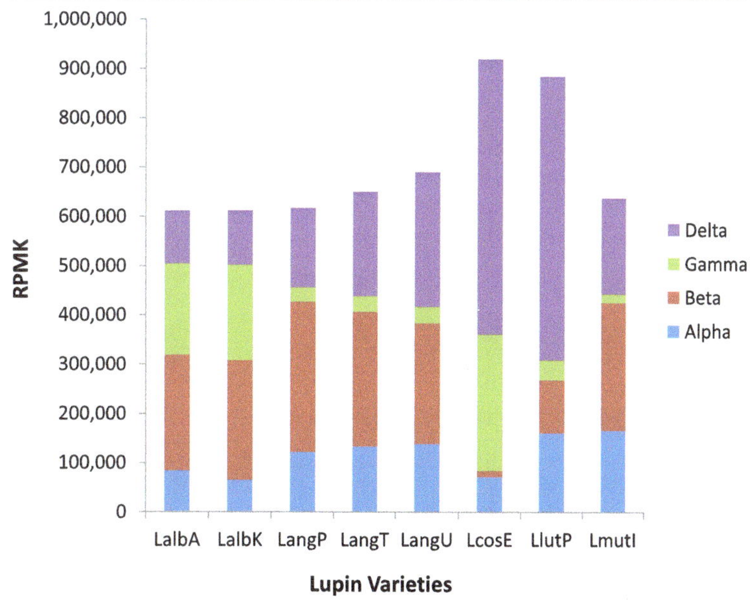

Figure 3 The reads per kilobase per million reads (RPKM) for each tested lupin species/variety for each family of conglutins (ALPHA, BETA, GAMMA and DELTA). The lupin varieties were *L. albus* Andromeda (LalbA), *L. albus* Kiev (LalbK), *L. angustifolius* P27255 (LangP), *L. angustifolius* Tanjil (LangT), *L. angustifolius* Unicrop (LangU), *L. cosentinii* Erregulla (LcosE), *L. luteus* Pootalong (LlutP) and *L. mutabilis* ID13 (LmutI).

of total conglutin transcripts. β conglutins also showed marked differences between lupin species, with β being the most abundant family (>35% of total conglutin transcripts) in *L. angustifolius*, *L. mutabilis* and *L. albus*, while in contrast there were extremely low levels in *L. cosentinii* (less than 2%). While α expression was quite similar across lupin species, γ expression was also quite variable, with high expression in *L. cosentinii* and *L. albus* (~30%) but low expression in the other species (<5%).

The conglutin expression profile can be further analysed to examine the expression of individual conglutin genes. In the case of the α family, ALPHA2 was the major α conglutin expressed in *L. albus* and *L. cosentinii*, but in the other lupin species examined, the different α conglutins were more evenly expressed (Figure 4A). Individual β family members shared substantial differences in their expression levels between lupin species (Figure 4B). For example, BETA7 while being the most abundant in *L. albus* and *L. mutabilis* was one of the least abundant in *L. angustifolius*. Interestingly, a *L. albus* β conglutin sequence has been previously submitted to Genbank as gi:62867685 and this shows greatest identity to BETA7 in *L. albus* Andromeda (98%) and the *L. albus* Kiev mutant (98%). As this is the most abundant β in this species, it was not surprising that this sequence was the first *L. albus* β conglutin sequence reported.

In the case of the γ conglutins, GAMMA1 was the most abundant in *L. mutabilis* and *L. angustifolius* while GAMMA2 was more abundant in the other lupin species (Figure 4C). In the case of the δ conglutins, with the exception of DELTA4, which was expressed at very low levels in all lupin species, the other δ on the whole showed relatively similar proportional expression levels in the different lupin species although DELTA1 expression was much lower in the *L. albus* Kiev mutant than in the *L. albus* Andromeda cultivar (Figure 4D).

Protein and transcript level comparison in mature seeds

A number of factors affect protein abundance, including transcription, translation and protein- turnover. To

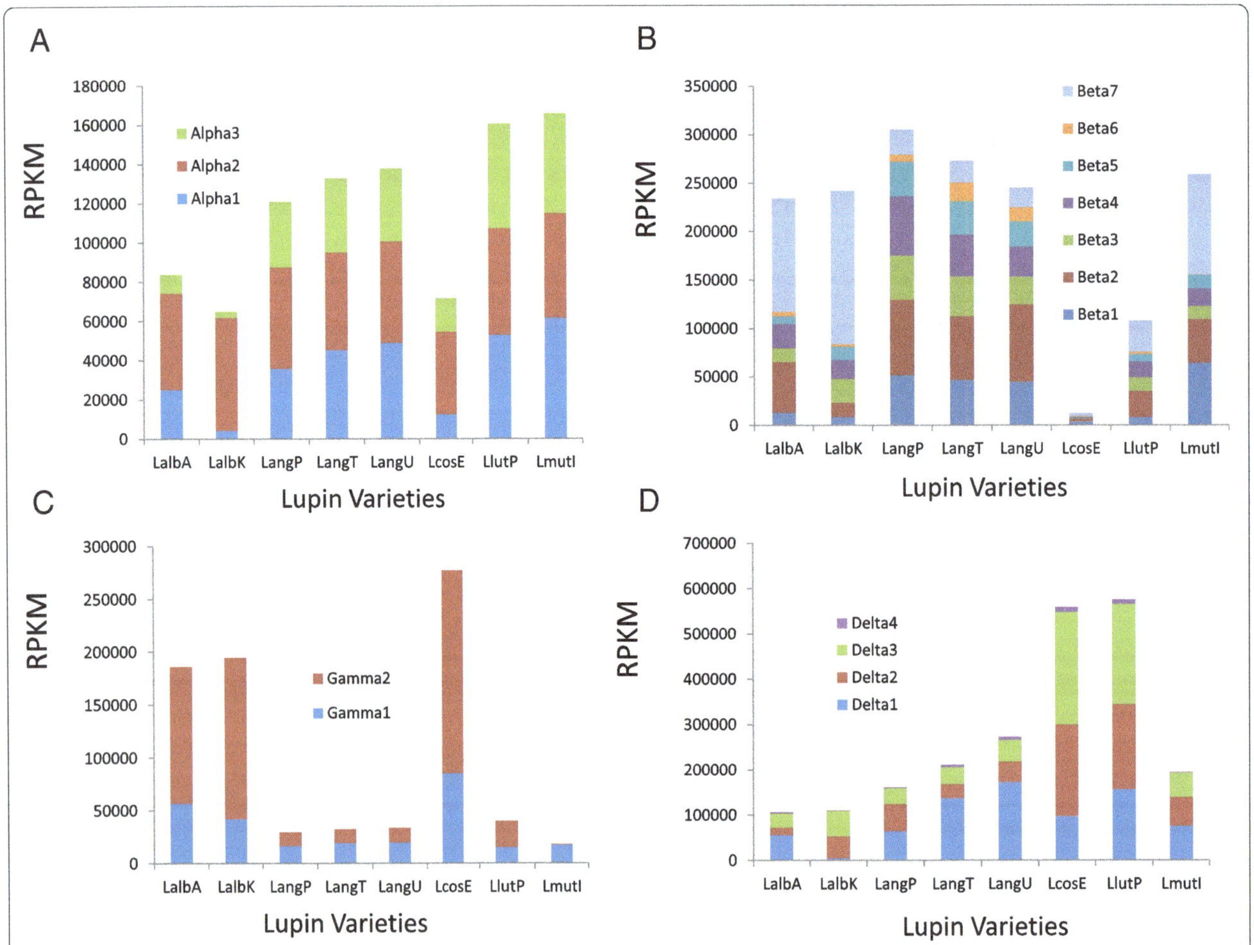

Figure 4 The RPKM for each tested lupin species/variety for each member of the conglutin families. RPKM for conglutin members **A)** Alpha, **B)** Beta, **C)** Gamma and **D)** Delta. The lupin varieties were *L. albus* Andromeda (LalbA), *L. albus* Kiev (LalbK), *L. angustifolus* P27255 (LangP), *L. angustifolius* Tanjil (LangT), *L. angustifolius* Unicrop (LangU), *L. cosentinii* Erregulla (LcosE), *L. luteus* Pootalong (LlutP) and *L. mutabilis* ID13 (Lmutl).

determine if the transcript numbers were correlated with the total protein abundance, the protein concentration in relation to the dry weight of the mature seed was measured. As shown in Figure 5, the cultivars ranged from 30-44% protein content. In the case of *L. albus, L. angustifolius* and *L. luteus*, there was a good correlation between the levels of conglutin transcripts in maturing seeds and the protein abundance in the mature seed, a stage where conglutin transcripts make up a major portion of the total seed transciptome in NLL [10]. However, this correlation was not maintained in *L. cosentinii* and *L.mutabilis* where *L. cosentinii* had the highest level of conglutin transcripts but the lowest percentage protein, while *L. mutabilis* was the opposite. There can be a number of explanations, for example translational regulation, post-translational modification and protein turnover. Moreover, a variety of seed components contribute to the weight of the seed (eg fibre, oil, etc), and these may differentially contribute to the final seed weight in some lupin species. For example, *L. mutabilis* it is known to have relatively high oil levels. Transcript expression of seed storage proteins in *Lotus japonicas* during seed development has been correlated to other genes including oleosins, seed maturation proteins, a late embryogenesis protein and 16 transcription factors [32].

Conglutin protein analysis during seed development in L. angustifolius, cv Tanjil

In *L. angustifolius*, cv Tanjil, transcription levels of the conglutin genes was demonstrated to increase during seed development quite dramatically from 4 to 30 *D*ays *A*fter *A*nthesis (DAA) [10]. It was therefore of interest to see if this was reflected in protein abundance, and the relative expression of conglutins in a variety of tissue types. Immunoblot analysis at different stages of seed development and for different tissues types were performed using polyclonal antibodies raised against synthetic peptides derived from each of the conglutin sequences in Tanjil. Comparison of protein expression of the four families of conglutins demonstrate strong tissue specificity with high expression in the seed and no detection in the leaf, petiole, root and stem tissue from two-month old plants (Figure 6A). In addition, protein from commercially available NLL lupin flour and wheat were also tested, demonstrating the proteins remain intact during lupin milling and emphasising the specificity of the antibodies, since no proteins were detected in wheat flour (Figure 6A). Seeds were collected from the pods at six time points that ranged from 4 DAA to mature seed. 20-26 DAA was the developmental stage used for EST sequencing and RNAseq reported in this study. Equal amounts of protein were loaded for each sample and the extracted protein profiles were separated by SDS-PAGE. Figure 6B illustrates that protein bands similar in size to the predicted pro-protein molecular weight range for α (56 kDa) and β (~70 kDa). In the case of BETA conglutins, the expression of pro-protein of predicted size of ~70 kD is higher at the mid-developmental range, and in later time points there is more expression of smaller polypeptides, suggesting post-translation processing is occurring within the maturing seed. GAMMA and DELTA proteins are expressed at higher levels in middle and later developmental seed stages (Figure 6b).

A 2D analysis of the *L. albus* seed proteome found that 37 out of 40 spots fell into the α, β or γ conglutin families, with many of the proteins having a large size range and several endoproteolytically cleaved subunits could be identified [33]. Our results are in agreement with this previous study, where we have observed a

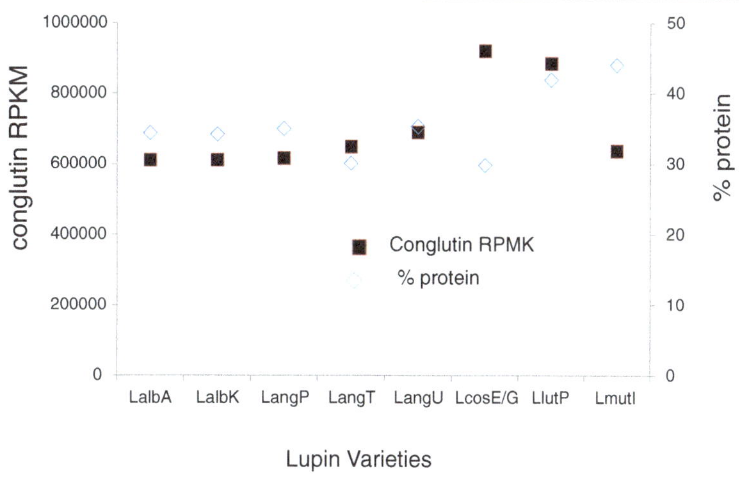

Figure 5 Protein concentrations (% dry weight) of mature seed and total conglutin transcript levels (RPKM) of *L. albus* Andromeda (LalbA), *L. albus* Kiev (LalbK), *L. angustifolius* P27255 (LangP), *L. angustifolius* Tanjil (LangT), *L. angustifolius* Unicrop (LangU), *L. cosentinii* Guss (LcosG), *L. luteus* Pootalong (LlutP) and *L. mutabilis* ID13 (Lmutl).

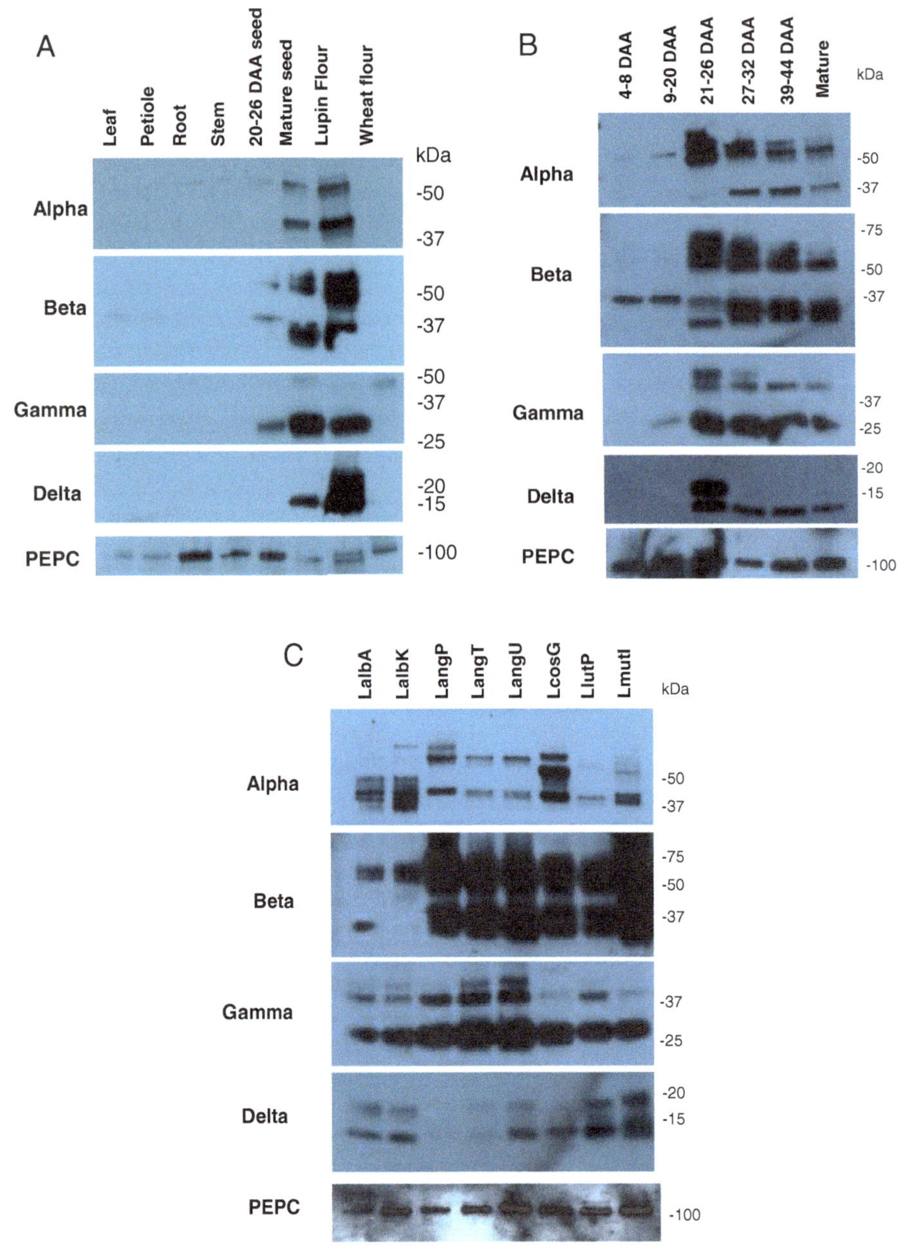

Figure 6 Western blot of 5 ug protein extracts from various lupin samples. The blots were probed using antibodies generated against ALPHA conglutin (expected size 56 kD), BETA conglutin (68-72 kD), GAMMA conglutin (49 kD – uncleaved, 33 kD cleaved), and DELTA conglutin (18 kD– uncleaved, 12 kD cleaved) peptides, and the commercially available PEPC antibody to illustrate similar protein loading. **A)** Protein expression of conglutins in *L. angustifolius* Tanjil and wheat flour. **B)** Protein expression in developmental stages of *L. angustifolius* Tanjil seeds. **C)** Protein expression of conglutin in mature seeds of various lupin species/varietie.

range of peptides with different molecular weights in our immunoblot analysis corresponding to the species, *L. angustifolius*. In addition, the heterogeneity of peptides corresponding to α and β conglutins have also been observed from purified conglutins in *L. albus* [34]. As the genomic and transcriptomic analyses have identified a fixed number of conglutin genes, this further supports that the majority of the heterogeneity of conglutin polypeptides is from post-translational modification from a limited number of precursor polypeptides rather than a large number of different genes.

Analysis of protein expression patterns in different lupin species

The expression profiles of conglutin protein were also analysed in mature seed for the different lupin species/ varieties used in the RNAseq studies. Sequence comparisons over the region of the target conglutin peptide used for antibody production, across the lupin species analysed, demonstrated high homology and therefore a high likely hood that the antibodies would recognise conglutins from all tested lupin species (data not shown). Similar levels of total protein were extracted from mature seeds and separated by SBS-PAGE as demonstrated by staining with Coomassie Brilliant blue (data not shown), and using an antibody against phosphoenolpyruvate caroxylase (PEPC). Figure 6C illustrates that differences occur in the protein profile for the different lupin species, although varieties from the same species often showed similar protein profiles, for example, *L. angustifolius* varieties Tanjil and Unicrop. There was no obvious correlation of conglutin transcript expression with final conglutin protein levels. For example, *L. mutabilis* exhibited the highest levels of proteins recognised by the β conglutin antisera but only had average β conglutin transcript expression levels, suggesting that post-transcriptional control is an important factor in conglutin expression in lupin species.

Conclusion

This study has provided substantial information on the complexity of the four conglutin families in a range of lupin species in terms of their gene structure, phylogenetic relationships and their relative RNA and protein abundance during seed development. These results will provide a good foundation for further research to elucidate the function and nutritional aspects of these seed storage proteins, that in turn will provide the knowledge to try and alter lupin seed protein composition to optimize the health properties of lupins. It was interesting to note that the different lupin species had large variations in conglutin RNA expression. These variations in RNA levels did not appear to have any obvious deleterious effects with regards to seed germination and development. There is evidence to support the plasticity of seed protein expression, for example the ability of sulphur and nitrogen supply to affect seed storage composition [30]. Interestingly, genetically manipulating seed storage proteins in soybean, using RNA interference or mutation results in rebalancing of the proteome thereby maintaining overall wild-type levels of overall seed storage proteins [35,36]. Future work in lupins may involve manipulating seed proteins for optimum nutrition and growth. For example, as β conglutins have been demonstrated to have some allergenic properties, it may be possible to reduce these proteins in favour of γ conglutins that may have higher nutritional benefits.

Methods

Plant growth

Lupin plants were grown in a growth cabinet at 22°C/18°C over a 14 h/10 h day/night schedule under 600 Photosynthetically Active Radiation μmol m^{-2} s^{-1} lights. For the transcriptomic comparisons, seeds were collected 20-26 *D*ays *A*fter *A*nthesis (DAA) for, *L. albus* cv Andromeda, *L. albus* cv Kiev mutant, *L. angustifolius* cv P27255, *L. angustifolius* Tanjil, *L. angustifolius* cv Unicrop, *L. cosentinii* cv Erregulla *L. luteus* cv Pootalong,and *L. mutabilis* cv ID13. For the protein level analysis, eight grams of mature seed was supplied to the ChemCentre (www.chemcentre.-wa.gov.au) of the above cultivars, except *L. cosentinii* cv Gus was used instead of *L. cosentinii* cv Erregulla, as the Erregulla variety lost its viability to germinate. Determination of protein levels was measured by the amount of nitrogen using thermal conductivity [37,38]. For the immunoblot analysis, seeds were collected from the pods 4 DAA to 53 DAA (mature seed) from *L. angustifolius* Tanjil. Commercially available lupin flour (Irwin Valley. *L. angustifolius*) was also obtained.

RNA isolation, library preparation, sequencing and quality control

RNA was isolated using the Trizol method [23] as followed: for each of the eight lupin samples approximately 150 mg of maturing seed (20-26 DAA) was ground to a fine powder under a stream of liquid nitrogen. The ground powder was homogenised in 1.5 mL Eppendorf tubes using two-times 500 μL Trizol reagents (Invitrogen, Carlsbad, CA), followed by a 15 min incubation period at room temperature. After centrifugation at 12000 g and 4°C for 10 min the supernatant was mixed with 200 μL chloroform, followed by another centrifugation step at 12000 g and 4°C for 15 min. The upper phase was mixed with 300 μL of high salt precipitation buffer (0.8 M sodium citrate/1.2 M NaCl) and 300 μL of isopropanol and incubated on ice for at least 10 min to selectively precipitate total RNA. Precipitated RNA was centrifuged (10 min 12000 g and 4°C) and washed twice in 75% ethanol. RNA was dissolved in diethylpyrocarbonate treated water. Quality and quantity was assessed by both BioAnalyzer (Agilent, Santa Clara, CA, USA) and Qubit assays (Invitrogen).

One μg of total RNA was used to generate TruSeq RNA libraries (Illumina, San Diego, CA, USA) according to the manufacturer's recommendations. The eight Lupin TruSeq RNA libraries were pooled evenly and sequenced on a single lane of an Illumina HiSeq1000 on a 2× 100 bp Paired End run.

Raw RNA-Seq data for libraries were trimmed for low quality (<Q30) and Illumina sequencing primer and adaptor sequences (Truseq v2) via Cutadapt 1.1 [39] (overlap 10, times 3, minimum length 25). Reads trimmed to less than 25 bp were discarded and reads with a discarded pair were retained as singleton reads. Transcriptome sequencing and subsequent trimming resulted in quality-controlled RNAseq data ranging from 1.38 to 2.69 Gb per

library (Additional file 1: Table S1). The RNAseq and conglutin EST sequences used in this study were submitted to the Sequence Read Archive and dbEST at NCBI under bioproject accession PRJNA271721.

Data analysis

The conglutin gene sequences identified [10] were used to search the Tanjil survey genome and transcriptome assembly [23], using BLASTN and tBLASTX in CLC Genomics WorkBench 6.0 (CLC Bio, Aarhus, Denmark) to confirm the presence of the conglutins in the genome and transcriptome assembly and to identify if any additional conglutin family members were present.

The Illumina trimmed sequencing reads were sorted into paired and unpaired groups and aligned to the 16 NLL conglutin genes, using CLC Genomics Workbench 6.0. Following the alignment to 16 NLL conglutin reference genes [10] the consensus generated homologous sequence was extracted for each conglutin gene from each variety. The extracted consensus sequences were aligned within each family using K-mer based tree construction (using CLC Genomics Workbench 6). The *r*eads *p*er *k*ilobase per *m*illion reads (RPKM) values were determined for each of the 16 consensus sequences for each lupin variety using the RNAseq datasets and the RNAseq analysis function in CLC Genomics WorkBench 6.0.

Antibodies

Synthetic peptides for production of the polyclonal serum against α conglutin, δ conglutin and β conglutin were from Mimotopes Pty Ltd, (www.mimotopes.com), and γ conglutin from Agrisera (www.agrisera.com). All peptides were conjugated to keyhole limpet hemocyanin as carrier with a maleimidocaproyl-N-hydroxysuccinimide linker and produced to immunograde purity. The sequences for each peptide synthesized were: α conglutin: (NH2)- CRETYEEPQEQEQG (-CONH2), β conglutin: (NH2-) CVDEGEGNYELVGIR (-CONH2), γ conglutin: (NH2)- CSSTYQAPFCHSTQCSRAN -amidated and δ conglutin: (NH2)- CRAILQQIYESQSEQ (-CONH2). Polyclonal antibodies were raised in rabbits at the following facilities: IMVS Pathology, Veterinary Services Division, University of Adelaide, Australia (β conglutin), Agrisera, (γ conglutin), and the Monoclonal Antibody Facility, Harry Perkins Institute of Medical Research (α conglutin, δ conglutin). Anti-PEPC antiserum was purchased from Agrisera.

Protein extraction, SDS-PAGE and immunoblot analyses

Seed material (0.5 g) was grounded into a fine powder in a mortar and pestle using liquid nitrogen, and resuspended in a solution containing 100 mM KPO_4 (pH 7.4), 1 mM EDTA, 1% Triton X-100, 10% glycerol, 1/200 proteinase inhibitor cocktail (Sigma: www.sigmaaldrich.com), 0.5 mM phenylmethanesulfonylfluoride (Sigma) and 1 mM DTT at a concentration of 50% (w/v). After a 15 min spin (14,000 rpm), the supernatant was recovered and measured for protein concentration using the Qubit Protein Assay kit (Invitrogen: Invitrogen.com.au).

Total protein (5 ug per lane) were loaded and separated by SDS-PAGE on a 4-20% acrylamide Mini Protean ®TGX gel (Biorad: www.bio-rad.com) using Mini Protean ®Tetra Cell apparatus (Biorad). To verify protein quantification, one gel was stained with Coomassie Brilliant Blue (Bio-Rad) according to manufacturer's procedures. The second gel was used for immunoblot analysis.

Proteins were electroblotted onto a PVDF member (Bio-Rad) using a Mini-Trans-Blot Electrophoretic Transfer Cell sysem (Bio-Rad). Western blots were performed using standard techniques [40] with an overnight incubation of 1/1000 dilution of rabbit serum containing polyclonal antibodies followed by TBS washes and a 2 h incubation of goat anti-rabbit IgG horseradish peroxidise conjugate (Bio-Rad). The chemiluminescence signal was detected using Western Lightning Plus-ECL substrate (Perkin Elmer: www.perkinelmer.com) following the manufacturers' instructions and Xray film (Kodak; www.kodak.com) detection.

Availability of supporting data

The RNAseq and conglutin EST sequences used in this study have been submitted to the Sequence Read Archive and dbEST at NCBI under bioproject accession PRJNA271721.

Additional file

Additional file 1: Table S1. RNAseq data information.

Abbreviations

DAA: Days after anthesis; EST: Expressed sequence tag; LalbA: *L. albus* Andromeda; LalbK: *L. albus* Kiev; LangP: *L. angustifolus* P27255; LangT: *L. angustifolius* Tanjil; LangU: *L. angustifolius* Unicrop; LcosE: *L. cosentinii* Erregulla; LlutP: *L. luteus* Pootalong; Lmutl: *L. mutabilis* ID13; NLL: Narrow-leafed lupin; RPKM: Reads per kilobase per million reads; RPM: Reads per million.

Competing interests

The authors declare they have no competing interests.

Authors' contributions

RF participated in experimental design, material preparation, bioinformatic analysis and interpretation, immunoblotting and drafting the manuscript. JJ assisted in immunoblotting and drafting the manuscript. LK participated in sequence preparation, sequencing, bioinformatics and drafting the manuscript. JH participated in bioinformatic analysis. SM provided the antibody raised against gamma conglutin. KS conceived the study, participated in the design and co-ordination and drafting the manuscript. All authors read and approved the final manuscript.

Acknowledgments
We thank Hayley Casarotto and Elaine Smith for valuable technical assistance and Linda Tabe and Ling-Ling Gao for helpful comments on the manuscript. The lupin seeds were kindly provided by Dr Jon Clements, Department of Agriculture and Food, Western Australia. This research was supported by the European Marie Curie Research Program (FP7-PEOPLE-2011-IOF), under the Grant Ref. Number PIOF-GA-2011-301550 to JL and KS and by CSIRO to RF, LK, JH, SM and KS.

Author details
[1]CSIRO Agriculture Flagship, Centre for Environment and Life Sciences, Floreat, WA, Australia. [2]The UWA Institute of Agriculture, University of Western Australia, Crawley, WA, Australia. [3]Current address: Curtin Institute for Computation and Centre for Crop & Disease Management, Department of Environment & Agriculture, Curtin University, Bentley, Western Australia, Australia.

References

1. Graham PH, Vance CP. Legumes: Importance and constraints to greater use. Plant Physiol. 2003;131(3):872–7.
2. Belski R, Mori TA, Puddey IB, Sipsas S, Woodman RJ, Ackland TR, et al. Effects of lupin-enriched foods on body composition and cardiovascular disease risk factors: a 12-month randomized controlled weight loss trial. Int J Obes. 2010;35(6):810–9.
3. Duranti M, Morazzoni P. Nutraceutical properties of lupin seed proteins A great potential still waiting for full exploitation. Agro Food Ind Hi-Tech. 2011;22(1):20–3.
4. Magni C, Sessa F, Accardo E, Vanoni M, Morazzoni P, Scarafoni A, et al. Conglutin gamma, a lupin seed protein, binds insulin in vitro and reduces plasma glucose levels of hyperglycemic rats. J of Nutritional Biochem. 2004;15(11):646–50.
5. Siddique KM, Johansen C, Turner N, Jeuffroy M-H, Hashem A, Sakar D, et al. Innovations in agronomy for food legumes. A review. Agron Sustain Dev. 2012;32(1):45–64.
6. Gladstones JS. Lupins as crop plants. Field Crop Abstracts. 1970;23:123–48.
7. Berger JD, Clements JC, Nelson MN, Kamphuis LG, Singh KB, Buirchell B. The essential role of genetic resources in narrow-leafed lupin improvement. Crop & Pasture Science. 2013;64(4):361–73.
8. Duranti M, Cucchetti E, Cerletti P. Changes in composition and subunits in the storage proteins of germinating lupine seeds. J Agric Food Chem. 1984;32(3):490–3.
9. Duranti M, Restani P, Poniatowska M, Cerletti P. The seed globulins of *Lupinus albus*. Phytochem. 1981;20(9):2071–5.
10. Foley RC, Gao LL, Spriggs A, Soo LYC, Goggin DE, Smith PMC, et al. Identification and characterisation of seed storage protein transcripts from *Lupinus angustifolius*. BMC Plant Biol. 2011;11:59.
11. Lopez-Pedrouso M, Alonso J, Zapata C. Evidence for phosphorylation of the major seed storage protein of the common bean and its phosphorylation-dependent degradation during germination. Plant Molecular Biology. 2014;84(4-5):415–28.
12. dosRamos PCR, Ferreira R, Franco E, Teixeira ARN. Accumulation of a lectin-like breakdown product of beta-conglutin catabolism in cotyledons of germinating *Lupinus albus* L seeds. Planta. 1997;203(1):26–34.
13. Duranti M, Gius C, Scarafoni A. Lectin-like activity of lupin seed conglutin-gamma, a gylcoprotein previously referred to as a storage protein. J Exp Bot. 1995;46(287):725–8.
14. Scarafoni A, Ronchi A, Duranti M. gamma-Conglutin, the Lupinus albus XEGIP-like protein, whose expression is elicited by chitosan, lacks of the typical inhibitory activity against GH12 endo-glucanases. Phytochem. 2010;71(2-3):142–8.
15. Gayler KR, Wachsmann F, Kolivas S, Nott R, Johnson ED. Isolation and Characterisation of Protein Bodies in Lupius angustifolius. Plant Physiol. 1989;91(4):1425–31.
16. Ogura T, Hernandez A, Aizawa T, Ogihara J, Sunairi M, Alcaino J, et al. Identification of a Low Digestibility delta-Conglutin in Yellow Lupin (*Lupinus luteus* L.) Seed Meal for Atlantic Salmon (*Salmo salar* L.) by Coupling 2D-PAGE and Mass Spectrometry. PLoS One. 2013;8:11.
17. Gao LL, Hane J, Kamphuis LG, Foley R, Shi B-J, Atkins CA, et al. Development of genomic resources for the narrow-leafed lupin (*Lupinus angustifolius*): construction of a Bacterial Artificial Chromosome (BAC) library and BAC-end sequencing. BMC Genomics. 2011;12:521.
18. Ksiazkiewicz M, Wyrwa K, Szczepaniak A, Rychel S, Majcherkiewicz K, Przysiecka L, et al. Comparative genomics of *Lupinus angustifolius* gene-rich regions: BAC library exploration, genetic mapping and cytogenetics. BMC Genomics. 2013;14:79.
19. Nelson MN, Moolhuijzen PM, Boersma JG, Chudy M, Lesniewska K, Bellgard M, et al. Aligning a New Reference Genetic Map of *Lupinus angustifolius* with the Genome Sequence of the Model Legume, Lotus japonicus. DNA Research. 2010;17(2):73–83.
20. Tian L, Peel GJ, Lei ZT, Aziz N, Dai X, He J, et al. Transcript and proteomic analysis of developing white lupin (*Lupinus albus* L.) roots. BMC Plant Biol. 2009;9:1.
21. Yang HA, Tao Y, Zheng ZQ, Shao D, Li ZZ, Sweetingham MW, et al. Rapid development of molecular markers by next-generation sequencing linked to a gene conferring phomopsis stem blight disease resistance for marker-assisted selection in lupin (*Lupinus angustifolius* L.) breeding. Theor Appl Genet. 2013;126(2):511–22.
22. Yang HA, Tao Y, Zheng ZQ, Zhang QS, Zhou GF, Sweetingham MW, et al. Draft genome sequence, and a sequence-defined genetic linkage map of the legume crop species *Lupinus angustifolius* L. PLoS One. 2013;8:5.
23. Kamphuis LG, Hane JK, Nelson MN, Gao L, Atkins CA, Singh KB. Transcriptome sequencing of different narrow-leafed lupin tissue types provides a comprehensive uni-gene assembly and extensive gene-based molecular markers. Plant Biotech J. 2015;13(1):14–25.
24. Parra-Gonzalez LB, Aravena-Abarzua GA, Navarro-Navarro CS, Udall J, Maughan J, Peterson LM, et al. Yellow lupin (*Lupinus luteus* L.) transcriptome sequencing: molecular marker development and comparative studies. BMC Genomics. 2012;13:425.
25. Ogura T, Ogihara J, Sunairi M, Takeishi H, Aizawa T, Olivos-Trujillo MR, et al. Proteomic characterization of seeds from yellow lupin (*Lupinus luteus* L.). Proteomics. 2014;14(12):1543–6.
26. Ainouche A-K, Bayer RJ. Phylogenetic Relationships in Lupinus (Fabaceae: Papilionoideae) based on Internal Transcribed Spacer Sequences (ITS) of nuclear ribosomal DNA. Am J of Bot. 1999;86(4):590–607.
27. Islam S, Ma W, Appels R, Yan G. Mass spectrometric fingerprints of seed protein for defining Lupinus spp. relationships. Genetic Resources and Crop Evolution. 2013;60(3):939–52.
28. Mortazavi A, Williams BA, McCue K, Schaeffer L, Wold B. Mapping and quantifying mammalian transcriptomes by RNA-Seq. Nature Methods. 2008;5(7):621–8.
29. Wagner GP, Kin K, Lynch VJ. Measurement of mRNA abundance using RNA-seq data: RPKM measure is inconsistent among samples. Theory in Biosci. 2012;131(4):281–5.
30. Tabe L, Hagan N, Higgins TJV. Plasticity of seed protein composition in response to nitrogen and sulfur availability. Curr Opin Plant Biol. 2002;5(3):212–7.
31. Bourgeois M, Jacquin F, Savois V, Sommerer N, Labas V, Henry C, et al. Dissecting the proteome of pea mature seeds reveals the phenotypic plasticity of seed protein composition. Proteomics. 2009;9(2):254–71.
32. Verdier J, Torres-Jerez I, Wang MY, Andriankaja A, Allen SN, He J, et al. Establishment of the Lotus japonicus Gene Expression Atlas (LjGEA) and its use to explore legume seed maturation. Plant J. 2013;74(2):351–62.
33. Magni C, Scarafoni A, Herndl A, Sessa F, Prinsi B, Espen L, et al. Combined 2D electrophoretic approaches for the study of white lupin mature seed storage proteome. Phytochem. 2007;68(7):997–1007.
34. Duranti M, Consonni A, Magni C, Sessa F, Scarafoni A. The major proteins of lupin seed: Characterisation and molecular properties for use as functional and nutraceutical ingredients. Trends in Food Science & Technology. 2008;19(12):624–33.
35. Schmidt MA, Barbazuk WB, Sandford M, May G, Song ZH, Zhou WX, et al. Silencing of soybean seed storage proteins results in a rebalanced protein composition preserving seed protein content without major collateral changes in the metabolome and transcriptome. Plant Physiology. 2011;156(1):330–45.
36. Kim WS, Gillman JD, Krishnan HB. Identification of a plant introduction soybean line with genetic lesions affecting two distinct glycinin subunits and evaluation of impacts on protein content and composition. Mol Breed. 2013;32(2):291–8.

37. Sosulski FW, Imafidon GI. Amino-acid-composition and nitrogen-to-protein conversion factors for animial and plant foods. J Agricl and Food Chem. 1990;38(6):1351–6.
38. AOAC. Official Methods of Analysis of AOAC International. 17th ed. Suite 500, 481 North Fredric Ave Gaithersburg, Maryland 20877-2417 USA: AOAC Int; 2000.
39. Martin M. Cutadapt removes adapter sequences from high-throughput sequencing reads. EMBnetjournal. 2011;17(1):10–2.
40. Foley RC, Singh KB. Western and south-western library screens. In: Molecular plant biology: A practical approach Volume One. 2002. p. 205–21.

Wheat homologs of yeast ATG6 function in autophagy and are implicated in powdery mildew immunity

Jieyu Yue[†], Hong Sun[†], Wei Zhang, Dan Pei, Yang He and Huazhong Wang[*]

Abstract

Background: Autophagy-related ATG6 proteins are pleiotropic proteins functioning in autophagy and the phosphatidylinositol 3-phosphate-signaling pathways. *Arabidopsis* ATG6 regulates normal plant growth, pollen development and germination, and plant responses to biotic/abiotic stresses. However, the ATG6 functions in wheat (*Triticum aestivum* L.), an important food crop, are lacking.

Results: We identified three members, *TaATG6a-6c*, of the *ATG6* family from common wheat. *TaATG6a*, *6b* and *6c* were localized on homeologous chromosomes 3DL, 3BL and 3AL, respectively, of the allo-hexaploid wheat genome, and evidence was provided for their essential role in autophagy. The TaATG6a-GFP fusion protein was found in punctate pre-autophagosomal structures. The expression of each *TaATG6* gene restored the accumulation of autophagic bodies in *atg6*-mutant yeast. Additionally, *TaATG6* knockdown plants showed impaired constitutive and pathogen-induced autophagy and growth abnormalities under normal conditions. We also examined the expression patterns of wheat *ATG6*s for clues to their physiological roles, and found that their expression was induced by the fungus *Blumeria graminis* f. sp. *tritici* (*Bgt*), the causal agent of powdery mildew, and by abiotic stress factors. A role for TaATG6s in wheat immunity to powdery mildew was further implied when knockdowns of *TaATG6*s weakly compromised the broad-spectrum powdery mildew resistance gene *Pm21*-triggered resistance response and, conversely and significantly, enhanced the basal resistance of susceptible plants. In addition, leaf cell death was sometimes induced by growth-retarded small *Bgt* mycelia on susceptible *TaATG6* knockdown plants after a long period of interaction. Thus, we provide an important extension of the previous characterization of plant ATG6 genes in wheat, and observed a role for autophagy genes in wheat immune responses to fungal pathogens.

Conclusions: Three wheat ATG6s were identified and shown to be essential for autophagy biogenesis. Wheat ATG6s are implicated in immunity to powdery mildew, playing a weak, positive role in the *Pm21*-triggered resistance response and a negative role in the basal resistance of susceptible plants.

Keywords: Autophagy, ATG6, Common wheat (*Triticum aestivum* L.), Powdery mildew

Background

Unlike animals, plants are sessile and must overcome or endure variable, sometimes severe environmental conditions. For example, plants have evolved two main defense mechanisms, pathogen-associated molecular pattern (PAMP)-triggered immunity (PTI) and effecter-triggered immunity (ETI), to combat pathogen infections [1,2]. ETI is a much stronger reaction than PTI and is often accompanied by a localized hypersensitive response (HR), a type of programmed cell death (PCD), to capture the pathogen at the site of infection. Autophagy is an evolutionarily conserved, eukaryotic process by which organelles and cytosolic macromolecules are consumed in lysosomes (vacuoles in yeast and plants) for nutrient recycling [3-5]. Recent evidence from autophagy-defective *Arabidopsis* plants suggests an important role for autophagy in regulating plant immune responses [6,7]. This role, however, remains unclear since both positive and negative effects of autophagy on plant immunity have been described in

* Correspondence: skywhz@mail.tjnu.edu.cn
[†]Equal contributors
Tianjin Key Laboratory of Animal and Plant Resistance, School of Life Sciences, Tianjin Normal University, Tianjin 300387, China

different *Arabidopsis*-pathogen systems as well as in the same *Arabidopsis*-pathogen system [8,9]. Likewise, both pro-survival and pro-death roles of autophagy in the regulation of HR-PCD or pathogen-induced cell death have been reported [8-10]. Therefore, the mechanisms behind the relationship between autophagy and plant immunity need to be elucidated in detail when considering plant species other than *Arabidopsis* and more plant-pathogen systems.

During autophagy, double-membrane structures called autophagosomes facilitate the transport of cellular cargos into lysosomes for degradation [3-5]. To date, more than 30 autophagy-related proteins (ATGs) functioning in autophagy induction and regulation, the initiation and biogenesis of autophagosomes, and the fusion of autophagosomes with lysosomes have been identified in yeast and some orthologs identified in animals and plants [3-5]. In yeast, a Class III phosphatidylinositol 3-kinase (PI3K) complex resides in the pre-autophagosomal structure/phagophore assembly site (PAS) where it catalyzes phosphatidylinositol 3-phosphate (PI3P) synthesis and recruits PI3P-binding proteins, especially the ATG18-ATG2 complex, for the initiation of autophagic membranes [11-13]. Yeast ATG6/vacuolar protein sorting 30 (VPS30), the ortholog of mammalian Beclin 1, is the core component of the PI3K complex along with VPS34, VPS15 and ATG14 [14,15] and is essential for auophagy. Plant homologs of yeast ATG6 play similar essential roles in autophagy, for ATG6-deficient plants produce a reduced number of autolysosomes under autophagy-inducing conditions [16,17], and the expression of plant *ATG6* restores autophagy in *atg6*-mutant yeast [18].

ATG6-deficient plants, like mutants of other *atg* genes, show an enhanced sensitivity to nutrient starvation and other stress factors [17,19-22]. This enhanced sensitivity may result from impaired autophagy because autophagy is responsible for nutrient recycling and distribution under severe environmental conditions. However, the ATG6-associated PI3K complex is not autophagy-specific. Its phospholipid product, PI3P, is a well-known second messenger involved in receptor signaling and protein sorting-related vesicle trafficking events [23,24]. Homozygous plant mutants of *atg6*, *vps34* or *vps15* cannot be obtained because of the requirement of PI3K in pollen development, germination and pollen tube growth [25-27]. Such effects may be attributed to the PI3P-, or its derivative PI(3,5)P2-, signaling pathways rather than to autophagy because homozygous mutants of the autophagy-specific *atg* genes can be routinely obtained by selfing heterozygous individuals. ATG6-deficient plants also show more severe growth abnormalities under normal conditions than mutants of the autophagy-specific *atg* genes, although they have similarly reduced autophagy levels [17,22,25,28]. ATG6, like autophagy-specific factors, is implicated in plant immune responses [16,17]. Yet the pleiotropy of ATG6 makes it difficult to interpret whether autophagy, the PI3P-signaling pathways, or their interplay underlines the roles of ATG6 in plant immunity. Well-established roles of ATG6, other than autophagy, also include its involvement in the vacuolar protein sorting pathway of yeast and in tumor suppression through interactions with the anti-apoptotic Bcl-2-like proteins in mammals [14,29].

So far, few studies on the mechanisms and physiological functions of autophagy and other ATG6-associated pathways have been reported in crop plants, and fewer still in common wheat (*Triticum aestivum* L.), which is an important food crop worldwide. Here, we report the identification of three members of the wheat ATG6 family. The conserved function of wheat ATG6s in autophagy was shown using yeast functional complementation, subcellular localization of green fluorescent protein (GFP) fusions and the determination of autophagy levels in *ATG6* knockdown plants. A role for ATG6 in the wheat immune response to the causal fungus, *Blumeria graminis* f. sp. *tritici* (*Bgt*), of powdery mildew was established by *Bgt*-induced *ATG6* expression profiles and by altered powdery mildew symptoms on *ATG6* knockdown plants.

Methods

Plant materials and fungal strain

Wheat used in this study included isogenic lines 92R137/Yangmai 158[7] and Yangmai 158, and isogenic lines Michigan Amber/Chancellor[8] and Chancellor. 92R137/Yangmai 158[7] carries the broad-spectrum powdery mildew resistance gene *Pm21*. Michigan Amber/Chancellor[8] carries the isolate-specific powdery mildew resistance gene *Pm3f*. Yangmai 158 and Chancellor are susceptible to powdery mildew.

The prevalent Chinese *Bgt* isolate E09, which is avirulent to *Pm21* and *Pm3f*, was maintained on seedlings of the susceptible wheat cultivar Sumai 3 under spore-proof conditions.

Plant growth conditions and treatments

All wheat plants were cultured hydroponically in one-fifth strength Hoagland's solution under normal conditions of 22°C/18°C (day/night) and a photoperiod of 16 h light/8 h dark in controlled climate chambers. Two-leaf stage seedlings were subjected to different treatments. Pathogen inoculation was conducted by vigorously shaking off fresh *Bgt* conidia from diseased Sumai 3 plants onto experimental seedlings. For abiotic stress treatments, seedlings were transferred to the same one-fifth strength Hoagland's solution containing 200 mM NaCl (high salinity) or 16% PEG6000 (drought). Additional abiotic stress treatments were nitrogen deprivation, and low temperature combined with darkness, in which seedlings were transferred to a 4°C refrigerator and kept in the dark. Leaves harvested at defined time points were frozen with liquid nitrogen and then stored at −80°C. These leaf samples were used for gene isolation and expression analyses.

Isolation of wheat ATG6 genes and expression analyses

Total RNA was prepared from each harvested leaf sample using TRIzol (Invitrogen, Life Technologies, CA, USA) and subjected to RNase-free DNase I digestion and purification. First-strand cDNA was synthesized using oligo(dT) primers and the Quantscript RT Kit (Tiangen, Beijing, China).

The protein sequence of *Arabidopsis ATG6* (At3g61710) was used as the input to search for wheat expressed sequence tags (ESTs) in the dbEST division of GenBank by TBLASTN. Both 5′- and 3′-terminal ESTs were iteratively used as queries to search for new terminal ESTs to cover full-length open reading frames (ORFs). EST contigs were assembled in Vector NTI Advance 11.5, and their sequences were used to design primers (Additional file 2: Table S1) flanking full-length ORFs. ATG6 genes were amplified from first-strand cDNA using the high-fidelity Pfu DNA polymerase. Amplified products were A-tailed and then cloned into the vector pGEM®-T Easy (Promega, Madison, WI, USA) for sequencing. Sequence data were deposited in the GenBank database.

The expression patterns of wheat *ATG6*s responding to biotic and abiotic treatments were quantified by quantitative real-time PCR (qRT-PCR) with gene-specific primers (Additional file 2: Table S1). qRT-PCR was conducted in a 20 μl RealMasterMix (SYBR Green) mixture (Tiangen, Beijing, China) containing 1 μl first-strand cDNA as the template, and the reaction was performed on an IQ5 (BioRad, Hercules, CA, USA). Each cDNA sample was analyzed in triplicate. Relative expression levels were calculated using the $2^{-\Delta\Delta CT}$ method [30] with the amplicon of the wheat β-tubulin gene as an internal control.

Chromosomal localization of wheat *ATG6*s

Wheat nullisomic-tetrasomic lines (NTLs) were used to map ATG6 genes on specific chromosomes. DNA of each line was isolated and used as the template for PCR amplification with gene-specific primers (Additional file 2: Table S1). Amplified products were analyzed by 12% polyacrylamide gel electrophoresis. Chromosome localization was inferred based on the absence of amplified fragments from lines lacking specific chromosomes.

Complementation test of yeast *atg6* mutant

The ORF of each wheat *ATG6* cDNA was amplified as a *Not*I- *Not*I fragment using the primers listed in Additional file 2: Table S1. Amplified products were digested and cloned into the yeast expression vector pFL61, in which the constitutive expression of the inserted gene was driven by the phosphoglycinecerate kinase (PGK) gene promoter. Constructed vectors were sequenced to confirm the correct cDNA insertion direction. Wild-type yeast strain BY4741 and a mutant strain, *atg6* (BY4741, *atg6Δ::kanMX MATa his3Δ1 leu2Δ0 met15Δ0 ura3Δ0*), were purchased from Open Biosystems (Thermo Scientific, Wilmington, DE, USA). Each constructed vector was introduced into yeast *atg6* cells according to the LiAc/SS-DNA/PEG TRAFO protocol [31]. Positive transformants were screened on SC-U medium. The yeast autophagy complementation test was performed as previously described [18]. The accumulation of autophagic bodies in vacuoles was observed under a differential interference contrast microscope (DM5000B, Leica).

Subcellular localization of wheat ATG6s

The ORF of each wheat *ATG6* cDNA lacking a stop codon was amplified as an *Xho*I-*Spe*I fragment using primers listed in Additional file 2: Table S1. Amplified products were digested, and ligated upstream and in-frame with the ORF of *GFP* in the vector pA7-GFP, in which the *GFP* fusion construct's expression was driven by two tandem 35S promoters. Prepared plasmids were introduced into onion epidermal cells by particle bombardment with PDS1000/He (BioRad, Hercules, CA, USA). GFP fluorescence was visualized by epifluorescence microscopy (DM5000B, Leica) 24 h after bombardment.

Virus-induced gene silencing (VIGS)

A barley stripe mosaic virus (BSMV)-based VIGS method [32,33] was used to create gene knockdown plants. To maximize knockdown efficiency, three $BSMV_{\gamma}$ vectors, each containing a 200–241 bp fragment from the upstream, middle, or downstream region of the *TaATG6a* coding sequence, were constructed, and *in vitro* transcribed RNA_{γ} molecules from these vectors were premixed for virus infection. To circumvent the potential functional redundancy among paralogs, we intended to produce simultaneous knockdowns of wheat *ATG6* members by choosing fragments from *TaATG6a* for the construction of $BSMV_{\gamma}$ vectors with 94.1–99.5% sequence identities to the corresponding fragments of *TaATG6b* and *6c* and multiple nucleotide stretches longer than 23 bp that are 100% identical among the three *ATG6* members. The three fragments were prepared by PCR using the primers containing *Nhe*I sites listed in Additional file 2: Table S1. Each amplified fragment was digested by *Nhe*I and then inserted in the reverse orientation into the vector $BSMV_{\gamma}$ to produce the vectors $BSMV_{\gamma:TaATG6as}$-1, $BSMV_{\gamma:TaATG6as}$-2 and $BSMV_{\gamma:TaATG6as}$-3. Plasmids of these three BSMVγ derivatives, as well as $BSMV_{\gamma:GFP}$ and $BSMV_{\alpha}$, were linearized by *Mlu*I, and $BSMV_{\beta}$ was linearized by *Spe*I. Linearized vectors were *in vitro* transcribed to produce 5′-capped infectious BSMV RNA molecules using the RiboMAX Large Scale RNA Production-T7 kit (Promega, Madison, WI, USA) with the addition of a cap analog (Promega, Madison, WI, USA) in the transcription mixture. $RNA_{\gamma:TaATG6as}$-1, $RNA_{\gamma:TaATG6as}$-2, and $RNA_{\gamma:TaATG6as}$-3 were premixed in a 1:1:1 ratio to produce $RNA_{\gamma:TaATG6as}$.

The second fully expanded leaves of wheat seedlings at the two-leaf stage were mechanically infected with a 1:1:1 mixture of RNA_{α}, RNA_{β} and $RNA_{\gamma:TaATG6as}$ (BSMV:ATG6 treatment) in 1× GKP buffer [33]. Seedlings infected with a 1:1:1 mixture of RNA_{α}, RNA_{β} and $RNA_{\gamma:GFP}$ (BSMV: GFP treatment) in 1× GKP buffer were used as controls. Three replications of the VIGS experiment were conducted on the two isogenic lines of 92R137/Yangmai 158[7] and Yangmai 158. In each replication, a total of 80 seedlings per line were used for each of the BSMV:ATG6 and BSMV:GFP treatments. The BSMV-treated plants were maintained on one-fifth strength Hoagland's solution under normal conditions. The success of the virus infection was judged by visual observations of mild chlorotic mosaic symptoms on the third and upper leaves of BSMV-treated plants. To confirm the knockdown efficiency, eight plants with virus symptoms were randomly selected from each of the two BSMV treatments, and their transcript levels of *TaATG6*s in the fourth leaves were determined using qRT-PCR as described above with a primer pair (Additional file 2: Table S1) designed to conserve regions in the three wheat *ATG6*s.

Determination of autophagy level by LysoTracker Red staining

BSMV-treated plants at the four-leaf stage were inoculated with *Bgt* conidia as described above. The fourth leaves were excised at 0 and 7 days after inoculation (dai) with *Bgt*, infiltrated with 100 μM E-64d (Sigma, St. Louis, MO, USA) and then stained with 2 μM LysoTracker Red DND-99 (Invitrogen, Life Technologies, CA, USA) according to the procedure described previously [16]. Fluorescence signals were observed under a confocal microscope (ECLIPSE 90i, Nikon) using a 543-nm excitation laser and 565 to 615-nm band pass.

Chlorophyll content determination

The fifth leaves of BSMV-treated plants at the five-leaf stage were excised, weighed and ground in 1 ml of 80% (v/v) acetone. The chlorophyll content was determined spectrophotometrically according to Patel and Dinesh-Kumar [17].

Leaf senescence assay

Green sections were excised from the fourth leaves of BSMV-treated plants at the four-leaf stage and floated on deionized water at room temperature in the dark. Leaf sections were photographed every day after dark incubation. The experiment was repeated in triplicate using leaves of three or more plants in each experiment.

Evaluation of powdery mildew resistance

To exclude the influence of early senescence/chlorosis on *Bgt* infections, only those BSMV:ATG6-treated plants without obvious symptoms of early senescence and chlorosis were selected for resistance evaluation. BSMV-treated plants at the six-leaf stage were inoculated with *Bgt* conidia as described above. For BSMV-treated Yangmai 158 plants, leaf discs with visible mildew colonies were gently harvested from different regions of the fourth leaves at 5 dai, immersed in 0.01% glycerol and subjected to an ultrasonic treatment for 1 min to disperse the conidia. The conidial production per mildew colony was counted by microscopic observation and expressed as the mean ± SD of at least 50 mildew colonies from five plants.

Fungal structures were stained by trypan blue [34] and visualized by light microscopy. For BSMV-treated 92R137/Yangmai 158[7] (*Pm21*) plants, the number of successfully penetrating conidium with secondary hyphae on the fourth leaves was counted at 7 dai and expressed as a percentage of at least 100 interaction sites on individual plants. The number of secondary hypha associated with each successfully penetrating conidium was also counted and expressed as the mean ± SD of the observed successfully penetrating conidia on all of the investigated plants. For BSMV-treated Yangmai 158 plants, the number of mycelium with mature chains of conidia on the fourth leaves was counted at 5 dai and expressed as a percentage of at least 100 mycelia on the leaves of five plants.

Statistical analysis

All experiments were repeated three times with similar results. Quantitative data were statistically analyzed using Student's *t* test ($P < 0.05$) using the IBM SPSS19.0 software package.

Results

Isolation of *ATG6* homologs from wheat

Using the sequence of *Arabidopsis* ATG6 as a query, 44 putative wheat ATG6-encoding ESTs were mined from the dbEST division of GenBank and assembled into contigs. Primers covering the full-length ORFs of these contigs were designed and used for gene isolation by PCR from a leaf cDNA pool of the 92R137/Yangmai 158[7] (*Pm21*) plants induced by *Bgt* for 48 h. Three genes, which encoded a typical ATG6 domain (Pfam PF04111) (Additional file 1: Figure S1A), were identified as *ATG6* homologs in wheat and designated as *TaATG6*a [GenBank: U49845], *6b* [GenBank: U49845] and *6c* [GenBank: U49845]. The ORF nucleotide sequences of

TaATG6a, *6b* and *6c* share identities of 96.1–98.4% and encode peptide sequences of 504, 500, and 503 amino acids, respectively, with 98–99% similarity. In the eukaryotic ATG6 phylogenetic tree (Additional file 1: Figure S1B), plant ATG6s are separately grouped into monocot and dicot clades. TaATG6a-6c showed 90–93% sequence similarities to OsATG6a-6c from rice and 75% to AtATG6 from *Arabidopsis*. The ATG6 domain region is more conserved across all plant ATG6s (80–89% similarity), whereas the N- and C-terminal flanking sequences show divergence between the monocot and dicot ATG6s (Additional file 1: Figure S1A).

Two methods, gene-specific PCR amplification of wheat NTL DNAs and BLAST searches against the mapped wheat genomic sequences (genome survey sequences and the completed reference sequence of chromosome 3B) [35], were performed to determine the chromosomal localizations of wheat *ATG6* members. *TaATG6b* and *6c* were localized on chromosomes 3BL and 3AL, respectively, by both methods (Figure 1). *TaATG6a* was localized on chromosome 3DL based only on the results of BLAST searches.

We also obtained the respective genomic sequences of *TaATG6a*, *6b* and *6c* by BLAST searches against wheat genomic sequences. Similar nine-intron structures of the *TaATG6a-6c* ORF sequences were revealed when comparing genomic sequences with the corresponding cDNA sequences (Additional file 1: Figure S2). However, *TaATG6b* has a much longer intron 3 than *TaATG6a* and *6c*. We compared the three genomic sequences and found a 900-bp insertion only in the intron 3 of *TaATG6b*, which is bordered by 19-bp direct short sequence repeats (TAGACTTAAATCATACTCC) (Additional file 1: Figure S2). As it encodes a truncated protein similar to RNA-directed DNA polymerases and polyproteins of mobile elements and because no long terminal repeats (LTR) were found around it, this inserted sequence most likely resulted from a transposition event of a non-LTR retrotransposon. A CpG island of 999 bp, covering the start codon, was detected in the *TaATG6b* genomic sequence, which has a 60.1% G + C content and a ratio of 1.07 of the observed versus expected presence of CpG dinucleotides (Figure 2A). This CpG island exists in other monocot *ATG6*s as well, including *OsATG6b* from rice, but not in dicot *ATG6*s, such as *AtATG6* from *Arabidopsis* (Figure 2A). Within the CpG island, there is a short GC-rich stretch just downstream of the start codon, and its length varies among paralogs as well as among homologs (Figure 2B). Only insertions or deletions that involve multiples of three bases are retained in the GC-rich stretch of monocot *ATG6*s because they do not result in translational frame shifts. Amino acids encoded in the GC-rich stretch are predominantly glycines and alanines.

Wheat ATG6s function in autophagy

Human and *Arabidopsis* ATG6 can rescue the autophagy activity of *atg6*-mutant yeast [16,18,36]. To determine if wheat ATG6s are functional homologs of yeast ATG6, yeast expression vectors were constructed for each *TaATG6* and transformed into *atg6*-mutant yeast cells. Autophagic bodies in vacuoles were monitored as an indicator of autophagy activity. After 5 h of nutrient starvation and treatment with the protease inhibitor phenylmethylsulfonyl fluoride, the wild-type yeast cells accumulated numerous autophagic bodies in vacuoles, which were rarely observed in the *atg6*-mutant yeast cells (Figure 3A). Cells of *atg6*-mutant yeast transformed with *TaATG6a*, *6b* or *6c* partially restored the accumulation of autophagic bodies (Figure 3A).

Punctate structures of ATG6-GFP fluorescence, considered as PAS for autophagosome initiation, have been observed in cells of yeast and *Arabidopsis* [18,37]. Here, *GFP* fusion constructs of wheat *ATG6*s were introduced into onion epidermal cells by particle bombardment. Punctate florescence signals were clearly observed in the cytoplasm of cells expressing TaATG6a-GFP and seemed not to occur in the central vacuole (Figure 3B). Fluorescent spots of TaATG6a-GFP were sometimes clustered at cytoplasmic and perinuclear membrane-like structures resembling the trans-Golgi network and endoplasmic reticulum (Figure 3C). Localizations of ATG6s to endosomes, the trans-Golgi network, endoplasmic reticulum,

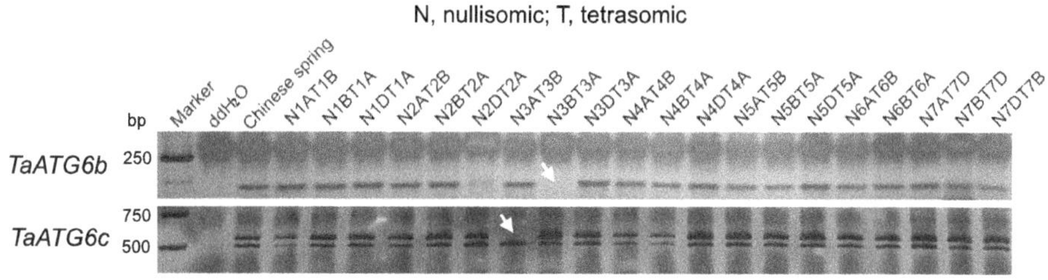

Figure 1 Chromosomal localization of wheat autophagy-related ATG6 genes. DNA of each wheat nullisomic-tetrasomic line was used as a template for PCR amplification with gene-specific primers. Amplified products were analyzed on 12% polyacrylamide gels. Chromosome localization was inferred based on the absence of amplified fragments (arrows) in lines lacking the specific chromosomes.

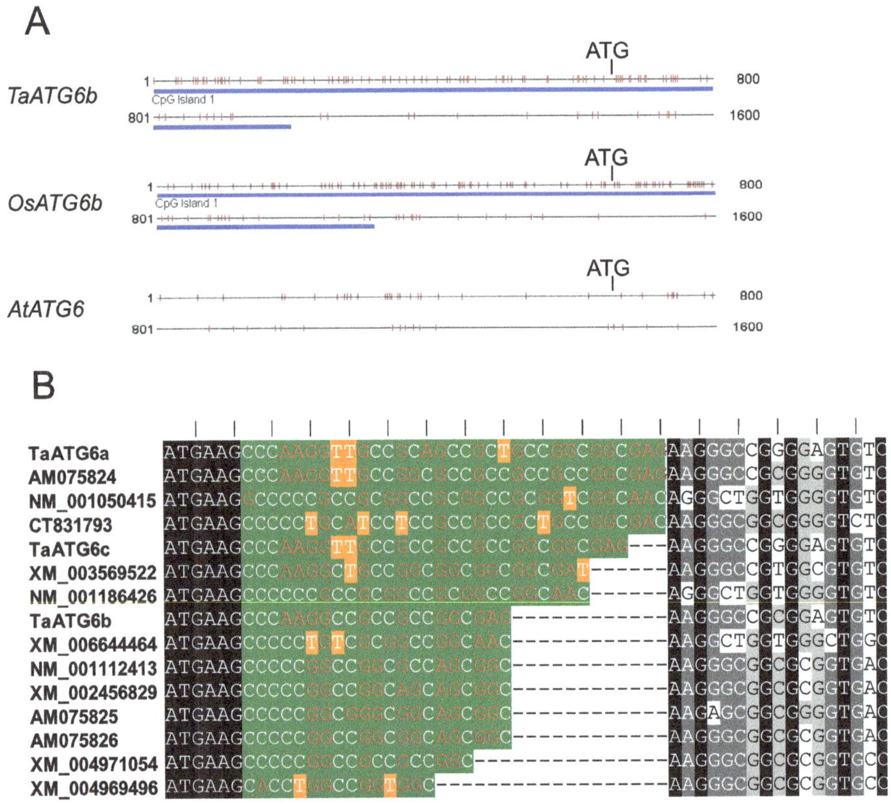

Figure 2 Characterization of wheat autophagy-related *ATG6* sequences. (A) The genomic sequences of monocot *ATG6*s contain a CpG island covering the start codon. The CpG island was predicted using the web tool CpG Island Searcher (http://cpgislands.usc.edu/) with a CpG island standard of longer than 500 bp in length, a minimum G + C content of 55% and a minimum 0.65 ratio of the observed versus the expected presence of CpG dinucleotides. Gray, horizontal lines represent the 1600-bp start codon-covering genomic sequences of *TaATG6b* from wheat, *OsATG6b* (Os03g0644000) from rice and *AtATG6* (At3g61710) from *Arabidopsis*. Short vertical red lines represent CpG dinucleotides. Blue horizontal thick lines represent the ranges of predicted CpG islands. **(B)** Short GC-rich stretch just downstream of the start codon in monocot *ATG6* sequences. Sequences shown are the 5′-terminal cDNA sequences beginning with the start codons of monocot *ATG6*s, including *TaATG6a*, *6b* and *6c* from wheat, AM075824 from *Hordeum vulgare*, NM_001050415, CT831793 and NM_001186426 from rice, XM_003569522 from *Brachypodium distachyon*, XM_006644464 from *Oryza brachyantha*, NM_001112413 from *Zea mays*, XM_002456829 and AM075825 from *Sorghum bicolor*, AM075826 from *Saccharum officinarum*, and XM_004971054 and XM_004969496 from *Setaria italic*. The GC-rich stretches (green background) are not properly aligned but the flanking sequences are aligned. The reading frames from ATG are indicated by the upper vertical lines.

mitochondria, cytoplasm, and nucleus have been reported in yeast and human [37-40]. Unexpectedly, TaATG6b- and TaATG6c-GFP florescence showed diffused distributions in the cytoplasm and nuclei, which were almost indistinguishable from those in cells expressing GFP alone (Figure 3B). Therefore, additional signals may be required to trigger the translocation of TaATG6b and 6c to PAS for their role in autophagy initiation.

To determine if an ATG6 deficiency can lead to autophagy impairment in wheat, BSMV:ATG6-treated *TaATG6* knockdown plants were generated by the VIGS method and their autophagy levels were compared with BSMV:GFP-treated control plants. To obtain the maximum knockdown efficiency, BSMV:ATG6 was designed to contain a mix of three BSMV RNA$_{\gamma}$ molecules, each containing an inserted antisense fragment corresponding to the upstream, middle, and downstream regions of the *TaATG6a* coding sequence. Taking into consideration the potential functional redundancy among paralogs in polyploid wheat, BSMV:ATG6 was also designed to knock down all three wheat *ATG6* members by inserting three fragments in the RNA$_{\gamma}$ molecules that are highly conserved (more than 94.1% identical) among the three genes. As determined by qRT-PCR using a primer pair designed based on identical regions of the three *TaATG6*s, the total constitutive transcript levels of *TaATG6a-6c* were greatly reduced in the BSMV:ATG6-treated plants compared with the transcript levels in BSMV:GFP-treated control plants (Figure 3D). This high knockdown efficiency was achieved in all eight BSMV:ATG6-treated plants investigated, which confirmed the effectiveness of our VIGS design. We then inoculated BSMV-treated plants with *Bgt* conidia and determined their constitutive autophagy activity at 0 dai and

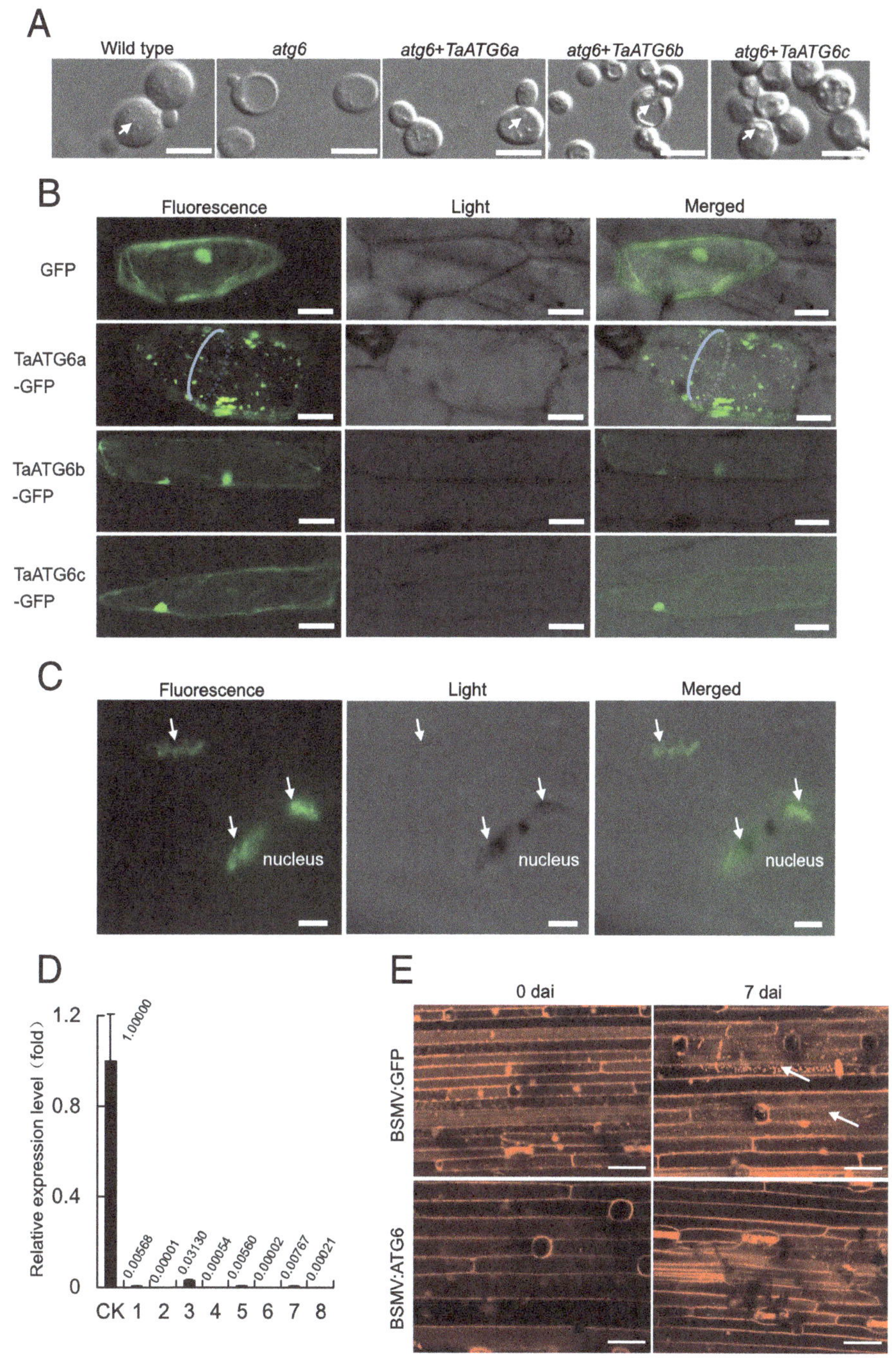

Figure 3 Autophagy-related ATG6 genes function in wheat autophagy. (A) Functional complementation of *atg6*-mutant yeast cells by wheat *ATG6*s. The autophagic bodies accumulated within vacuoles are indicated by arrows. Scale bars represent 5 μm. **(B)** Subcellular localization of GFP-TaATG6 fusion structures in onion epidermal cells. Scale bars represent 50 μm. **(C)** Localization of GFP-TaATG6a in cytoplasmic and perinuclear membrane-like structures in onion epidermal cells. Florescence spots localized in cytoplasmic and perinuclear membrane-like structures are indicated by arrows. Scale bars represent 10 μm. **(D)** Relative transcript levels of wheat *ATG6*s in leaves of barley stripe mosaic virus (BSMV)-treated plants. The fourth leaves were sampled from BSMV-treated 92R137/Yangmai 158[7] (*Pm21*) plants under normal conditions and used in the quantification of *TaATG6* transcripts by qRT-PCR with a primer pair designed based on identical regions of *TaATG6a*, *6b*, and *6c*. Amplification of the wheat β-tubulin gene served as an internal control. CK is from BSMV: GFP-treated control plants and Arabic numbers indicate individual BSMV:ATG6-treated plants. **(E)** Constitutive and *Blumeria graminis* f. sp. *tritici* (*Bgt*)-induced autophagy are impaired in BSMV:ATG6-treated 92R137/Yangmai 158[7] (*Pm21*) plants. The fourth leaves were excised from BSMV-treated plants at 0 and 7 days after inoculation (dai) with *Bgt* conidia, infiltrated with E-64d and then stained with LysoTracker Red. LysoTracker Red-stained punctate autolysosome-like structures are indicated by arrows. The results were reproduced from three independent experiments using three or more plants in each experiment. Scale bars represent 5 μm.

Bgt-induced autophagy activity at 7 dai using the LysoTracker Red staining method. LysoTracker Red is a dye that stains acidic organelles, such as lysosomes and endosomes, as well as autophagosomes/autolysosomes (vacuoles and small lytic organelles). Punctate structures labeled by this dye are generally considered indicative of plant autophagy activity [16,41-44]. Before LysoTracker Red staining, leaf samples were pre-infiltrated with the cysteine protease inhibitor E-64d to enhance the accumulation of autophagic bodies inside autolysosomes [16,45,46]. For the BSMV:GFP-treated control plants of the resistant line 92R137/Yangmai 158^7 (*Pm21*), low levels of a LysoTracker Red-stained punctate pattern were observed at 0 dai, and enhanced levels were observed at 7 dai in *Bgt*-challenged leaf epidermal cells (Figure 3E). These *Bgt*-enhanced levels implied that the TaATG6s-regulating autophagy process was induced and implicated in the *Pm21*-triggered immune response to powdery mildew. However, because of the knocking down of *TaATG6*s, the LysoTracker Red-stained punctate structures were not visible in leaf samples of BSMV:ATG6-treated plants regardless of *Bgt* challenge (Figure 3E). This result of impaired autophagy in *TaATG6* knockdown plants, together with the complementation of yeast autophagy-defective cells with each *TaATG6* and the observation of PAS localization of TaATG6a-GFP, demonstrates the essential role of TaATG6s in wheat autophagy.

Wheat *ATG6* knockdown plants exhibited growth abnormalities

BSMV:ATG6-treated plants, when compared with control plants, showed varying degrees of growth defects, including stunted growth and chlorosis under normal conditions (Figure 4A–C). Some were thin and weak, and a higher percentage of BSMV:ATG6-treated plants shifted into early senescence and accelerated death before the five-leaf stage than of the controls (Figure 4D). To further monitor senescence, green leaf sections from plants at the same developmental stage were excised and kept floating on water in the dark to induce senescence. Leaf sections from BSMV:ATG6-treated plants began senescing within 3 d post-detachment, as indicated by yellowing (Figure 4E). This was significantly earlier than leaf sections from control plants, which began senescing by 7 d post-detachment. Thus, TaATG6s are essential for normal plant growth and to prevent premature senescence in wheat.

Expression of wheat *ATG6s* in response to abiotic stress factors and fungal infection

Yangmai 158 plants were studied using qRT-PCR to determine the effects of abiotic stress factors on the expression of wheat *ATG6*s. The expression levels of *TaATG6b* and *6c* were upregulated by high salinity, drought, low temperature/darkness and nitrogen deprivation (Figure 5A). The expression of *TaATG6a* was also upregulated by high salinity, drought and nitrogen deprivation, but downregulated by low temperature/darkness (Figure 5A). Among the four investigated abiotic stress factors, wheat *ATG6*s were most sensitive to the two osmotic stress factors, high salinity and drought, with up to 63-fold (*TaATG6a* responding to high salinity) and 230-fold (*TaATG6a* responding to drought) increases in the accumulation of transcripts. These upregulated expression modes indicate that TaATG6s are closely related to the wheat's responses to abiotic stress factors.

Two powdery mildew resistant lines, 92R137/Yangmai 158^7 (*Pm21*) and Michigan Amber/Chancellor^8 (*Pm3f*), and their respective isogenic susceptible lines were included in the expression analysis, using qRT-PCR, of wheat *ATG6*s responding to *Bgt* infection. In the broad-spectrum resistance gene *Pm21*- and the isolate-specific resistance gene *Pm3f*-triggered resistance responses, the expression of wheat *ATG6*s was upregulated by *Bgt*, producing bimodal profiles over a time interval of 0 to 36 h after inoculation (hai) with *Bgt* (Figure 5B). The two peaks of transcript accumulation occurred at 10 and 24 hai. Upon *Bgt* challenge, *Pm21* activated the three *ATG6*s to about the same extent, while *Pm3f* activated *TaATG6a* to a greater extent than *TaATG6b* and *6c*. In the susceptible response of Yangmai 158 to *Bgt*, however, a single instance of upregulated expression within 16 to 24 hai was detected for the three *ATG6*s (Figure 5B). This corresponded to, but was higher than, the second instance of upregulated expression detected in the two resistance responses. In the susceptible response of Chancellor to *Bgt*, only one member, *TaATG6a*, showed a *Bgt*-induced expression profile (Figure 5B). These *Bgt*-upregulated expression patterns implied that TaATG6s are involved in the wheat immune response to powdery mildew.

Wheat ATG6s play a weak, positive role in the *Pm21*-triggered resistance response to *Bgt*

We used *TaATG6* knockdown plants to test if TaATG6s have any effect on *Bgt* infection. To exclude the influence of early senescence/chlorosis on *Bgt* infection, only those knockdown plants that did not show obvious early senescence and chlorosis were selected for resistance evaluation. BSMV-treated plants were heavily inoculated with *Bgt* conidia, and disease symptoms were determined when mildew colonies were apparently visible on susceptible plant leaves. The knocking down of *TaATG6*s did not result in visible mildew colonies on leaves of BSMV:ATG6-treated 92R137/Yangmai 158^7 (*Pm21*) plants by 7 dai, which seemed to retain a resistant phenotype similar to BSMV:GFP-treated control plants. We then stained leaf samples with trypan blue and, under a microscope, counted the number of successfully penetrating *Bgt*

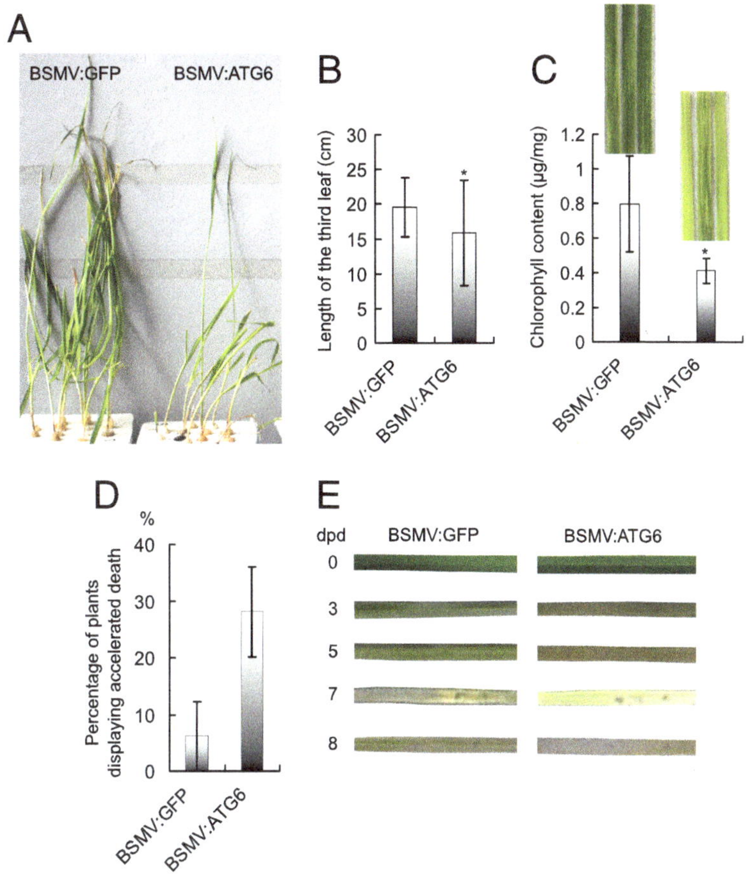

Figure 4 Wheat autophagy-related *ATG6* knockdown plants display growth abnormalities. All barley stripe mosaic virus (BSMV)-treated plants were cultured hydroponically in one-fifth strength Hoagland's solution and kept under normal conditions of 22°C/18°C (day/night) and a photoperiod of 16 h light/8 h dark in controlled climate chambers. **(A)** The growth morphology of BSMV-treated Yangmai 158 plants. Photographs were taken at the three-leaf stage. The second leaves, which had been mechanically infected with BSMV, were generally withered by this stage and thus were excised before photographing. **(B)** Leaf length of BSMV-treated Yangmai 158 plants. The lengths of the third leaves were measured at the three-leaf stage. Values represent the mean ± SD (standard deviation) of data measured from 30 individual plants. Asterisks indicate a significant difference ($P < 0.05$, Student's *t* test) between the BSMV:ATG6-treated plants and BSMV:GFP-treated control plants. **(C)** Chlorophyll content of BSMV-treated Yangmai 158 plants. The chlorophyll content of the fifth leaves was determined at the five-leaf stage. Values represent the mean ± SD of data determined from five individual plants. Asterisks indicate a significant difference ($P < 0.05$) between the BSMV:ATG6-treated plants and BSMV:GFP-treated control plants. **(D)** Percentage of BSMV-treated Yangmai 158 plants displaying accelerated death before the five-leaf stage. Values represent the mean percentage ± SEM (standard error of the mean) of at least 30 plants. **(E)** Detached leaves from BSMV:ATG6-treated plants display accelerated dark-induced senescence. Green sections of the fourth leaves were excised from BSMV-treated Yangmai 158 plants at the four-leaf stage and were floated on deionized water at room temperature in the dark. Leaf sections were photographed at the indicated days post-detachment (dpd). The results are reproduced from three independent experiments using the leaves of three or more plants in each experiment.

conidium with secondary hyphae. As shown in Figure 6A and B, the percentage of successfully penetrating *Bgt* conidia (penetration frequency) was very low (0.5%) on leaves of the highly resistant BSMV:GFP-treated control plants, but significantly increased to 1.7–29.1% on leaves of different BSMV:ATG6-treated plants. Furthermore, successfully penetrating *Bgt* conidia on leaves of BSMV: ATG6-treated plants developed an average of 4.16 ± 2.91 prolonged secondary hyphae, whereas only one short secondary hypha was associated with successfully penetrating *Bgt* conidia on leaves of BSMV:GFP-treated control plants (Figure 6B). These results indicated that knocking down *TaATG6*s resulted in a compromised *Bgt* resistance on 92R137/Yangmai 158[7] (*Pm21*) plants. Therefore, we propose that TaATG6s play a positive role in *Pm21*-triggered wheat resistance response to *Bgt* and that this positive role is weak since no large, sporulating mycelia were observed on *TaATG6* knockdown plants of 92R137/Yangmai 158[7] (*Pm21*) by 7 dai. In *ATG6, VPS34, ATG3*, or *ATG7* knockdown tobacco plants and in *ATG6* knockdown or *atg5* mutant *Arabidopsis* plants, avirulent viral or bacterial pathogen-induced HR-PCD was unrestricted and spread into uninfected cells, suggesting that autophagy has a pro-survival function in the pathogen-induced cell

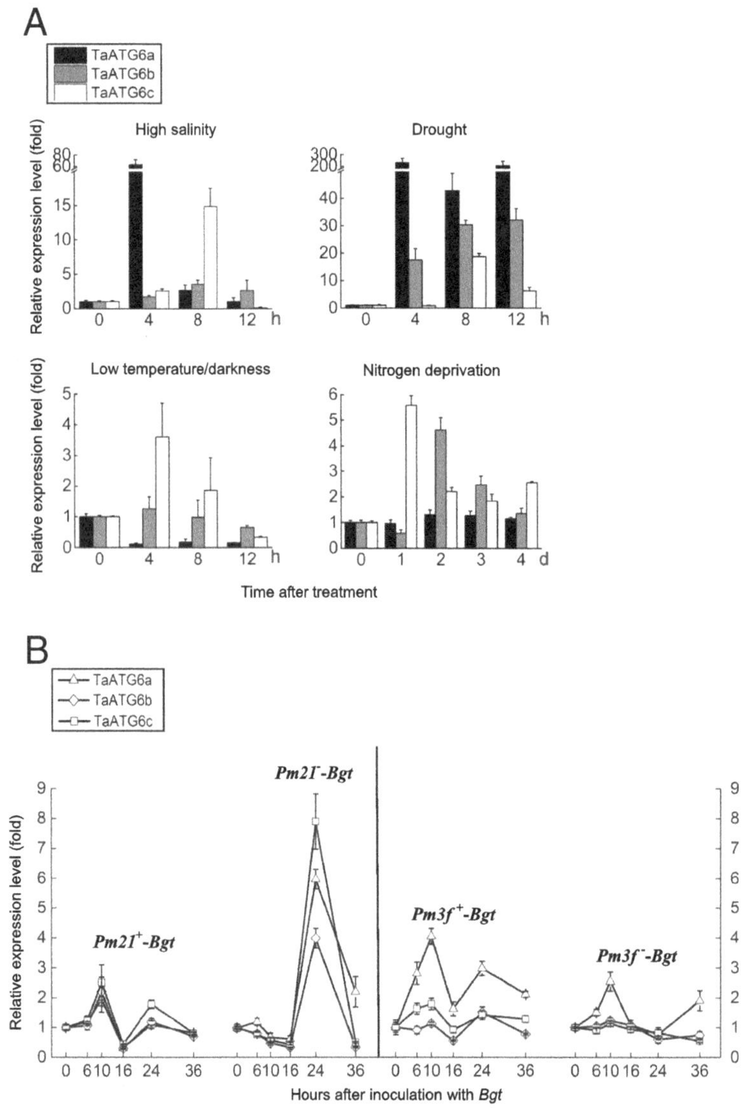

Figure 5 Expression patterns of wheat autophagy-related ATG6 genes. Two-leaf stage seedlings were subjected to different treatments. Samples were collected at the indicated time points after the initiation of treatment. Total RNA was extracted and qRT-PCR was performed with gene-specific primers. The relative expression levels were normalized by the wheat β-tubulin gene and relative to the control value measured at 0 h. Data represent the mean of three independent experiments ± SD. **(A)** Expression of wheat *ATG6s* in response to abiotic stress factors. Yangmai 158 seedlings were transferred to the one-fifth strength Hoagland's solution containing 200 mM NaCl (high salinity) or 16% PEG6000 (drought). Additional seedling treatments were nitrogen deprivation and exposure to 4°C in the dark (low temperature/darkness). **(B)** Expression of wheat *ATG6s* in response to *Blumeria graminis* f. sp. *tritici* (*Bgt*) infection. Pathogen inoculations were conducted by vigorously shaking fresh *Bgt* conidia from diseased plants onto seedlings of the isogenic lines 92R137/Yangmai 158[7] (*Pm21*$^+$) and Yangmai 158 (*Pm21*$^-$), and the isogenic lines Michigan Amber/Chancellor[8] (*Pm3f*$^+$) and Chancellor (*Pm3f*$^-$).

death response [16,17,44,47]. In our study, however, such an unrestricted spread of cell death was not induced by the biotrophic fungal pathogen *Bgt* on *TaATG6* knock-down plants of 92R137/Yangmai 158[7] (*Pm21*).

Wheat ATG6s also play a negative role in the basal resistance of plants susceptible to *Bgt*

Although the BSMV:ATG6-treated plants and BSMV: GFP-treated control plants of Yangmai 158 were both colonized by *Bgt* at 5 dai, the sizes of visible mycelia and their sporulation (conidial production per mildew colony) were significantly reduced on both lower leaves (the fourth leaves, Figure 7A and B) and upper leaves (the sixth leaves, Figure 8A) of BSMV:ATG6-treated plants. *Bgt* mycelia were also examined under a microscope after trypan blue staining, and a larger portion of small mycelia and a significantly reduced percentage of mycelia with mature chains of conidia were observed on

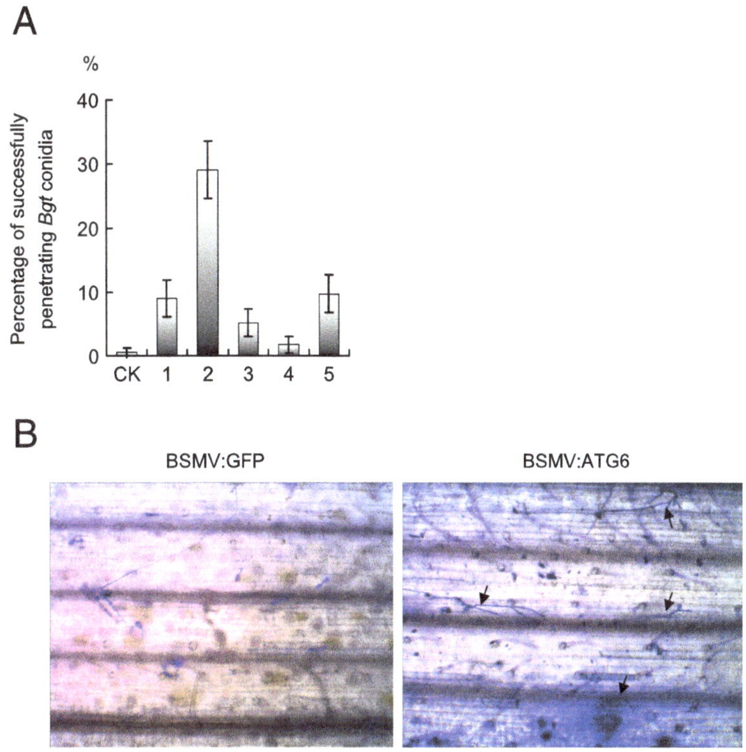

Figure 6 Knocking down wheat autophagy-related ATG6 genes weakly compromises the broad-spectrum resistance gene *Pm21*-triggered resistance response to *Blumeria graminis* f. sp. *tritici* (*Bgt*). Barley stripe mosaic virus (BSMV)-treated 92R137/Yangmai 158^{7} (*Pm21*) plants of the six-leaf stage were heavily inoculated with *Bgt* conidia. Fungal structures were stained by trypan blue and visualized by microscopy. **(A)** *Bgt* penetration frequency. The number of successfully penetrating conidium with secondary hyphae on the fourth leaves was counted at 7 days after inoculation (dai) with *Bgt* conidia. Values represent the mean percentage of successfully penetrating *Bgt* conidia ± SEM calculated from at least 100 interaction sites. CK is from BSMV:GFP-treated 92R137/Yangmai 158^{7} (*Pm21*) control plants. Arabic numbers indicate individual BSMV:ATG6-treated 92R137/Yangmai 158^{7} (*Pm21*) plants. **(B)** Representative images of *Bgt* infection on the fourth leaves of BSMV-treated 92R137/Yangmai 158^{7} (*Pm21*) plants at 7 dai. Successfully penetrating conidia with secondary hyphae are indicated by arrows.

leaves of BSMV:ATG6-treated plants at 5 dai (Figure 7C). Clearly, a significantly enhanced resistance should be responsible for the retarded mycelial spread and sporulation on *TaATG6* knockdown plants. We thus proposed that TaATG6s play negative roles in the basal resistance of the susceptible line Yangmai 158 to powdery mildew. Moreover, growth-retarded small mycelia, after a prolonged infection time of 9–10 dai, induced lesions and chlorosis around lesions on the middle-to-tip leaf regions of BSMV:ATG6-treated Yangmai 158 plants (Figure 8A, left panel of 8B). The exhibition of enhanced resistance preceded the occurrence of leaf lesions and not all small mycelia were finally associated with leaf cell death (Figure 8B), although leaf cell death may further impair the biotrophic lifestyle of *Bgt*.

Discussion

As a key component of the Class III PI3K kinase complex, ATG6 is essential for autophagy and the PI3P-signaling pathways. Previously, plant ATG6 orthologs had been identified in *Nicotiana benthamiana*, *Arabidopsis* and rice, and their involvement in autophagy had been demonstrated [16,17,48]. Here, we characterized three wheat ATG6 genes, *TaATG6a*, *6b* and *6c*, which were localized on the homeologous chromosomes 3DL, 3BL and 3AL, respectively, of the allo-hexaploid wheat genome (AABBDD). Heterologous expression of *TaATG6a*, *6b*, or *6c* can restore autophagy in *atg6*-mutant yeast cells. The TaATG6a-GFP fusion protein was located in cytoplasmic punctate PAS structures. Constitutive and pathogen-induced autophagy processes were impaired in *TaATG6* knockdown plants. Collectively, these results support the essential role of TaATG6s in wheat autophagy biogenesis. Consistent with *Arabidopsis* plants deficient in the PI3K components ATG6 or VPS34 [17,22,49], *TaATG6* knockdown wheat plants exhibit abnormal growth phenotypes, such as dwarfing, chlorosis, and accelerated senescence/death, under normal conditions. However, the role of PI3K on normal plant growth may be beyond autophagy, as mutants of

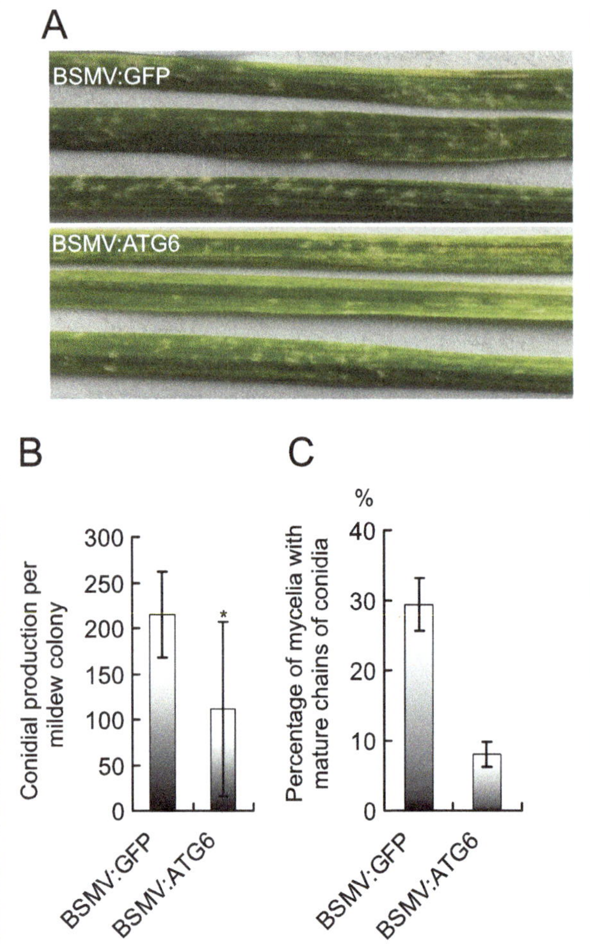

Figure 7 Knocking down wheat autophagy-related ATG6 genes enhances the basal resistance of susceptible plants to *Blumeria graminis* f. sp. *tritici* (*Bgt*). Barley stripe mosaic virus (BSMV)-treated Yangmai 158 plants of the six-leaf stage were heavily inoculated with *Bgt* conidia. **(A)** Representative images of mildew colonies on the fourth leaves of BSMV-treated Yangmai 158 plants. Images were taken at 5 days after inoculation (dai) with *Bgt* conidia. **(B)** Sporulation of *Bgt* colonies on BSMV-treated Yangmai 158 plants. Leaf discs with visible mildew colonies were gently harvested from different regions of the fourth leaves at 5 dai, immersed in 0.01% glycerol and subjected to ultrasonic treatment for 1 min to disperse the conidia. The conidial production per mildew colony was counted by microscopic observation. Values represent the mean conidial production per mildew colony ± SD of at least 50 mildew colonies from five plants. Asterisks indicate a significant difference ($P < 0.05$, Student's *t* test) between the BSMV:ATG6-treated plants and BSMV:GFP-treated control plants of Yangmai 158. **(C)** Percentage of mycelia with mature chains of conidia on BSMV-treated Yangmai 158 plants. Fungal structures were stained by trypan blue and visualized by light microscopy. The number of mycelium with mature chains of conidia on the fourth leaves was counted at 5 dai. At least 100 mycelia from five individual plants were counted and values represent the mean percentage of mycelia with mature conidia ± SEM.

autophagy-specific *atg* generally do not show such extensive growth abnormalities under normal conditions [19-21,28,50].

Pathogen-induced expression patterns and positive roles of ATG6 in plant innate immunity have been reported in plant pathosystems, such as *N. benthamiana*-TMV [16] and *Arabidopsis-Pseudomonas syringae* [17]. Here, two transcript accumulation events, an early and a late one, of wheat *ATG6s* were induced in the *Pm21*- and *Pm3f*-triggered wheat resistance responses to *Bgt* within 0 to 36 hai. *Bgt*-induced bimodal expression profiles have also been characterized for other powdery mildew resistance-related genes in wheat [51,52]. Two pathogen-induced oxygen burst processes reported previously also match the two time points [53]. The early wheat *ATG6* transcript accumulation event occurred within 0 to 10 hai, which is when *Bgt* appressorium germ tubes contact and attempt to invade wheat epidermal cells. This transcript accumulation may be pivotal for the *Pm21*-triggered resistance response because it was not detected in the isogenic susceptible plants of Yangmai 158. Although knocking down *TaATG6*s did not result in a striking change in the *Pm21*-triggered resistance phenotype, an increased penetration frequency of inoculated *Bgt* conidia was observed by microscopy on knockdown plants of 92R137/Yangmai 158[7] (*Pm21*). Since successfully penetrating conidia on knockdown plants were inevitably arrested before extending into large mycelia, we hypothesize that TaATG6s play only a weak positive role, probably in the early phase of *Bgt* invasion, in the *Pm21*-triggered resistance response. Recent studies on autophagy-specific *atg* gene mutants support the involvement of autophagy in plant immune responses [6,8]. The *Bgt*-enhanced occurrence of LysoTracker Red-stained autolysosome-like structures in control plants of 92R137/Yangmai 158[7] (*Pm21*), which "disappeared" in *TaATG6* knockdown plants, also links autophagy to ATG6 during the wheat immune response to *Bgt*. Positive roles of autophagy in plant resistance responses have been reported [16,17,43,44,54]. Similar to our finding of a weak, positive role of TaATG6s in the *Pm21*-triggered resistance response to *Bgt*, *Arabidopsis atg7* and *atg9* mutants are compromised in the early restriction of avirulent *P. syringae* growth, but retain the capacity to induce other defense responses [42].

Contrary to their positive roles in the *Pm21*-triggered resistance response, TaATG6s had negative roles in the basal resistance of the susceptible line Yangmai 158. Knocking down *TaATG6*s enhanced the resistance of Yangmai 158 plants to *Bgt* mycelial extension and sporulation. Our expression data showed that, compared with their performances in resistance plants of 92R137/Yangmai 158[7] (*Pm21*), TaATG6s were absent for the first response to *Bgt*, which occurred within 0–10 hai, but were intensely involved in the second response to *Bgt*, which occurred after 16 hai, in the susceptible plants of Yangmai 158. The large quantity of successfully penetrating

Figure 8 Lesions and chlorosis are induced by *Blumeria graminis* f. sp. *tritici* (*Bgt*) on wheat autophagy-related *ATG6* knockdown susceptible plants.. (A) Representative images of mildew colonies, leaf lesions and chlorosis on the sixth leaves of BSMV-treated Yangmai 158 plants at 10 days after inoculation (dai) with *Bgt* conidia. **(B)** Microscopic observation of growth-retarded small mycelia associated (left panel) or not associated (right panel) with leaf cell death on BSMV:ATG6-treated Yangmai 158 plants at 10 dai.

Bgt conidia on susceptible plants may induce the highly upregulated expression of *TaATG6*s after 16 hai. The retarded mycelial spread and sporulation on *TaATG6* knockdown susceptible plants suggested that, in wild-type susceptible plants, the ATG6 recruitment after 16 hai might be beneficial for the further development of *Bgt* mycelium. Thus, high levels of TaATG6s, and certain TaATG6-regulated pathways, in susceptible host cells might be hijacked by *Bgt* after the penetration stage to facilitate its pathogenicity, leading them to function negatively in the basal resistance of susceptible plants. The autophagy process in host cells is a potential target because of its function in the remobilization of cellular nutrients, which may be exploited by *Bgt* to meet the nutrient supply for fungus proliferation. Negative roles of autophagy in plant basal resistance have been reported in the susceptible responses of *Arabidopsis* to the powdery mildew fungus *Golovinomyces cichoracearum* [55] and to the bacterial pathogen *P. syringae* [44,54]. However, we failed to observe autophagy activity in *Bgt*-infected cells of wild-type Yangmai 158 plants using the LysoTracker Red staining method, probably because of the very low induction level (data not shown). Considering this and the pleiotropy of ATG6, we cannot exclude other ATG6- or PI3P-regulated pathways, or their interplay with autophagy, in underpinning the negative role of ATG6 in the basal resistance of susceptible wheat plants to powdery mildew.

Cell death was induced by growth-retarded small mycelia of *Bgt* on the leaves of *TaATG6* knockdown plants of Yangmai 158. However, cell death does not seem to be the reason for the enhanced resistance because it occurred after the exhibition of enhanced resistance and was not induced by all small mycelia. Similarly, powdery mildew induced lesions, but they were not necessarily coupled with the enhanced resistance of susceptible *Arabidopsis atg2* mutants [55]. The enhanced resistance of *Arabidopsis pmr* mutants to the powdery mildew pathogen *Erysiphe cichoracearum* and autophagy-defective mutants to *P. syringae* is also not associated with cell death [54,56]. Since ATG2 is autophagy-specific, the powdery mildew-induced cell death of susceptible *Arabidopsis atg2* mutants indicates a pro-survival role of autophagy in this compatible interaction [55]. However, in view of the weak *Bgt*-induced autophagy level and the pleiotropy of ATG6, whether autophagy alone or combined with other ATG6-regulated pathways underlies the suppression function of TaATG6s on *Bgt*-induced cell death of susceptible wheat plants needs to be determined.

Conclusions

In this study we provide evidence from yeast mutant complementation, subcellular localization of GFP fusions, and detection of autophagy levels in knockdown plants to show that the three identified wheat ATG6s are essential for autophagy biogenesis. *Bgt*-induced expression patterns, and the altered symptoms of powdery mildew on knockdown plants, suggest that TaATG6s are implicated in wheat immunity to *Bgt*, playing a weak, positive role in the *Pm21*-triggered resistance response and a negative role in the basal resistance of the susceptible line Yangmai158. These findings lay the foundations for a better understanding of autophagy mechanisms in crop plants and the role of ATG6 in wheat immunity.

Additional files

Additional file 1: Figure S1. Sequence comparison and phylogenetic relationship of eukaryotic autophagy-related ATG6 proteins. (A) Sequence comparison between wheat ATG6 proteins and their homologs. The predicted amino acid sequences of wheat TaATG6a, 6b and 6c were aligned with homologous ATG6s from *Arabidopsis* (AtATG6, AAK62668), yeast (ScATG6, Q02948) and human (HsATG6, Q14457). The alignment was generated in ClustalX 2.1 and viewed in the GeneDoc 2.7 program. Numbers on the right indicate amino acid residue positions. Similarity was coded as follows: 100%, black; 80 to 100%, dark gray; 60 to 80%, light gray; and < 60%, white. The ATG6 domain (Pfam PF04111) region is under the straight lines. (B) Phylogenetic relationship of plant ATG6s. The phylogenetic tree was generated with MEGA 5 using the neighbor-joining method. The reliability of internal branches was assessed by bootstrapping, with 1000 bootstrap replicates, and the values are shown in percentages, with a branching cut-off at 50%. Wheat ATG6s are indicated by diamonds. Figure S2. Diagram of the exon-intron structures of wheat *ATG6* genomic ORF sequences. Exons (boxes) and introns (lines between boxes) are drawn to scale, except introns indicated by question marks, which represent there are gaps between available genomic sequences. Two vertical short lines in intron 3 of *TaATG6b* represent direct short sequence repeats (TAGACTTAAATCATACTCC) delimiting the non-LTR retrotransposon (RT) sequence.

Additional file 2: Table S1. List of primers used in this paper.

Abbreviations

ATG: Autophagy-related gene; BSMV: Barley stripe mosaic virus; GFP: Green fluorescent protein; PAS: Pre-autophagosomal structure/phagophore assembly site; PI3K: Phosphatidylinositol 3-kinase; VIGS: Virus-induced gene silencing.

Competing interests

The authors declare that they have no competing interests.

Authors' contributions

JY and HW designed the research. JY, HS, WZ, DP and YH conducted the experiments and analyzed the data. HW wrote the manuscript. All authors read and approved the final manuscript.

Acknowledgements

We thank the other members of our laboratory for help with the research and for insightful remarks. This research was supported by the Natural Science Foundation of Tianjin, grants 12JCQNJC09700 and 12JCZDJC23000, and the Fok Ying-Tong Education Foundation for Young Teachers in the Higher Education Institutions of China, grant 131026.

References

1. Jones JD, Dangl JL. The plant immune system. Nature. 2006;444:323–9.
2. Chisholm ST, Coaker G, Day B, Staskawicz BJ. Host-microbe interactions: shaping the evolution of the plant immune response. Cell. 2006;124:803–14.
3. Liu Y, Bassham DC. Autophagy: pathways for self-eating in plant cells. Annu Rev Plant Biol. 2012;63:215–37.
4. Li F, Vierstra RD. Autophagy: a multifaceted intracellular system for bulk and selective recycling. Trends Plant Sci. 2012;17:526–37.
5. Klionsky DJ, Abdalla FC, Abeliovich H, Abraham RT, Acevedo-Arozena A, Adeli K, et al. Guidelines for the use and interpretation of assays for monitoring autophagy. Autophagy. 2012;8:445–544.
6. Talbot NJ, Kershaw MJ. The emerging role of autophagy in plant pathogen attack and host defence. Curr Opin Plant Biol. 2009;12:444–50.
7. Yoshimoto K, Takano Y, Sakai Y. Autophagy in plants and phytopathogens. FEBS Lett. 2010;584:1350–8.
8. Hayward AP, Dinesh-Kumar SP. What can plant autophagy do for an innate immune response? Annu Rev Phytopathol. 2011;49:557–76.
9. Zhou J, Yu JQ, Chen Z. The perplexing role of autophagy in plant innate immune responses. Mol Plant Pathol. 2014;15:637–45.
10. Teh OK, Hofius D. Membrane trafficking and autophagy in pathogen-triggered cell death and immunity. J Exp Bot. 2014;65:1297–312.
11. Kihara A, Noda T, Ishihara N, Ohsumi Y. Two distinct Vps34 30 phosphatidylinositol 3-kinase complexes function in autophagy and carboxypeptidase Y sorting in *Saccharomyces cerevisiae*. J Cell Biol. 2001;152:519–30.
12. Suzuki K, Kirisako T, Kamada Y, Mizushima N, Noda T, Ohsumi Y. The pre-autophagosomal structure organized by concerted functions of APG genes is essential for autophagosome formation. EMBO J. 2001;20:5971–81.
13. Obara K, Sekito T, Niimi K, Ohsumi Y. The Atg18-Atg2 complex is recruited to autophagic membranes via phosphatidylinositol 3-phosphate and exerts an essential function. J Biol Chem. 2008;283:23972–80.
14. Kametaka S, Okano T, Ohsumi M, Ohsumi Y. Apg14p and Apg6/Vps30p form a protein complex essential for autophagy in the yeast, *Saccharomyces cerevisiae*. J Biol Chem. 1998;273:22284–91.
15. Yang Z, Klionsky DJ. An overview of the molecular mechanism of autophagy. Curr Top Microbiol Immunol. 2009;335:1–32.
16. Liu Y, Schiff M, Czymmek K, Tallóczy Z, Levine B, Dinesh-Kumar SP. Autophagy regulates programmed cell death during the plant innate immune response. Cell. 2005;121:567–77.
17. Patel S, Dinesh-Kumar SP. *Arabidopsis* ATG6 is required to limit the pathogen-associated cell death response. Autophagy. 2008;4:20–7.
18. Fujiki Y, Yoshimoto K, Ohsumi Y. An *Arabidopsis* homolog of yeast ATG6/VPS30 is essential for pollen germination. Plant Physiol. 2007;143:1132–9.
19. Doelling JH, Walker JM, Friedman EM, Thompson AR, Vierstra RD. The APG8/12-activating enzyme APG7 is required for proper nutrient recycling and senescence in *Arabidopsis thaliana*. J Biol Chem. 2002;277:33105–14.
20. Hanaoka H, Noda T, Shirano Y, Kato T, Hayashi H, Shibata D, et al. Leaf senescence and starvation-induced chlorosis are accelerated by the disruption of an *Arabidopsis* autophagy gene. Plant Physiol. 2002;129:1181–93.
21. Xiong Y, Contento AL, Bassham DC. AtATG18a is required for the formation of autophagosomes during nutrient stress and senescence in *Arabidopsis thaliana*. Plant J. 2005;42:535–46.
22. Harrison-Lowe NJ, Olsen LJ. Autophagy protein 6 (ATG6) is required for pollen germination in *Arabidopsis thaliana*. Autophagy. 2008;4:339–48.
23. Petiot A, Faure J, Stenmark H, Gruenberg J. PI3P signaling regulates receptor sorting but not transport in the endosomal pathway. J Cell Biol. 2003;162:971–9.
24. Xue HW, Chen X, Mei Y. Function and regulation of phospholipid signalling in plants. Biochem J. 2009;421:145–56.
25. Qin G, Ma Z, Zhang L, Xing S, Hou X, Deng J, et al. *Arabidopsis* AtBECLIN 1/AtAtg6/AtVps30 is essential for pollen germination and plant development. Cell Res. 2007;17:249–63.
26. Lee Y, Kim ES, Choi Y, Hwang I, Staiger CJ, Chung YY, et al. The *Arabidopsis* phosphatidylinositol 3-kinase is important for pollen development. Plant Physiol. 2008;147:1886–97.
27. Xu N, Gao XQ, Zhao XY, Zhu DZ, Zhou LZ, Zhang XS. *Arabidopsis* AtVPS15 is essential for pollen development and germination through modulating phosphatidylinositol 3-phosphate formation. Plant Mol Biol. 2011;77:251–60.
28. Kurusu T, Koyano T, Hanamata S, Kubo T, Noguchi Y, Yagi C, et al. OsATG7 is required for autophagy-dependent lipid metabolism in rice postmeiotic anther development. Autophagy. 2014;10:1–11.
29. Liang XH, Kleeman LK, Jiang HH, Gordon G, Goldman JE, Berry G, et al. Protection against fatal Sindbis virus encephalitis by beclin, a novel Bcl-2-interacting protein. J Virol. 1998;72:8586–96.
30. Livak KJ, Schmittgen TD. Analysis of relative gene expression data using real-time quantitative PCR and the 2 −ΔΔCT method. Methods. 2001;25:402–8.
31. Gietz RD, Schiestl RH. Quick and easy yeast transformation using the LiAc/SS carrier DNA/PEG method. Nat Protoc. 2007;2:35–7.
32. Holzberg S, Brosio P, Gross C, Pogue GP. Barley stripe mosaic virus-induced gene silencing in a monocot plant. Plant J. 2002;30:315–27.
33. Zhou H, Li S, Deng Z, Wang X, Chen T, Zhang J, et al. Molecular analysis of three new receptor-like kinase genes from hexaploid wheat and evidence for their participation in the wheat hypersensitive response to stripe rust fungus infection. Plant J. 2007;52:420–34.
34. Van Wees S: Trypan Blue Stain for Fungi, oomycetes, and dead plant cells. In Arabidopsis: a laboratory manual. Chapter 4. Edited by Weigel D, Glazebrook J. Cold Spring Harbor: Cold Spring Harbor Laboratory Press; 2002: 86–87.
35. A database for wheat genomic sequences and physical and genetic maps [http://wheat-urgi.versailles.inra.fr]

36. Liang XH, Jackson S, Seaman M, Brown K, Kempkes B, Hibshoosh H, et al. Induction of autophagy and inhibition of tumorigenesis by beclin 1. Nature. 1999;402:672–6.
37. Obara K, Sekito T, Ohsumi Y. Assortment of phosphatidylinositol 3-kinase complexes-Atg14p directs association of complex I to the pre-autophagosomal structure in *Saccharomyces cerevisiae*. Mol Biol Cell. 2006;17:1527–39.
38. Kihara A, Kabeya Y, Ohsumi Y, Yoshimori T. Beclin-phosphatidylinositol 3-kinase complex functions at the trans-Golgi network. EMBO Rep. 2001;2:330–5.
39. Li BX, Li CY, Peng RQ, Wu XJ, Wang HY, Wan DS, et al. The expression of beclin 1 is associated with favorable prognosis in stage IIIB colon cancers. Autophagy. 2009;5:303–6.
40. Kang R, Zeh HJ, Lotze MT, Tang D. The Beclin 1 network regulates autophagy and apoptosis. Cell Death Differ. 2011;18:571–80.
41. Moriyasu Y, Hattori M, Jauh G, Rogers JC. Alpha tonoplast intrinsic protein is specifically associated with vacuole membrane involved in an autophagic process. Plant Cell Physiol. 2003;44:795–802.
42. Hofius D, Schultz-Larsen T, Joensen J, Tsitsigiannis DI, Petersen NH, Mattsson O, et al. Autophagic components contribute to hypersensitive cell death in *Arabidopsis*. Cell. 2009;137:773–83.
43. Lai Z, Wang F, Zheng Z, Fan B, Chen Z. A critical role of autophagy in plant resistance to necrotrophic fungal pathogens. Plant J. 2011;66:953–68.
44. Kwon SI, Cho HJ, Kim SR, Park OK. The Rab GTPase RabG3b positively regulates autophagy and immunity-associated hypersensitive cell death in *Arabidopsis*. Plant Physiol. 2013;161:1722–36.
45. Inoue Y, Suzuki T, Hattori M, Yoshimoto K, Ohsumi Y, Moriyasu Y. *AtATG* genes, homologs of yeast autophagy genes, are involved in constitutive autophagy in *Arabidopsis* root tip cells. Plant Cell Physiol. 2006;47:1641–52.
46. Bassham DC. Plant autophagy-more than a starvation response. Curr Opin Plant Biol. 2007;10:587–93.
47. Yoshimoto K, Jikumaru Y, Kamiya Y, Kusano M, Consonni C, Panstruga R, et al. Autophagy negatively regulates cell death by controlling NPR1-dependent salicylic acid signaling during senescence and the innate immune response in *Arabidopsis*. Plant Cell. 2009;21:2914–27.
48. Rana RM, Dong S, Ali Z, Huang J, Zhang HS. Regulation of ATG6/Beclin-1 homologs by abiotic stresses and hormones in rice (*Oryza sativa* L.). Genet Mol Res. 2012;11:3676–87.
49. Welters P, Takegawa K, Emr SD, Chrispeels MJ. ATVPS34, a PtdIns 3-kinase of *Arabidopsis thaliana* is an essential protein with homology to a calcium-dependent lipid-binding domain. Proc Natl Acad Sci U S A. 1994;91:11398–402.
50. Yoshimoto K, Hanaoka H, Sato S, Kato T, Tabata S, Noda T, et al. Processing of ATG8s, ubiquitin-like proteins, and their deconjugation by ATG4s are essential for plant autophagy. Plant Cell. 2004;16:2967–83.
51. Liu GS, Sheng XY, Greenshields DL, Ogieglo A, Kaminskyj S, Selvaraj G, et al. Profiling of wheat class III peroxidase genes derived from powdery mildew-attacked epidermis reveals distinct sequence-associated expression patterns. Mol Plant Microbe Interact. 2005;18:730–41.
52. Bhuiyan NH, Selvaraj G, Wei Y, King J. Gene expression profiling and silencing reveal that monolignol biosynthesis plays a critical role in penetration defense in wheat against powdery mildew invasion. J Exp Bot. 2009;60:509–21.
53. Hückelhoven R, Kogel KH. Reactive oxygen intermediates in plant-microbe interactions: who is who in powdery mildew resistance? Planta. 2003;216:891–902.
54. Lenz HD, Haller E, Melzer E, Kober K, Wurster K, Stahl M, et al. Autophagy differentially controls plant basal immunity to biotrophic and necrotrophic pathogens. Plant J. 2011;66:818–30.
55. Wang Y, Nishimura MT, Zhao T, Tang D. ATG2, an autophagy-related protein, negatively affects powdery mildew resistance and mildew-induced cell death in *Arabidopsis*. Plant J. 2011;68:74–87.
56. Vogel J, Somerville S. Isolation and characterization of powdery mildew-resistant *Arabidopsis* mutants. Proc Natl Acad Sci U S A. 2000;97:1897–902.

Carotenoid accumulation during tomato fruit ripening is modulated by the auxin-ethylene balance

Liyan Su[1,7], Gianfranco Diretto[3], Eduardo Purgatto[4], Saïda Danoun[5], Mohamed Zouine[1,2], Zhengguo Li[6], Jean-Paul Roustan[1,2], Mondher Bouzayen[1,2], Giovanni Giuliano[3] and Christian Chervin[1,2*]

Abstract

Background: Tomato fruit ripening is controlled by ethylene and is characterized by a shift in color from green to red, a strong accumulation of lycopene, and a decrease in β-xanthophylls and chlorophylls. The role of other hormones, such as auxin, has been less studied. Auxin is retarding the fruit ripening. In tomato, there is no study of the carotenoid content and related transcript after treatment with auxin.

Results: We followed the effects of application of various hormone-like substances to "Mature-Green" fruits. Application of an ethylene precursor (ACC) or of an auxin antagonist (PCIB) to tomato fruits accelerated the color shift, the accumulation of lycopene, α-, β-, and δ-carotenes and the disappearance of β-xanthophylls and chlorophyll *b*. By contrast, application of auxin (IAA) delayed the color shift, the lycopene accumulation and the decrease of chlorophyll *a*. Combined application of IAA + ACC led to an intermediate phenotype. The levels of transcripts coding for carotenoid biosynthesis enzymes, for the ripening regulator *Rin*, for chlorophyllase, and the levels of ethylene and abscisic acid (ABA) were monitored in the treated fruits. Correlation network analyses suggest that ABA, may also be a key regulator of several responses to auxin and ethylene treatments.

Conclusions: The results suggest that IAA retards tomato ripening by affecting a set of (i) key regulators, such as *Rin*, ethylene and ABA, and (ii) key effectors, such as genes for lycopene and β-xanthophyll biosynthesis and for chlorophyll degradation.

Keywords: Auxin, Ethylene, Abscisic acid, Tomato, Carotenoids, Chlorophyll, Lycopene, Rin, Ripening

Background

Auxin and ethylene are hormones known to impact plant development, often with antagonistic roles. Auxin exerts pleiotropic effects, on the development of roots, shoots, flowers and fruits [1]. Ethylene is one of the plant hormones regulating the ripening of fruits, the opening of flowers, and the abscission of leaves. Tomato is a model plant for the study of climacteric fruit development, which is promoted by ethylene [2]. Observations of tomato fruits and some non-climacteric fruits, like grape berry and strawberry, have suggested that ripening is also regulated by auxin, since they can delay ripening and regulate gene expression [3-6]. However, the impact of auxin on tomato ripening has not been extensively studied, as previous works using exogenous auxin [3,6] do not study carotenoid accumulation and related gene expression. Moreover in the plant kingdom, the crosstalk between auxin and ethylene is not yet deciphered [7].

Color change from green to red is a very important indicator of tomato ripening and can easily be measured by chromametry [8]. This change is associated with the degradation of chlorophylls and the shift of the carotenoid composition from leaf-like xanthophylls (mainly lutein and neoxanthin) to carotenes (mainly phytoene, lycopene and β-carotene) as described by Fraser et al.

* Correspondence: chervin@ensat.fr
[1]Université de Toulouse, INP-ENSA Toulouse, UMR990 Génomique et Biotechnologie des Fruits, Avenue de l'Agrobiopole, CS 32607, F-31326 Castanet-Tolosan, France
[2]INRA, UMR990 Génomique et Biotechnologie des Fruits, 24 Chemin de Borde Rouge, CS 52627, F-31326 Castanet-Tolosan, France
Full list of author information is available at the end of the article

[9]. In the fruit tissues, the degradation of chlorophylls is slow, while the accumulation of red carotenoids is rapid [10] when checked by time lapse imaging. The carotenoid biosynthetic pathway in tomato is well described [11,12] and is detailed on Figure 1. The first committed step is the condensation of two molecules of geranylgeranyl diphosphate (GGPP) to form the colorless carotene 15-*cis*-phytoene, a reaction catalyzed by phytoene synthases (PSY); 15-*cis*-phytoene is then desaturated and isomerized to all-*trans*-lycopene through the action of two desaturases and two isomerases: phytoene desaturase (PDS), ζ-carotene desaturase (ZDS), prolycopene isomerase (CRTISO) and ζ-carotene isomerase (ZISO). The formation of δ-carotene and γ-carotene from lycopene are catalyzed by lycopene ε-cyclase (ε-LCY) and β-cyclases (β-LCY and CYC-β), and then the orange α - carotene and β-carotene are synthetized by β-cyclases. Finally, these carotenes are transformed into lutein and zeaxanthin by heme and non-heme β-carotene hydroxylases (CYP97 and CRTR-b). Zeaxanthin is converted to violaxanthin by the action of zeaxanthin epoxidase (ZEP) and further to neoxanthin by the action of the NXD and ABA4 proteins. These two xanthophylls are cleaved by 9-*cis*-epoxycarotenoid dioxygenase (NCED), a key enzyme in the biosynthesis of ABA [13].

For the purpose of this article, the pathway will be divided into two parts, upstream of lycopene and downstream of lycopene (Figure 1). In the upstream part, the key rate-limiting steps are catalyzed by PSY1, PDS, ZDS, ZISO and CRTISO [9,14,15]. The expression of *Psy1, Ziso, Crtiso* is directly regulated by the ripening inhibitor (RIN) protein, which is a member of the MADS-box family of transcription factors [16,17]. In the downstream part, lycopene cyclases (ε-LCY, β-LCY/CYC-β) are also key enzymes, catalyzing the transformation of lycopene to δ- and β- carotene [18-21].

To study the role of cross-talk between auxin and ethylene in the accumulation of carotenoid pigments in tomato fruits, we treated mature green fruits with the auxin indole acetic acid (IAA) and the ethylene precursor aminocyclopropane carboxylic acid (ACC), alone or in combination, and also with p-chlorophenoxy isobutyric acid (PCIB). The latter compound is an antagonist of auxin action, although its mechanism of action is not well characterized [22]. The effects of these treatments on color change, pigment content and on the levels of transcripts involved in carotenoid biosynthesis were studied.

Results and Discussion

Contrasting effects of ethylene and auxin on tomato fruit color

The hormonal treatments induced significant color changes within 96 hours (Figure 2). Treatment with ACC accelerated significantly the transition from green to orange/red compared to controls. On the contrary, treatment with IAA induced a significant delay in the transition from green to orange/red compared to controls. After 96 h, IAA-treated fruits began to turn orange and then never became red (data not shown).

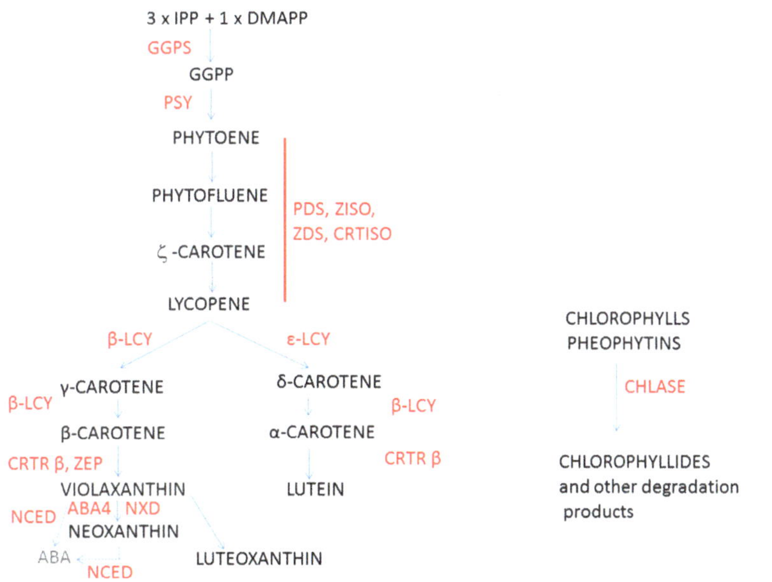

Figure 1 Carotenoid biosynthetic pathway based on a previous study [12]. Names of intermediate compounds are in black and names of enzymes are in red. IPP = isopentenyl diphosphate, GGPS = GGPP synthase, GGPP = geranyl-geranyl pyrophosphate, PSY = phytoene synthase, PDS = phytoene desaturase, ZISO = zeta-carotene isomerase, ZDS = zeta-carotene desaturase, CRTISO = carotenoid isomerase, ε-LCY = lycopene ε-cyclase, β-LCY = lycopene β-cyclase, CRTR-β = β-carotene hydroxylase, ZEP = zeaxanthin epoxydase, NXD = neoxanthin synthase, CHL = chlorophyllases, ABA = abscisic acid.

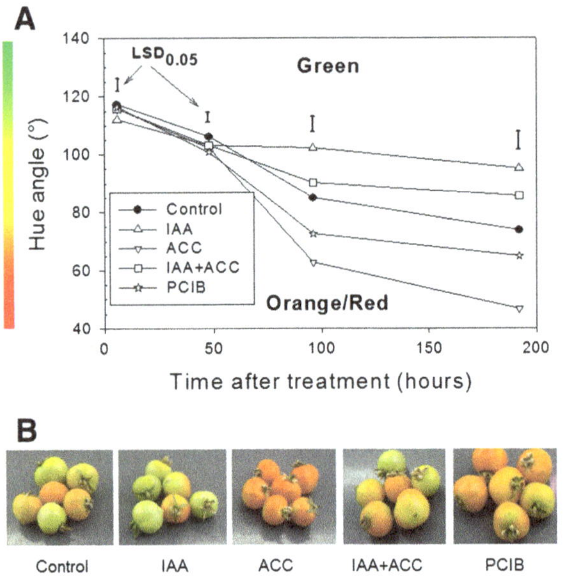

Figure 2 A) Changes of tomato color as a function of time after hormonal treatments. IAA: indole acetic acid, ACC: aminocyclopropane carboxylic acid (ethylene precursor), PCIB: p-chlorophenoxy isobutyric acid (auxin antagonist). The color bar next to the Y axis gives an indication of the relation between Hue angle and fruit color, but it is not the exact color of the fruit on the CIELab scale. n = 6 biological replicates, LSD bars calculated at 0.05 level. **B)** Pictures of tomatoes 96 h after hormonal treatments.

In fruits treated with a combination of ACC and IAA, color evolution was slower than in controls, but faster than the fruits treated by IAA alone, indicating that IAA treatment is epistatic over ACC treatment. In the presence of the auxin antagonist PCIB, fruits turned red faster than control ones and the color change kinetics were very similar to those treated with ACC (Figure 2A). These results confirmed previous studies showing that IAA slows down ripening of tomato fruits [2,6], and that ACC accelerates it [2].

Effects of hormonal treatments on carotenoid, chlorophyll and ABA accumulation

To further investigate the influence of hormonal treatments on fruit pigment composition, fruit extracts were analyzed. At 96 hours, the main carotenoids in control fruits were lutein and β-carotene (Figure 3). Large amounts of chlorophylls *a* and *b* were observed, together with trace amounts of lycopene, violaxanthin, neoxanthin, luteoxanthin, ζ-, δ- and α-carotene. The upstream compounds phytoene and phytofluene were not detectable. This composition is typical of a ripening stage between the "Breaker" and "Orange" stages of ripening [41].

The ACC and PCIB treatments induced large changes in carotenoid composition at 96 hours (Figure 3). Lycopene was greatly induced, becoming a major pigment, together with β-carotene which was also induced and lutein which was unaffected. The upstream compounds phytoene, phytofluene and ζ-carotene and the downstream compounds δ- and α-carotene were also induced, while the β-xanthophylls, neoxanthin and violaxanthin were reduced.

The IAA treatment reduced significantly lycopene accumulation compared to controls while it did not affect α-, β- or δ-carotene accumulation. It also led to higher levels of neoxanthin, violaxanthin and chlorophyll *a* than in the controls (Figure 3).

The 9-*cis* forms of neoxanthin and violaxanthin are the precursors of abscisic acid (ABA) [23,24], a phytohormone known to control ripening of many fruits, including tomato, in which it triggers ethylene biosynthesis and thus accelerates ripening [25]. ABA levels were decreased by the ACC and PCIB treatments and increased by the IAA treatment (Figure 4), mimicking the evolution of neoxanthin/violaxanthin, thus suggesting that the accumulation of these compounds might be directly correlated. This observation is consistent with the idea that in the tomato fruit, levels of neoxanthin and violaxanthin are rate-limiting for ABA accumulation [26]. Finally, the ACC and PCIB treatments led to an increased degradation of chlorophyll *b* (Figure 3).

Our results detail the auxin effects on carotenoid accumulation, thus completing preliminary observations that were not detailing this aspect [6]. Our results also detail carotenoid changes induced by ACC, following previous studies showing that ethylene treatments accelerated chlorophyll degradation, the appearance of orange color [10,27] and the accumulation of lycopene [28]. It is noticeable that PCIB, which acts as an auxin antagonist, induced the same effects as ACC.

Effects of hormonal treatments on gene expression

In order to investigate if the above hormone-induced phenotypes were controlled at least partially at the gene expression level, we determined the levels of all transcripts involved in carotenoid biosynthesis by quantitative Real Time PCR (qPCR) at two different times after the hormonal treatments (Figure 5).

As observed in Figure 5A, IAA treatment resulted in lower transcript levels for most of the genes upstream of lycopene (*Psy1, Psy3, Pds, Ziso* and *Crtiso*). With the exception of *Psy3* which has been reported to be mainly expressed in roots, all these genes are rate-limiting for lycopene accumulation [15]. Thus, these changes in transcript levels match well the slower color change and the decreased accumulation of lycopene after treatment with IAA (Figures 2 and 3). Regarding the downstream part of the pathway (Figure 5B), the transcript levels of *β-Lcy1* and *Crtr-β1* genes were induced by IAA treatment, concomitant with the higher amounts of violaxanthin and neoxanthin, while *Aba4* showed a biphasic

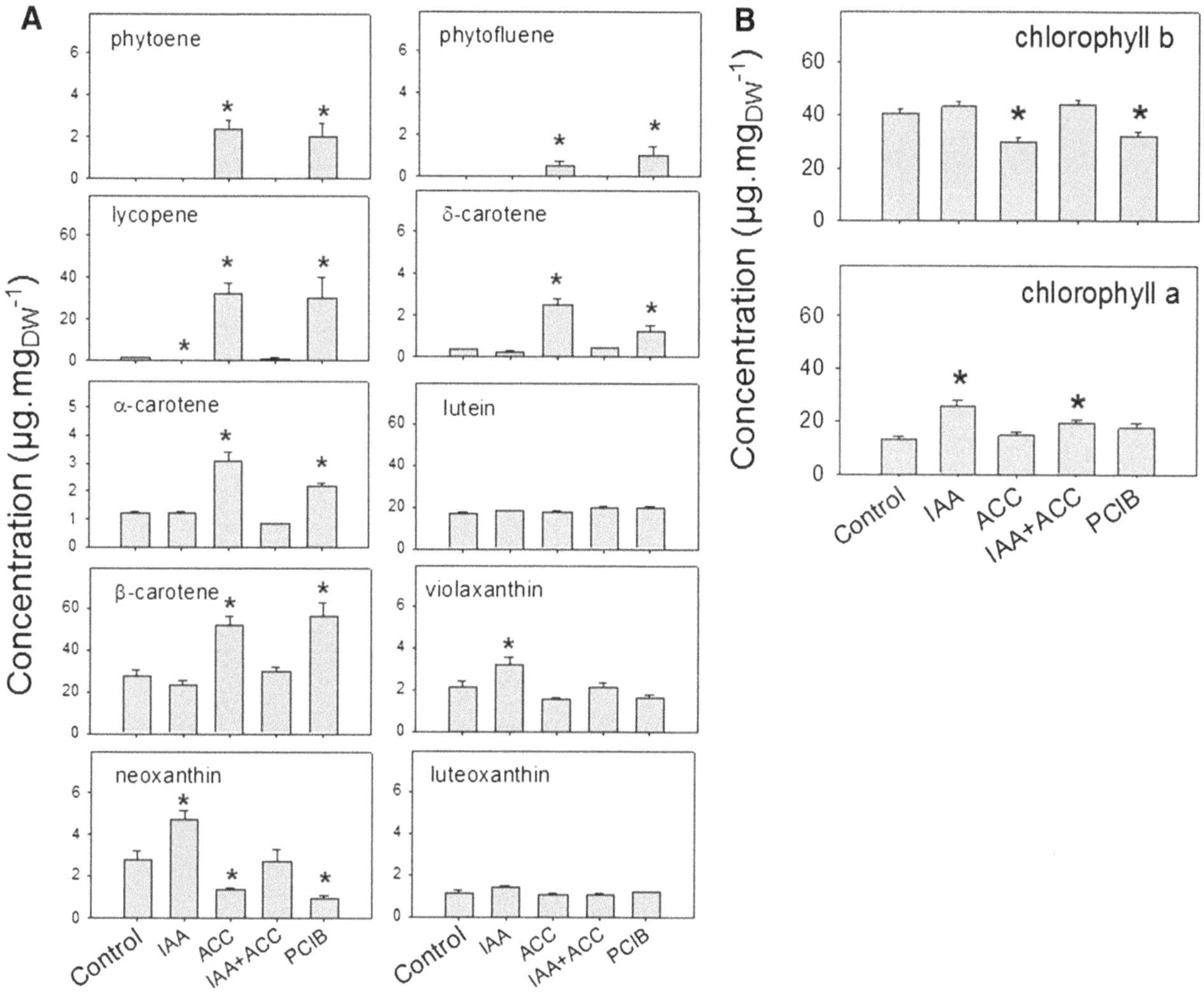

Figure 3 Carotenoid **[A]** and chlorophyll **[B]** contents 96 h after hormonal treatments. Abbreviations are as in Figure 2. n = 3 biological replicates, error bars are standard errors. An asterisc (*) shows a significant difference at 0.05 level using t-test between control and the corresponding treatment.

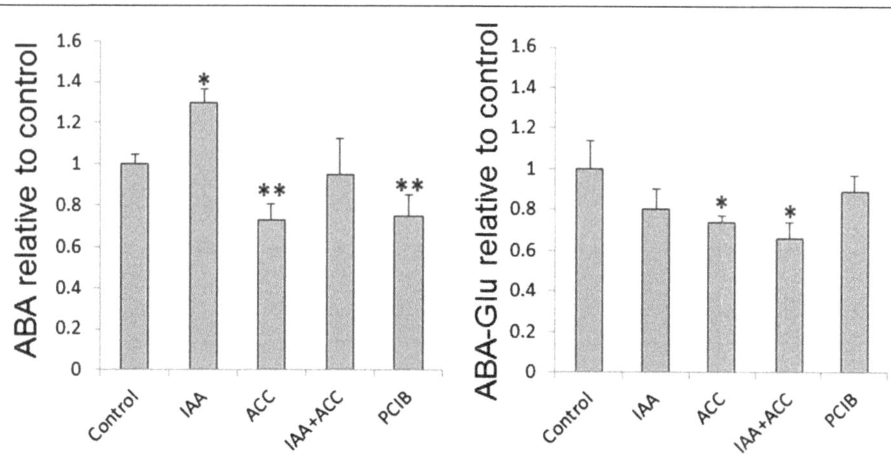

Figure 4 Variations of free ABA levels and ABA glucoside 96 h after hormonal treatments. Abbreviations are as in Figure 2. n = 5 biological replicates, error bars are standard errors. Asteriscs, * or ** show significant differences at the 0.05 or 0.01 levels compared to controls, respectively (t-test).

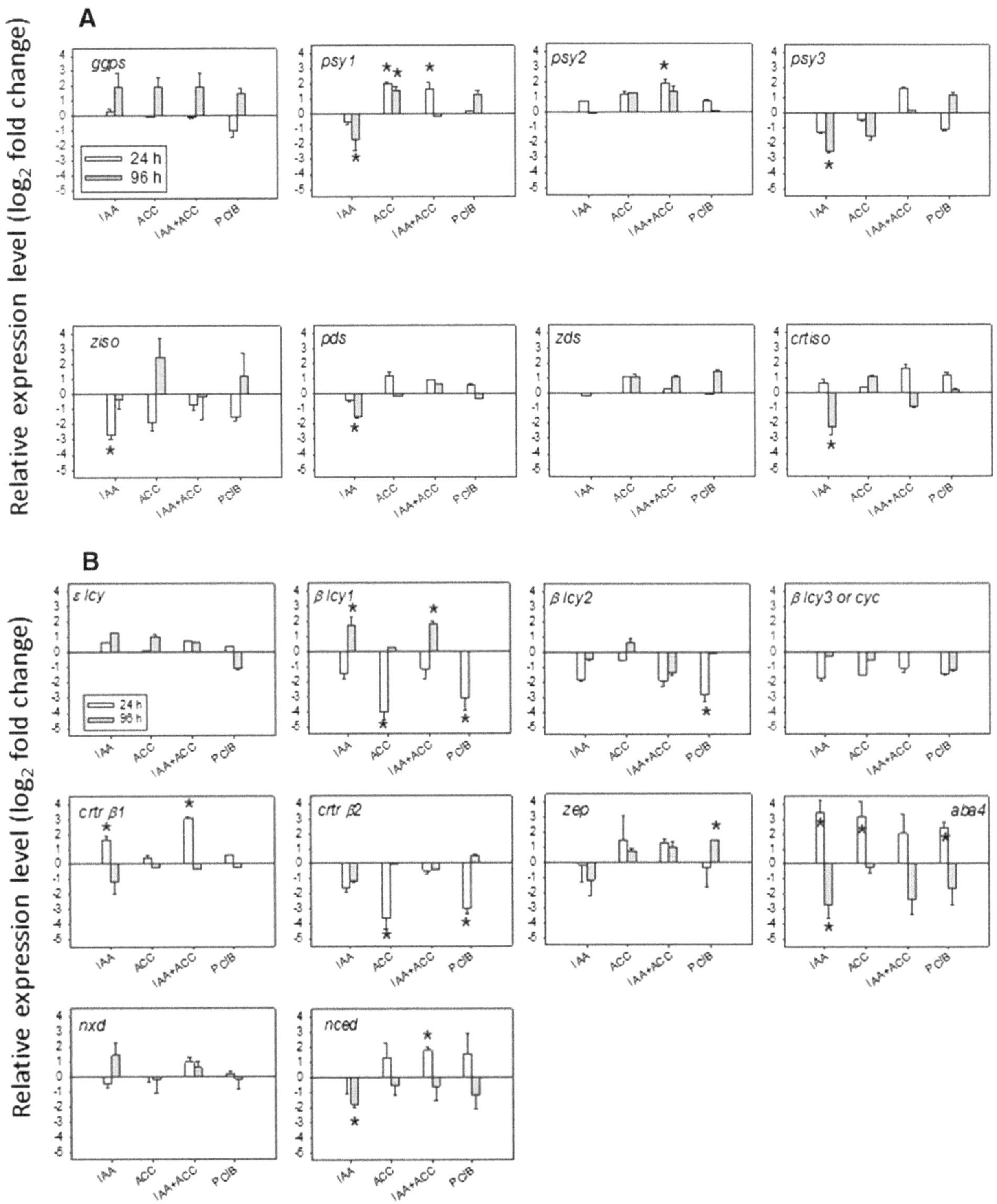

Figure 5 Modulation of transcript accumulation related to carotenoid pathway, **A)** upstream of lycopene, **B)** downstream of lycopene, 24 h or 96 h after hormonal treatments (see abbreviations in legend of Figure 2). n = 3 biological replicates, bars = std errors. Expression relative to controls (set at 0). An asterisc (*) shows significant differences at the 0.05 level with controls (t-test).

response (induction at 24 h and repression at 96 h) and *Nced1* a repression at 96 h. Together, these observations indicate that the ABA increase after IAA treatment is a fast response, probably due to an increase in the synthesis of its precursors violaxanthin and neoxanthin, mediated by an activation of the *β-Lcy1*, *Crtr-β1* and *Aba4* genes. The repression of *Aba4* and *Nced1* at 96 h may be due to a negative feedback regulation exerted by the increased ABA levels on these genes. ABA is known to increase in tomatoes prior to the ethylene peak [25].

ACC treatment led to higher levels of *Psy1* and *Psy2* transcripts, and also, to a lesser extent, of the *Ziso, Pds, Zds* and *Crtiso* ones (Figure 5A). All these genes encode rate-limiting steps for lycopene biosynthesis [15] and thus the observed changes in gene expression are in agreement with the faster color change and accelerated lycopene accumulation (Figures 2 and 3). Moreover, ACC treatment decreased *β-Lcy1* transcript levels (Figure 5B) with unexpected increase of α-, β- and δ- carotenes, indicating that the *β-Lcy1* repression was possibly offset by the unaltered levels of the other cyclase transcripts. ACC also repressed *Crtr-β2* expression that was not offset by the unaltered *Crtr-β1* levels, reducing the further conversion of carotene compounds into β-xanthophylls. This was confirmed by the reduced neoxanthin and ABA levels after ACC treatment (Figures 3 and 4), in spite of an induction of *Aba4*. It is also worth noticing that IAA and ACC affected the expression of two different hydroxylase paralogs, *Crtr-βi* being stimulated by IAA and *Crtr-β2* being inhibited by ACC, respectively. Overall, these data explain the faster accumulation of lycopene and β-carotene, and also the lower accumulation of β-xanthophylls and ABA in ACC treated fruits than in controls.

Similar changes in transcript levels occurred in PCIB-treated fruits (Figure 5), which showed an additional repression of *β-Lcy2* and an induction of *Zep*, as well as a very similar carotenoid profile (Figure 3) to the ACC-treated samples. There was no significant effect of any treatment on *Ggps* expression (Figure 5A and Additional file 1: Figure S1).

The combined IAA + ACC treatment resulted in a visual and carotenoid phenotype intermediate between those of each treatment alone and more similar to that of IAA alone, with the exception of violaxanthin, neoxanthin and ABA induction, which was less pronounced than in IAA alone (Figures 2, 3 and 4). At the transcriptional level, IAA + ACC was less inhibitory of upstream transcripts than IAA alone. Although the significance of these observations awaits clarification, it confirms the antagonistic effects of the two hormones at the biochemical and transcriptional levels.

Chlorophyll degradation in *Citrus* fruits is an active process mediated by chlorophyllase (Chlase) [29]. In tomato, chlorophyll degradation was affected by hormonal treatments, with IAA treatment retarding chlorophyll *a* degradation, both alone and in combination with ACC treatment, while chlorophyll *b* degradation was accelerated by both ACC and PCIB treatments (Figure 3). We measured the levels of the three *Chlase* transcripts identified in the tomato genome. Repression of all three transcripts was obvious 96 h after the IAA treatment (Figure 6). This correlates well with the higher levels of chlorophyll *a* and to a lesser extent of chlorophyll *b*, in both treatments with IAA (Figure 3). However, the marked decrease of chlorophyll *b* in the ACC and PCIB treatments does not correlate with increased *Chlase* transcript accumulation (Figure 6). This suggests that, in contrast to Citrus [29], tomato *Chlase* gene expression is not under ethylene control and that, as observed in Citrus [30], posttranscriptional mechanisms may also regulate *Chlase* activity in tomato.

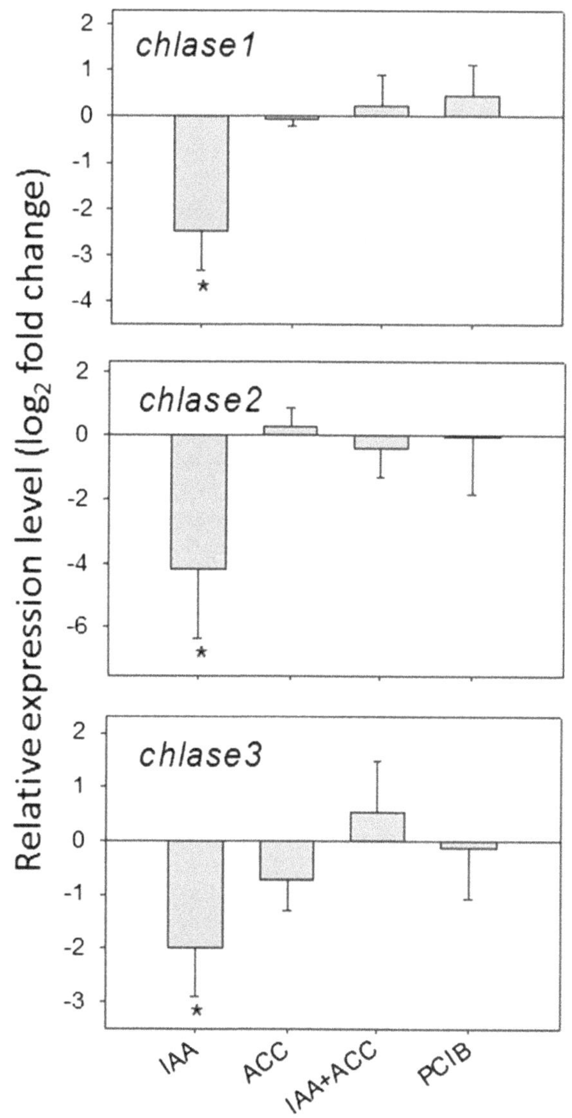

Figure 6 Modulation of chlorophyllase transcripts, 96 h after hormonal treatments (see abbreviations in legend of Figure 2). n = 3 biological replicates. Expression relative to controls (set at 0). Error bars are standard errors. An asterisc (*) shows significant differences the 0.05 level with controls (t-test).

Effects of hormonal treatments on the Rin transcript and on transcripts of the carotenoid/ABA pathway

Several genes in the carotenoid pathway are regulated by the Rin transcription factor [16,17]: *Psy1*, *Ziso* and *Crtiso* display direct positive regulation, *Zds* indirect positive regulation, and *ε-Lcy* and *β-Lcy2* indirect negative regulation. Analyses carried out by qPCR (Figure 7A) showed that the transcript levels of *Rin* were stimulated by ACC and inhibited by IAA, even if the sole significant difference was noticed for ACC 96 h. The qPCR profiles of *Rin* (Figure 7A) and *Psy1* (Figure 5A) seem to match quite well. Indeed, in keeping with the findings of Fujisawa et al. [17], high positive correlations ($\rho > 0.60$, and in some cases $\rho > 0.80$) were observed between transcript levels of *Rin* and *Psy1* at both 24 h and 96 h, *Ziso* and *Crtiso* at 96 h, and *ZDS* at 24 h (Figure 7B).

In contrast, *ε-Lcy* did not show high correlations with *Rin* neither at 24 h nor at 96 h, while *β-Lcy2* showed strong positive correlations at both time points. This contrasts with the findings of Fujisawa et al. [17] and suggests that lycopene cyclase transcripts are subject to additional layers of regulation. Strong positive correlations with *Rin* were identified for *Pds* and *Zep* at 24 h and for *ABA4* at 96 h. The latter two genes mediate the biosynthesis of the ABA precursors, violaxanthin and neoxanthin (Figure 1), and thus their positive correlations with *Rin* may be indicative of the fact that Rin activates two hormonal cascades: one acting through ethylene [16], and one acting through ABA. Finally, *Ggps4* showed a negative correlation with *Rin* levels at 96 h. This gene is unrelated to fruit carotenoid biosynthesis and may control the biosynthesis of other isoprenoid compounds (Falcone et al., unpublished).

Effects of hormonal treatments on fruit ethylene production

Ethylene is assumed to be a "master switch" controlling tomato fruit ripening. Therefore, it is interesting to verify if the hormonal treatments described above alter ethylene production. We measured ethylene production in hormone-treated fruits at various times after treatments (Figure 8). As expected, ACC treatment accelerated the appearance of the climacteric ethylene peak by about 2 days whereas IAA treatment repressed the ethylene production, and this repression was only partially reversed by combined IAA + ACC treatments. PCIB treatment had little effect up to 100 hours after treatment, while it slightly decreased ethylene production around 200 hours. So it seems that PCIB enhancement of carotenoid accumulation in comparison to controls (Figure 2) is not mediated by a variation in ethylene production. The IAA decrease of carotenoid accumulation in comparison to controls could be partially mediated by the repression of ethylene production.

Factorial and network analyses show associations between hormonal treatments and carotenoid levels

Factorial analyses are used to determine and describe the dependencies within sets of variables. In this study the treatments, and many observed variations, in this study the transcript levels (Figure 9A) or the carotenoid levels (Figure 9B). These factorial correspondence analyses clearly show strong positive correlation between

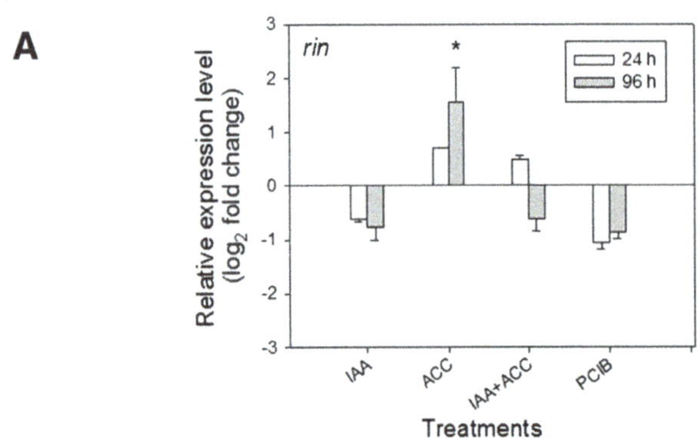

B

	GGPS1	GGPS2	GGPS3	GGPS4	PSY1	PSY2	PSY3	ZISO	PDS	ZDS	CRTISO	E-LCY	B-LCY1	B-LCY2	CYC	CRTR-B1	CRTR-B2	ZEP	NCED	ABA4	NXD
24H	0.259	-0.031	0.366	-0.126	0.966	0.645	0.512	0.372	0.809	0.829	-0.012	-0.209	-0.029	0.78	0.4	0.314	0.224	0.997	0.361	-0.102	0.412
96H	0.412	0.0404	0.305	-0.771	0.699	0.547	-0.459	0.942	0.0198	0.124	0.863	0.346	-0.488	0.866	-0.119	0.378	0.0713	-0.032	0.599	0.971	-0.483

Figure 7 A) Modulation of the Rin transcript 24 h or 96 h after various hormonal treatments (see abbreviations in legend of Figure 2). n = 3 biological replicates. Expression relative to controls (set at 0). Error bars are standard errors. **B)** Correlation coefficients between Rin and other transcripts shown in Figure 5.

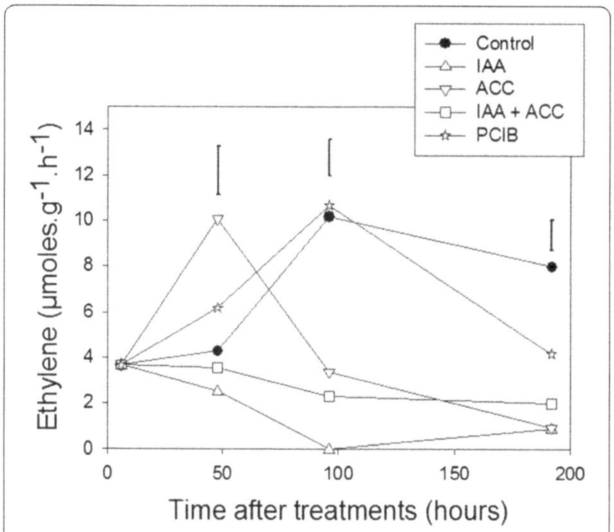

Figure 8 Variations in ethylene production after the hormonal treatments (see abbreviations in legend of Figure 2). n = 3 biological replicates, error bars are LSD at the 0.05 level.

the effects of ACC and PCIB, and their negative correlation to the effects of IAA treatment, whatever the regulatory level measured: transcript accumulation or carotenoid accumulation. It is noticeable that, at the transcript level, the IAA + ACC treatment is strongly correlated with the ACC and PCIB ones (Figure 9A), while at the carotenoid composition level - which matches the fruit phenotype more closely - it is correlated with the IAA treatment (Figure 9B). This may be due to the fact that changes in transcript accumulation occur ahead of those in metabolite accumulation, or to the fact that some of the latter changes are due to post-transcriptional events, or to both.

The transcripts correlating with the ripening delay associated to IAA treatment are lycopene cyclases (*ε and β-Lcy*) and, to a lesser extent, carotene hydroxylases (*Crtr-β*) (Figure 9A). These results confirm previous studies [18-21]. All transcripts mediating lycopene biosynthesis in tomato fruits: *Psy1, Pds, Ziso, Zds,* and *Crtiso* [15] correlate well with the accelerated ripening

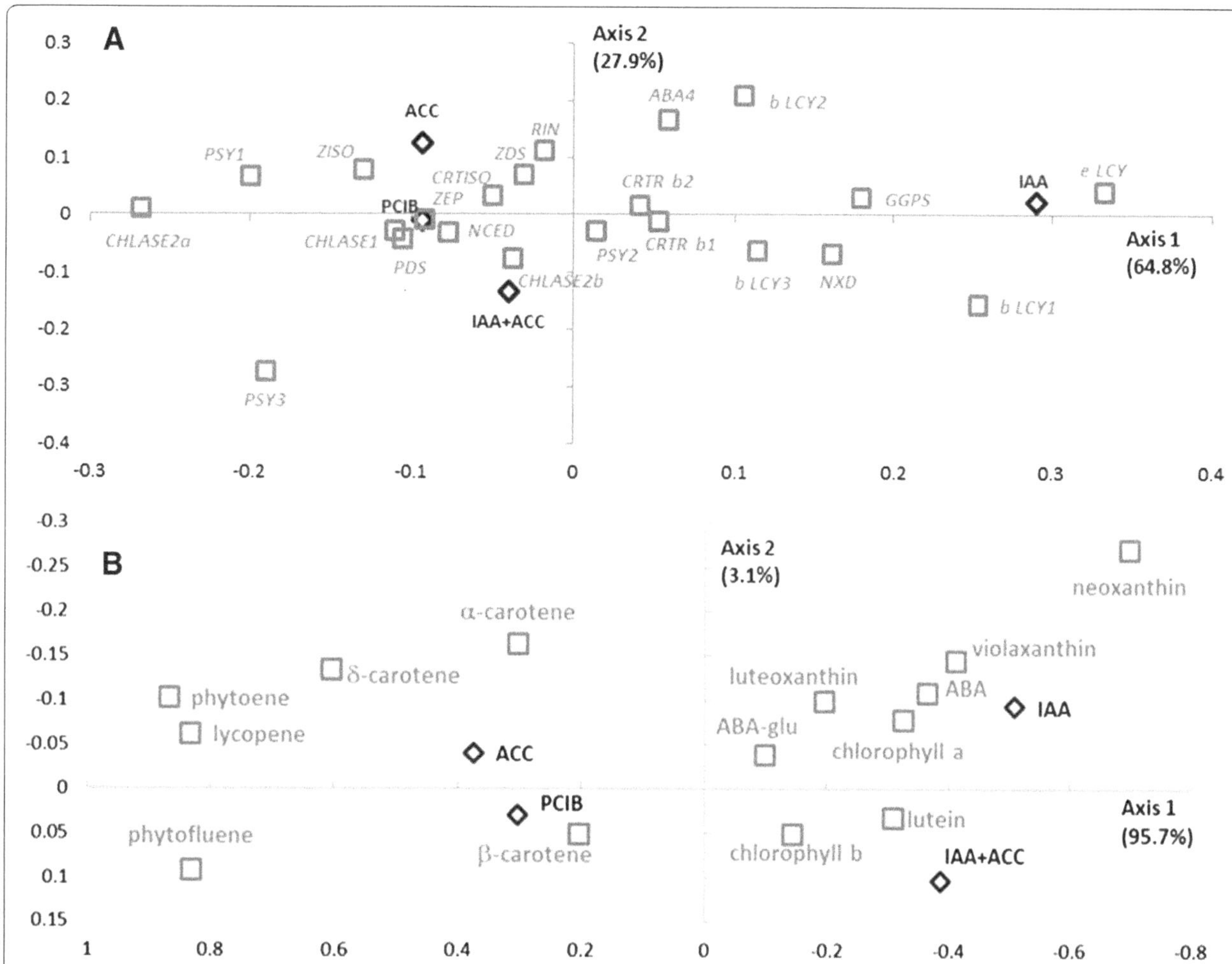

Figure 9 Factorial correspondence analyses with data 96 h after hormonal treatments, **A)** transcript accumulation and **B)** carotenoid and metabolite content. Abbreviations are as in Figures 1 and 2.

induced by ACC or PCIB. Also *Psy3*, which is much less expressed and is non-essential for lycopene biosynthesis, shows a position opposed to IAA treatment (Figure 9A) as it was strongly repressed by IAA at 96 h (Figure 5A). Same case for the position of *Chlase* transcripts in Figure 9A which is mainly due to the strong inhibition by IAA, rather than to a stimulation by ACC. Regarding carotenoids, the accumulation of upstream intermediates and lycopene and, to a lesser extent, of α-, β-, and δ-carotene is correlated directly with ACC and PCIB treatments. Inversely IAA and IAA + ACC treatments correlate well with chlorophylls and xanthophylls, (especially violaxanthin and neoxanthin) and their product ABA (Figure 9B). This is consistent with the fact that ripening is associated with the accumulation of cyclic carotenes and with the decrease of chlorophylls and xanthophylls.

We also applied correlation network analysis based on transcript-metabolite data integration (Figure 10). The time spent after treatments increased the strength in the network [31], and at 96 h the network shows four nodes with strong correlation values ($|\rho| > 0.60$) (Additional file 2: Table S2): ABA, its metabolic precursors violaxanthin and neoxanthin and *Nxd*, a gene essential for neoxanthin biosynthesis [33]. All four nodes exhibited a prevalence of negative correlations with the other ripening-specific variables in the network.

Conclusions

Our results suggest that ACC treatment induces lycopene and α-, β- and δ-carotene accumulation by inducing *Psy1*, and repressing *β-Lcy1* and *Crtr-β2*. These transcriptional responses are fast, reaching a peak at 24 h.

On the other hand, treatment with IAA represses several upstream carotenoid transcripts (*Psy, Ziso, Pds, Crtiso*) as well as *Chlases 1-3* and promotes the accumulation of *β-Lcy1* and *Crtr-β1* transcripts, leading to higher levels of chlorophyll *a*, neoxanthin, violaxanthin and ABA. These responses show a temporal curve: *Ziso* and some downstream transcripts (*Crtr-β1* and *ABA4*) respond already at 24 h, while most other transcripts (*Psy1* to *β-Lcy1*) respond later, at 96 h. This response could be due to the fact that downstream transcripts respond directly to auxin, while upstream transcripts respond to the repression of ethylene production induced by IAA treatment (Figure 8). Treatment with PCIB (an auxin antagonist) led to responses similar to those obtained after ACC treatment, confirming the antagonism

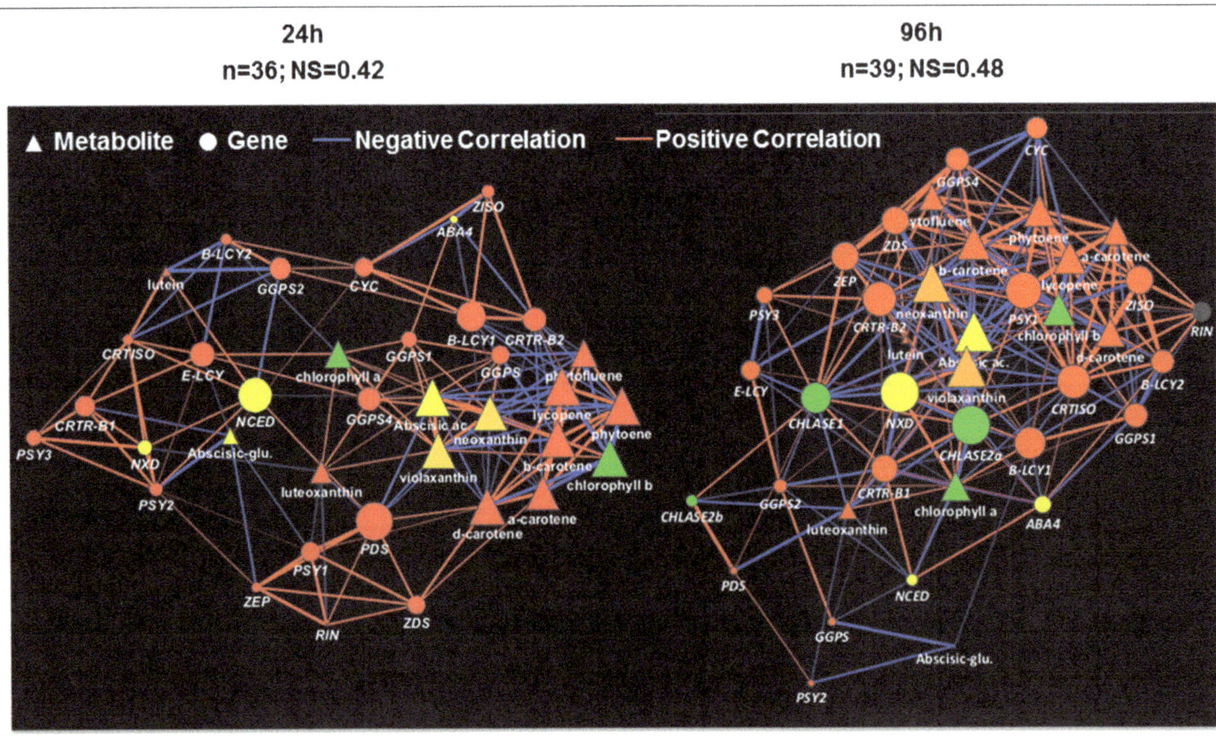

Figure 10 Correlation networks at 24 h and 96 h, generated as previously described [31]. In all network diagrams, nodes of different shape represent genes and metabolites. Direct and inverse corrlations ≥ |0.60| are shown as red and blue edges, respectively. Edge thickness is proportional to the absolute values of the Pearson correlation coefficient ($|\rho|$), while node sizes are proportional to node strengths [31] (Additional file 2: Table S2). n = number of nodes, NS = network strength [31]. Nodes related to carotenoids are shown in red, to chlorophyll in green, to ABA in yellow, neoxanthin and violaxanthin (ABA precursors) in orange, Rin in grey. The "organic layout" was used for network visualization with Cytoscape 2.6.3 (www.cytoscape.org) [32].

between ethylene and auxin. Interestingly, while IAA completely repressed ethylene production, PCIB did not increase it (Figure 8) indicating that endogenous auxin does not play a major role in regulating ethylene production during normal ripening. The repression of ethylene production and the induction of *Crtr-β1* by exogenous IAA supplementation were epistatic over ACC supplementation when both treatments were given together, while the final phenotype of the fruits did not show a clear epistasis of IAA over ACC supplementation.

Factorial and correlation network analyses allowed the identification, at 96 h, of transcriptional and metabolite "hubs" which may represent central regulators; these comprised ABA, its carotenoid precursors (violaxanthin and neoxanthin) and the *Nxd* gene, leading to neoxanthin biosynthesis. Overall, these data suggest a central role for ABA as a negative intermediate regulator in the perturbation of tomato fruit ripening following auxin and ethylene treatments.

Methods

Plant materials and growth conditions

Tomato plants (*Solanum lycopersicum* cv. *MicroTom*) were grown under standard greenhouse conditions. The culture chamber room was set as follows: 14-h day/10-h night cycle, 25/20°C day/night temperature, 80% relative humidity, 250 μmol m^{-2} s^{-1} light intensity. Tomato seeds were first sterilized 5 min in sterile water and sown in Magenta vessels containing 50 ml 50% Murashige and Skoog (MS) culture medium and 0.8% (w/v) agar, pH 5.9 [34].

Treatments of tomato fruits

Tomato fruits were harvested at the mature green stage of development and injected with a buffer solution contained 10 mM MES, pH 5.6, sorbitol (3% w/v) and 100 μM of ACC, or IAA, or IAA + ACC (100 μM each), or PCIB (all Sigma-Aldrich products). Preliminary tests were performed with concentrations ranging from 1 μM to 1 mM, in order to choose the minimal concentration impacting the ripening kinetics without showing toxic effects. Buffer injection was performed as described previously [35]. Briefly, tomato fruits were infiltrated using a 1 ml syringe with a 0.5 mm needle, inserted 3 to 4 mm into the fruit tissue through the stylar apex. The infiltration solution was gently injected into the fruit until the solution ran off the stylar apex and the hydathodes at the tip of the sepals. Only completely infiltrated fruits were used in the experiments. Controls were treated with buffer only. After the treatment, fruits were incubated in a culture room at 26°C, under 16 h light/8 h dark cycle with a light intensity of 100 μmol s^{-1} m^{-2}. After 24 h and 96 h, fruits pericarp was collected and frozen at -80°C until further analysis. For each condition, 27 fruits were sampled arising from 9 different plants.

Color and pigment measurement

Surface color was assessed with a Chromameter (CR400, Konica Minolta), using the D65 illuminant and the L*, a*, b* space, and the data were processed to obtain Hue as previously described [8]. In the culture room, the fruit color was measured after 6 h, 48 h, 96 h and some fruit were kept up to 8 days for assessing this parameter. Three measures were taken at the equator of each fruit, before being averaged. The Hue angle (in degrees) was calculated according to the following equations: Hue = $\tan^{-1}$ (b*/a*) if a > 0 and 180 + $\tan^{-1}$ (b*/a*) if a < 0. For pigment analysis, fruit samples were chosen at 96 h after treatment with IAA, ACC, IAA + ACC, PCIB and ground to a fine powder in liquid nitrogen. Pigments (chlorophylls/carotenoids) were extracted from freeze-dried tissues and analyzed as described previously [36] using an Accela U-HPLC system coupled to an Orbitrap high-resolution mass spectrometer (HRMS) operating in positive mode-atmospheric pressure chemical ionization (APCI) (Thermo Fischer Scientific, Waltham, MA).

ABA and ethylene assays

The ABA assays were performed as described previously [37]. Briefly, 110 mg of frozen tissue, sampled at 96 h after treatments, were extracted at 4°C for 30 min with 400 μl of H_2O with 10% methanol + 1% acetic acid. The internal standard was [^{2}H6] labelled ABA. The extract was centrifuged at 13,000 *g* for 10 min at 4°C. The supernatant was carefully removed and the pellet re-incubated for 30 min with 400 μl of methanol-acetic acid mix. Following the centrifugation, the supernatants were pooled. Extracts were then analysed by LC-MS using an Acquity UPLC coupled to a XevoQtof (Waters, Massachusetts, USA). Analysis parameters were described in Jaulneau et al. [38]. Fruit ethylene production was assayed as previously described [36]. The fruit ethylene production was measured after 6 h, 48 h, 96 h and some fruit were kept up to 8 days in the culture room for assessing this parameter.

RNA isolation and quantitative PCR (qPCR)

Total fruit RNA was extracted using the PureLink™ Plant RNA Reagent (Invitrogen) according to the manufacturer's instructions. On fruit sampled at 24 and 96 h, total RNA was treated by DNase I to remove any genomic DNA contamination. First-strand cDNA was reverse transcribed from 2 μg of total RNA using an Omniscript kit (Qiagen). qPCR analyses were performed as previously described [39]. The primer sequences are listed in Additional file 3: Table S1. Relative fold changes were calculated using *Sl-actin* as housekeeping gene. As for pigment analyses, three independent RNA isolations were used for cDNA synthesis. Efficiency of DNAse was assessed by PCR with actin primers designed on both size of a zone with an intron, thus giving two bands if genomic DNA is still present.

Factorial analyses of correspondence, correlation networks and statistics

We used transcript accumulation relative to controls under the ΔΔCt format to get only positive values, and the carotenoid accumulation levels were calculated relative to controls. The factorial analyses of correspondence were calculated with the explore.xla Excel macro developed previously [40]. Correlation networks were built as previously described [31]. Networks were visualized as organic layouts with Cytoscape version 2.6.3 (www.cytoscape.org) [32]. When LSD are presented they were calculated using Tukey's HSD.

Additional files

Additional file 1: Figure S1. Modulation of *ggps* transcript accumulation, by various hormonal treatments, 24th or 96h after treatment (see abbreviations in legend of Figure 2), n=3 biological replicates, bars = std errors. Expression relative to controls (set at 0).

Additional file 2: Table S2. Node strengths (ns) of the network in Figure 10, calculated as avg(|ρ|) (Diretto et al, 2010).

Additional file 3: Table S1. List of primers used in qPCR experiments.

Abbreviations

ABA: Abscisic acid; ACC: Aminocyclopropane carboxylic acid; CHLASE: Chlorophyllases; CRTISO: Carotenoid isomerase; CRTR-β: β-carotene hydroxylase; IAA: Indole acetic acid; IPP: Isopentenyl diphosphate; GGPS: GGPP synthase; GGPP: Geranyl-geranyl pyrophosphate; ε-LCY: Lycopene ε-cyclase; β-LCY: Lycopene β-cyclase; NXD: Neoxanthin synthase; PCIB: *p*-chlorophenoxy isobutyric acid; PDS: Phytoene desaturase; PSY: Phytoene synthase; ZEP: Zeaxanthin epoxydase; ZDS: Zeta-carotene desaturase; ZISO: Zeta-carotene isomerase*.

Competing interests

The authors declare that they have no competing interests.

Authors' contributions

LS participated in the design of the study, performed the qRT-PCR experiments and participated to the redaction, GD and GG performed carotenoid assays and correlation network and contributed to the writing of the manuscript, EP contributed to the design of the study, performed the fruit treatments and carried out the color and ethylene measurements, SD performed ABA assays, MZ contributed in the transcriptomic data treatment, ZL participated to the design of the project, MB and JPR conceived the study, participated to its design and contributed to the writing of the manuscript, CC participated to experiment design, wrote the body of the paper and coordinated the writing of the final manuscript. All the authors read and approved the manuscript.

Acknowledgements

We are grateful to Dr Bo Burla, University Hospital Zürich, for the generous gift of some ABA glucosyl ester. The work benefited from the networking activities within the European COoperation in Science and Technology Action FA1106. We thank Dominique Saint-Martin (Université de Toulouse, Institut National Polytechnique-Ecole Nationale Supérieure Agronomique de Toulouse (INP-ENSAT) Laboratoire de Génomique et Biotechnologie des Fruits (GBF) for tomato cultures, and the staff of the Toulouse Genomics plateform (GenoToul) for their assistance with qPCR. This work was supported by the French Laboratory of Excellence project "TULIP" (ANR-10-LABEX-41; ANR-11-IDEX-0002-02), by the Italian Ministry of Research ("Integrated Knowledge for the Sustainability and Innovation of Italian Agri-Food" Project) and by the European Commission (FP7 Project "From discovery to products: A next generation pipeline for the sustainable generation of high-value plant products", contract n. 613513; H2020 Project "Traditional tomato varieties and cultural practices: a case for agricultural diversification with impact on food security and health of European population", contract n. 634561). LS was supported by a CSC PhD grant.

Author details

[1]Université de Toulouse, INP-ENSA Toulouse, UMR990 Génomique et Biotechnologie des Fruits, Avenue de l'Agrobiopole, CS 32607, F-31326 Castanet-Tolosan, France. [2]INRA, UMR990 Génomique et Biotechnologie des Fruits, 24 Chemin de Borde Rouge, CS 52627, F-31326 Castanet-Tolosan, France. [3]Italian National Agency for New Technologies, Energy, and Sustainable Development, Casaccia Research Centre, 00123 Rome, Italy. [4]Department Food and Experimental Nutrition; NAPAN/FoRC - Food Research Center, Universidade de São Paulo, School of Pharmaceutical Sciences, Av. Prof. Lineu Prestes 580, Butantã, CEP 05508-000 São Paulo, SP, Brazil. [5]Université de Toulouse; UPS; UMR 5546; Laboratoire de Recherche en Sciences Végétales (LRSV), 24 Chemin de Borde Rouge, F-31326 Castanet-Tolosan, France. [6]Genetic Engineering Research Centre, Bioengineering College, Chongqing University, Chongqing 400044, PR China. [7]Actual address: Department of Life Sciences, Xi'an University of Arts and Science, Xi'an 710065, PR China.

References

1. Tromas A, Perrot-Rechenmann C. Recent progress in auxin biology. Comptes Rendus Biologies. 2010;333:297–306.
2. Giovannoni JJ. Fruit ripening mutants yield insights into ripening control. Current Opinion Plant Biology. 2007;10:283–9.
3. Cohen J. In vitro tomato fruit cultures demonstrate a role for indole-3-acetic acid in regulating fruit ripening. J Am Soc Hortic Sci. 1996;121:520–4.
4. Davies C, Boss P, Robinson S. Treatment of grape berries, a nonclimacteric fruit with a synthetic auxin, retards ripening and alters the expression of developmentally regulated genes. Plant Physiol. 1997;115:1155–61.
5. Given N, Venis M, Grierson D. Hormonal regulation of ripening in the strawberry, a non-climacteric fruit. Planta. 1988;174:402–6.
6. Vendrell M. Dual effect of 2,4-D on ethylene production and ripening of tomato fruit tissue. Physiol Plant. 1985;64:559–63.
7. Muday GK, Rahman A, Binder BM. Auxin and ethylene: collaborators or competitors? Trends Plant Science. 2012;17:181–95.
8. Ecarnot M, Baczyk P, Tessarotto L, Chervin C. Rapid phenotyping of the tomato fruit model, Micro-Tom, with a portable VIS-NIR spectrometer. Plant Physiol Biochem. 2013;70:159–63.
9. Fraser PD, Truesdale MR, Bird CR, Schuch W, Bramley PM. Carotenoid biosynthesis during tomato fruit development evidence for tissue-specific gene expression. Plant Physiol. 1994;105:405–13.
10. Egea I, Bian W, Barsan C, Jauneau A, Pech JC, Latche A, et al. Chloroplast-to-chromoplast transition in tomato fruit: spectral confocal microscopy analyses of carotenoids and chlorophylls in isolated plastids and time-lapse recording on intact live tissue. Ann Bot. 2011;108:291–7.
11. Hirschberg J. Carotenoid biosynthesis in flowering plants. Current Opinion Plant Biology. 2001;4:210–8.
12. Giuliano G. Plant carotenoids: genomics meets multi-gene engineering. Current Opinion Plant Biology. 2014;19:111–7.
13. Ji K, Kai W, Zhao B, Sun Y, Yuan B, Dai S, et al. SlNCED1 and SlCYP707A2: key genes involved in ABA metabolism during tomato fruit ripening. J Exp Bot. 2014;65:5243–55.
14. Giuliano G, Bartley GE, Scolnik PA. Regulation of carotenoid biosynthesis during tomato development. The Plant Cell. 1993;5:379–87.
15. Fantini E, Falcone G, Frusciante S, Giliberto L, Giuliano G. Dissection of tomato lycopene biosynthesis through virus-induced gene silencing. Plant Physiol. 2013;163:986–98.
16. Martel C, Vrebalov J, Tafelmeyer P, Giovannoni JJ. The tomato MADS-box transcription factor RIPENING INHIBITOR interacts with promoters involved

in numerous ripening processes in a COLORLESS NONRIPENING-dependent manner. Plant Physiol. 2011;157:1568–79.
17. Fujisawa M, Nakano T, Shima Y, Ito Y. A large-scale identification of direct targets of the tomato MADS box transcription factor RIPENING INHIBITOR reveals the regulation of fruit ripening. The Plant Cell. 2013;25:371–86.
18. Ronen G, Cohen M, Zamir D, Hirschberg J. Regulation of careteniod biosynthesis during tomato fruit development: expression of the gene for lycopene epsilon-cyclase is down-regulated during ripening and is elevated in the mutant Delta. The Plant Journal. 1999;17:341–51.
19. Ronen G, Carmel-Goren L, Zamir D, Hirschberg J. An alternative pathway to beta -carotene formation in plant chromoplasts discovered by map-based cloning of beta and old-gold color mutations in tomato. ProcNatlAcadSci U S A. 2000;97:11102–7.
20. Rosati C, Aquilani R, Dharmapuri S, Pallara P, Marusic C, Tavazza R, et al. Metabolic engineering of beta-carotene and lycopene content in tomato fruit. The Plant Journal. 2000;24:413–9.
21. Ma C, Ma B, He J, Hao Q, Lu X, Wang L. Regulation of carotenoid content in tomato by silencing of lycopene β/ε-cyclase genes. Plant Mol Biol Report. 2011;29:117–24.
22. Oono Y, Ooura C, Rahman A, Aspuria ET, Hayashi K, Tanaka A, et al. p-Chlorophenoxyisobutyric acid impairs auxin response in Arabidopsis root. Plant Physiol. 2003;133:1135–47.
23. Liotenberg S, North H, Marion-Poll A. Molecular biology and regulation of abscisic acid biosynthesis in plants. Plant Physiol Biochem. 1999;37:341–50.
24. Taylor IB, Burbidge A, Thompson AJ. Control of abscisic acid synthesis. J Exp Bot. 2000;51:1563–74.
25. Zhang M, Yuan B, Leng P. The role of ABA in triggering ethylene biosynthesis and ripening of tomato fruit. J Exp Bot. 2009;60:1579–88.
26. Galpaz N, Wang Q, Menda N, Zamir D, Hirschberg J. Abscisic acid deficiency in the tomato mutant high-pigment 3 leading to increased plastid number and higher fruit lycopene content. The Plant Journal. 2008;53:717–30.
27. Saltveit ME. Effect of ethylene on quality of fresh fruits and vegetables. Postharvest Biology and Technology. 1999;15:279–92.
28. Frenkel C, Garrison SA. Initiation of lycopene synthesis in the tomato mutant rin as influenced by oxygen and ethylene interactions. HortSci. 1976;11:20–1.
29. Jacob-Wilk D, Holland D, Goldschmidt EE, Riov J, Eyal Y. Chlorophyll breakdown by chlorophyllase: isolation and functional expression of the Chlase1 gene from ethylene-treated Citrus fruit and its regulation during development. The Plant Journal. 1999;20:653–61.
30. Harpaz-Saad S, Azoulay T, Arazi T, Ben-Yaakov E, Mett A, Shiboleth YM, et al. Chlorophyllase is a rate-limiting enzyme in chlorophyll catabolism and is posttranslationally regulated. The Plant Cell. 2007;19:1007–22.
31. Diretto G, Al-Babili S, Tavazza R, Scossa F, Papacchioli V, Migliore M, et al. Transcriptional-metabolic networks in β-carotene-enriched potato tubers: the long and winding road to the golden phenotype. Plant Physiol. 2010;154:899–912.
32. Cline MS, Smoot M, Cerami E, Kuchinsky A, Landys N, Workman C, et al. Integration of biological networks and gene expression data using Cytoscape. Nat Protoc. 2007;2:2366–82.
33. Neuman H, Galpaz N, Cunningham Jr FX, Zamir D, Hirschberg J. The tomato mutation nxd1 reveals a gene necessary for neoxanthin biosynthesis and demonstrates that violaxanthin is a sufficient precursor for abscisic acid biosynthesis. The Plant Journal. 2014;78:80–93.
34. Pirrello J, Jaimes-Miranda F, Sanchez-Ballesta MT, Tournier B, Khalil-Ahmad Q, Regad F, et al. Sl-ERF2, a tomato ethylene response factor involved in ethylene response and seed germination. Plant Cell Physiol. 2006;47:1195–205.
35. Orzaez D, Mirabel S, Wieland WH, Granell A. Agroinjection of tomato fruits. A tool for rapid functional analysis of transgenes directly in fruit. Plant Physiol. 2006;140:3–11.
36. Liu M, Diretto G, Pirrello J, Roustan JP, Li Z, Giuliano G, et al. The chimeric repressor version of an Ethylene Response Factor (ERF) family member, Sl-ERF.B3, shows contrasting effects on tomato fruit ripening. New Phytol. 2014;203:206–18.
37. Forcat S, Bennett MH, Mansfield JW, Grant MR. A rapid and robust method for simultaneously measuring changes in the phytohormones ABA, JA and SA in plants following biotic and abiotic stress. Plant Methods. 2008;4:16.
38. Jaulneau V, Lafitte C, Jacquet C, Fournier S, Salamagne S, Briand X, et al. Ulvan, a sulfated polysaccharide from green algae, activates plant immunity through the jasmonic acid signaling pathway. J Biomed Biotechnol. 2010;2010:525291.
39. Chervin C, Deluc L. Ethylene signalling receptors and transcription factors over the grape berry development: gene expression profiling. Vitis. 2010;49:129–36.
40. Vaillé J. La Statistique au service des Données: quelques macros Excel pour faire de l'analyse exploratoire des données. 2010, Modulad n°43, http://www.modulad.fr/index.htm
41. Bramley PM. Regulation of carotenoid formation during tomato fruit ripening and development. J Exp Bot. 2002;53:2107–13.

Comparative study of the protein profiles of Sunki mandarin and Rangpur lime plants in response to water deficit

Tahise M Oliveira[1], Fernanda R da Silva[2], Diego Bonatto[2], Diana M Neves[1], Raphael Morillon[3,4], Bianca E Maserti[5], Mauricio A Coelho Filho[6], Marcio GC Costa[1], Carlos P Pirovani[1] and Abelmon S Gesteira[6*]

Abstract

Background: Rootstocks play a major role in the tolerance of citrus plants to water deficit by controlling and adjusting the water supply to meet the transpiration demand of the shoots. Alterations in protein abundance in citrus roots are crucial for plant adaptation to water deficit. We performed two-dimensional electrophoresis (2-DE) separation followed by LC/MS/MS to assess the proteome responses of the roots of two citrus rootstocks, Rangpur lime (*Citrus limonia* Osbeck) and 'Sunki Maravilha' (*Citrus sunki*) mandarin, which show contrasting tolerances to water deficits at the physiological and molecular levels.

Results: Changes in the abundance of 36 and 38 proteins in Rangpur lime and 'Sunki Maravilha' mandarin, respectively, were observed via LC/MS/MS in response to water deficit. Multivariate principal component analysis (PCA) of the data revealed major changes in the protein profile of 'Sunki Maravilha' in response to water deficit. Additionally, proteomics and systems biology analyses allowed for the general elucidation of the major mechanisms associated with the differential responses to water deficit of both varieties. The defense mechanisms of Rangpur lime included changes in the metabolism of carbohydrates and amino acids as well as in the activation of reactive oxygen species (ROS) detoxification and in the levels of proteins involved in water stress defense. In contrast, the adaptation of 'Sunki Maravilha' to stress was aided by the activation of DNA repair and processing proteins.

Conclusions: Our study reveals that the levels of a number of proteins involved in various cellular pathways are affected during water deficit in the roots of citrus plants. The results show that acclimatization to water deficit involves specific responses in Rangpur lime and 'Sunki Maravilha' mandarin. This study provides insights into the effects of drought on the abundance of proteins in the roots of two varieties of citrus rootstocks. In addition, this work allows for a better understanding of the molecular basis of the response to water deficit in citrus. Further analysis is needed to elucidate the behaviors of the key target proteins involved in this response.

Keywords: Citrus rootstock, Water deficit, Proteomics, Protein network

* Correspondence: abelmon.gesteira@embrapa.br
[6]Embrapa Mandioca e Fruticultura, Rua Embrapa, s/n, Cruz das Almas 44380-000, Bahia, Brazil
Full list of author information is available at the end of the article

Background

Among potential abiotic stresses, water deficit is considered to have the largest effect on agricultural productivity and is one of the main factors limiting the distribution of species worldwide [1]. When plants are subjected to water deficit, numerous morphological and physiological responses are observed, and the amplitude of these responses depends on the plant genotype as well as the duration and severity of the stress [2,3].

The plant response to water deficit involves several processes, beginning with the perception of stress, followed by modulation of the expression of specific genes, and finally, the appearance numerous transcriptomic, proteomic and metabolomic changes. These changes result in the regulation of metabolism and the generation of regulatory networks that are involved in plant defense against the harmful effects of stress [4,5].

Transcriptomic studies have revealed that the expression of a wide range of genes is regulated in response to water deficit in citrus plants. Analysis of 2,100 expressed sequence tags (ESTs) in the roots of Rangpur lime (*Citrus limonia* Osbeck) subjected to osmotic stress resulted in the identification of genes involved in the water stress response, including those encoding aquaporins, dehydrins, sucrose synthase and enzymes related to the synthesis of proline [6]. Using a microarray containing 6,000 genes, Gimeno et al. [7] investigated the response of the transcriptome of 'Clementine' mandarin (*C. clementina* Ex Tanaka) grafted onto 'Cleopatra' mandarin (*C. reshni* hort. Ex Tanaka) to water deficit conditions. As observed in other species, genes encoding proteins involved in lysine, proline and raffinose catabolism, hydrogen peroxide reduction, vacuolar malate transport, and defense (including osmotins, dehydrins and chaperones) were induced. Analysis of the NAC family of transcription factors resulted in the identification of one member, *CsNAC1*, that was strongly induced by water deficit in the leaves of 'Cleopatra' mandarin and Rangpur lime and by salt stress, cold and abscisic acid (ABA) only in the leaves and roots of 'Cleopatra' mandarin [8]. In 'Cleopatra' mandarin, Xian et al. [9] isolated a gene encoding *CrNCED1*, which is an enzyme involved in ABA synthesis, and produced transgenic plants that constitutively overexpressed this gene. The transgenic lines displayed tolerance to dehydration, drought, salt, and oxidative damage compared with wild-type plants. Furthermore, low levels of reactive oxygen species (H_2O_2 and O_2^-) were detected in the transgenic plants under salt stress and dehydration.

In addition to studies addressing the effects of water deficit on the transcriptome, proteomic studies have revealed the role of proteins involved in the complex mechanisms underlying the stress responses of plants [4,10]. Indeed, many proteins related to stress defense, detoxification, carbohydrate metabolism and photosynthesis that participate in the process of adaptation and tolerance to stress have been identified [11,12]. In a study that evaluated changes in the leaves of two contrasting populations of *Populus cathayana* in response to water deficit, 40 drought-responsive proteins were identified: several of the proteins showing altered abundance were involved in transcriptional regulation, secondary metabolism, redox homeostasis and stress defense [13]. An investigation of soybean (*Glycine max* L.) roots subjected to short-term water deficit revealed changes in the abundance of proteins involved in carbohydrate and nitrogen metabolism, cellular defense and programmed cell death [14]. Zadražnik et al. [5] identified drought-responsive proteins in the leaves of two bean cultivars with differing responses to drought stress. These proteins are primarily involved in energy metabolism, ATP conversion, photosynthesis, protein synthesis and proteolysis and stress defence. Changes in protein levels in the leaves of 'Willow leaf' and 'Cleopatra' mandarin plants subjected to salt stress were analyzed by Podda et al. [15]. Significant variations in the abundance of 44 protein spots were detected. These salt-responsive proteins play roles in photosynthetic processes, ROS scavenging, stress defense, and signaling. However, there are few studies of the root proteome. Analysis of the root proteome of wild watermelon (*Citrullus lanatus* sp.) has revealed that proteins involved in root morphogenesis, carbon/nitrogen metabolism, lignin synthesis and molecular chaperones are differentially regulated under drought stress [16].

In the present study, we used proteomic approaches to analyses changes in the protein profiles of the roots of two citrus rootstock cultivars with contrasting responses to water deficit. Proteins showing significantly altered abundance were selected for identification via mass spectrometry and bioinformatics analysis. In Rangpur lime, the abundance of various proteins involved in protein metabolism, the stress response and proteolysis were modulated under water deficit conditions. In contrast, repair-related proteins contributed more specifically to the response of 'Sunki Maravilha' mandarin to this stress. This is the first report to examine the effects of water deficit on the abundance of proteins in citrus roots.

Results

In the present study, root samples of Rangpur lime and 'Sunki Maravilha' mandarin collected in a previous study by Neves et al. [17] were used. Considering the soil moisture data from the previous report [17], two sampling points were selected for proteomic analysis as follows: 1) plants grown in soil with moisture ranging from 0.29-0.28 m^3m^{-3} were defined as 'control' plants, whereas 2) the soil moisture for 'drought-stressed' plants ranged from 0.15-0.14 m^3m^{-3}. According to a previous report by Neves et al. [17], stomatal resistance is more pronounced in both varieties at the selected drought stress sampling points. In addition, they have reported that the leaf water potential decreases in water-stressed plants, reaching −1.43 MPa and −1.3 MPa in Rangpur lime and 'Sunki Maravilha' mandarin, respectively. Interestingly, Rangpur lime shows a higher growth rate when grown under water deficit compared with the rate observed for 'Sunki Maravilha'. When subjected to water deficit, the leaves and roots of 'Sunki Maravilha' display a progressive increase in the ABA concentration. The lower leaf growth rate that has been recorded for 'Sunki Maravilha' mandarin may be associated with its greater leaf ABA concentration. In contrast, in Rangpur lime, alternations between high and low ABA concentrations were observed [17].

Analysis of root protein profiles in response to water deficit

To elucidate the changes in protein abundance in response to water deficit, comparative analysis of the protein profiles of roots of Rangpur lime and 'Sunki Maravilha' mandarin was performed via 2D gel electrophoresis. The root protein profiles of both varieties that were grown under control conditions and subjected to water stress are shown in Figure 1. More than 350 spots were detected in both varieties via image analysis. A total of 81 spots showed significant changes in abundance ($P < 0.05$) in the Rangpur lime roots. These spots were subjected to mass spectrometry (MS) analysis, and 36 proteins were identified. Among these proteins, 11 were increased and 18 were decreased in abundance, and seven proteins were unique to this genotype. In 'Sunki Maravilha' mandarin, 72 spots showed significant changes in abundance. Among these spots, 38 proteins were identified, 14 of which increased and 12 of which decreased in abundance, and nine were unique to this genotype.

To understand the relationship between the two plant varieties as a function of water stress, multivariate analysis and principal component analysis (PCA) were performed (Figure 2). PC1 represented 75% of the variance, suggesting that there were differences between Rangpur lime and 'Sunki Maravilha' mandarin in response to water deficit. PC2 accounted for 17% of the variance, indicating that in 'Sunki Maravilha', there were differences between the well-watered plants and those under water stress. Interestingly, we observed only minor changes in the protein profiles of Rangpur lime under control versus drought-stressed conditions, suggesting that protein abundance was less affected by water deficit in this variety.

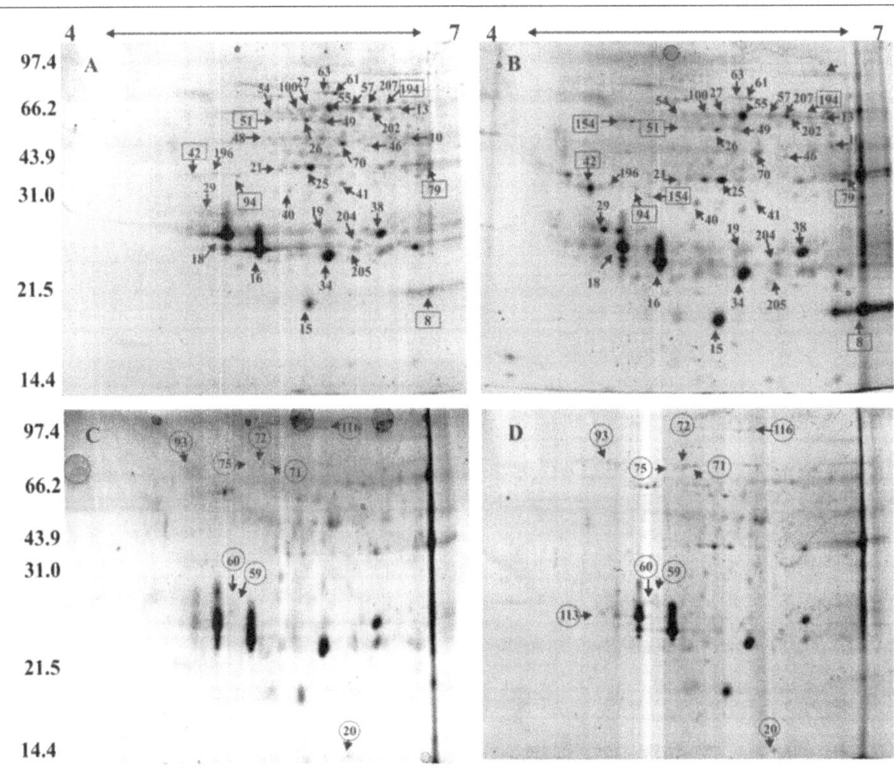

Figure 1 2-DE analysis of root proteins in Rangpur lime under control conditions (A) and following water deficit (B) and in 'Sunki Maravilha' mandarin under control (C) and water deficit conditions (D). The proteins indicated by the arrows were differentially expressed under the applied treatment. The proteins in the squares are unique to Rangpur lime, and those in the circles are exclusive to 'Sunki Maravilha' mandarin.

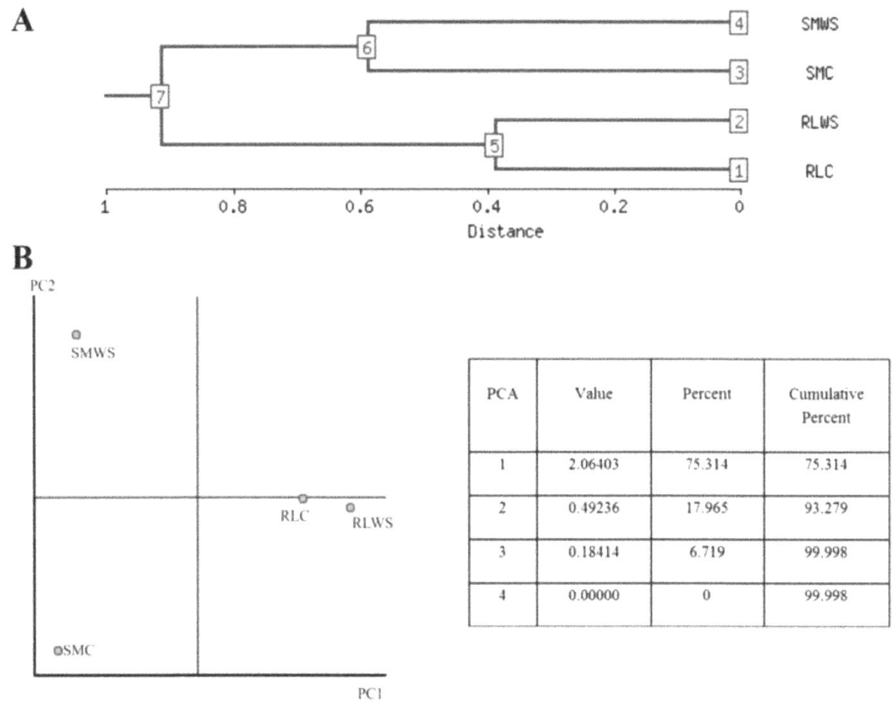

PCA	Value	Percent	Cumulative Percent
1	2.06403	75.314	75.314
2	0.49236	17.965	93.279
3	0.18414	6.719	99.998
4	0.00000	0	99.998

Figure 2 Principal component analysis (PCA) and evaluation of variance under control conditions and drought in Rangpur lime and 'Sunki Maravilha' mandarin. (A) Hierarchical clustering of the experiments and **(B)** PCA and eigenvalues table in control and water stress-treated samples from Rangpur lime and 'Sunki Maravilha'.

Identification and analysis of differentially expressed proteins

Spots showing differential intensities under water deficit were excised from the two-dimensional polyacrylamide gel electrophoresis (2D-PAGE) gels and identified via MS (detailed MS/MS results are provided in Additional file 1: Table S1). Some proteins were identified more than once in different spots, reflecting different isoforms, post-translational modifications or alternative mRNA splice forms [18]. Two spots were identified as epidermis-specific secreted glycoprotein EP1-like (8 and 13), five as germin-like (16, 18, 19, 34 and 38), four as 2-phospho-D-glycerate hydrolase (21, 48, 54 and 79), three as mitochondrial processing peptidase alpha-1 subunit (26, 55 and 57), two as putative mitochondrial processing peptidase (202 and 207), four as annexin 1 (51, 59, 60), two as annexin D2 (25 and 94), two as heat shock protein 70 (46 and 154), two as fructokinase (116 and 196) and two as lactoylglutathione lyase (61 and 63).

In addition, some of the proteins that were represented by different spots on the 2D gel showed opposite expression patterns (one spot showed an increase in abundance, whereas the other exhibited a decrease in abundance). The germin-like proteins, which were represented by five spots (16, 18, 19, 34, and 38), and putative mitochondrial processing peptidase (202 and 207) exhibited opposite patterns of accumulation in 'Sunki Maravilha' mandarin (Table 1). In contrast, mitochondrial processing peptidase alpha 1 subunit (26, 55, and 57) showed opposite accumulation patterns in Rangpur lime.

The functions of the identified proteins were inferred using the UniProt database (http://www.uniprot.org). The identified proteins were classified into the following seven major groups according to their possible biological functions: stress and defense response (36% and 35%), metabolism (25% and 21%), transport (9% and 8%), energy (13% and 14%), signal transduction (4% and 5%), protein metabolism (10% and 10%) and unknown (2% and 1%) for Rangpur lime and 'Sunki Maravilha' mandarin (Figure 3A and B), respectively. Although the protein groups did not differ significantly between the two studied varieties, an additional class of proteins involved in DNA repair was observed for 'Sunki Maravilha' mandarin (Figure 3B).

Analysis of protein-protein interactions

Analysis of interactomic data of *A. thaliana* orthologous proteins corresponding with the protein interaction profiles of Rangpur lime and 'Sunki Maravilha' mandarin (both with and without water deficit) allowed us to draw an interactome network. The network developed for 'Sunki Maravilha' mandarin included 723 proteins and 10,430 connectors, and that constructed for Rangpur

Table 1 Identification of differentially expressed proteins in the roots of Rangpur lime and 'Sunki Maravilha' mandarin subjected to water deficit

ID spot[a]	Identified protein reference organism[b]	Accession number[c]	Mascot score/*P*-value[d]	Mr Theor/Exp[e]	pI Theor/Exp[f]	Expression level[g] A B C D	Fold change ($P < 0.05$)[h] RL Sk
8	Epidermis-specific secreted glycoprotein EP1-like *Citrus sinensis*	XP_006477736	155/1e-08	48.8/18	6.26/6.92		**−3.81 np**
10	Dihydrolipoyllysine-residue succinyltransferase component of 2-oxoglutarate dehydrogenase complex 1, mitochondrial-like *Citrus sinensis*	XP_006475040	67/4e-15	51.1/42	9.13/6.55		−2.11 **2.17**
13	epidermis-specific secreted glycoprotein EP1-like *Citrus sinensis*	XP_006477736	130/0.0	48.8/59	6.27/6.61		−1.82 **1.50**
15	miraculin-like protein 1 *Citrus maxima*	AEK31192	176/1e-20	18.9/15	8.18/5.80		**2.32** *1*
16	Germin-like protein subfamily T member 2-like *Citrus sinensis*	XP_006477534	235/2e-169	25.9/24	5.74/5.35		−1.65 **1.2**
18	Germin-like protein subfamily T member 2-like *Citrus sinensis*	XP_006477534	235/2e-169	25.9/24	5.74/5.15		−1.57 -1.38
19	Germin-like protein subfamily T member 2-like *Citrus sinensis*	XP_006477534	235/2e-169	25.9/27	5.74/5.96		−1.39 **2.27**
20	Nucleoside diphosphate kinase *Citrus sinensis*	XP_006464834	60/2e-09	16.3/13	5.91/6.21		Np *1*
21	2-phospho-D-glycerate hydrolase *Citrus trifoliata*	ADD12953	911/2e-18	47.77/39	5.42/5.57		−1.86 **1.56**
25	Annexin D2 *Arabidopsis thaliana*	NP_174810	160/6e-173	36.20/38	5.21/5.83		**1.60 2.47**
26	Mitochondrial processing peptidase subunit alpha-1 *Arabidopsis thaliana*	NP_175610	189/4e-14	48.20/50	6.08/5.79		−1.47 *1*
27	Lipase class 3 family protein *Arabidopsis thaliana*	NP_567515	150/0.0	59.06/59	9.33/5.86		−1.33 ∞
29	TIR-NBS-LRR type disease resistance protein *Citrus trifoliata*	AAN62351	93/2e-37	41.4/30	7.10/4.91		**3.17** -1.62
34	Germin-like protein 3–3 like *Citrus sinensis*	XP_006477531	222/3e-31	43.3/24	5.73/6.07		*1* -2.79
38	Germin-like protein 3–3 like *Citrus sinensis*	XP_006477531	222/3e-31	43.3/27	5.73/6.45		*1* -3.45
40	Glyoxalase *Theobroma cacao*	XP_007026102	413/4e-144	27.06/33	6.52/5.64		**1.73** -3.26
41	Mitochondrial malate dehydrogenase *Citrus sinensis*	AET22414	285/4e-98	30.89/34	5.2/6.10		−1.71 ∞
42	Chitinase *Citrus sinensis*	CAA93847	58/0.0045	32.45/35	5.06/4.80		**3.44** np
46	Heat shock protein 70 *Arabidopsis thaliana*	CAA05547	77/0.0	71.4/46	5.14/6.32		**1.64 1.76**
48	2-phospho-D-glycerate hydrolase *Citrus trifoliata*	ADD12953	911/0.0	47.77/46	5.42/5.45		**2.63** *1*
49	Methyl-CPG-binding domain 6 protein *Arabidopsis thaliana*	NP_200746	150/2e-43	24.44/49	9.03/6.0		**1.35 2.07**
51	Annexin 1 *Theobroma cacao*	NP_174810	109/0.0	35.8/48	6.34/5.72		**2.16** np
54	2-phospho-D-glycerate hydrolase *Citrus trifoliata*	ADD12953	911/0.0	47.77/51	5.42/5.7		−1.94 -2.53
55	Mitochondrial processing peptidase subunit alpha-1 *Arabidopsis thaliana*	NP_175610	545/0.0	54.4/56	5.94/5.9		**1.42** ∞
57	Mitochondrial processing peptidase subunit alpha-1 *Arabidopsis thaliana*	NP_175610	545/0.0	60.79/57	7.06/6.26		**1.83 1.82**
59	Annexin 1 *Theobroma cacao*	EOY16019	566/0.0	35.8/36.1	6.34/5.42		np **2.09**
60	Annexin 1 *Theobroma cacao*	EOY16019	566/0.0	35.8/36.1	6.34/5.42		np **2.38**
61	Lactoylglutathione lyase *Citrus X paradisi*	CAB09799	598/0.0	32.63/66	5.28/6.05		−1.81 **1.66**
63	Lactoylglutathione lyase S-transferase *Ricinus communis*	XP_002518470	52/8e-146	31.5/66	7.63/5.99		−1.90 **1.65**
70	Peroxidase *Citrus maxima*	ABG49115	517/0.0	37.88/44	4.52/6.10		−3.21 *1*
71	Histone ubiquitination proteins group *Populus trichocarpa*	XP_002302510	188/0.0	48.1/67	5.56/5.71		np −3.61

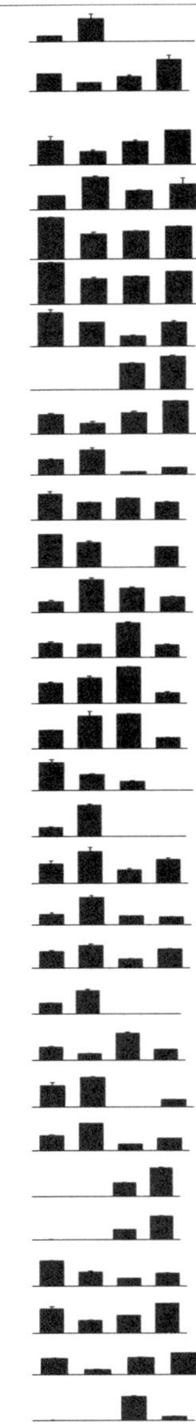

Table 1 Identification of differentially expressed proteins in the roots of Rangpur lime and 'Sunki Maravilha' mandarin subjected to water deficit *(Continued)*

72	Acyl-CoA -N-acetyltransferase *Arabidopsis thaliana*	NP_196882	46.4/7e-29	20.39/22	7.8/5.48		np *1*
75	5-formyltetrahydrofolate cyclo-ligase *Arabidopsis thaliana*	NP_565139	355/1e-119	39.55/39	9.41/6.63		np −2.20
79	2-phospho-D-glycerate hydrolase *Citrus trifoliata*	ADD12953	62.6/0.0	47.78/34	5.54/6.39		−2.68 np
93	mRNA-capping enzyme *Arabidopsis thaliana*	NP_974263	74.4/0.0	78.7/75.57	6.74/5.52		Np *1*
94	Annexin D2 *Citrus sinensis*	CAB09799	116/4e-122	19.8/36	5.30/5.16		−1.79 np
100	ATP synthase beta subunit *Citrus macroptera*	ABM74441	69.4/3e-152	37.07/58	5.01/5.74		−2.68 -4.18
113	2-dehydro-3-deoxyphosphooctonate aldolase *Medicago truncatula*	ABN05924	427/2e-13	31.9/28	6.61/4.91		np ∞
116	Fructokinase *Oryza sativa*	A2WXV8	70/1e-34	30.3/87	5.50/6.19		np ∞
154	Heat shock protein-70 cognate protein *Arabidopsis thaliana*	NP_176036	73/0.0	71.4/65	5.10/5.27		∞ np
194	F-box family protein *Vitis vinifera*	XP_002279122	414/4e-139	47.2/55	9.4/6.39		−1.57 np
196	Fructokinase *Citrus unshiu*	AAS67872	219/2e-71	37.5/36	5.11/4.97		−2.54 **1.64**
202	Putative mitochondrial processing peptidase *Arabidopsis thaliana*	BAE98412	202/0.0	51.53/53	5.71/6.49		**2.88** -1.95
205	Putative L-galactose dehydrogenase *Citrus unshiu*	ADV59927	294/1e-18	37.62/25	6.03/6.23		**1.75** ∞
207	Putative mitochondrial processing peptidase *Arabidopsis thaliana*	BAE98412	480/0.0	51.53/55	5.71/6.33		**1.56 1.47**

[a]Spot ID corresponding to the position in the 2D gel illustrated in Figure 1. [b]Protein accession number according to the NCBI database (http://www.ncbi.org). [c]Best matching protein identified by pBLAST analysis of the non-redundant (NCBInr) database. [d]Mascot score *P* value of the homology between citrus proteins and orthologous, homologous, or paralogous proteins, as annotated in NCBInr. [e]Theoretical and experimental masses (KDa) of identified proteins. [f]Theoretical and experimental *pIs* of identified proteins. [g]Expression levels, presented as the % normalised volume, in the control and water deficit-stressed roots. Vertical bars indicate the mean ± SE. **Rangpur lime**: (A) control; and (B) water deficit. **'Sunki Maravilha'**: (C) control; and (D) water deficit. [h]Fold change (water deficit-treated normalised volume/control normalised volume): **bold** = increased protein abundance; **underlined** = decreased protein abundance; *italics* = no significant difference; np = protein not found in gel; ∞ = present in one treatment in the genotype.

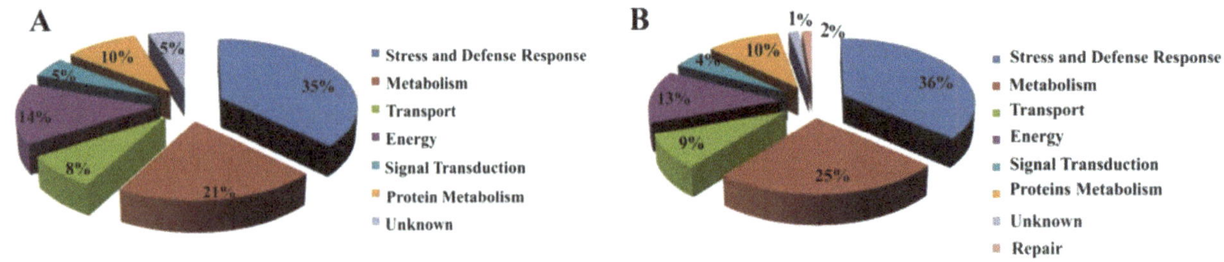

Figure 3 Functional classifications of drought-responsive proteins. (A) Functional categorization of proteins that showed significant changes in abundance in Rangpur lime. **(B)** Functional categorization of proteins with significantly altered levels in 'Sunki Maravilha' mandarin.

lime included 566 proteins and 5,954 connectors (Additional file 2: Figure S1 and S2).

Based on analysis of the intersection between the networks of the two varieties, 190 proteins specific to the Rangpur lime network, 347 proteins unique to the 'Sunki Maravilha' mandarin network, and 376 proteins shared between the two networks were observed (Additional file 2: Figure S1). The interactome networks obtained for each variety could be divided into several functional clusters. Evaluation of the highly connected regions and gene ontologies of each cluster revealed the presence of 21 clusters in the Rangpur lime interactome network and 22 clusters in the 'Sunki Maravilha' mandarin interactome network (Additional file 3: Table S2).

To evaluate the proteins forming the most relevant network, centrality analysis was performed by sorting the proteins into hubs and/or bottlenecks. Among the 45 proteins showing altered abundance in Rangpur lime and 'Sunki Maravilha' mandarin, mitochondrial malate dehydrogenase (AT1G53240), mitochondrial processing peptidase (MPPBETA), 2-phospho-D-glycerate hydrolase (LOS2) and nucleoside diphosphate kinase 1 (NDPK1) were considered hubs/bottlenecks. Glyoxalase (ATGLX1), annexin 1 (ANNAT1), glutathione S-transferase (ATG5TF12) and putative L-galactose dehydrogenase (L-Gadh) were only considered to be bottlenecks (Additional file 2: Figure S1A).

Protein-protein interactions in Rangpur lime

Among the clusters of *A. thaliana* orthologous proteins corresponding with the differentially abundant proteins identified in the two studied citrus varieties, eight were exclusively related to Rangpur lime (Figure 4, clusters A-H). Fructokinase, which was a bottleneck protein in Rangpur lime, was present in a sub-functional network involved in growth, development and the stress response (Figure 4, cluster A, Additional file 3: Table S2) that contained the proteins CYP96A4, CYP71A, CYP70, and CYP76 and representatives of the cytochrome P450 superfamily. This cluster also included WRKY transcription factors (WRKY6 and WRKY75). The WRKY75 transcription factors were associated with osmotin 34 (ATOSM34), which interacted with chitinase (ATHCHIB), which is an enzyme involved in the response to various environmental stresses. Moreover, ATHCHIB was present in the clusters corresponding to metabolism and ethylene-dependent systemic resistance (Figure 4, cluster B) and was related to beta-hexosaminidase (HEXO1), which was in turn associated with galactosidases (BGAL and AtAGAL), which are involved in carbohydrate metabolism. The cluster related to the metabolism of amino acids and protein modification (Figure 4, cluster C) contained cinnamyl alcohol dehydrogenases (CAD2, CAD3, and CAD6), which were linked to numerous peroxidases (Figure 4, cluster H) associated with the oxidative stress response. In the cluster related to the methylation and transposition of DNA, methyl-CpG-binding proteins (MBD6 and MBD3) and chromatin remodelling 1 (CHR1) were found.

Tyrosine aminotransferase 3 (TAT3) was present in a cluster associated with hormone biosynthesis and responses to jasmonic acid and ABA (Figure 4, cluster F), and this protein interacted with coronatine-induced 3 (CORI3), which is involved in methyl jasmonate signalling in guard cells (Figure 4, cluster F). CORI3 was related to superroot 1 (SUR1) and bisphosphate nucleotidase/inositol (SAL1). SAL1 was also associated with a cluster of proteins involved in amino acid metabolism (Figure 4, cluster G) and interacted with proteins involved in the biosynthesis of myoinositol, which is a signalling molecule involved in the stress response (IMPL1, VTC4, MIPS1, and MIPS3), and with phospholipase C (PLC), which is an important protein in the adaptation of plants to environmental stresses. These proteins constituted the cluster related to the response to water deficit stress (Figure 4, cluster D).

Interactome network analysis for 'Sunki Maravilha' mandarin

Gene ontology (GO) analysis allowed us to identify the most representative biological processes in the protein interaction network of *A. thaliana* related to 'Sunki Maravilha' mandarin (Additional 3: Table S2). The interactome networks were divided into several sub-functional networks (Figure 5), and six of these clusters were found

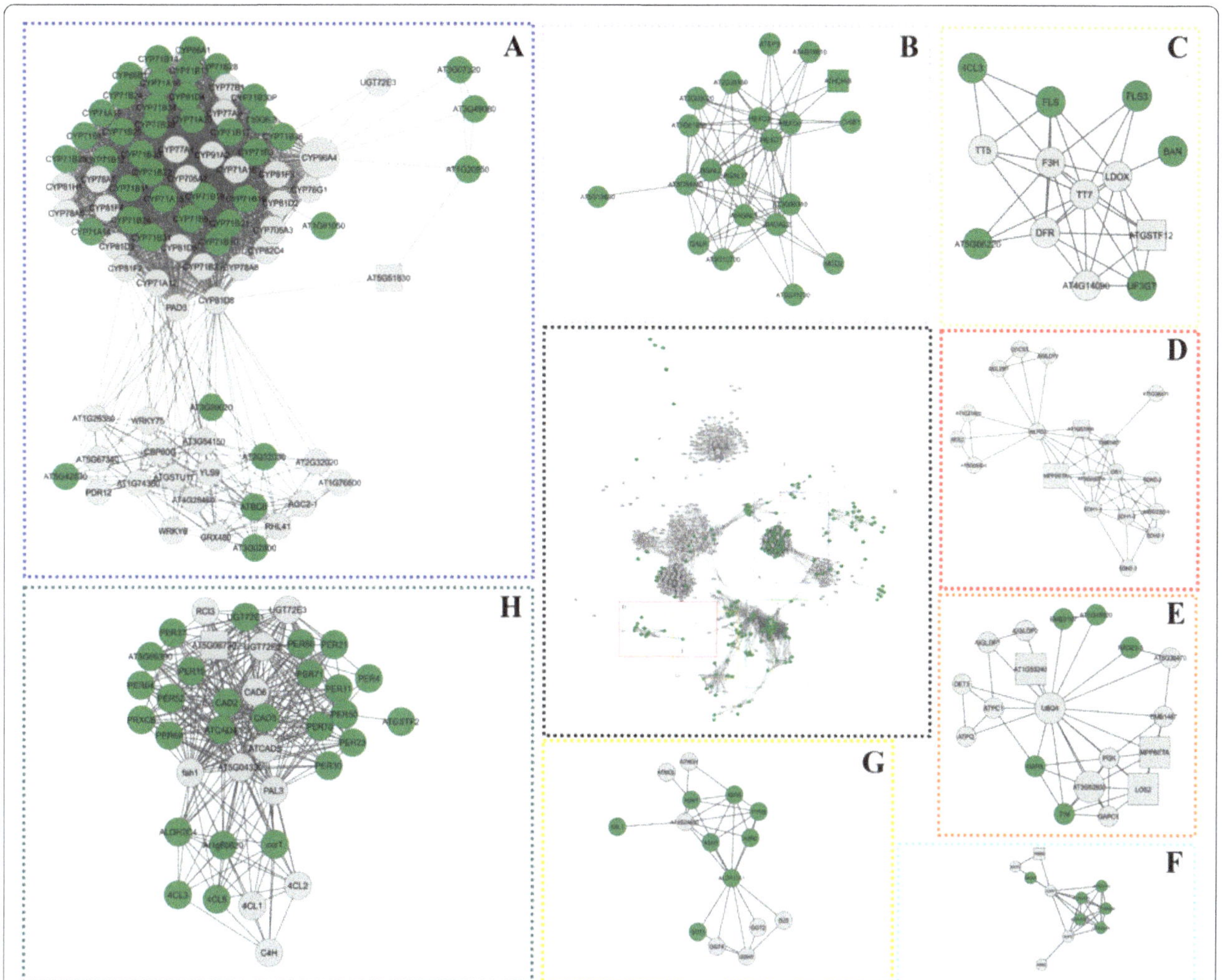

Figure 4 Interactome network of *A. thaliana* orthologous proteins related to water stress response of Rangpur lime. General network with inserts (in colour) that represent clusters (detailed in **A-H**). **(A)** A cluster (in blue) corresponding with proteins related to metabolism, development and the abiotic stress response. **(B)** A cluster (in purple) corresponding with proteins involved in carbohydrate metabolism and systemic responses that are dependent on ethylene. **(C)** A cluster (in light green) containing proteins related to protein modification. **(D)** A cluster (in red) of proteins involved in the response to drought stress. **(E)** A cluster (in orange) corresponding with proteins related to DNA methylation. **(F)** A cluster (in light blue) related to proteins involved in amino acid metabolism. **(G)** A cluster (in yellow) comprising amino acid precursor proteins. **(H)** A cluster (in dark green) corresponding with proteins related to oxidative stress. The circles indicate proteins involved in biological processes corresponding to the network, and the squares indicate proteins that were also differentially expressed during treatment. The green nodes indicate proteins that were unique to the Rangpur lime.

to be unique to 'Sunki Maravilha' mandarin orthologous proteins. The main clusters were related to DNA repair and amino acid and nucleic acid metabolism (Figure 5, clusters A-F). A large number of RNA polymerases (NRPB1, NRPB5C, NRPB11, NRPE5, RPB5E and AT1G61700) were found in a cluster corresponding with DNA repair and methylation (Figure 5, cluster F). These proteins were related to the NDPK1 protein, which was exclusive to 'Sunki Maravilha' and was considered to be a hub/bottleneck (HB) of this genotype (Additional file 2: Figure S1b). In turn, NDPK1 was associated with LOS2 and MPPBETA (Figure 5, cluster D), which were also considered to be HB proteins in the 'Sunki Maravilha' mandarin network (Additional file 2: Figure S1b). This cluster was related to metabolism and biotic and abiotic stresses (Figure 5, cluster D) and included several proteins that showed altered abundance in 'Sunki Maravilha' following exposure to water deficit that are known to be involved in the stress response. These proteins included mitochondrial malate dehydrogenase (AT1G53240), glyoxalase (ATGLX1), 2-dehydro-3-deoxyphosphooctonate aldolase (KDSA), fructokinase (AT3G54090) and the dihydrolipoyllysine-residue succinyltransferase component of 2-oxoglutarate dehydrogenase complex 2 (AT4G26910).

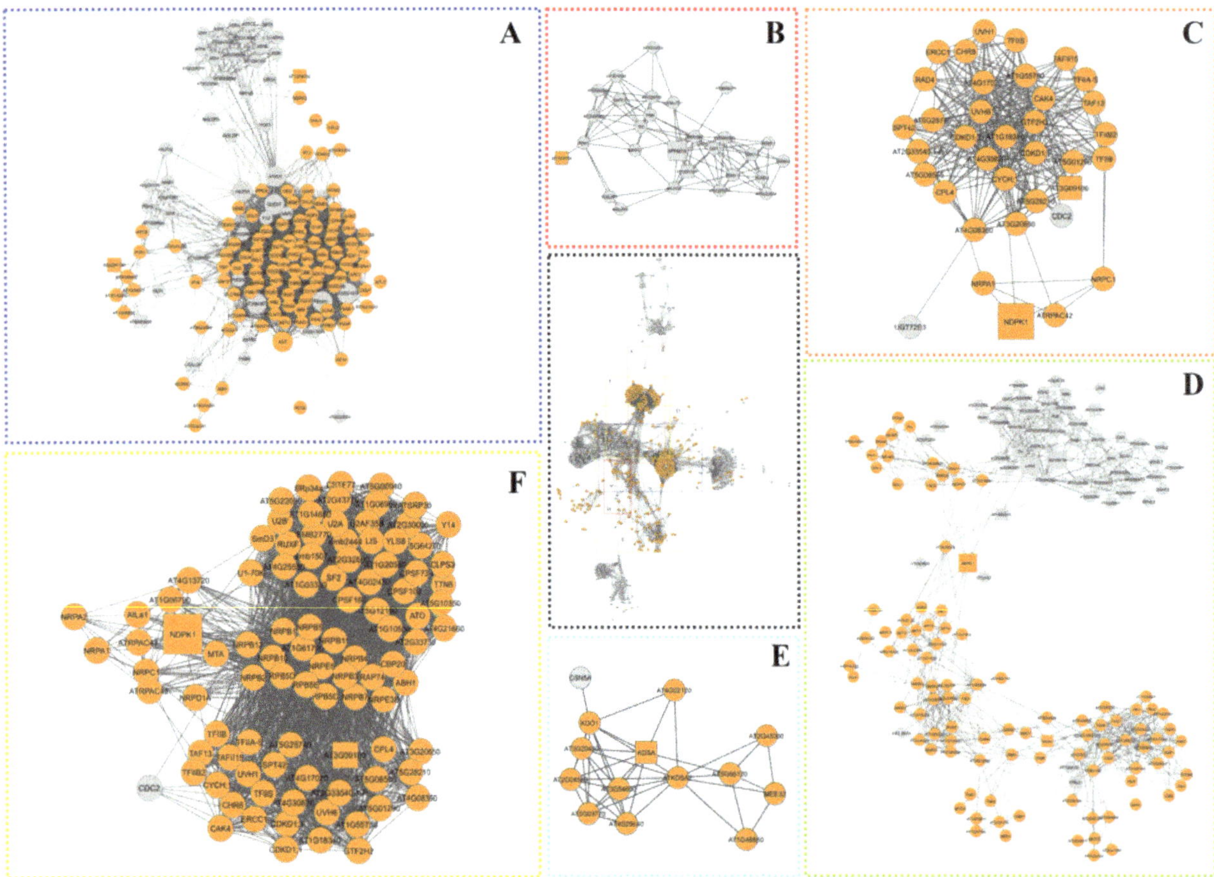

Figure 5 Interactome network of *A. thaliana* orthologous proteins related to the water stress response of 'Sunki Maravilha' mandarin. General network with inserts (in colour) that represent clusters (detailed **A**-**F**). **(A)** A cluster (in dark blue) corresponding with proteins related to the metabolism of nucleotides and ubiquitination. **(B)** A cluster (in red) comprising proteins involved in nucleotide metabolism. **(C)** A subgraph (in orange). **(D-F)** Two clusters (in green and yellow in **D** and **F**, respectively) corresponding with proteins involved in DNA repair. **(E)** A cluster (in light blue) containing proteins related to cell division. The circles indicate proteins involved in biological processes corresponding to the network, and the network squares indicate proteins that were also differentially expressed during treatment. The orange nodes indicate proteins that were unique to 'Sunki Maravilha' mandarin.

The cluster related to DNA repair (Figure 5, cluster C) contained RAD4, 5′-flap endonuclease (ERCC1), chromatin remodelling 8 (CHR8), and ultraviolet hypersensitive (UVH) proteins, which are involved in nucleotide excision from damaged DNA, the regulation of replication and the response to UV light, respectively. A large number of transcription factors were also found (GTF2H2, SPT42, TAF13, TFII-S, TFIIB, TFIIS, and TAFI15). A cluster (Figure 5, cluster E) related to cell division included the following proteins: cytidylyltransferase family protein (KDO1), 3-deoxy-D-manno-octulosonic acid transferase (AT5G03770), 3-deoxy-8-phosphooctulonate synthase (ATKDSA2), and maternal effect embryo arrest 32 (MEE32).

A cluster related to energy, the metabolism of nucleic acids and ubiquitination (Figure 5, cluster A) included an ubiquitin (UBQ4) protein that binds to ATPC1, which was in turn associated with the following proteins: glyceraldehyde 3-phosphate dehydrogenase subunit 2a (GAPA-2), glycine dehydrogenase (GDH), serine transhydroxymethyltransferase 1 (SHM1), A. *thaliana* glycine decarboxylase P-protein 2 (AtGLBP2) and sedoheptulose bisphosphatase (SBPASE), which is involved in the metabolism of osmoprotectants.

Discussion

Proteins showing altered abundance in response to water deficit in the roots of Rangpur lime and 'Sunki Maravilha' mandarin were identified using a proteomic approach. Rangpur lime and 'Sunki Maravilha' were selected because in a previous study, these cultivars have been shown to differ in their use of available soil water, ABA accumulation and expression of ABA biosynthesis genes, suggesting that they use different systems to adapt to water restriction [17].

Protein changes associated with water deficit

Water deficit caused alterations in protein abundance in Rangpur lime and 'Sunki Maravilha' mandarin. Approximately 40% of the identified proteins were detected in multiple spots and had different isoelectric points (pIs) and molecular weights (MWs), suggesting the presence of isoforms and post-translational modifications or that these proteins were translated from different products of paralogous genes within a multigene family (Table 1) [19]. The observed changes were related to phenotypic responses that determined the plant tolerance to water deficit [4].

Major changes in protein abundance caused by water stress were observed in 'Sunki Maravilha' mandarin compared with Rangpur lime (Figure 2). Under control conditions, Neves et al. [17] have measured higher ABA concentrations in the roots of unstressed 'Sunki Maravilha' mandarin compared with Rangpur lime. This finding can be considered to indicate increased physiological responsiveness to biotic and abiotic stresses, which enables better stomatal regulation and consequently reduces water use by 'Sunki Maravilha' mandarin plants subjected to water stress. This physiological responsiveness leads to a series of changes at the protein level as an adaptive response to water stress. The proteins identified in Rangpur lime and 'Sunki Maravilha' mandarin were functionally categorized in terms of their roles in the response to water restriction. Comparative analysis of protein accumulation in Rangpur lime and 'Sunki Maravilha' mandarin, together with the use of a systems biology approach, allowed us to establish a general profile of the biological processes involved in the response to water deficit in these plant varieties (Figure 4). The main functional groups of proteins were examined in relation to water deficit.

Proteins involved in metabolism and energy

The energy metabolism of proteins is often affected by water deficit. In the present study, the abundance of some enzymes involved in the tricarboxylic acid (TCA) cycle and glycolysis was altered following water deficit in both evaluated varieties. In Rangpur lime and 'Sunki Maravilha' mandarin, the mitochondrial malate dehydrogenase levels (spot 41) declined in response to water deficit. This enzyme, which was considered to represent an HB in Rangpur lime (Figure 4A), is a key component of the TCA cycle [20] that is involved in central metabolism and redox homeostasis between organelle compartments [21]. Another protein in the energy metabolism class, ATP synthase, showed decreased abundance in both varieties, which suggested that damage had occurred to the mitochondria and chloroplasts exposed to water deficit. In addition, energy metabolism may have been weakened, which is a disadvantage due to the resulting decreases in the syntheses of ATP and metabolites and feedback signaling. Thus, plants require additional energy to repair the damage caused by water stress.

Mitochondrial processing protein accounted for approximately 11% of the proteome-level changes observed (Table 1), suggesting that mitochondrial function and, hence, plant metabolism were altered and that the integrity of these processes must be protected from oxidative stress induced by drought [22]. Indeed, it is known that a key function of the mitochondria is defense against an excess of ROS. In fact, in plant cells, the mitochondria represent major sources of ROS production and subsequent oxidative damage, as indicated in other proteomic studies [23-25]. These findings suggest that plant tolerance to water deficit may be associated with efficient defense responses against oxidative stress at the cellular and subcellular levels.

LOS2 is an essential glycolytic enzyme that catalyses the interconversion of 2-phosphoglycerate and phosphoenolpyruvate and is induced by several types of abiotic stress, including water deficit and salinity [26]. In the present study, LOS2 showed different patterns of expression and isoforms and was classified as an HB protein in both varieties. In addition, it was identified in a unique 'Sunki Maravilha' mandarin cluster that was involved in the stress response and metabolism (Figure 5, cluster D). The opposite expression patterns observed for LOS2 indicate that this protein may play different roles during the water stress response in the two varieties. Systems biology analysis identified interactions between LOS2 and other proteins with important roles in the stress response and energy metabolism, such as NDPK1 (Figure 5, cluster D), which was detected exclusively in 'Sunki Maravilha' mandarin and may be involved in the acclimation of this variety to water deficit due to its relationships with the general homeostasis of cellular nucleoside triphosphate [27], oxidative stress responses [28] and water deficit tolerance in bean [5].

Stress and defense proteins

Approximately 37% of the proteins identified in this study were related to stress and defense (Figure 3). Plants have evolved antioxidant defense pathways to protect cells against the damage caused by high levels of ROS under stress conditions [29]. Several enzymes involved in redox homeostasis were differentially regulated in the responses of Rangpur lime and 'Sunki Maravilha' mandarin to water stress, including peroxidase (spot 70, Figure 4, cluster H), lactoylglutathione lyase (spot 61 and 63) and glyoxalase (spot 40, Figure 4, cluster H, Figure 5, cluster D). In the two plant varieties, glyoxalase (Figure 5D) showed opposite abundance patterns compared with those observed for lactoylglutathione lyase. The excessive production of ROS in stressed plants contributes to the accumulation of other toxic compounds, such as methylglyoxal, which is regulated by the glyoxalase system. This system plays a role in tolerance to oxidative stress through the recycling of reduced glutathione (GSH) and specific changes in the absolute concentrations of ROS [5].

Another protein that showed contrasting abundance between the two evaluated varieties was peroxidase, which, together with other enzymes, participates in the removal of H_2O_2 from cells. In Rangpur lime, we observed a decrease in peroxidase abundance under water stress conditions, whereas in 'Sunki Maravilha' mandarin, the abundance of this enzyme increased. In addition, systems biology analysis identified a cluster related to the oxidative stress response that was unique to Rangpur lime, in which peroxidase was directly associated with phenylalanine, aldehyde dehydrogenase and glutathione S-transferase, which are also involved in the detoxification of ROS (Figure 4, cluster H). The changes in the abundance of several antioxidant enzymes observed in the Rangpur lime and 'Sunki Maravilha' mandarin roots may also be necessary to balance the antioxidant system during water deficit. Choi and Hwang [30] have studied the expression of a peroxidase ($CaPO_2$) in response to different biotic and abiotic stresses in pepper, demonstrating that its expression is strongly induced by drought, cold, salinity and osmotic stresses. They also have shown that treatment with ABA induces the expression of this enzyme. The increased expression of peroxidase observed in 'Sunki Maravilha' mandarin may be associated with the accumulation of ABA, which has been previously observed by Neves et al. [17] during water stress in this variety. Based on our results, it appears that the response of ROS-scavenging enzymes to water deficit in citrus roots is genotype-dependent.

Another protein involved in the stress response is annexin. Two isoforms of this protein with altered abundance following exposure to water stress were identified in our study (ANNAT1 and ANNAT2). The ANNAT1 protein was up-regulated in 'Sunki Maravilha' mandarin but not in Rangpur lime, whereas the abundance of the ANNAT2 isoform was altered in both varieties (Figure 5). Annexins are directly involved in the regulation of signaling pathways that are activated by stress, and changes in the abundance of these proteins can alter plant tolerances to various types of abiotic stresses [31,32]. In addition, the protein concentration of annexin in *Arabidopsis* is increased by ABA treatment, and this hormone may be a general regulator of annexin expression in several varieties of plant species [33-35]. Thus, based on the high levels of ABA found in 'Sunki Maravilha' mandarin [17], it can be speculated that this protein may function as a positive regulator of the accumulation of annexin in this variety. Moreover, ANNAT1 was classified as a bottleneck protein in 'Sunki Maravilha' mandarin (Figure 4B), and it is therefore an important candidate for future experiments designed to provide new insights regarding signaling pathways that are modulated by water stress.

Repair and processing proteins

Proteins related to the repair and processing of nucleotides, such as mRNA-capping enzyme (spot 93, Figure 5, cluster C and F) and the histone ubiquitination protein group (spot 71, Figure 5, cluster A), demonstrated altered abundance in response to water deficit. These proteins were identified exclusively in 'Sunki Maravilha' mandarin, and interestingly, they were present in the interactome sub-network exclusive to this variety. Systems biology analysis identified the interaction of mRNA-capping protein with nucleotide excision repair (NER). The NER pathway is one of the most versatile repair pathways considering the diversity of DNA lesions, including those induced by environmental factors, such as UV radiation [36,37]. DNA repair mechanisms are essential for maintaining genomic stability and integrity under stress conditions, allowing for greater genomic plasticity in response to environmental changes [38].

Protein modification represents a potential target in the engineering of plants with the goal of increasing tolerance to multiple stresses. Post-translational modifications of signaling proteins, such as phosphorylation and ubiquitination, are important for the regulation of gene expression in response to stress. The ubiquitination of proteins can modulate stress response mechanisms involving the regulation and detection of hormone biosynthesis and the control of the abundance of proteins in signaling pathways, particularly transcription factors [39,40].

Studies have shown that ABA is involved in controlling the processing of proteins and RNA [41]. Analyses of the expression of genes in response to ABA treatment in *A. thaliana* have revealed that this hormone alters the expression of ribosomal proteins and genes involved in proteolysis [42]. In addition, Liu and Stone [43] have found that the presence of ABA promotes the self-ubiquitination of multi-domain ubiquitin E3 ligase (KEG) to increase the level of the transcription factor ABSCISIC ACID-INSENSITIVE5 (ABI5) in *A. thaliana*. The high levels of ABA that have been found in 'Sunki Maravilha' mandarin by Neves et al. [17] may have induced the post-translational modifications of some proteins that are involved in the response to water restriction, contributing to the improved adaptation to drought of this variety.

Putative mechanisms in Rangpur lime and 'Sunki Maravilha' mandarin in response to water stress

The results of this study demonstrated that the dynamic stress response observed in Rangpur lime included changes in the metabolism of carbohydrates and amino acids as well as alterations in the activation of ROS detoxification and in the abundance of proteins involved in water stress defense (Figures 3 and 4). For instance, L-galactose dehydrogenase (L-GALDH) (spot 205), which exhibited increased abundance in Rangpur lime, held an important position in the network (corresponding to a bottleneck with a high capacity for interacting with other proteins; Figure 4D). This enzyme was directly related to an ascorbate 5'-biphosphate nucleotide (VTC4) that was associated with several proteins in cluster D (such as SAL1 [inositol phosphatase] and PLC1 [phospholipase C]) and is involved in the water deficit response, signal transduction and the response to ABA (Figure 4). SAL1 acts as a negative regulator of drought tolerance in *A. thaliana*, and its inactivation increases the relative water content, improves water-use efficiency, reduces gas exchange and maintains viable tissues during prolonged water stress [44,45]. In addition, PLC1 is strongly induced under various environmental stresses, such as dehydration, salinity, and low temperature, and it plays a role in the inhibition of stomatal opening by ABA [46,47]. We hypothesize that the interactions of these proteins may be related to the faster response of Rangpur to water stress, as demonstrated by its increased soil water collection efficiency. The ABA level increases in the roots of this plant during water deficit conditions in correlation with the activation of defense mechanisms [17].

Protein interaction network analysis suggested that the adaptation of 'Sunki Maravilha' plants to stress was aided by the activation of DNA repair and processing proteins (Figure 5). Furthermore, our systems biology analysis revealed that the NDPK1 protein, which was considered to represent an HB in 'Sunki Maravilha' mandarin, interacted with proteins in the three clusters involved in the repair and processing of nucleotides (Figure 5, clusters 2, 5 and 6). NDPKs are ubiquitous housekeeping enzymes involved in the response to heat stress [48], UV-B light signaling, growth [49], ROS signaling [27] and phytochrome signaling [50]. Changes in protein abundance and interactions with important proteins involved in DNA repair pathways could be important factors determining the tolerance of 'Sunki Maravilha' mandarin to water deficit.

Conclusions

The present study revealed that the levels of a number of proteins involved in various cellular pathways are affected during water deficit in citrus roots. The results showed that acclimatization to water deficit involved specific responses that differed between Rangpur lime and 'Sunki Maravilha' mandarin. Proteins involved in metabolism, energy and signal transduction were downregulated in 'Sunki Maravilha' mandarin. In addition, proteins involved in the repair and processing of nucleotides were identified exclusively in this variety and showed higher levels in the drought-stressed roots compared with the control roots. In contrast, the response to water deficit in Rangpur lime included an increased abundance of proteins involved in transport, protein metabolism, the stress response and proteolysis and a decreased abundance of proteins related to metabolism and energy.

This study provides insights into the effects of drought on the abundance of proteins in the roots of two varieties of citrus rootstocks. In addition, this work allows for a better understanding of the molecular basis of the response to water deficit in citrus. Further analyses are needed to elucidate the behaviors of the key target proteins that appear to be involved in this response.

Methods

Plant materials and growth conditions

Rangpur lime (*Citrus limonia* Osbeck) and 'Sunki Maravilha' mandarin (*C. sunki* hort. Ex Tanaka) were obtained from the Active Germplasm Bank (BAG) of Embrapa Cassava and Tropical Fruits (Cruz das Almas, Bahia, Brazil). Rootstocks of nucellar and diploid Rangpur lime and 'Sunki Maravilha' were used in our proteomics experiments. The varieties were selected according to their contrasting responses to water deficit; Rangpur lime is considered to be more tolerant than 'Sunki Maravilha' mandarin [51,17]. We used the same plants that were subjected to water deficit analyses in the study conducted by Neves et al. [17]. The plants were divided into two groups as follows: (i) a control treatment group, in which the plants were constantly irrigated to near field capacity, and (ii) a drought stress group, in which the plants were subjected to a complete suspension of irrigation. The applied water deficit lasted for 40 days, and during this period, soil moisture, transpiration, stomatal conductance and leaf water potential were evaluated. Based on soil moisture data, Neves et al. have selected three sampling periods for plant materials; accordingly, plants with moisture values ranging from 0.29 to 0.28 m^3 m^{-3} were selected for the control group, and those with values ranging from 0.20 to 0.19 and 0.17 to 0.16 m^3 m^{-3} were selected for the water deficit group [17]. The relationship between transpiration and soil moisture was estimated for each variety by relating the normalized transpiration rate (NTR) to the fraction of transpirable soil water (FTSW) according to Sinclair and Ludlow [52]. The NTR was calculated by dividing the daily transpiration rate of each plant in the water-deficient group by the average transpiration rate of the control plants of each variety. The FTSW was calculated for each plant daily by subtracting the lower limit of the soil moisture from the soil moisture determined for each plant daily and then dividing that value by the total transpirable soil water for the plant. Rangpur lime seedlings subjected to water deficit displayed decreases in the FTSW and NTR of approximately 20% with respect to the available water (FTSW of 0.19 and NTR of 0.24), and these values in 'Sunki Maravilha' mandarin plants were significantly reduced to approximately 30% of the available water (FTSW of 0.33 and NTR of 0.27). The collection of root samples from water-deficient plants was based on the observation of soil water humidity levels of 0.15 to 0.14 m^3m^{-3}. For the control plants, soil water humidity ranged from 0.29-0.28 m^3m^{-3} [17]. Three independent biological replicates from the drought-stressed plants and control plants of each variety were analyzed. First, the root samples were frozen in liquid nitrogen and stored at −80°C, after which they were lyophilized and stored at −20°C until analysis.

Protein extraction

Total root proteins were extracted using the protocol described by Bertolde et al. [53]. The proteins were dissolved in rehydration buffer (7 M urea, 2 M thiourea, 4% CHAPS, 5 mM tributylphosphine, 0.5% IPG buffer 4–7, and a trace amount of bromophenol blue). The amount of protein in each extract was quantified using a 2-D Quant Kit according to the manufacturer's recommendations (GE Healthcare - Brazil).

Two-dimensional gel electrophoresis

Approximately 350 μg of protein was dissolved in 250 μl of rehydration buffer and loaded onto 13-cm Immobiline DryStrips, pH 4–7 (GE Healthcare). The proteins were separated in the first dimension using an Ettan IPGPhor II unit (GE Healthcare) with the following program: 500 V for 1 hour, 1,000 V gradient for 1 hour, 8,000 V gradient for 2.5 hours and 8,000 V for 55 min for a total of 80 kV hours. Following isoelectric focusing, the gel strips were equilibrated for 15 min in equilibration buffer containing 7.5 M Tris–HCl, pH 8.8, 6 M urea, 30% (v/v) glycerol, 2% SDS and 1% DDT, followed by another 15 min equilibration with 7.5 M Tris–HCl, pH 8.8, 6 M urea, 29.3% (v/v) glycerol, 2% SDS and 2.5% iodoacetamide. In the second dimension, the proteins were resolved via 12.5% SDS-PAGE using a Ruby SE600 system (GE Healthcare) at 30 mA/gel for 15 min, 40 mA/gel for 30 min, and then 50 mA/gel for 3 hours. After electrophoresis, the 2D gels were stained with 0.08% w/v colloidal Coomassie blue G-250 [54].

Image acquisition and statistical analysis

The gels were scanned using an ImageScanner II (Amersham) at 300 dpi, and the images were analysed using ImageMaster 2D Platinum 7.0 software (GE Healthcare). To compare spot quantities between gels accurately, the spot volumes were normalised as percentages of the total volume of all spots in the gel. The normalised percentage volumes (volume %) of the protein spots were then subjected to statistical analysis via analysis of variance (ANOVA) using the web-based NIA array analysis tool [55]. The entire data set was log_2-transformed and loaded onto NIA, and after determining the biological replicates and log_{10} transformation values, the data were statistically analysed using the following settings: error model "max (average, actual)", 0.01 for the proportion of the highest variance values removed before variance averaging, 10° of freedom for the Bayesian error model, 0.05 false discovery rate (FDR) threshold and zero permutations. PCA was conducted to assess genotype and experimental differences using the following settings: covariance matrix type, three principal components, 1.5-fold change threshold for clusters, and 0.5 correlation threshold for clusters. The PCA results were represented as biplot graphs, with proteins that exhibited higher tissue expression located in the same area of the graph. A comparison of the abundance of proteins was performed using the following settings: 0.05 FDR and 1.5-fold change threshold.

In-gel digestion and mass spectrometry analysis

Selected protein spots were manually excised from the gel and subjected to in-gel trypsin digestion according to Shevchenko et al. [56]. Selected gel plugs were washed extensively with 50% (v/v) acetonitrile to remove dye and SDS impurities. Colourless gel plugs were completely dried with 100% acetonitrile and then vacuum-dried, digested with 25 ng/μl Promega Trypsin Gold (MS grade) in 25 mM ammonium bicarbonate, and incubated overnight at 37°C. The tryptic fragments were eluted from the gel with 50% acetonitrile and 5% formic acid. The extracts were dried under a vacuum to a volume of 15 μl. The resulting peptides from the digests were purified with Reversed-Phase ZipTip C18 pipette tips according to the manufacturer's specifications (Millipore™). The obtained peptides were subjected to online nanoflow liquid chromatography tandem mass spectrometry (LC/MS/MS) using a nanoACQUITY system (Waters, Milford, MA, USA) coupled to a Q-TOF micro mass spectrometer (Waters, Milford, MA, USA). Peptide mixtures were loaded onto a 1.7 μm × 100 mm nanoACQUITY BEH300 column packed with C18 resin (Waters, USA) and separated at a flow rate of 0.6 μl min^{-1} using a linear gradient of up to 50% of solvent B (95% acetonitrile and 0.1% formic acid) over 23 min, followed by an increase to 85% of solvent B in 4 min and maintenance at 85% of solvent B for an additional 3 min. Solvent A was 0.1% formic acid in water. The eluent from the high-performance liquid chromatography (HPLC) column was directly electrosprayed into the mass spectrometer, which was operated in data-dependent acquisition mode to switch automatically between full-scan MS and MS/MS acquisition. The MS and MS/MS raw data were processed using ProteinLynx Global server v2.3 (Waters), and the resulting pkl files were subjected to searches against the NCBI non-redundant (NCBInr) database with the taxonomy parameter set to green plants, using the Mascot server v2.4 (http://www.matrixscience.com). The applied search criteria were as follows: trypsin digestion, carbamidomethyl (Cys) as a fixed modification and oxidation (Met) as a variable modification, a maximum of one missed cleavage event and peptide mass tolerances of ± 0.3 Da for the parent ion and 0.10 Da for the fragment ions. Ion scores of greater than 44 were considered significant ($P < 0.05$).

Systems biology analysis

To obtain information about protein-protein interactions (PPIs) based on the proteomic profiles of Rangpur lime and 'Sunki Maravilha' mandarin, we searched for orthologous proteins in *Arabidopsis thaliana*. To achieve this goal, a list of 45 proteins (including 9 proteins exclusive to 'Sunki Maravilha' mandarin and 7 exclusive to Rangpur lime) was used to perform reciprocal BLASTp searches [http://blast.ncbi.nlm.nih.gov], after which the initial networks of orthologous proteins of *A. thaliana* were prospected in STRING 9.05 [57]. Using this online search tool, PPIs were downloaded based on the following parameters: no more than 50 interactions, a high confidence score (0.400), and a network depth equal to 2. All of the active prediction methods were enabled, excluding text mining. The true ortholog sequences were obtained in the *A. thaliana* by performing best-reciprocal BLASTP hits from the in silico translation of *Citrus sinensis* and *Citrus clementine* peptides sequences .The BLAST analyses were performed locally on LINUX operating system using the model *C. sinensis* and *C. clementine* databanks extracted from NCBI, the filter soft option (-F "m S") with the local Smith–Waterman alignment algorithm (-s T) and 10- as cut-off e-value. The best-reciprocal BLASTP and the true ortholog identification were supported by the development of a specific script in PEARL programming language (Additional file 4: Table S3).

The interactome networks obtained from the orthologous proteins of *A. thaliana* for each variety were combined in the Rangpur lime network and 'Sunki Maravilha' mandarin network using the union function in the Cytoscape 2.8.2 [58] Advanced Merge Network plugin. The Rangpur lime and 'Sunki Maravilha' mandarin networks were then analyzed using the Cytoscape 2.8.2 Molecular Complex Detection (MCODE) plugin [59] to detect modules/clusters (densely connected regions) that were suggestive of functional protein complexes. The parameters used for MCODE to generate the clusters were as follows: loops included, degree cut-off of 2, deletion of single connected nodes from the cluster (haircut option enabled), expansion of the cluster by one neighbor shell (fluff option enabled), node density cut-off of 0.1, node score cut-off of 0.2, k-core of 2, and maximum depth of the network equal to 100.

To demonstrate the overlaps of nodes and exclusive regions between the Rangpur lime and 'Sunki Maravilha' mandarin networks, the Cytoscape 3.0 plugin Venn and Euler diagrams [60] was used. The major biological processes associated with the clusters generated using MCODE were analyzed with the Cytoscape 2.8.2 Biological Network Gene Ontology program (BiNGO) plugin [61]. The degree of functional enrichment for a given cluster and category was quantitatively computed (*p* value) using hypergeometric distribution, and multiple test corrections were assessed by applying the FDR algorithm [62], which was fully implemented using the BiNGO software at a significance level of $P < 0.05$. Node centrality analysis was computed using the Cytoscape 2.8.2 plugin Centiscape 1.2 [63] to identify nodes (proteins) with central positions within the networks. The implemented centralities were degree and betweenness. Nodes with relatively higher degrees were termed hubs, and those with higher betweenness were termed bottlenecks. Hubs were nodes that were highly connected, whereas bottlenecks were nodes with a higher probability of joining different clusters [64,65]. HB nodes can be considered to be key regulators of biological processes and to be essential for successful information transfer throughout the network. Furthermore, non-hub/bottleneck (NH-B) nodes were also identified due to the essential nature of the bottleneck nodes in the network dynamics.

Edge centrality analysis was performed using the WERW-Kpath algorithm [66], and data input generation was added as a network attribute in Cytoscape 2.8.2. In addition, quantitative protein expression data were distributed in the protein interaction networks, which included data input as a network attribute. Three networks were generated based on the overlap between the functional and interaction information. The first network compared the proteomic profile of Rangpur lime under water stress (e.g., *treatment a*) to the Rangpur lime control (e.g., *treatment b*), the second compared the proteomic profile of 'Sunki Maravilha' under water stress (e.g., *treatment a*) to the 'Sunki Maravilha' control (e.g., *treatment b*), and the third compared the proteomic profile of Rangpur lime under water stress (e.g., *treatment a*) to 'Sunki Maravilha' under water stress (e.g., *treatment b*), using the color gradient $Z = a/(a + b)$ [67].

Availability of supporting data

The data sets supporting the results of this article are included within the article and its additional files.

Additional files

Additional 1: Table S1. Detailed MS/MS results of 46 differentially expressed proteins in *Citrus*.

Additional 2: Figure S1. Interactome network of protein involved in water stress response in 'Rangpur' lime and Sunki 'Maravilha' mandarin. **Figure S2.** Distribution of the quantitative differential expression of proteins in water stress conditions, the interactomic in Rangpur lime and Sunki 'Maravilha' mandarin.

Additional 3: Table S2. Specific Gene Ontology (GO) classes derived from protein-protein interactions observed in different clusters.

Additional 4: Table S3. IDs orthologous proteins *Arabidopsis thaliana-Citrus*.

Abbreviations

ROS: Reactive oxygen species; LC/MS/MS: Liquid nanoflow chromatography tandem mass spectrometry; PCA: Principal component analyses; ABA: Abscisic acid; MW: Molecular weight; TCA: Tricarboxylic acid cycle; NER: Nucleotide excision repair; NTR: Normalized transpiration soil water; FTSW: Fraction of transpirable soil water; DTT: Dithiothreitol; SDS: Sodium dodecyl sulfate; FDR: False discovery rate; HPLC: High-performance liquid chromatography; PPI: Protein-protein interaction.

Competing interests

The authors declare that they have no competing interest.

Authors' contributions

TMO and ASG were responsible for the conception and design of the experiments, the analysis of the data and the redaction of the manuscript; TMO was responsible for the execution of all the proteomics experiments and for bioinformatics. FRS was responsible for the preparation and analysis of protein-protein interaction networks. DB supervised FRS in preparation and analysis of protein-protein interaction networks and helped to write and edit the manuscript. DMN and MACF were responsible for the water deficit experiment, physiological analysis and sample collection. BEM and RM helped in data analysis and write the manuscript. MGCC helped to interpretation of data, write and edit the manuscript. CPP helped to interpretation of data and proteomics experiments, LC/MS/MS analysis, data analysis and edit the manuscript. ASG coordinated the project design, and edited the manuscript. All authors read and approved the final manuscript.

Acknowledgements

The work of TMO was supported by the Coordenação de Aperfeiçoamento de Pessoal de Nível Superior (CAPES) and Conselho Nacional de Desenvolvimento Científico e Tecnológico (CNPq). This study supported by the CNPq (project 470156/2010-4 coordinated by ASG). The authors thank Fabienne Florence Lucienne Michelli for critical Reading of the manuscript.

Author details

[1]Universidade Estadual de Santa Cruz-UESC, Rodovia Ilhéus-Itabuna, Km 16, Salobrinho, Bahia, Brazil. [2]Centro de Biotecnologia, Departamento de Biologia Molecular e Biotecnologia, Universidade Federal do Rio Grande do Sul-UFGRS, Avenida Bento Goncalves, 9500 Porto Alegre, Rio Grande do Sul, Brazil. [3]IVIA; Centro de Genomica, Ctra. Moncada-Náquera Km 5, 46113 Moncada, Valencia, Spain. [4]CIRAD, UMR AGAP, Avenue Agropolis - TA A-75/02 – 34398, Montpellier Cedex 5, France. [5]Dipartimento di Scienze BioAgroAlimentari, CNR-IPSP, Istituto per la Protezione Sostenibile delle Piante, Area della Ricerca CNR, Via Madonna del Piano 10, Via Madonna del Piano n 10, 50019 Sesto Fiorentino, FI, Italy. [6]Embrapa Mandioca e Fruticultura, Rua Embrapa, s/n, Cruz das Almas 44380-000, Bahia, Brazil.

References

1. Xu C, Jing R, Mao X, Jia X, Chang X. A wheat (*Triticum aestivum*) protein phosphatase 2A catalytic subunit gene provides enhanced drought tolerance in tobacco. Ann Bot. 2007;99:439–50.
2. Nepomuceno AL, Neumaier N, Farias JRB, Oya T. Tolerância à seca em plantas. Biotecnologia Ciência e Desenvolvimento. 2001;23:12–8.
3. Kausar R, Arshad M, Shahzad A, Komatsu S. Proteomics analysis of sensitive and tolerant barley genotypes under drought stress. Amino Acids. 2013;44:345–59.
4. Kosová K, Vítámvás P, Prásil IT, Renaut J. Plant proteome changes under abiotic stress- Contribution of proteomics studies to understanding plant stress response. J Proteomics. 2011;71:1301–22.
5. Zadražnik T, Hollung K, Egge-Jacobsen W, Meglič V, Šuštar-Vozlič J. Differential proteomic analysis of drought stress response in leaves of common bean (*Phaseolus vulgaris* L.). J Proteomics. 2013;78:254–72.
6. Boscariol-Camargo RL, Berger IJ, Souza AA, Amaral AM d, Carlos EF, Freitas Astúa J, et al. In silico analysis of ESTs from roots of Rangpur lime (*Citrus limonia* Osbeck) under water stress. Genet Mol Biol. 2007;30:906–16.
7. Gimeno J, Gadea J, Forment J, Pérez-Valle J, Santiago J, Martínez-Godoy MA, et al. Shared and novel molecular responses of mandarin to drought. Plant Mol Biol. 2009;70:403–20.
8. Oliveira TM, Cidade LC, Gesteira AS, Coelho Filho MA, Soares Filho WS, Costa MGC. Analysis of the *NAC* transcription factor gene family in citrus reveals a novel member involved in multiple abiotic stress responses. Tree Genet Genomes. 2011;7:1123–34.
9. Xian L, Sun P, Hu S, Wu J, Liu JH. Molecular cloning and characterization of *CrNCED1*, a gene encoding 9-cis-epoxycarotenoid dioxygenase in *Citrus reshni*, with functions in tolerance to multiple abiotic stresses. Planta. 2013;239:61–77.
10. Abreu IA, Farinha AP, Negrão S, Gonçalves N, Fonseca C, Rodrigues M, et al. Coping with abiotic stress: Proteome changes for crop improvement. J Proteomics. 2013;93:145–68.
11. Shao HB, Liang MA, Wang BC. Changes of antioxidative enzymes and membrane peroxidation for soil water deficits among 10 wheat genotypes at seedling stage. Colloids Surf B Biointerfaces. 2008;42:107–13.
12. Ge P, Ma C, Wang S, Gao L, Guo G. Comparative proteomic analysis of grain development in two spring wheat varieties under drought stress. Anal Bioanal Chem. 2012;402:1297–313.
13. Xiao X, Yang F, Zhang S, Korpelainen H, Li C. Physiological and proteomic responses of two contrasting *Populus cathayana* populations to drought stress. Physiol Plant. 2009;136:150–68.
14. Alam I, Sharmin SA, Kim KH, Yang JK, Choi MS, Lee BH. Proteome analysis of soybean roots subjected to short-term drought stress. Plant Soil. 2010;333:491–505.
15. Podda A, Checcucci G, Mouhaya W, Centeno D, Rofidal V, Del Carratore R, et al. Salt-stress induced changes in the leaf proteome of diploid and tetraploid mandarins with contrasting Na^+ and Cl^- accumulation behaviour. J Plant Physiol. 2013;170:952–9.
16. Yoshimura K, Masuda A, Kuwano M, Yokota A, Akashi K. Programmed Proteome Response for Drought Avoidance/Tolerance in the Root of a C3 Xerophyte (Wild Watermelon) Under Water Deficits. Plant Cell Physiol. 2008;49:226–41.
17. Neves DM, Coelho Filho MA, Bellete BS, Silva MFGF, Souza DT, Soares Filho WS, et al. Comparative study of putative 9-cis-epoxycarotenoid dioxygenase and abscisic acid accumulation in the responses of Sunki mandarin and Rangpur lime to water deficit. Mol Biol. 2013;40:5339–49.
18. Rossignol M, Peltier JB, Mock HP, Matros A, Maldonado AM, Jorrin JV. Plant proteome analysis: a 2004–2006 update. Proteomics. 2006;6:5529–48.
19. Gao F, Zhou Y, Zhu W, Li X, Fan L, Zhang G. Proteomic analysis of cold stress-responsive proteins in Thellungiella rosette leaves. Planta. 2009;230:1033–46.
20. Caruso G, Cavaliere C, Guarino C, Gubbiotti FP, Laganà A. Identification of changes in *Triticum durum* L. leaf proteome in response to salt stress by two-dimensional electrophoresis and MALDI-TOF mass spectrometry. Analytical Bioanal Biochem. 2008;391:381–90.
21. Tomaz T, Bagard M, Pracharoenwattana I, Lindén P, Lee CP, Carroll AJ, et al. Mitochondrial Malate Dehydrogenase Lowers Leaf Respiration and Alters Photorespiration and Plant Growth in *Arabidopsis*. Plant Physiol. 2010;154:1143–57.
22. Tanou G, Job C, Rajjou L, Arc E, Belghazi M, Diamantidis G, et al. Proteomics reveals the overlapping roles of hydrogen peroxide and nitric oxide in the acclimation of citrus plants to salinity. Plant J. 2009;60:795–804.
23. Sweetlove LJ, Heazlewood JL, Herald V, Holtzapffel R, Day DA, Leaver CJ, et al. The impact of oxidative stress on *Arabidopsis* mitochondria. Plant J. 2002;32:891–904.
24. Bartoli CG, Gómez F, Martínez DE, Guiamet JJ. Mitochondria are the main target for oxidative damage in leaves of wheat (*Triticum aestivum* L.). J Exp Bot. 2004;55:1663–9.

25. Taylor NL, Heazlewood JL, Day DA, Millar AH. Differential impact of environmental stresses on the pea mitochondrial proteome. Mol Cell Proteomics. 2005;4:1122–3.
26. Kollipara KP, Saab IN, Wych RD, Lauer MJ, Singletary GW. Expression profiling of reciprocal maize hybrids divergent for cold germination and desiccation tolerance. Plant Physiol. 2002;129:974–92.
27. Moon H, Lee B, Choi G, Shin D, Prasad DT, Lee O. NDP kinase 2 interacts with two oxidative stress-activated MAPKs to regulate cellular redox state and enhances multiple stress tolerance in transgenic plants. Proc Natl Acad Sci U S A. 2003;100:358–63.
28. Kim YH, Kim M, Choi YI, Park SC, Yun DJ, Noh EW, et al. Transgenic poplar expressing Arabidopsis NDPK2 enhances growth as well as oxidative stress tolerance. Plant Biotechnol J. 2011;9:334–47.
29. Selote DS, Khanna-Chopra R. Antioxidant response of wheat roots to drought acclimation. Protoplasma. 2010;245:153–63.
30. Choi HW, Hwang BK. The pepper extracellular peroxidase *CaPO2* is required from salt, drought and oxidative stress tolerance as well as resistance to fungal pathogens. Planta. 2012;235:1369–82.
31. Cantero A, Barthakur S, Bushart T, Morgan RO, Fernandez P, Chou S. Expression profiling of the Arabidopsis annexin gene family during abiotic stress, germination and de-etiolation. Plant Physiol Biochem. 2006;44:13–24.
32. Clark G, Konopka-Postupolska D, Hennig J, Roux S. Is annexin 1 a multifunctional protein during stress responses? Plant Signal Behav. 2010;5:303–7.
33. Hoshino D, Hayashi A, Temmei Y, Kanzawa N, Tsuchiya T. Biochemical and immunohistochemical characterization of Mimosa annexin. Planta. 2004;219:867–75.
34. Lee H, Guo Y, Ohta M, Xiong L, Atevenson B, Zhu J-K. LOS2, a genetic locus required for cold-responsive gene transcription encodes a bi-functional enolase. EMBO J. 2002;21:2692–702.
35. Konopka-Postupolska D, Clark G, Goch G, Debski J, Floras K, Cantero A, et al. The role of annexin 1 in drought stress in *Arabidopsis*. Plant Physiol. 2009;150:1394–410.
36. Balajee AA, Bohr VA. Genomic heterogeneity of nucleotide excision repair. Gene. 2000;250:15–30.
37. Tuteja N, Singh MB, Misra MK, Bhalla PL, Tuteja R. Molecular Mechanisms of DNA Damage and Repair: Progress in Plants. Crit Rev Biochem Mol Biol. 2001;36:337–97.
38. Waterworth WM, Drury GE, Bray CM, West CE. Repairing breaks in the plant genome: the importance of keeping it together. New Phytol. 2011;192:805–22.
39. Stone SL, Callis J. Ubiquitin ligases mediate growth and development by promoting protein death. Curr Opin Plant Biol. 2007;10:624–32.
40. Lyzenga WJ, Stone SL. Protein Ubiquitination: An Emeerging Theme in Plant Abiotic Stress Tolerance. Am J Plant Sci Biotechnol. 2011;5:1–11.
41. Himmelbach A, Yang Y, Grill E. Relay and control of abscisic acid signaling. Plant Biol. 2003;6:470–9.
42. Hoth S, Morgante M, Sanchez M, Hanafey MK, Tingey SV, Chua NH. Genome-wide gene expression profiling in *Arabidopsis thaliana* reveals new targets of abscisic acid and largely impaired gene regulation in the *abi1-1* mutant. J Cell Sci. 2002;115:4891–900.
43. Liu H, Stone SL. Abscisic Acid Increases Arabidopsis ABI5 Transcription Factor Levels by Promoting KEG E3 Ligase Self-Ubiquitination and Proteosomal Degradation. Plant Cell. 2010;22:2630–41.
44. Wilson PB, Estavillo GM, Field KJ, Porsiriwong W, Carroll AJ, Howell KA, et al. The nucleotidase/phosphatese SAL1 is a negative regulator of drought tolerance in Arabidopsis. Plant J. 2009;58:299–317.
45. Manmathan H, Shaner D, Sneling J, Tisserat N, Lapitan N. Virus-induced gene silencing of Arabidopsis thaliana gene homologues in wheat identifies genes conferring improved drought tolerance. J Exp Bot. 2013;64:1381–92.
46. Tasma M, Brendel V, Whitham SA, Bhattacharyya MK. Expression and evolution of the phosphoinositide-specific phospholipase C gene family in Arabidopsis thaliana. Plant Physiol Biochem. 2008;46:627–37.
47. Singh A, Kanwar P, Pandey A, Tyagi AK, Sopory SK, Kapoor S, et al. Comprehensive Genomic Analysis and Expression Profiling of Phospholipase C Gene Family during Abiotic Stresses and Development in Rice. PLoS One. 2013;30:1–14.
48. Escorbar-Galvis ML, Marttila S, Hakansson G, Forsberg J, Knorpp C. Heat stress response in pea involves interaction of mitochondrial nucleoside diphosphate kinase with a novel 86-kilodalton protein. Plant Physiol. 2001;126:69–77.
49. Pan L, Kawai M, Yano A, Uchimiya H. Nucleoside diphosphate kinase required for coleoptile elongation in rice. Plant Physiol. 2000;122:447–52.
50. Uhring JF. Protein interaction networks in plants. Planta. 2006;224:771–81.
51. Soares Filho WDS, Cunha Sobrinho APD, Passos OS, Moitinho EDB. Maravilha: uma nova seleção de "Sunki". Rev Bras Frutic. 2003;25:268–71.
52. Sinclair TR, Ludlow MM. Influence of soil water supply on the plant water balance of four tropical grain legumes. Aust J Plant Physiol. 1986;13:329–41.
53. Bertolde FZ, Almeida AAF, Silva FAC, Oliveira TM, Pirovani CP. Efficient method of protein extraction from *Theobroma cacao* L. roots for two-dimensional gel electrophoresis and mass spectrometry analyses. Genet Mol Res. 2014;13:5036–47.
54. Neuhoff V, Arold N, Taube D, Ehrhardt W. Improved staining of proteins in polyacrylamide gels including isoelectric focusing gels with clear background at nanogram sensitivity using Coomassie Brilliant Blue G-250 and R-250. Electrophoresis. 1988;9:255–62.
55. Sharov AA, Dudeka DB, Ko MS. A web-based tool for PC and significance of microarray data. Bioinformatics. 2005;21:2548–9.
56. Shevchenko A, Tomas H, HavlīsOlsen JV, Mann M. In-gel digestion for mass spectrometric characterization of proteins and proteomes. Nat Protoc. 2007;1:2856–60.
57. Snel B, Lehmann G, Bork P, Huynen MA. STRING: a web-server to retrieve and display the repeatedly occurring neighborhood of a gene. Nucleic Acids Res. 2000;28:3442–4.
58. Shannon P, Markiel A, Ozier O, Baliga NS, Wang JT, Ramage D, et al. Cytoscape: a software environment for integrated models of biomolecular interaction networks. Genome Res. 2003;13:2498–504.
59. Bader GD, Hogue CW. An automated method for finding molecular complexes in large protein interaction networks. BMC Bioinf. 2003;4:2.
60. Saito R, Smoot ME, Ono K, Ruscheinski J, Wang PL, Lotia S, et al. A travel guide to Cytoscape plugins. Nat Methods. 2012;9:1069–76.
61. Maere S, Heymans K, Kuiper M. BINGO: a Cytoscape plugin to assess over representation of gene ontology categories in biological networks. Bioinformatics. 2005;21:3448–9.
62. Benjamin Y, Hochberg Y. Controlling the false discovery rate: a practical and powerful approach to multiple testing. J R Stat Soc B Stat Methodol. 1995;57:289–300.
63. Scardoni G, Petterlini M, Laudanna C. Analyzing biological network parameters with CentiScaPe. Bioinformatics. 2009;25:2857–9.
64. Barabási AL, Oltvai ZN. Network biology: understanding the cell's functional organization. Nat Rev Genet. 2004;5:101–13.
65. Yu H, Kim PM, Sprecher E, Trifonov V, Gerstein M. The importance of bottlenecks in protein networks: correlation with gene essentiality and expression dynamics. PLoS Comput Biol. 2007;3:713–20.
66. De Meoa P, Ferrara E, Fiumara G, Ricciardello A. A novel measure of edge centrality in social networks. Knowl-Based Syst. 2012;30:136–50.
67. Castro MA, Filho JL, Dalmolin RJ, Sinigaglia M, Moreira JC, Mombach JC, et al. ViaComplex: software for landscape analysis of gene expression networks in genomic context. Bioinformatics. 2009;25:1468–9.

The lycopene β-cyclase plays a significant role in provitamin A biosynthesis in wheat endosperm

Jian Zeng†, Cheng Wang†, Xi Chen, Mingli Zang, Cuihong Yuan, Xiatian Wang, Qiong Wang, Miao Li, Xiaoyan Li, Ling Chen, Kexiu Li, Junli Chang, Yuesheng Wang, Guangxiao Yang* and Guangyuan He*

Abstract

Background: Lycopene β-cyclase (LCYB) is a key enzyme catalyzing the biosynthesis of β-carotene, the main source of provitamin A. However, there is no documented research about this key cyclase gene's function and relationship with β-carotene content in wheat. Therefore, the objectives of this study were to clone *TaLCYB* and characterize its function and relationship with β-carotene biosynthesis in wheat grains. We also aimed to obtain more information about the endogenous carotenoid biosynthetic pathway and thus provide experimental support for carotenoid metabolic engineering in wheat.

Results: In the present study, a lycopene β-cyclase gene, designated *TaLCYB*, was cloned from the hexaploid wheat cultivar Chinese Spring. The cyclization activity of the encoded protein was demonstrated by heterologous complementation analysis. The *TaLCYB* gene was expressed differentially in different tissues of wheat. Although *TaLCYB* had a higher expression level in the later stages of grain development, the β-carotene content still showed a decreasing tendency. The expression of *TaLCYB* in leaves was dramatically induced by strong light and the β-carotene content variation corresponded with changes of *TaLCYB* expression. A post-transcriptional gene silencing strategy was used to down-regulate the expression of *TaLCYB* in transgenic wheat, resulting in a decrease in the content of β-carotene and lutein, accompanied by the accumulation of lycopene to partly compensate for the total carotenoid content. In addition, changes in *TaLCYB* expression also affected the expression of several endogenous carotenogenic genes to varying degrees.

Conclusion: Our results suggest that *TaLCYB* is a genuine lycopene cyclase gene and plays a crucial role in β-carotene biosynthesis in wheat. Our attempt to silence it not only contributes to elucidating the mechanism of carotenoid accumulation in wheat but may also help in breeding wheat varieties with high provitamin A content through RNA interference (RNAi) to block specific carotenogenic genes in the wheat endosperm.

Keywords: Lycopene, Lycopene β-cyclase, β-carotene, Provitamin A, RNA interference, Wheat

Background

Carotenoids are important natural isoprenoid pigments synthesized in plants that have essential roles in protecting against excess light energy and oxidative damage, and in light-harvesting [1,2]. Their provitamin A activity and antioxidant properties are their most attractive functions. β-carotene is the major and most effective vitamin A precursor among carotenoids, and plays a crucial role in human health, protecting against age-related degenerative diseases, cardiovascular disease, certain cancers and vitamin A deficiency (VAD) [3-5]. Generally, β-carotene is the most attractive target product for metabolic engineering.

In higher plants, although the main pathway of carotenoid biosynthesis has been studied extensively [6-8], the regulatory mechanisms of carotenoid biosynthesis are still not well known. Lycopene cyclization is the first branch point of the carotenoid biosynthetic pathway, and is hypothesized to regulate the proportion of carotenes through two competing lycopene cyclases, LCYB and lycopene ε-cyclase (LCYE). In general, lycopene is cyclized by LCYE and LCYB to introduce ε and β-ionone end groups and produce α- and β-carotene,

* Correspondence: ygx@hust.edu.cn; hegy@hust.edu.cn
†Equal contributors
The Genetic Engineering International Cooperation Base of Chinese Ministry of Science and Technology, The Key Laboratory of Molecular Biophysics of Chinese Ministry of Education, College of Life Science and Technology, Huazhong University of Science and Technology, Wuhan, China

respectively (Figure 1). Only a small number of species such as *Lactuca sativa* produce ε,ε-carotene [9]. Because of the special position of lycopene cyclization, researchers have focused on the function of LCYB and its relationship with carotenoid accumulation in plants [10-13]. Plant LCYBs share similar and highly conserved functional domains, which are involved in many reactions in β-ionone catalysis [10,14].

Through the deeper understanding of the benefits of carotenoids for human health, scientists have been prompted to explore effective methods to increase the carotenoid composition and content in plants, especially in staple crops. However, precise carotenoid metabolic engineering in crop plants has been hindered by limited data about the endogenous regulation of carotenogenic genes despite recent progress in staple crops [15-17]. Thus, the first step to understanding how carotenoids are biosynthesized is to identify the related key enzymes and clone the relevant genes.

Wheat is one of the most important cereal crops in the world [18]. Given the huge daily consumption of wheat-based products in populations worldwide, increasing the β-carotene content in wheat grains could significantly impact VAD. Although carotenoids are one of the major pigments that affect the nutritional value of wheat [19], wheat grains have a very low carotenoid content and mainly accumulate lutein, which lacks provitamin A activity. To improve the carotenoid or provitamin A content in wheat, the detailed regulation of carotenoid biosynthesis must be clarified. So far, the polyploid characteristics and huge size of the wheat genome have been substantial barriers to identifying and cloning key carotenoid biosynthetic genes. Only a few carotenoid biosynthetic enzyme genes such as phytoene synthase (*PSY*) and *LCYE* have been identified [20-23]. Therefore, identifying and cloning more genes in the wheat carotenoid biosynthetic pathway will provide more information about carotenoid biosynthesis and its regulatory mechanism. According to the latest research, about 50% of the genome of hexaploid wheat has now been sequenced [24]. Although gene cloning will become easier and more precise after sequencing is completed in the future, there is still plenty of work to be done and many difficulties to be overcome. Recently, we found that endogenous *LCYB* was up-regulated by the co-expression of *CrtB* and *CrtI* in transgenic wheat, which resulted in an increase in the total carotenoid and provitamin A contents [25]. However, there is no documented research about this key cyclase gene's function and its relationship with β-carotene content in wheat. Therefore, the objectives of this study were to clone *TaLCYB* and characterize its function and relationship with β-carotene biosynthesis in the wheat grain. We also aimed to obtain more information about the endogenous

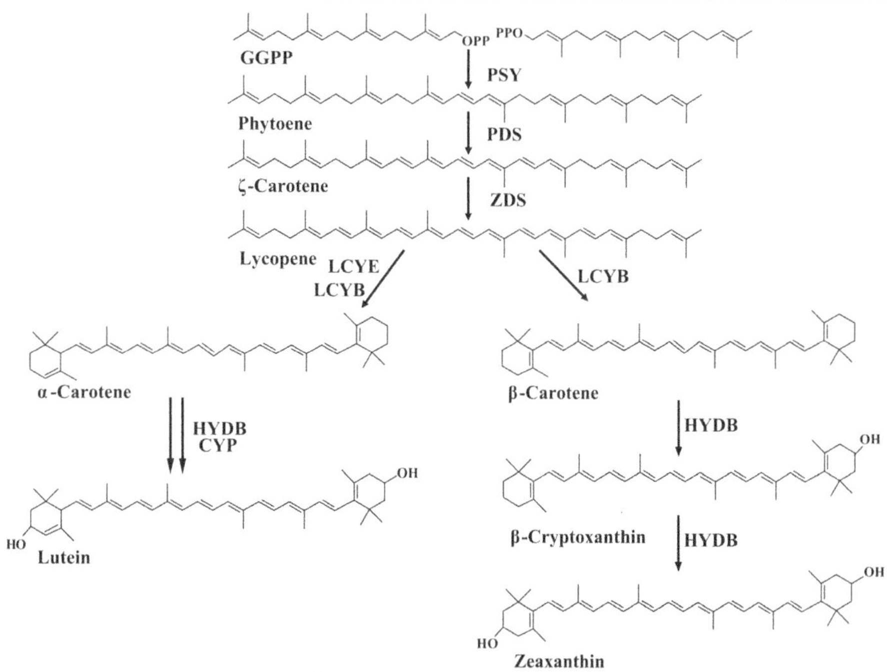

Figure 1 Carotenoid biosynthetic pathway in wheat. PSY, phytoene synthase; PDS, phytoene desaturase; ZDS, zeta-carotene desaturase; CRTISO, carotene isomerase; LCYB, lycopene β-cyclase; LCYE, lycopene ε-cyclase; HYDB, β-carotene hydroxylase; CYP, carotenoid ε-hydroxylase (cytochrome P450 type).

carotenoid biosynthetic pathway and thus provide experimental support for carotenoid metabolic engineering in wheat.

Results

Cloning and sequence analysis of *TaLCYB*

A 1,455 bp full-length cDNA of *LCYB* from common wheat was isolated through an *in silico* cloning strategy. The full-length cDNA of the *LCYB* gene was designated *TaLCYB* (GenBank Accession No.: JN622196.1). Comparison of the obtained cDNA sequence with the gDNA sequences of wheat revealed an intronless structure. Based on the latest database of the International Wheat Genome Sequencing Consortium, *TaLCYB* was localized on 6AS and 6DS (https://urgi.versailles.inra.fr/blast/blast.php). The ORF encoded a polypeptide of 484 amino acid residues with a predicted relative molecular mass of 53.3 kDa containing a predicted plastid transit peptide of 30 amino acids. Multiple alignment showed that TaLCYB shared a significant degree of sequence identify with other LCYB proteins in monocots (86.3% sequence identity with OsLCYB from *O. sativa*, 86.2% with ZmLCYB from *Z. mays*), and relatively lower homology with LCYB proteins from dicot species, such as *C. annuum*, *A. thaliana*, *S. lycopersicum* (67.4%, 67.1% and 66.5% respectively). Conserved motifs analysis (Figure 2A) showed a conserved β-LCY region, a dinucleotide-binding signature, a LCY-specific motif, cyclase motifs I and II, a charged region, two predicted TM helices and three β-LCY CAD regions (Catalytic Activity Domain). These domains were shown to be essential for LCYB activity [10,26,27]. A phylogenetic tree was constructed based on the amino acid sequence alignment of TaLCYB and five other plant LCYBs from GenBank (Figure 2B). These results suggested that TaLCYB isolated from wheat was a genuine member of the plant lycopene β-cyclase family.

Functional characterization of TaLCYB in *E. coli*

While the LCYB conserved motifs gave an indication of the enzymatic function of the encoded protein, they could not fully determine or reflect its cyclic function *in vivo*. To investigate the function of TaLCYB, an *in vivo* analysis using *E. coli* BL21 was conducted. *TaLCYB* was cloned into pET32α(+), which was then designated pET32-LCYB. *E. coli* strain BL21 was co-transformed with pAC-LYC, which contains genes for lycopene biosynthesis, and pET32-LCYB. Carotenoids were extracted from the bacterial cells and analyzed by High Performance Liquid Chromatography (HPLC). As shown in Figure 3, HPLC analysis of BL21 extracts showed that the strains containing pAC-LYC or pAC-LYC + pET32α(+) exhibited a single peak, whose retention time and absorbance spectrum corresponded to lycopene, and the cultures appeared pink. In contrast, extracts from pAC-LYC + pET32-LCYB cells mainly accumulated β-carotene, the cultures turned yellow with an undefined peak (maybe an isomer of β-carotene), and lycopene was virtually undetectable. These results demonstrated that TaLCYB was a functional β-cyclase in the carotenoid biosynthetic pathway in *E. coli*, which could convert lycopene to β-carotene.

TaLCYB is expressed in different tissues including developing grains of common wheat

To assess the spatial and temporal expression patterns of *TaLCYB* in different wheat tissues, quantitative PCR (qPCR) was carried out with RNA extracted from leaves, stems, roots, pistils, stamens and five developmental stages of grains: Grain 1 (4–10 days after pollination (DAP)), Grain 2 (10–16 DAP), Grain 3 (16–20 DAP), Grain 4 (20–25 DAP) and Grain 5 (25–35 DAP). As shown in Figure 4, *TaLCYB* was expressed in all of these tissues. The highest expression of the *TaLCYB* gene was observed in the leaf followed by the stamen, pistil, stem and root. In developing grains, it was interesting that the expression of *TaLCYB* always remained at a relatively high level, particularly at the later stages. *TaLCYB* expression peaked (15-fold) at 20–25 DAP (Grain 4) in common wheat and then decreased, but still remained at 4.1-fold in Grain 5 when compared with Grain 1 (Figure 4B).

Common wheat carotenoid accumulation presents a decreasing tendency in developing grains

The carotenoid composition of wheat grains at different developmental stages was determined by HPLC analysis. As shown in Additional file 1: Figure S1, detailed HPLC analysis revealed a decreasing tendency in total carotenoid content during grain development. The β-carotene content in the wheat grains also showed a decreasing tendency, despite *TaLCYB* expression remaining at a high level during later developmental stages. Carotenoid pigments, encompassing lutein, zeaxanthin, β-cryptoxanthin, α-carotene and β-carotene, were measured and their concentrations all decreased progressively during grain development. At the last stage, the main carotenoids were lutein, zeaxanthin and β-carotene (Table 1).

In parallel with the carotenoid content analysis, the expression of carotenogenic genes in developing wheat grains was also analyzed. As shown in Figure 4B, *TaPSY*, *TaPDS*, *TaZDS* and *TaHYD2* showed similar expression patterns in developing grains, presenting a declining tendency. The expression pattern of *TaLCYE* was similar to that of *TaLCYB*; both maintained a relatively high expression level in all development stages (Grains 1–5). *TaHYD1* expression showed a decline in Grains 1–3, but was up-regulated in Grains 4–5. In the last two stages, carotenogenic gene expression was dramatically reduced in

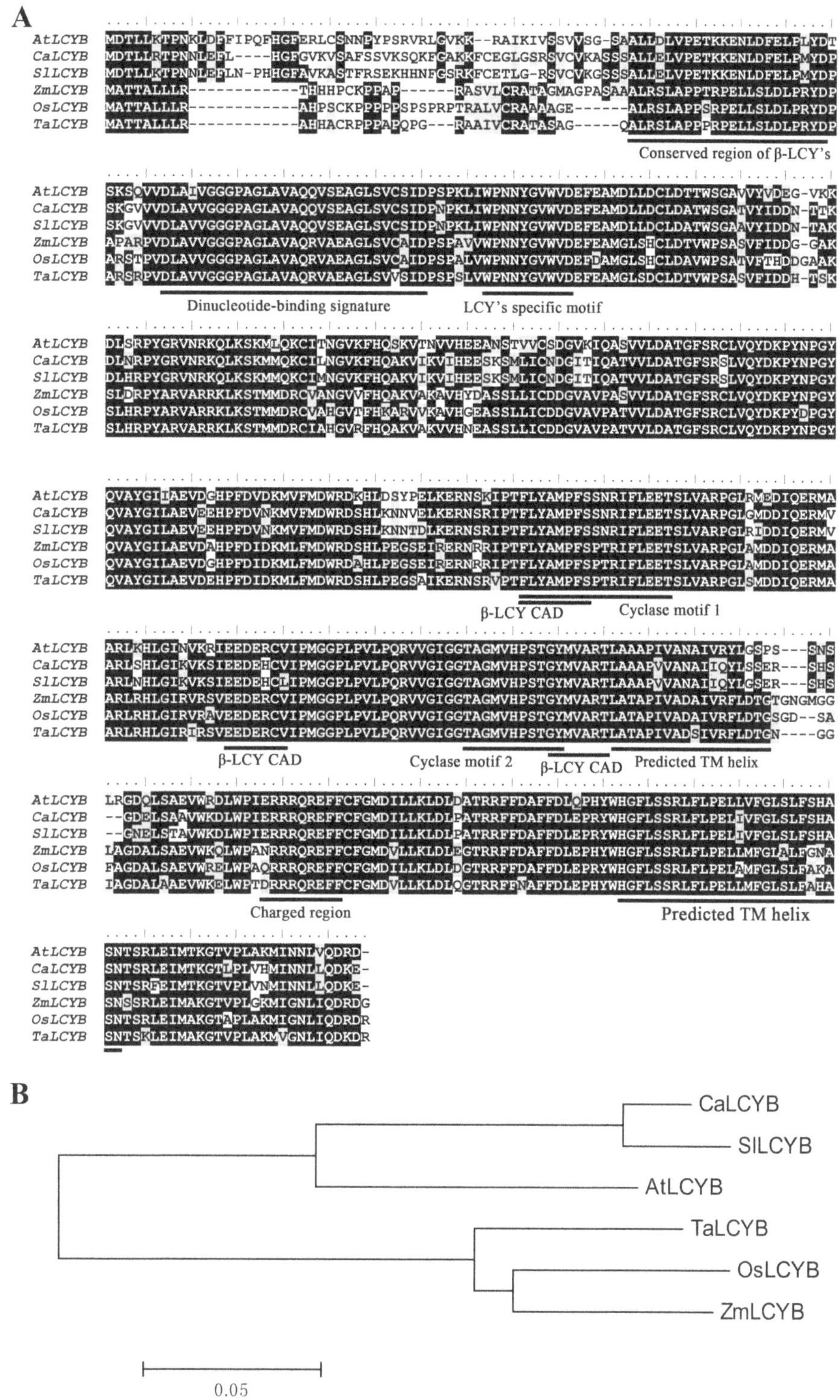

Figure 2 Comparative alignment and phylogenetic tree of lycopene β-cyclase. **(A)** The alignment was created using ClustalW. The amino acid residues which are identical in all sequences are shown in white text on a black background, whereas different residues are shown in black text on a white background. Characteristic regions of plant LCYBs are indicated under the LCYB sequence: Conserved LCYB region, Di-nucleotide binding site, LCY's specific motif, Cyclase motifs (CM) I and II, Predicted TM helices, Charged region and β-LCY CAD (Catalytic Activity Domain). AtLCYB: *A. thaliana* lycopene β-cyclase; CaLCYB: *C. annuum* lycopene β-cyclase; SlLCYB: *S. lycopersicum* lycopene β-cyclase; ZmLCYB: *Z. mays* lycopene β-cyclase; OsLCYB: *O. sativa* lycopene β-cyclase. **(B)** The multiple alignments were generated by ClustalW and the phylogenetic tree was constructed with MEGA4.0 using a bootstrap test of phylogeny with minimum evolution test and default parameters.

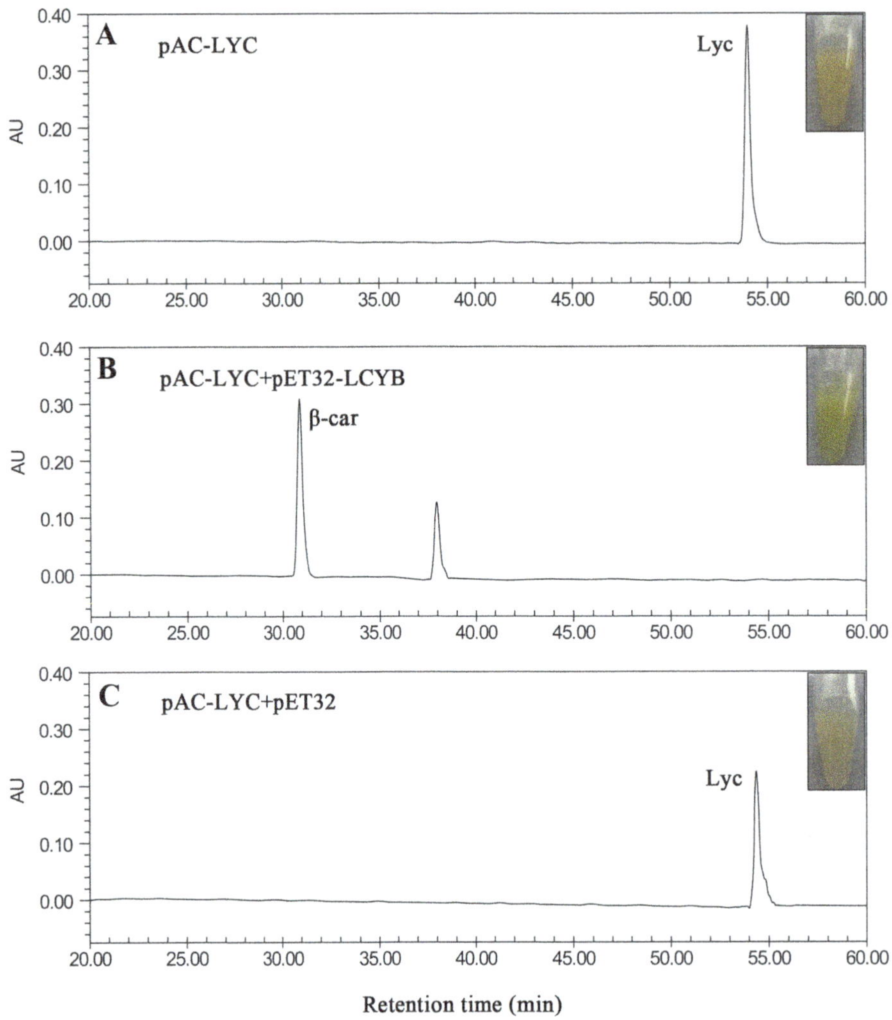

Figure 3 Reverse phase HPLC analysis of carotenoids accumulated in *E. coli* BL21 strain complemented with *TaLCYB*. Carotenoids were extracted from suspension cultures of cells with different plasmids **(A)** plasmids pAC-LYC; **(B)** pAC-LYC + pET32-LCYB; **(C)** pAC-LYC + pET32.

comparison with the early stages, except that *TaLCYE* and *TaLCYB* were slightly down-regulated and *TaHYD1* was up-regulated. Overall, the carotenogenic genes had relatively stable expression levels in the early stages (Grains 1–3). This suggested that carotenoids were synthesized at a stable rate during the early stages.

Expression patterns of *TaLCYB* under different abiotic stresses and their effects on β-carotene accumulation

qPCR was performed to analyze the expression level of *TaLCYB* under different abiotic stresses, such as strong light, darkness and cold. As shown in Additional file 1: Figure S2, *TaLCYB* transcripts were up-regulated by strong light and cold, and inhibited by darkness. Under strong light conditions, dramatic and fast changes in *TaLCYB* expression were observed. The expression of *TaLCYB* reached a peak (about 8.5-fold) after 4 h under strong light treatment. By contrast, changes in *TaLCYB* expression were more gradual under cold, with the highest expression (1.4-fold) at 8 h after treatment. As the expression of *TaLCYB* was dramatically induced by strong light, the carotenoid profiles of leaves at different treatment times were analyzed by HPLC (Additional file 1: Figure S3). Notably, the β-carotene content variation seemed concurrent with changes of *TaLCYB* expression; the highest expression level corresponded to the maximum β-carotene content (Additional file 1: Figure S2B). The expression of other upstream genes in the pathway such as *TaPSY, TaPDS, TaZDS* and *TaLCYE* was also up-regulated by strong light (Additional file 1: Figure S4). These results suggested a correlation between *TaLCYB* expression and β-carotene content in wheat.

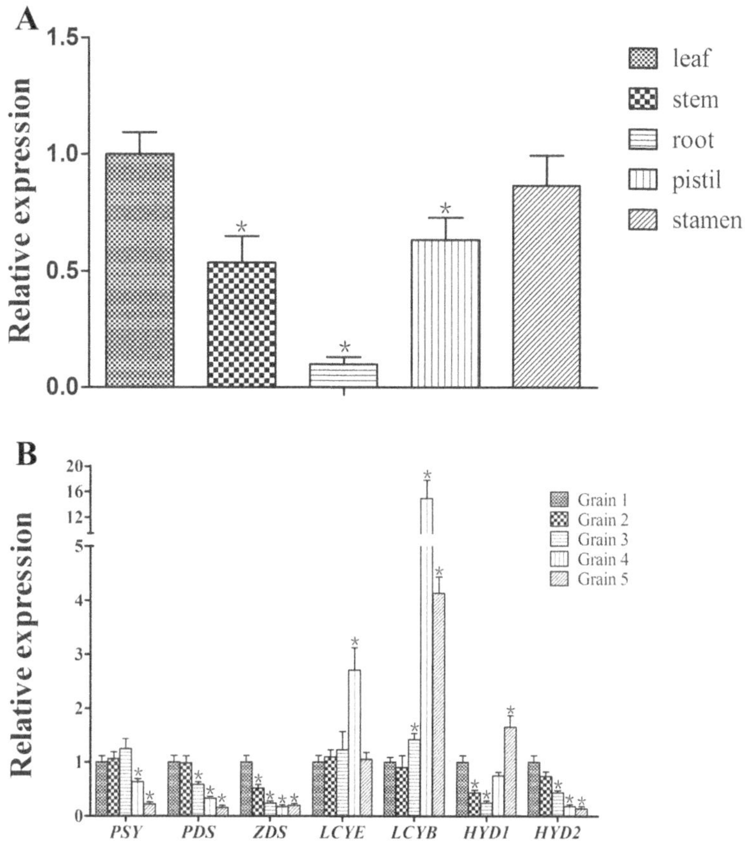

Figure 4 Expression patterns of wheat *TaLCYB* revealed by qRT-PCR analysis. **(A)** Expression patterns of *TaLCYB* in different tissues. **(B)** Expression levels of the endogenous carotenoid biosynthetic genes in developing grains of common wheat. Poly(A)$^+$ mRNA of 200 ng was subjected to reverse transcription, and served as the qPCR template. qPCR results for each gene were performed in three biological replicates with three technical repeats each and all data are shown as Mean ± SEM. Single asterisk indicates significant differences in the expression levels between controls at $P = 0.05$ probability level.

TaLCYB RNAi increases lycopene by decreasing β-carotene accumulation in the seeds

To explore the function of *TaLCYB* in the wheat carotenoid biosynthetic pathway, an RNAi vector was constructed and transformed into wheat (cv. Chinese Spring). After herbicide-selective regeneration, positive transgenic wheat lines were screened out in the T_0 generation by specific PCR-amplification of both the *bar* gene sequence and vector (sense-intron) sequence. Three transgenic lines and several lines only transformed with the pAHC25 plasmid were obtained; the latter lines were regarded as vector control lines (VC). Self-pollination of the PCR-positive transgenic plants in subsequent generations led to the identification of non-segregant RNAi transgenic lines. HPLC analysis of carotenoids showed no distinction in the carotenoid composition between the VC and wild-type. The carotenoid profiles and total carotenoid content of the transgenic lines differed from the wild-type. However, transgenic line BI-2 did not show any changes in carotenoid content or profile compared with the wild-type. Several novel carotenoids were observed in the transgenic wheat lines including lycopene, β-cryptoxanthin and α-carotene (Additional file 1: Table S1). To further analyze the carotenoid profiles, detailed HPLC analysis was carried out on the T_3 generation, which showed significant differences in carotenoid content and composition in seeds between transgenic and control lines, implying profound changes in the carotenoid biosynthetic pathways of the transgenic lines (Figure 5). The total carotenoid content slightly decreased to 0.84 μg g^{-1} seed dry weight in BI-6 and 0.75 μg g^{-1} seed dry weight in BI-9 compared with the wild-type (0.96 μg g^{-1}). In these two lines, consistent with the hypothesized silencing of *TaLCYB* genes, the β-carotene content decreased to 0.16 μg g^{-1} and 0.09 μg g^{-1} compared with the wild-type (0.22 μg g^{-1}). Lycopene is the immediate precursor of lycopene β-cyclase, and was accumulated to 0.22 μg g^{-1} and 0.39 μg g^{-1} in BI-6 and BI-9, respectively. Because

Table 1 Cartenoids content and compositions in developing grains

Grain	Lutein (μg g^{-1})	Zeaxanthin (μg g^{-1})	β-cryptoxanthin (μg g^{-1})	α-carotene (μg g^{-1})	β-carotene (μg g^{-1})	Total carotenoid (μg g^{-1})
1	5.52 ± 0.61a (60%)	0.60 ± 0.09a (7%)	0.27 ± 0.03a (3%)	0.36 ± 0.05a (4%)	2.35 ± 0.31a (26%)	9.10 ± 1.82a
2	4.70 ± 0.42b (59%)	0.56 ± 0.08a (7%)	0.23 ± 0.02a (3%)	0.33 ± 0.04a (4%)	2.12 ± 0.29b (27%)	7.94 ± 0.95b
3	3.80 ± 0.88c (62%)	0.40 ± 0.06b (6%)	0.17 ± 0.02b (3%)	0.32 ± 0.04a (5%)	1.47 ± 0.21c (24%)	6.16 ± 0.80c
4	1.93 ± 0.29d (50%)	0.22 ± 0.03c (6%)	0.08 ± 0.01c (2%)	0.31 ± 0.03a (8%)	1.32 ± 0.15d (34%)	3.86 ± 0.54d
5	0.52 ± 0.08e (53%)	0.17 ± 0.03c (17%)	ND	ND	0.29 ± 0.03e (30%)	0.98 ± 0.16e

Data represent the average carotenoid content (±SEM) of grains from five individual ears per line. Different letters indicate significant differences ($P = 0.05$) in carotenoid pigment content were determined by Tukey's HSD test. Values in parentheses represent the percentages of each carotenoid composition relative to the total content. ND = not detected.

LCYB participates in the biosynthesis of lutein, the lutein content was also decreased to 0.22 μg g^{-1} and 0.18 μg g^{-1} in BI-6 and BI-9, respectively, compared with 0.59 μg g^{-1} in the wild-type (Table 2).

Decreased β-carotene content in transgenic wheat was due to down-regulation of *TaLCYB*

Transcriptional regulation of carotenogenic genes is a crucial regulatory mechanism of carotenoid accumulation in plants. A post-transcriptional gene silencing strategy was used to verify the function of *TaLCYB*. The β-carotene and lutein contents were demonstrated to be reduced through HPLC analysis. The expression levels of the endogenous carotenogenic genes were thus analyzed in both endosperms and leaves from transgenic and control lines to investigate whether the decrease of β-carotene and lutein content in transgenic lines was related with carotenogenic gene expression. As shown in Figure 6, transgenic line BI-2, VC and the wild-type showed similar expression levels for all carotenogenic genes in the endosperm. In transgenic lines BI-6 and BI-9, the expression of *TaLCYB* showed a 70% and 84% reduction, respectively. This was consistent with the HPLC results that lower expression of *TaLCYB* accompanied decreased β-carotene content. In these transgenic lines, the expression levels of *TaZDS*, *TaLCYE* and *TaHYD1* were up-regulated. The expression of *TaPSY* showed slight suppression, which was possibly correlated with the decrease in total carotenoids. The remaining carotenogenic genes such as *TaPDS* and *TaHYD2* appeared to be unaffected by the reduced *TaLCYB* expression.

Endogenous carotenogenic genes from the transgenic lines were much less affected in leaves than in the endosperm. In transgenic lines BI-6 and BI-9, the expression of *TaLCYB* was down-regulated, the other cyclase *TaLCYE* was up-regulated, and *TaHYD2* was slightly down-regulated. In the leaves of transgenic line BI-2, the expression of all carotenogenic genes showed the same transcriptional levels as the VC line and wild-type (Figure 6B). The carotenoid composition and content in the leaves of transgenic and control lines were also analyzed by HPLC; there was no distinction between the transgenic lines and control lines (data not shown).

Discussion

Because of the nutritional value and health benefits of carotenoids, there have been many attempts to improve the carotenoid content in staple crops by metabolic engineering, especially the β-carotene content. For example, the transgenic cereal 'Golden Rice 2' was developed because of the deficiency of β-carotene in rice grains [28], in which the endogenous LCYB plays a crucial role in β-carotene biosynthesis. However, owing to the complexity of the wheat genome, there are limited reports on carotenoid metabolic engineering to improve the carotenoid content in wheat. One of the major limitations to metabolic engineering in wheat is the lack of a fully elucidated carotenoid biosynthetic pathway [25,28-30]. Additionally, only a few related genes have been cloned and characterized in wheat because of its complicated and huge genome, which seriously hinders the understanding of carotenoid biosynthesis in wheat. Therefore, cloning and analyzing carotenoid biosynthetic genes in wheat is very important to elucidate the carotenoid biosynthesis pathway and to improve its nutritional value by metabolic engineering. In this study, a novel wheat gene, *TaLCYB*, was identified and characterized to function as a lycopene β-cyclase. Its relationship to carotenoid biosynthesis was also investigated, in particular to β-carotene biosynthesis.

TaLCYB has β-lycopene cyclase function according to bioinformatics analysis and heterologous complementation in *E. coli*

To provide more information about *TaLCYB*, the TaLCYB protein was analyzed by comparing its amino acid sequence with other LCYBs from monocot (OsLCYB and ZmLCYB) and dicot species (CaLCYB, AtLCYB and SlLCYB) [11,31-33]. The amino acid sequence analysis

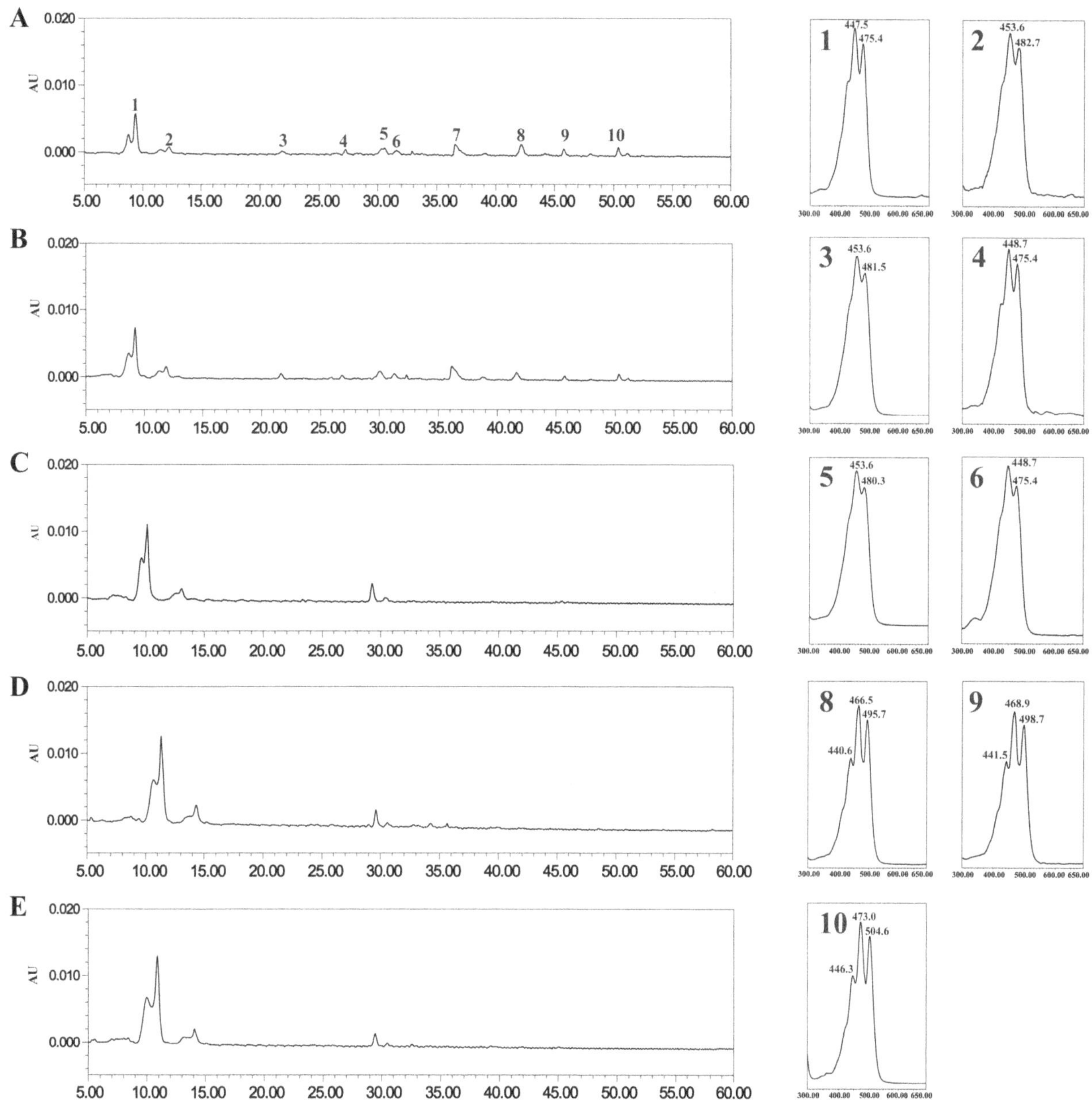

Figure 5 HPLC characterization of carotenoids extracted from grains of T_3 transgenic and control wheat. **(A)** BI-9; **(B)** BI-6; **(C)** BI-2; **(D)** VC-10 (transgenic vector control); **(E)** Chinese Spring (wild-type); Peak 1, lutein; Peak 2, zeaxanthin; Peak 3, β-cryptoxanthin; Peak 4, α-carotene; Peak 5, *trans*-β-carotene; Peak 6, 9-*cis*-β-carotene; Peak 7, undefined carotene; Peak 8,9, *cis*-lycopene; Peak 10, *trans*-lycopene.

revealed that TaLCYB contains all the conserved domains of plant LCYBs (Figure 2A). For instance, β-LCY CAD regions, which have been reported as crucial to LCYB catalytic activity, were found in TaLCYB. The "Conserved region β-LCY" was also found in TaLCYB, which is regarded as a crucial factor for the association of LCYB with membrane components and for its catalytic activity. These conserved motifs showed a high degree of conservation in amino acids between monocot and dicot species (Figure 2B) [10,26,27]. However, the sequence analysis only gave a preliminary indication of the function of TaLCYB. Therefore, a heterologous complementation system was used to verify its function *in vivo*. This method has been proved to be efficient for functional characterization of carotenoid biosynthetic genes [8]. Consistent with the results of sequence analysis, TaLCYB was demonstrated to be a functional β-cyclase enzyme *in vivo*, converting lycopene to β-carotene (Figure 3).

Table 2 Cartenoids content and compositions in grains from transgenic wheat

	Lutein (μg g^{-1})	Zeaxanthin (μg g^{-1})	Lycopene (μg g^{-1})	α-carotene (μg g^{-1})	β-carotene (μg g^{-1})	β-cryptoxanthin (μg g^{-1})
Bobwhite	0.59 ± 0.07a	0.15 ± 0.017a	ND	ND	0.22 ± 0.025a	ND
VC	0.57 ± 0.08a	0.17 ± 0.015a	ND	ND	0.24 ± 0.03a	ND
BI-2-9	0.51 ± 0.06a	0.18 ± 0.02a	ND	ND	0.26 ± 0.03a	ND
BI-6-1	0.22 ± 0.02b	0.13 ± 0.018a	0.22 ± 0.028b	0.06 ± 0.016a	0.16 ± 0.017b	0.04 ± 0.015
BI-9-5	0.18 ± 0.015c	0.05 ± 0.014b	0.39 ± 0.06a	0.05 ± 0.016a	0.09 ± 0.022c	ND

Carotenoid composition in wheat grains from transgenic and control lines in T_3 generation. Average of each carotenoid species are determined from five individual plants ears per line. Data represent the average carotenoid content (±SEM) of grains from five individual ears per line. Different letters indicate significant differences ($P = 0.05$) in carotenoid pigment content were determined by Tukey's HSD test. ND = not detected.

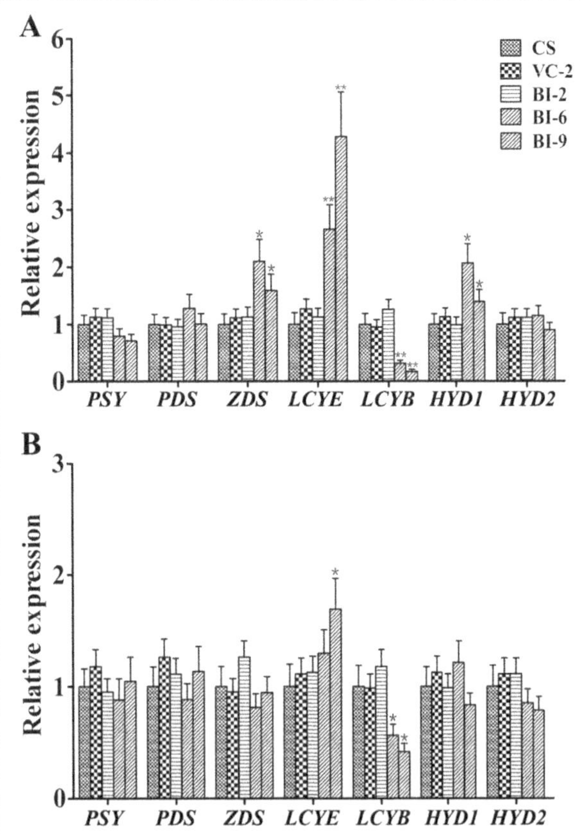

Figure 6 Expression levels of the endogenous carotenoid biosynthetic genes in endosperms and leaf from transgenic and control wheat lines. Gene expression levels were measured by qPCR and are determined relative to the transcript levels of the constitutively expressed β-actin gene in the same samples. Expression levels of these genes for the transformed lines are given as expression levels relative to the values for non-transformed control line Chinese Spring (CS). qPCR results for each gene were performed in three biological replicates with three technical repeats each and all data are shown as Mean ± SEM. Single asterisk and double asterisk indicate significant differences in the expression levels between control CS and transgenic lines at $P = 0.05$ or $P = 0.01$ probability level, respectively.

The endogenous expression level of *TaLCYB* is not positively correlated with β-carotene accumulation in developing grains

The tissue specificity of gene expression usually mirrors the function of the corresponding gene products in plant development. Maximal expression in leaves should be associated with the photosynthetic system and photoprotective function where carotenoids are a key component [34,35]. In developing grains, it was found that *TaLCYB* reached its highest expression at 20–25 DAP (Grain 4), suggesting that at later developmental stages grains may also have high ability to synthesize β-carotene compared with the early stages. The decreasing tendency of β-carotene content meant that increased β-ring cyclization capacity did not present as a large amount of β-carotene accumulation in developing wheat grains (from 2.35 μg g^{-1} to 0.29 μg g^{-1}). This phenomenon might be explained by the β-carotene produced by increased synthesis being rapidly transformed into downstream compounds, resulting in a net decrease in β-carotene content. The total carotenoid content in developing wheat grains also showed a decreasing trend, which was the same as in Qin *et al.* [20]. The high expression of *TaHYD1* in later stages also suggested high downstream synthetic capacity to some degree, which is more preferred to the β,β-branch [25]. In addition, the accumulation of carotenoids is inversely determined by the rate of carotenoid turnover, in which the activities of various carotenoid cleavage dioxygenases (*CCDs*) play a crucial role [36]. The CCD family catabolizes the turnover of different carotenoids to apo-carotenoids in various crops, such as rice and maize [37]. Experimental evidence from the expression of carotenoid cleavage dioxygenase 1 and the carotenoid content in maize endosperm demonstrates that high expression of *CCD1* accompanies lower carotenoid accumulation [37,38]. LCYB is supposed to be a key step for β,ε- and β,β-branch biosynthesis [10]. In kiwifruit and papaya, the major carotenoid was controlled by the expression level of *LCYB* [12,13]. In transgenic wheat with introduced

CrtB and *CrtI*, constant lutein content in mature grains was still maintained despite high expression of *TaLCYB*. The relatively higher expression of *TaLCYB* and *TaLCYE* in developing grains did not translate into accumulation of the corresponding carotenoids. All of these phenomena suggest that coordination between *TaLCYB* and *TaLCYE* expression would regulate carbon flux through different branches in the wheat carotenoid pathway. They also show that there is a mechanism resulting in a net decrease due to more carotenoid compounds entering into turnover relative to the biosynthetic capacity.

TaLCYB plays a crucial role in the β-carotene biosynthesis of wheat

LCYB and *LCYE* determine the flux towards the β,β- and β,ε-carotenoid branches. Thus, a strategy of modulating the levels of these two competing cyclases should enable the control of carotenoid composition [39]. Our results showed that the *TaLCYB* transcript could be regulated by silencing, which resulted in decreased β-carotene content and lycopene accumulation. In the transgenic lines BI-6 and BI-9, the β-carotene content showed obvious reduction (Table 2). The most dramatic down-regulation of the endogenous *TaLCYB* gene was coincident with the lowest metabolic flux into the β,β-carotene branch in BI-9. The same results were also observed in transgenic carrot with *DcLCYB1* silencing, while the β-carotene content was increased by overexpression of *DcLCYB1* [40]. Under strong light treatment, qPCR and HPLC analysis results showed that the expression level of *TaLCYB* and the β-carotene content in leaves presented the same change tendency. The increased biosynthesis of β-carotene was due mainly to the combined effects of *TaLCYB* and *TaPSY* up-regulation. In the present study, lycopene and α-carotene were the carotenoid composition in the silenced transgenic lines, but these were undetectable in the VC and wild-type. Lycopene and α-carotene are direct substrates or products of LCYB, which indicates that the silencing of *TaLCYB* simultaneously affects its upstream and downstream products. TaLCYB is also required for lutein synthesis; thus, one of the possible reasons for the decreased lutein content could be that TaLCYB is the key enzyme of lutein biosynthesis in wheat seeds. Additionally, it is also possible that changing the pigment composition could regulate the expression of enzymes. For example, the expression of *TaHYD1*, which is more related with β,β-branch synthesis, was up-regulated in the transgenic lines while the expression of *TaHYD2*, which is more related with β,ε-branch synthesis, maintained stable expression. In addition, cyclases and hydroxylases are thought to form a protein complex to function; thus, the down-regulation of *TaLCYB* may impair protein complex formation and influence the biosynthesis of lutein or other carotenoids [41]. The accumulation of lycopene partly compensated for the decrease of β-carotene and lutein in the total carotenoid content, and also showed that carotenoid flux was a whole, opening the possibility for the metabolic engineering of compounds in the carotenoid pathway through an appropriate strategy to modulate the expression of carotenogenic genes in the carotenoid biosynthetic pathway. Generally, all these results suggested that TaLCYB acts as the key enzyme in the downstream carotenoid biosynthetic pathway and determines the β-carotene synthesis capacity.

Down-regulation of *TaLCYB* transcripts affects the expression of other genes in the carotenoid biosynthetic pathway of wheat

The expression levels of related endogenous carotenogenic genes are often altered when introducing exogenous genes, and simultaneously alter the levels of carotenoids in the biosynthetic pathway. This phenomenon has been documented in tomato leaves, potato tubers and maize kernels [29,30,42]. In the present study, the expression of endogenous carotenogenic genes was analyzed in endosperms and leaves from transgenic and control lines. In endosperms, the expression level of *TaLCYB* was reduced, *TaHYD1* and *TaLCYE* were up-regulated, *TaHYD2* maintained stable expression, and *TaPSY* was slightly down-regulated (Figure 6A). In previous research, *TaHYD1* was shown to be more related with β,β-branch synthesis, while *TaHYD2* was more related to β,ε-branch synthesis. This implies that the down-regulation of *TaLCYB* led to more flux to the β,ε-branch, accompanied by the occurrence of α-carotene and the accumulation of lycopene. Other intermediates such as phytoene, phytofluene or ζ-carotene upstream of the carotenoid pathway were not detected, which was probably due mainly to the expression of *TaZDS* and *TaPDS* without significantly altering compared with the wild-type. Because of insufficient TaLCYB, the expression of *TaHYD2* in the transgenic lines was stable. Although *TaHYD1* and *TaLCYE* were up-regulated, the repression of *TaLCYB* could explain the reduction of zeaxanthin and lutein. Since ubiquitous feedback or forward regulation exists in the carotenoid biosynthetic pathway and *PSY* has frequently been reported as the rate-limiting gene in non-green plant tissues [28,43-45], the accumulation of lycopene may lead to feedback regulation to suppress *TaPSY* accompanied by a decrease in total carotenoid content, which might suggest an attenuated phytoene synthesis capacity as a consequence of *TaLCYB* down-regulation. These results are consistent with a previous report on the endosperm of transgenic lines that were transformed with the exogenous genes *CRTB* or/and *CRTI* [25]. However, in the leaves of our transgenic lines (Figure 6B), carotenogenic gene transcripts and carotenoid content and composition did not show dramatic changes. There are four possible explanations for this phenomenon. (1) It is possible that the silencing efficiency was too low or that hexaploid wheat has other *LCYB* gene copies that might not be totally silenced by *LCYB* RNAi, and which might have compensated for β-ring formation. For example, there are two types of LCYBs expressed differently in the diploid species tomato; LCYB1 is active in green tissues, while LCYB2/CYCB functions only in chromoplast-containing tissues such as ripening fruit [46]. The same phenomenon has been observed with the other wheat carotenoid biosynthetic genes, such as the different expression patterns of *TaHYD1* and *TaHYD2* in vegetative tissues and developing grains of wheat. Additionally, our results suggest that *TaLCYB* may be located on 6AS and 6DS according to the latest sequencing of hexaploid wheat [24]; these results indicate that *TaLCYB* might have other copies in hexaploid wheat. (2) The regulatory mechanism of carotenoid biosynthesis in leaves is more stringent to prevent disruption of photosynthesis, because carotenoids are an important part of the photosynthetic apparatus [35,47]. (3) The enzyme activity of LCYB is sufficient for product synthesis to maintain normal photosynthesis in leaves, so incomplete silencing might not affect carotenoid biosynthesis in leaves. (4) The effects in leaves are very weak, which correlates with the low reduction in co-suppressed *TaLCYB*. Thus, it also appears likely that the lines with stronger co-suppression of *TaLCYB* than those we obtained were not viable as a critical level of photosynthetically relevant carotenoids could not be attained.

Conclusions

In summary, this study demonstrates that *TaLCYB* is a genuine carotenoid biosynthetic gene. The silencing of *TaLCYB* led to a decrease of β-carotene content and altered the carotenoid profile and accumulation, accompanied by changes in the expression of endogenous carotenogenic genes to varying degrees. This provides new ideas and means for improving the total carotenoid content or specific carotenoid products by metabolic engineering in wheat. For example, the combination of RNAi-induced gene silencing with overexpression of upstream synthetic genes constitutes a strategy to improve specific carotenoid products in wheat. Generally, our data demonstrate that LCYB is a key enzyme of β-carotene biosynthesis and plays an essential role in the regulation of provitamin A biosynthesis in wheat, controlling flux to the downstream carotenogenic pathway. Although the precise regulatory mechanisms of carotenoid biosynthesis in wheat need to be investigated in future, these findings increase our knowledge of carotenoid biosynthesis in wheat and provide novel implications for wheat carotenoid bioengineering.

Methods

Plant materials and treatments

Wheat (*Triticum aestivum* L. cv. Chinese Spring) plants were grown in the experimental field of Huazhong University of Science and Technology in Wuhan, China. Developing grains were collected between 5 and 35 DAP at 5-day intervals. Leave, stem, root, stamen and pistil tissues were collected from wheat plants in the field. Abiotic stresses including cold, darkness and strong light (800 $\mu mol \cdot m^{-2} \cdot s^{-1}$) were used to examine their effect on the expression of *TaLCYB*. The 10-day-old seedlings were transferred respectively into a dark box, cold room (4°C) or in a growth chamber at constant temperature of 25°C as a control. Strong light stress was imposed by increasing light intensity to 800 $\mu mol \cdot m^{-2} \cdot s^{-1}$ photosynthetic photon flux density.

RNA and genomic DNA isolation

Total RNA was extracted from different wheat tissues using Trizol reagent (Invitrogen, Carlsbad, CA) according to the manufacturer's instructions. RNA concentration and purity were analyzed by Nanodrop ND-2000 spectrophotometer (Thermo Scientific, Wilmington, DE). The integrity of RNA sample was assessed by a non-denaturing agarose gel analysis. Genomic DNA was isolated from wheat leaves by cetyltrimethyl ammonium bromide (CTAB) extraction method [48].

Cloning and bioinformatics analysis of *TaLCYB*

Total RNA extracted from wheat seedlings was used to synthesize cDNAs using RevertAid™ first-strand cDNA synthesis kit (Fermentas, Lithuania). In order to identify putative lycopene cyclase genes in wheat, a BLASTN search was performed with the sequences of *OsLCYB* from *O. sativa* (GenBank Accession No.: AP005849) [49], and *ZmLCYB* from *Z. mays* to identify the putative *LCYB* (GenBank Accession No.: AAO18661) [11]. A *Triticum aestivum* cDNA clone (WT009_F16) from cultivar Chinese Spring showed high identity with *OsLCYB* and *ZmLCYB*. Sequence analysis by ORF Finder showed that WT009_F16 contained the full-length ORF and amplified from wheat cDNA by PCR using specific primer pairs (Additional file 1: Table S2). Cycling parameters for RT-PCR were: 94°C for 3 min, 30 cycles of 94/58/72°C for 30/30/90 s, respectively, and 72°C for 10 min. Purified PCR products were cloned into pMD18-T simple vector (Takara, Dalian, China) then sequenced.

Prediction of transit peptide of TaLCYB was performed using ChloroP 1.1 Prediction Server program [50]. LCYBs sequences were searched at the NCBI (Bethesda, USA) and five amino acid sequences of LCYB were used for phylogenetic analysis. A phylogenetic tree was constructed by the Neighbor–Joining method [51] included in the ClustalW program [52] and bootstrap re-sampling analysis (1000 replicates) was performed.

Functional characterization of wheat TaLCYB in *E. coli*

Full-length of the *TaLCYB* cDNA was cloned into pET-32a+. The plasmid pET-LCYB with pAC-LYC was used to transform *E. coli* strain. Plasmid pAC-LYC is a pACYC184 derived vector including several carotenoid biosynthesis genes, such as geranylgeranyl pyrophosphate synthase (*CrtE*), phytoene synthase (*CrtB*), and phytoene desaturase (*CrtI*) [8]. *E. coli* colonies containing pAC-LYC accumulate lycopene and appear pink. The co-transformants of plasmid pET-LCYB with pAC-LYC were plated onto LB agar medium added with chloramphenicol (50 μg ml^{-1}) and ampicillin (100 μg ml^{-1}). Colonies were incubated for 24 h at 37°C.

Plasmid constructs

PCR primers were designed on the sequence of *TaLCYB* using the Primer 5. The pAHC25 containing the maize *ubi-1* promoter and the nopaline synthase terminator were used to construct RNAi vector. Fragments of 204 bp corresponding to *TaLCYB* were isolated by RT-PCR using specific primers with incorporated restriction sites (Additional file 1: Table S2). The selected fragments followed the selection strategies for RNAi in wheat [53]. Cycling parameters for PCR amplification were: 35 cycles of 94/62/72°C for 30/45/30 s, respectively, and 72°C for 10 min. The amplified fragments were then subcloned into pBluescript SK plus and sequenced. The fragments recovered by *Sma*I and *Not*I digestion were cloned in the same restriction site of plasmid pAHC25. RNAi construct contained a cDNA fragment derived from *TaLCYB* and oriented in the sense and antisense directions at the 3' and 5' ends of the construct separated by an intron sequence, respectively, and the resulting plasmid was named pAHC25-LCYB-RNAi (Additional file 1: Figure S5). The intron was derived from the wheat *TAK14* gene (AF325198).

Wheat transformation and plant regeneration

Wheat genetic transformation was according to the bombardment method reported by Sparks [54]. Wheat immature scutella (14 DAP) from Chinese Spring were transformed with the plasmids of pAHC25-LCYB-RNAi or pAHC25 as a control. The regenerated plants were screened by the herbicide phosphinotricin medium (3 mg L^{-1}). The surviving plants were transferred to soil and grown to maturity under growth chamber conditions (22°C/16°C day/night, 16/8-h light/dark cycle and 300 $\mu mol \cdot m^{-2} \cdot s^{-1}$ photosynthetic photon flux density). The regenerated plants were continued to screen by PCR amplification using gene-specific primers (Additional file 1: Table S2). The PCR-positive transgenic plants were self pollinated and the non-segregant lines were selected to analyze the carotenoid profiles and expression levels of carotenoid biosynthetic genes (Additional file 1: Figure S6).

qPCR analysis

The qPCR analysis was performed with the Realtime System (Bio-Rad, CFX Connect Optics Module, USA) using SuperReal PreMix Plus (SYBR Green) (FP205, Tiangen, Beijing, China). The amplification was performed with the following programme: 40 cycles of 95°C for 15 s, and 60°C for 60 s. Fluorescence was acquired at 60°C. The specificity of the unique amplification product was determined by a melt curve analysis from 55-99°C. Data were analyzed using the Lightcycler software version 4 and normalized to the expression of wheat β-*actin* gene as its relatively constitutive expression levels throughout wheat developmental process. The quality of the cDNA templates and PCR amplifications were verified by the analysis of negative controls without template and no-reverse transcription for each primer pair. Dissociation curve analysis was performed following qPCR and a single peak was observed for each primer pair. A portion of qPCR products was separated on agarose gels and single band at expected sizes were detected.

Analysis of carotenoid composition by HPLC

Carotenoids in the mature wheat seeds were extracted according to Wang *et al.* [25] with some modifications. Seed samples were ground into fine powder. One gram powder was added with 15 mL extracting solvent (hexane: acetone: ethanol, 50:25:25, v/v/v) containing 0.01% (w/v) 2,6-di-tert-butyl-methylphenol (BHT, Sigma, Shanghai, China) following sonication for 30 min (SB-5200DTN, Scientz, China), and centrifuged for 10 min at 10,000 g under 4°C (CR-21G, Himac, Japan). The colored supernatant was collected, and the residue was re-extracted several times with extraction operation until colorless. The combined supernatant was then washed three times with saturated NaCl solution until neutral, and the aqueous phase was discarded. The solvent was evaporated under nitrogen stream, the pigments were redissolved in 0.3 mL methyltert-butyl ether (MTBE) containing 0.01% (w/v) BHT. After centrifuging at 12,000 rpm under 4°C for 30 min, the sample was filtered through a 0.22 μm filter before HPLC analysis. For quantitative purpose, β-apo-8'-carotenal was added to each sample as an internal standard prior to extraction (10 μg g^{-1} of freeze-dried sample). Carotenoids in the *E. coli* cells were extracted according to Alquezar [27].

The sample was injected into the HPLC system. The HPLC system included a model 2996 photodiode array detection (DAD) system, a 1525 solvent delivery system, and a Breeze2 Chromatography Manager (Waters Corpora-tion, Milford, MA. Carotenoids were separated by an YMC C30 carotenoid column (150 × 4.6 mm, packing 3 μm) (Wilmington, NC, USA) at 25°C. All the eluate was under 200 to 700 nm monitoring. The solvent A was acetonitrile/methanol (3:1, v/v), containing 0.01% BHT and 0.05% triethylamine (TEA, Sigma, Shanghai, China), and solvent B was 100% MTBE, containing 0.01% BHT. The parameters of mobile-phase gradient were programmed as follows: 0–10 min, A-B (95:5); 10–19 min, A-B (86:14); 19–29 min, A-B (75:25); 29–54 min, A-B (50:50); 54–66 min, A-B (26:74) and back to the initial condition for re-equilibration. All solvents were HPLC grade (J.T. Baker, Phillipsburg, USA). Carotenoid standards for lutein, zeaxanthin, β-cryptoxanthin, α-carotene, *trans*-β-carotene, β-apo-8'-carotenal, *trans*-lycopene calibration were purchased from Sigma-Aldrich (Shanghai, China); *9-cis*-β-carotene was purchased from Carotenature (Lupsingen, Switzerland). These standards, and the β-apo-8'-carotenal internal standard, were used to generate standard calibration curves. Carotenoids and chlorophylls were identified by comparing the retention time and spectra with published data and then quantified from their peak areas [55-58].

Additional file

Additional file 1: Figure S1. HPLC characterization of carotenoids extracted from developing grains. **Figure S2.** Expression patterns of wheat *TaLCYB* in leaves with different treatments and β-carotene content varied with strong light. **Figure S3.** HPLC characterization of wheat leaf's carotenoids composition under strong light treatment. **Figure S4.** Expression levels of the endogenous carotenoid biosynthetic genes in leaves with strong light treatment. **Figure S5.** The structures of transformation plasmids (pAHC25-LCYB-RNAi and pAHC25) used in this study. **Figure S6.** Propagation of transgenic wheat and selection of non-segregant lines of *TaLCYB* silencing. **Table S1.** Cartenoids content and compositions in T_2 seeds from the transgenic and control wheat plants. **Table S2**. Primer sequences used in this study.

Abbreviations

BHT: 2,6-di-tert-butyl-methylphenol; β-LCY CAD region: β-LCY Catalytic Activity Domain region; CCDs: Carotenoid cleavage dioxygenases; CrtB: Phytoene synthase; CrtE: Geranylgeranyl pyrophosphate synthase; CrtI: Phytoene desaturase; CTAB: Cetyltrimethyl ammonium bromide; DAD: Photodiode array detection; DAP: Days after pollination; HPLC: High Performance Liquid Chromatography; HYD: β-carotene hydroxylase; LCYB: Lycopene β-cyclase; LCYE: Lycopene ε-cyclase; MTBE: Methyl tert-butyl ether; PDS: Phytoene desaturase; PSY: Phytoene synthase; qPCR: Quantitative PCR; RNAi: RNA interference; TEA: Triethylamine; VAD: Vitamin A deficiency; VC: Vector control; ZDS: Zeta-carotene desaturase..

Competing interests

The authors declare that they have no competing interests.

Authors' contributions

GYH and GXY conceived the study. XC, MLZ, CHY, XTW, QW, ML, XYL and LC performed the experiments. KXL, JLC and YSW carried out the analysis. JZ, CW, GYH and GXY designed the experiments and wrote the manuscript. All authors read and approved this submitted manuscript.

Authors' information

The Genetic Engineering International Cooperation Base of Chinese Ministry of Science and Technology, The Key Laboratory of Molecular Biophysics of Chinese Ministry of Education, College of Life Science and Technology, Huazhong University of Science and Technology, Wuhan, China

Acknowledgements

We thank Dr. F. X. Cunningham, University of Maryland, for providing the complementation plasmid pAC-LCY. This work was supported by International S & T Cooperation Key Projects of MoST (Grant no. 2009DFB30340), National Genetically Modified New Varieties of Major Projects of China (Grant no. 2013ZX08002004-007 and 2014ZX08010004), the National Natural Science Foundation of China (no. 31071403 and no. 31371614), Research Fund for the Doctoral Program of Higher Education of China (Grant no. 20120142110 75) and Open Research Fund of State Key Laboratory of Hybrid Rice in Wuhan University (Grant no. KF201302).

Reference

1. Bramley PM. Regulation of carotenoid formation during tomato fruit ripening and development. J Exp Bot. 2002;53(377):2107–13.
2. Britton G. Functions of intact carotenoids. In: *carotenoids*. Springer. 2008. 189–212.
3. Fraser PD, Bramley PM. The biosynthesis and nutritional uses of carotenoids. Prog Lipid Res. 2004;43(3):228–65.
4. Clagett-Dame M, Knutson D. Vitamin A in reproduction and development. Nutrients. 2011;3(4):385–428.
5. Abdel-Aal E-SM, Akhtar H, Zaheer K, Ali R. Dietary sources of lutein and zeaxanthin carotenoids and their role in eye health. Nutrients. 2013;5(4):1169–85.
6. Hirschberg J. Carotenoid biosynthesis in flowering plants. Curr Opin Plant Biol. 2001;4(3):210–8.
7. Ruiz-Sola MA, Rodriguez-Concepcion M. Carotenoid biosynthesis in Arabidopsis: a colorful pathway. Arabidopsis Book. 2012;10, e0158.
8. Cunningham FX, Sun Z, Chamovitz D, Hirschberg J, Gantt E. Molecular structure and enzymatic function of lycopene cyclase from the cyanobacterium Synechococcus sp strain PCC7942. The Plant Cell Online. 1994;6(8):1107–21.
9. Cunningham FX, Gantt E. One ring or two? Determination of ring number in carotenoids by lycopene ε-cyclases. Proc Natl Acad Sci. 2001;98(5):2905–10.
10. Cunningham FX, Pogson B, Sun Z, McDonald KA, DellaPenna D, Gantt E. Functional analysis of the beta and epsilon lycopene cyclase enzymes of Arabidopsis reveals a mechanism for control of cyclic carotenoid formation. The Plant Cell Online. 1996;8(9):1613–26.
11. Singh M, Lewis PE, Hardeman K, Bai L, Rose JK, Mazourek M, et al. Activator mutagenesis of the pink scutellum1/viviparous7 locus of maize. The Plant Cell Online. 2003;15(4):874–84.
12. Ampomah-Dwamena C, McGhie T, Wibisono R, Montefiori M, Hellens RP, Allan AC. The kiwifruit lycopene beta-cyclase plays a significant role in carotenoid accumulation in fruit. J Exp Bot. 2009;60(13):3765–79.
13. Devitt LC, Fanning K, Dietzgen RG, Holton TA. Isolation and functional characterization of a lycopene beta-cyclase gene that controls fruit colour of papaya (Carica papaya L.). J Exp Bot. 2010;61(1):33–9.
14. Bouvier F, d'Harlingue A, Hugueney P, Marin E, Marion-Poll A, Camara B. Xanthophyll biosynthesis cloning, expression, functional reconstitution, and regulation of β-cyclohexenyl carotenoid epoxidase from pepper (Capsicum annuum). J Biol Chem. 1996;271(46):28861–7.
15. Sandmann G, Römer S, Fraser PD. Understanding carotenoid metabolism as a necessity for genetic engineering of crop plants. Metab Eng. 2006;8(4):291–302.
16. Zhu C, Naqvi S, Gomez-Galera S, Pelacho AM, Capell T, Christou P. Transgenic strategies for the nutritional enhancement of plants. Trends Plant Sci. 2007;12(12):548–55.
17. Fraser PD, Enfissi E, Bramley PM. Genetic engineering of carotenoid formation in tomato fruit and the potential application of systems and synthetic biology approaches. Arch Biochem Biophys. 2009;483(2):196–204.
18. Shewry P. Wheat. *J Exp Bot* 2009. 60(6):1537–1553.
19. Mares DJ, Campbell A. Mapping components of flour and noodle colour in Australian wheat. Crop and Pasture Science. 2001;52(12):1297–309.
20. Qin X, Zhang W, Dubcovsky J, Tian L. Cloning and comparative analysis of carotenoid β-hydroxylase genes provides new insights into carotenoid metabolism in tetraploid (Triticum turgidum ssp. durum) and hexaploid (Triticum aestivum) wheat grains. Plant Mol Biol. 2012;80(6):631–46.

21. Cong L, Wang C, Li ZQ, Chen L, Yang GX, Wang YS, et al. cDNA cloning and expression analysis of wheat (Triticum aestivum L.) phytoene and zeta-carotene desaturase genes. Mol Biol Rep. 2010;37(7):3351–61.
22. Wang J, He X, He Z, Wang H, Xia X. Cloning and phylogenetic analysis of phytoene synthase 1 (Psy1) genes in common wheat and related species. Hereditas. 2009;146(5):208–56.
23. Howitt CA, Cavanagh CR, Bowerman AF, Cazzonelli C, Rampling L, Mimica JL, et al. Alternative splicing, activation of cryptic exons and amino acid substitutions in carotenoid biosynthetic genes are associated with lutein accumulation in wheat endosperm. Funct Integr Genomics. 2009;9(3):363–76.
24. Mayer KF, Rogers J, Doležel J, Pozniak C, Eversole K, Feuillet C, et al. A chromosome-based draft sequence of the hexaploid bread wheat (Triticum aestivum) genome. Sci. 2014;345(6194):1251788.
25. Wang C, Zeng J, Li Y, Hu W, Chen L, Miao Y, et al. Enrichment of provitamin A content in wheat (Triticum aestivum L.) by introduction of the bacterial carotenoid biosynthetic genes CrtB and CrtI. J Exp Bot. 2014;65(9):2545–56.
26. Hugueney P, Badillo A, Chen H-C, Klein A, Hirschberg J, Camara B, et al. Metabolism of cyclic carotenoids: A model for the alteration of this biosynthetic pathway in Capsicum annuum chromoplasts. Plant J. 1995;8(3):417–24.
27. Alquezar B, Zacarias L, Rodrigo MJ. Molecular and functional characterization of a novel chromoplast-specific lycopene beta-cyclase from Citrus and its relation to lycopene accumulation. J Exp Bot. 2009;60(6):1783–97.
28. Paine JA, Shipton CA, Chaggar S, Howells RM, Kennedy MJ, Vernon G, et al. Improving the nutritional value of Golden Rice through increased pro-vitamin A content. Nat Biotechnol. 2005;23(4):482–7.
29. Aluru M, Xu Y, Guo R, Wang Z, Li S, White W, et al. Generation of transgenic maize with enhanced provitamin A content. J Exp Bot. 2008;59(13):3551–62.
30. Diretto G, Al-Babili S, Tavazza R, Papacchioli V, Beyer P, Giuliano G. Metabolic Engineering of Potato Carotenoid Content through Tuber-Specific Overexpression of a Bacterial Mini-Pathway. PLoS One. 2007;2(4), e350.
31. Scolnik PA, Bartley G. Nucleotide sequence of Lycopene Cyclase (Genbank L40176) from Arabidopsis. Plant Physiol. 1995;108:1342.
32. Guzman I, Hamby S, Romero J, Bosland PW, O'Connell MA. Variability of carotenoid biosynthesis in orange colored *Capsicum* spp. Plant Sci. 2010;179(1):49–59.
33. Pecker I, Gabbay R, Cunningham Jr FX, Hirschberg J. Cloning and characterization of the cDNA for lycopene beta-cyclase from tomato reveals decrease in its expression during fruit ripening. Plant Mol Biol. 1996;30(4):807–19.
34. Demmig-Adams B, Gilmore AM, Adams III WW. In vivo functions of carotenoids in higher plants. FASEB J. 1996;10(4):403–12.
35. Dall'Osto L, Fiore A, Cazzaniga S, Giuliano G, Bassi R. Different roles of α-and β-branch xanthophylls in photosystem assembly and photoprotection. J Biol Chem. 2007;282(48):35056–68.
36. Nisar N, Li L, Lu S, Khin NC, Pogson BJ. Carotenoid Metabolism in Plants. Mol Plant. 2015;8(1):68–82.
37. Vallabhaneni R, Bradbury LM, Wurtzel ET. The carotenoid dioxygenase gene family in maize, sorghum, and rice. Arch Biochem Biophys. 2010;504(1):104–11.
38. da Silva MR. Galli V, dos Anjos e Silva SD, Rombaldi CV: Carotenoid Biosynthetic and Catabolic Pathways: Gene Expression and Carotenoid Content in Grains of Maize Landraces. Nutrients. 2014;6(2):546–63.
39. Diretto G, Tavazza R, Welsch R, Pizzichini D, Mourgues F, Papacchioli V, et al. Metabolic engineering of potato tuber carotenoids through tuber-specific silencing of lycopene epsilon cyclase. BMC Plant Biol. 2006;6:13.
40. Moreno JC, Pizarro L, Fuentes P, Handford M, Cifuentes V, Stange C. Levels of Lycopene β-Cyclase 1 Modulate Carotenoid Gene Expression and Accumulation in Daucus carota. PLoS One. 2013;8(3), e58144.
41. Cunningham FX, Gantt E. Genes and enzymes of carotenoid biosynthesis in plants. Annu Rev Plant Physiol Plant Mol Biol. 1998;49:557–83.
42. Rosati C, Aquilani R, Dharmapuri S, Pallara P, Marusic C, Tavazza R, et al. Metabolic engineering of beta-carotene and lycopene content in tomato fruit. Plant J. 2000;24(3):413–9.
43. Fraser PD, Romer S, Shipton CA, Mills PB, Kiano JW, Misawa N, et al. Evaluation of transgenic tomato plants expressing an additional phytoene synthase in a fruit-specific manner. Proc Natl Acad Sci U S A. 2002;99(2):1092–7.
44. Rodríguez-Villalón A, Gas E, Rodríguez-Concepción M. Phytoene synthase activity controls the biosynthesis of carotenoids and the supply of their metabolic precursors in dark-grown Arabidopsis seedlings. PIJ. 2009;60(3):424–35.
45. Welsch R, Arango J, Bär C, Salazar B, Al-Babili S, Beltrán J, et al. Provitamin A accumulation in cassava (Manihot esculenta) roots driven by a single nucleotide polymorphism in a phytoene synthase gene. The Plant Cell Online. 2010;22(10):3348–56.
46. Ronen G, Carmel-Goren L, Zamir D, Hirschberg J. An alternative pathway to β-carotene formation in plant chromoplasts discovered by map-based cloning of Beta and old-gold color mutations in tomato. Proc Natl Acad Sci. 2000;97(20):11102–7.
47. Davison P, Hunter C, Horton P. Overexpression of β-carotene hydroxylase enhances stress tolerance in Arabidopsis. Nature. 2002;418(6894):203–6.
48. Sambrook J, Fritsch E. Maniatis. 1989. Molecular cloning: a laboratory manual. In.: Cold Spring Harbor Laboratory, Cold Spring Harbor. NY. 1997.
49. Yu Q, Schaub P, Ghisla S, Al-Babili S, Krieger-Liszkay A, Beyer P. The lycopene cyclase CrtY from Pantoea ananatis (formerly Erwinia uredovora) catalyzes an FADred-dependent non-redox reaction. J Biol Chem. 2010;285(16):12109–20.
50. Emanuelsson O, Nielsen H, Heijne GV. ChloroP, a neural network-based method for predicting chloroplast transit peptides and their cleavage sites. Protein Sci. 1999;8(5):978–84.
51. Saitou N, Nei M. The neighbor-joining method: a new method for reconstructing phylogenetic trees. Mol Biol Evol. 1987;4(4):406–25.
52. Thompson JD, Higgins DG, Gibson TJ. CLUSTAL W: improving the sensitivity of progressive multiple sequence alignment through sequence weighting, position-specific gap penalties and weight matrix choice. Nucleic Acids Res. 1994;22(22):4673–80.
53. Fu D, Uauy C, Blechl A, Dubcovsky J. RNA interference for wheat functional gene analysis. Transgenic Res. 2007;16(6):689–701.
54. Sparks C, Jones H. Transformation of wheat by biolistics. In: Transgenic Crops of the World. Springer; 2004. p. 19–34.
55. Sander LC, Sharpless KE, Pursch M. C < sub > 30</sub > Stationary phases for the analysis of food by liquid chromatography. J Chromatogr A. 2000;880(1):189–202.
56. Fraser PD, Pinto MES, Holloway DE, Bramley PM. Application of high-performance liquid chromatography with photodiode array detection to the metabolic profiling of plant isoprenoids. PIJ. 2000;24(4):551–8.
57. Rodriguez-Amaya D, Kimura M: Harvest plus handbook for carotenoid analysis. international Food Policy Research institute and international Center for Tropical agriculture. 2004.
58. Liu Q, Xu J, Liu Y, Zhao X, Deng X, Guo L, et al. A novel bud mutation that confers abnormal patterns of lycopene accumulation in sweet orange fruit (Citrus sinensis L. Osbeck). J Exp Bot. 2007;58(15–16):4161–71.

18

Identification and analysis of novel salt responsive candidate gene based SSRs (cgSSRs) from rice (*Oryza sativa* L.)

Kutubuddin Ali Molla, Ananda Bhusan Debnath, Showkat Ahmad Ganie and Tapan Kumar Mondal*

Abstract

Background: Majority of the Asian people depend on rice for nutritional energy. Rice cultivation and yield are severely affected by soil salinity stress worldwide. Marker assisted breeding is a rapid and efficient way to develop improved variety for salinity stress tolerance. Genomic microsatellite markers are an elite group of markers, but there is possible uncertainty of linkage with the important genes. In contrast, there are better possibilities of linkage detection with important genes if SSRs are developed from candidate genes. To the best of our knowledge, there is no such report on SSR markers development from candidate gene sequences in rice. So the present study was aimed to identify and analyse SSRs from salt responsive candidate genes of rice.

Results: In the present study, based on the comprehensive literature survey, we selected 220 different salt responsive genes of rice. Out of them, 106 genes were found to contain 180 microsatellite loci with, tri-nucleotide motifs (56%) being most abundant, followed by di-(41%) and tetra nucleotide (2.8%) motifs. Maximum loci were found in the coding sequences (37.2%), followed by in 5′UTR (26%), intron (21.6%) and 3′UTR (15%). For validation, 19 primer sets were evaluated to detect polymorphism in diversity analysis among the two panels consisting of 17 salt tolerant and 17 susceptible rice genotypes. Except one, all primer sets exhibited polymorphic nature with an average of 21.8 alleles/primer and with a mean PIC value of 0.28. Calculated genetic similarity among genotypes was ranged from 19%-89%. The generated dendrogram showed 3 clusters of which one contained entire 17 susceptible genotypes and another two clusters contained all tolerant genotypes.

Conclusion: The present study represents the potential of salt responsive candidate gene based SSR (cgSSR) markers to be utilized as novel and remarkable candidate for diversity analysis among rice genotypes differing in salinity response.

Keywords: Microsatellite, Genic-SSR, cgSSR, Salt tolerance, Salt responsive gene, Rice, Molecular diversity, Candidate gene, Rice genotype

Background

Rice (*Oryza sativa* L.) is the most widely consumed staple food by over half of the world's population and it provides 27 percent of dietary energy supply worldwide [1]. The burgeoning world population growth and shrinkage of agricultural land are the two main reasons of an estimated food shortage in the coming days. Rice production must increase at least 25% by 2030 in order to feed the estimated world population [2]. The situation is more aggravated due to the huge loss of crop yield as a result of different abiotic stresses. Soil salinity, one of the top most abiotic stresses, imposes limitation to the growth and development of rice plant causing yield losses of more than 50 percent [3]. Rice being a natural glycophyte, for every unit of excess salinity (deciSiemens/metre), rice yields are estimated to reduce over 12 percent [4]. In contrast to animal, plants, the creature of nature, are unable to move from one place to other compelling them to endure the stress in standing condition. In this scenario, rice genetic improvement is one of the top priority areas to increase yield overcoming those constraints to meet the future demand.

* Correspondence: mondaltk@yahoo.com
Division of Genomic Resources, National Bureau of Plant Genetic Resources, IARI Campus, Pusa, New Delhi 110012, India

Marker assisted selection remarkably speeds up the efficiency and preciseness of breeding programme over the traditional breeding. Availability of high quality genome sequence [5] further eases up the mining of DNA markers to facilitate marker assisted breeding programme in rice. With the advancement of molecular techniques, a diverse group of molecular markers like restriction fragment length polymorphism (RFLP), random amplification of polymorphic DNA (RAPD), variable number tandem repeat (VNTR), amplified fragment length polymorphisms (AFLP), microsatellites polymorphism or simple sequence repeats (SSR) and single nucleotide polymorphism (SNP) have been developed. Among all, SSR markers are outstanding in application because of their high reproducibility, multi-allelic nature, codominant inheritance, good uniform genome coverage, simplicity and inexpensive developmental methodology [6]. SSRs are present in the genome as tandem arrays of short nucleotide repeats usually 1–5 bases per unit. SSR markers have been extensively used in phylogenetic relationship cum diversity analysis among rice genotypes [7-9], association mapping [10,11] and identification and characterization of important trait related QTL [12-14].

Traditional SSR markers developed from random genomic sequences have uncertainty of linkage with the transcribed regions (genes) of the genome, whereas genic SSR derived from expressed sequence tag (EST) or candidate gene sequences based SSR have far better possibility of linkage to important loci conferring desired phenotypes [15]. Genic SSR markers are highly valuable by virtue of their high transferability to related species, usefulness in functional diversity analysis and utilization as anchor markers for comparative mapping and evolutionary studies [16]. Genic SSR markers were developed from EST sequences available from public database in different crop species like rice [17,18], wheat [19], barley [20], date palm [21], common bean [22] and many others. As another approach, development of SSR markers based on important candidate genes related to a particular trait may greatly expedite marker assisted breeding programme for the trait. Moreover, looking for SSR in candidate genes may attain many unanswered question about the regulation of those genes as increasing evidences are being reported about the regulatory roles of microsatellites in gene sequences [23-25]. However, report on the development of genic SSR marker based on candidate gene sequences (cgSSR) are scanty [26,27].

From literature survey, around 220 different genes in rice were found to be salt responsive as evidenced by forward/reverse genetics study. To the best of our knowledge, there is no report on the development of candidate gene based SSR markers in rice. In this study, we report an exclusive identification of novel salt responsive candidate gene based SSR markers (cgSSRs) from rice. We extensively investigated all characterized salt responsive rice genes from published reports. When those gene sequences were subjected for mining SSR, 106 genes were found to contain simple sequence repeats. Among those cgSSRs, 19 primer sets were evaluated and validated for their extent of polymorphism in 17 salt tolerant and 17 salt sensitive rice genotypes. The originated dendrogram revealed their remarkable ability to distinguish rice genotype on the basis of salinity response. This is the first report of salt responsive candidate gene based SSR (cgSSR) marker identification and validation in rice.

Methods

Plant materials

A total of 34 rice genotypes including two contrasting panel (salt tolerant and salt susceptible), each of which contains 17 genotypes, were subjected for the polymorphism survey in this study. Details of rice genotypes along with their salt sensitivity level are given in Additional file 1.

Isolation of genomic DNA

Fresh green leaves were collected, weighed (100 mg) and immediately used for DNA isolation or stored at −80°C after snap freezing in liquid N_2. DNA was isolated following a previously described protocol [28]. Leaf tissues were grinded to fine powder employing liquid N_2 in a pre-chilled morter. Prewarmed CTAB buffer (2.0% CTAB (w/v); 0.1 M Tris Cl, pH 8; 0.02 M of EDTA, pH 8; 1.4 M NaCl) was added to the powder for extraction and the mixture was incubated at 60°C for 20 min. Supernatant was collected after centrifugation and a solution of Chloroform: Isoamyl alcohol (24:1) was mixed. After centrifugation, aqueous phase was collected, mixed with equal volume of isopropanol and incubated for 20 min at −20°C. Centrifugation was done to pellet down DNA. Pellet was washed with 70% (v/v) ethanol, air dried and dissolved in nuclease free water. The sample was treated with RNase enzyme at 37°C and subsequently purified by phenol-chloroform method [28]. Concentration and quality of purified DNA were checked in Nanodrop spectrophotometer (Thermo scientific, USA) employing 260/280 and 260/230 ratio and also by 1% (w/v) agarose gel electrophoresis.

Salt tolerant genes, SSRs mining and Primer designing

An extensive search of literature was performed manually to identify the rice candidate genes conferring salt tolerant phenotype. All rice genes which have been reported elsewhere to confer stable salt tolerance in transgenic plants on homologous and heterologous over expression and which showed either enhanced or suppressed expression upon salt stress were considered in this study. The gene bank locus numbers were retrieved and subsequently sequences of all those genes were downloaded from the web (http://rice.plantbiology.msu.edu/) resources of Rice Genome Annotation Project [29]. The gene sequences were used to mine SSRs in SSR identification tool [30]. Respective references of those candidate genes which have been found to contain microsatellite repeats were given in Additional file 2: Table S2. We designed primers from the flanking sequences of the identified microsatellite repeat region. Primers were designed manually with the following parameters: primer length 20–25 bp, melting temperature 55–60°C, GC percentage- 45–60 and product size- 130–250 bp. Details of the primers, melting temperature and the anticipated amplification product length are given in Table 1.

PCR amplification and 6% polyacrylamide gel electrophoresis

PCR amplification was done from 34 genotypes with 19 pairs of SSR primers in a total volume of 25 μl using a C1000 Thermal Cycler (Bio Rad, USA). Each 25 μl volume of reaction mixture contained 50 ng of genomic DNA as template, 1X Taq polymerase buffer, 2 mM $MgCl_2$, 0.2 mM dNTPs mix, 0.4 pM each of the forward and reverse primer, 1 U of Taq polymerase. The optimized condition was initial 5 minutes incubation at 97°C for complete denaturation, followed by 38 cycles consisting of 94°C for 1 min, 55°C- 60°C (vary with the primer pair) for 1 min, 72°C for 2 min and finally 72°C for 10 min. The experiments were repeated twice.

Resolving of all PCR products were performed in a vertical 6% non denaturing Polyacrylamide gel electrophoresis (PAGE) system at constant 140 V with 1X TAE (Tris acetate EDTA) buffer (pH-8.0). The gel was stained with ethidium bromide solution and visualized in gel documentation system (Protein Simple, USA).

Allele scoring and sequencing

Molecular weights of the amplified bands were determined based on the relative migration of standard 100 bp DNA ladder (Thermo scientific, USA) in the gel. The molecular weight of each allele was determined using the Alpha View software (Protein Simple, USA). Presence or absence of a particular allele was denoted as 1 or 0 respectively and the data was plotted to generate a data matrix for further analysis. When an allele was found exclusively in one genotype, it was designated as unique allele. Alleles found in less than 5% of genotypes were designated as rare.

DNA was eluted from the bands of selected alleles and purified using QIAEX II Gel Extraction Kit (Qiagen, Germany). The purified DNA was sequenced. The obtained sequences were aligned with the original target sequence using NCBI blastn tool.

Data analysis

Analysis of data was performed according to the method described in a previous report [31]. Polymorphism information content (PIC) value of each primer pairs was calculated according to the formula: $PIC = 1 - \Sigma\, pi^2$, where *pi* is equal to the frequency of the *i*th allele of a particular locus [32]. DARwin v5 software was used to draw the phylogenetic relationship among the rice genotypes [33]. Euclidean distance matrix was computed for evaluation of genetic distances between genotypes and further utilized to construct a dendrogram using the neighbour joining method [34]. Bootstrapping data over a locus for 1000 replications of the original matrix (1/0 data matrix) was used to evaluate the significance of each node. Principal coordinate analysis (PCoA) was carried out in DARwin v5 for differentiating the genotypes.

Result

Frequency and distribution of salt responsive cgSSRs

A total of 220 different salinity responsive candidate genes were screened for the presence of SSR which yielded a total of 180 SSR loci from 106 (48.18%) candidate genes. List of those genes harbouring SSR loci with their respective gene bank LOC number, function, number, types and location of motif found were detailed in Additional file 2. The study included only di-tetra nucleotide repeats and reiteration of motifs less than 5 times was excluded. Tri-nucleotide motifs were found to be the largest (56.11%) and tetra-nucleotide motifs formed the smallest group (2.8%) (Figure 1A). A total of 50 different kinds of motifs were found, of which, CT/TC motifs (12.8%) were most frequent, followed by AT/TA (10%) and CGG (7.7%) motif (Additional file 3). Among the trinucleotide repeat motifs, CGG (coding for arginine) and GCC (coding for alanine) were more abundant than others (Additional file 3). The number of repetition of a motif varied from 5–40, among which, motif with 5 reiterations were the highest in frequency, followed by six, seven and eight repetition indicating that there is an inverse relationship between number of reiteration of a SSR motifs and its frequency. To survey the trend of distribution of SSR loci in candidate gene sequences, the location of motifs were thoroughly investigated. Our results showed that maximum percentage of SSR loci were found in CDS (37.22%) followed by 5′UTR (26.11%), intron (21.66%) and 3′UTR (15%) (Figure 1B). We classified the all 106 candidate genes into seven broad functional groups. Among the groups, 28.3% of total cgSSRs were found in transcription factor genes, while antioxidant genes contained 9.43% of total cgSSRs (Figure 1C). Further we have analyzed the location of SSR loci in each individual functional group. The result revealed that most SSR loci were found in CDS region in case of transcription factor genes and of genes

Table 1 Details of salt tolerance gene, respective genbank LOC number, motifs with repeat number and location in sequence, primers with Tm and molecular weight of expected band

Gene bank LOC No.	Gene	Forward (Tm)	Reverse (Tm)	Expected amplicon size	Number of alleles	PIC value	Function	(Motif) repeat*	Location of motif
LOC_Os11g08210	OsNac5	ATGTGATTAGAGTCGCTTTCAGTTGG (56.9C)	CCAGCTTGTACTTGTGCCAGCC (58.4C)	238 bp	10	0.110	TF	$(TAA)_{18}$	3′UTR
LOC_Os10g25010	OsCML8	GAGAATCAGAGCAAGAGTCTGAACCAGC (61C)	GTCAGCCGCTTCTTCCTCACCTG (61C)	209 bp	21	0.275	Signaling	$(CGG)_{9}$	CDS
LOC_Os01g32120	OsCML11	CATGCAAGCCTGCGGAGACG (61.2C)	CGGTCGAAGGAGCGGAAGATCT (60.5C)	154 bp	42	0.221	Signaling	$(CAG)_{10}$	CDS
LOC_Os02g17500	OsGMST1	AGGAACCAACAGAAGCAAAGGTG (56.3C)	GAGGTGATTTGATGCTGTGAGGC (57.4C)	194 bp	25	0.289	Sugar Transporter	$(AG)_{10}$	5′UTR
LOC_Os07g06740.2	OsCPK17	TTGCCTTTTGATCTAGTGCATTGG (57.2C)	GTCTTCGTCCTTTACTAAATAGCACTCC (55.8C)	267 bp	21	0.353	Kinase	$(CT)_{9}/(CT)_{11}$	3′UTR
LOC_Os02g04630	OsCAX (D)	CTGTTTGGCAATCTGCCAGC (55.6C)	CGTCTCGGCAAAATGTTCCTC (56.1C)	139 bp	20	0.326	Ion transporter	$(CT)_{18}$	Intronic
LOC_Os02g04630	OsCAX (T)	CTTTGGTTGGTTCAGGACGATG (55.9)	GAATTGGAAGCTGTTGGCTCATTC (57.9)	163 bp	18	0.169	"	$(TTA)_{26}$	Intronic
LOC_Os07g38090	OsC3H50	GAGGAATTAGACCATTTAACTCGTCGC (58.7)	GAATCCGACCCAATCCAATCAAG (58.3)	214 bp	24	0.199	RNA processing	$(TC)_{9}$	5′UTR
LOC_Os06g48590.1	OsMAPK4	GACATCTAAGTGCCGCGTGTTC (56.2)	TACATGCAGCGTCGAATCGAAG (57.6)	254 bp	17	0.358	Kinase	$(CT)_{12}$	5′UTR
LOC_Os01g54600	OsWRKY13	CCATGCGTACATACACGTTCATGTG (57C)	GATGGGTGCAGCTTTCAATGATC (57.3C)	246 bp	20	0.370	TF	$(AG)_{16}/(GA)_{9}$	5′UTR
LOC_Os01g72530.1	OsCML31	GTTGATGGATCTGTAAATGCTTCATGG (58.8)	GGCACCATGGAGCACCAAAC (57.4)	167 bp	Not amplified	Not amplified	Signaling	$(AT)_{40}$	3′UTR
LOC_Os01g45274.1	OsCA1	CCATCGAGTACGCCGTCTGC (57.9)	CTTCACCATGAATGTTACACACCCTAC (56.8)	281 bp	35	0.296	Chloroplast photosynthesis	$(CT)_{9}$	Intronic
LOC_Os02g02840.1	OsRacB (D)	GCTCCTCCTTCAACCTTCTTCTTTC (57.1C)	GTGACGCACTTTATGAACCTGGAC (56.5C)	176 bp	30	0.318	Signaling, GTPase	$(GA)_{21}$	5′UTR
LOC_Os02g02840.1	OsRacB (T)	CAAGACCTGCATGCTCATCTCC (56.1C)	CCAGATCAAGAACCATAATCCTAGCTC (56.9C)	202 bp	14	0.386	"	$(TTC)_{9}$	Intronic
LOC_Os05g51670.1	OsUGE1	CACAACGCCAACAACCTCGAC (57.7C)	GCTTATCGAGATGGGAATGGTTG (56.5)	154 bp	11	0.087	Nucleotide sugar metabolism	$(TC)_{9}$	Intronic
LOC_Os06g48590.1	OsMSRMK3	CACCTCCATTTCCCATTCCACC (58.9C)	CGAATCGAAGGCGGCAGCTATAG (60.9)	201 bp	27	0.340	Signaling, Kinase	$(CT)_{12}$	5′UTR
LOC_Os02g35190.2	OsCLC-1	CAGAGAAGCCAAGCAAAGAAAGTCTC (58.1C)	CCGTGCTCTCGATGTCGTAGTTG (59.2)	179 bp	24	0.322	Ion trasport	$(AGA)_{11}$	5′UTR
LOC_Os09g13570	OsbZIP71	CTCAGTAAGCTCCCTGTAGTTGTAGCC (57.3)	GTTCAGGTCATCTTCCGACCTGG (58.5)	259 bp	13	0.323	TF	$(TA)_{12}$	5′UTR
LOC_Os03g02590	OsPEX11-1	GCTGCTCTCGACTTTCTTGTTCC (56.2)	ACTAGCCCTGCACAGACTGAAGAG (55.8)	276 bp	21	0.261	Peroxisomal biogenesis	$(TG)_{19}$	Intronic

D- di-nucleotide and T- tri-nucleotide. *subscript denotes the number of repeats.

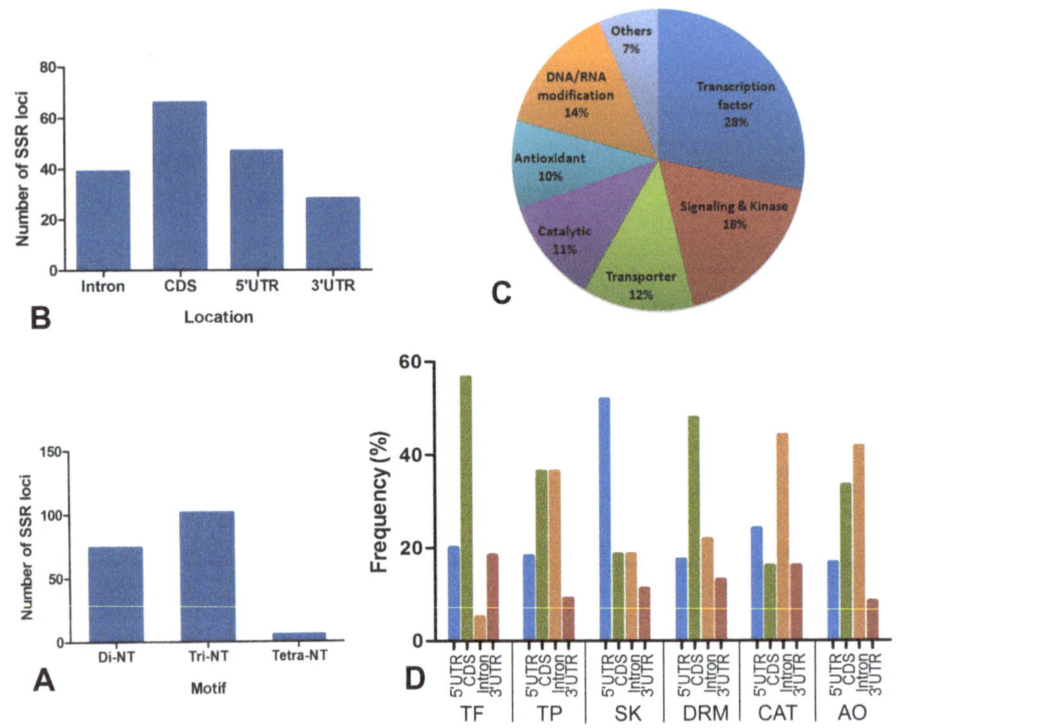

Figure 1 Frequency and distribution of salt responsive cgSSRs in rice. **A)** Number of different SSR motifs found, **B)** number of motifs found in different locations of salt responsive gene sequences, **C)** Percentage of different functional classes of salt responsive genes harbouring SSR loci. **D)** Location of SSR loci in each functional class of salt responsive genes. TF- transcription factor, TP- transporter, SK- signaling & kinase, DRM- DNA/RNA modifying, CAT- catalytic and AO- Antioxidant.

involved in DNA/RNA modification and in intronic region in case of catalytic and antioxidant genes, whereas the genes showing kinase activity and involved in signaling showed highest frequency of SSRs in 5′UTR (Figure 1D). Equal percentage of SSR loci were found in CDS and intron of transporter genes (Figure 1D). Although salt responsive cgSSRs are present on all 12 chromosomes of rice, yet their distribution were not equal among the chromosomes (Figure 2). For example, maximum frequency (23.88%) of salt responsive cgSSR loci were found on chromosome 1, whereas the least (2.22%) was found on chromosome 10. Chromosome 2, 3 and 5 were found to contain more than 10% cgSSR loci.

Development and validation of salt responsive candidate gene based SSR (cgSSR) markers

Out of 180 cgSSRs, primers were designed for 19 loci (NCBI probe- Pr032302526- Pr032302544) from 17 different salt responsive candidate genes (Table 1) for validation. Among 19 different loci, only one designed from CML31 gene failed to amplify. Therefore, we used finally 18 different cgSSR loci to study polymorphism in 34 rice genotypes containing two contrasting panels (17 tolerant and 17 susceptible genotypes). All 18 primer sets generated clear distinct polymorphic profiles as evident from the 6% agarose gel profile (Figure 3) and PIC values. A total of 393 alleles were detected including 32 rare alleles and 32 unique alleles. The average number of alleles produced per primer was 21.8. There was also a degree of stutter bands associated with the main alleles of almost half of the markers used. The cgSSR from *Nac5* gene produced the lowest number of alleles (10), whereas the cgSSR from *CML11* gave rise to the highest number (42) of alleles. Although SSR markers are multiallelic in nature, in order to avoid erroneous calculation and to

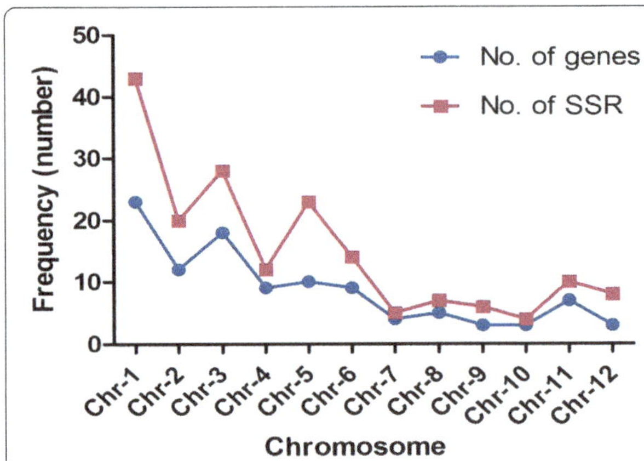

Figure 2 Frequency and distribution of salt responsive cgSSR loci in different rice chromosomes.

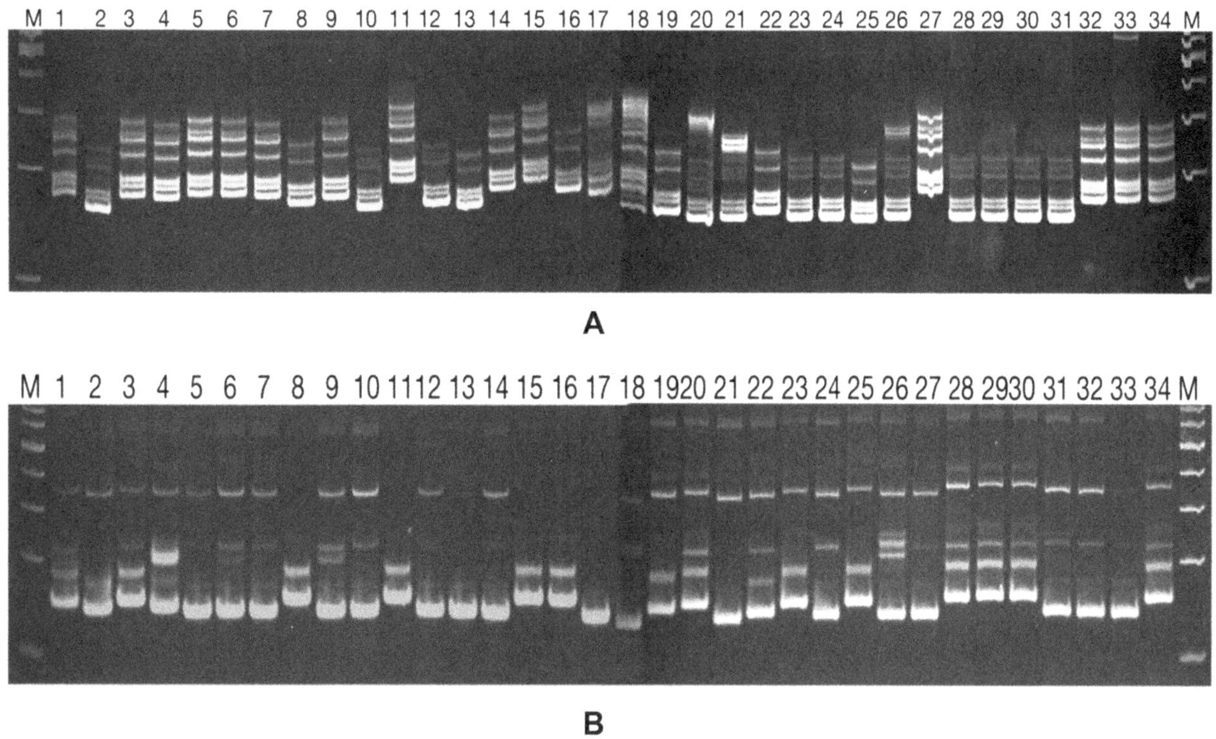

Figure 3 Representative images of 6% Polyacrylamide gel profile of amplified product from 34 genotypes using salt responsive cgSSR primer. **A**- Gel picture with marker- OsRacB (2)-SSR and **B**- with marker OsCML11-SSR. Image was taken in gel documentation system after staining with EtBr. Lane M- 100 bp DNA ladder (Thermo scientific), 1-34- different rice genotypes as defined in Additional file 1.

ascertain the nature of amplicons, we have sequenced all the amplified bands for a particular genotype with a specific marker as a representative case. A total of 20 randomly chosen alleles were sequenced and aligned with the original target sequence. The alignment results confirmed the similarity of each bands with its particular original target sequence (data not shown).

The PIC value denotes the allelic diversity and frequency among genotypes. In our study, an average of about 0.278 PIC value was obtained per cgSSR. The lowest PIC value (0.087) was exhibited by the cgSSR from *UGE1* gene, while highest value (0.386) was obtained with the cgSSR from tri-nucleotide motif of *RacB* gene. Primer designed from di-nucleotide motif loci of *RacB* had a bit lower PIC value of 0.318. On the contrary, cgSSR primers based on di-nucleotide and tri-nucleotide motif loci of the gene *CAX* showed a PIC value of 0.326 and 0.169 respectively. Details of primers and their corresponding PIC values were depicted in the Table 1.

Genetic diversity analysis using salt responsive cgSSR

The data matrix generated from 18 cgSSRs profiling of 34 genotypes were utilized to study the genetic diversity by dissimilarity analysis, factorial analysis through PCoA (principal coordinates analysis) and cluster analysis. The dendrogram generated through unweighted pair group method of arithmetic mean (UPGMA) showed the similarity among the rice accessions ranging from 19% to 89%. The dendrogram exhibited 3 distinct clusters of which two containing all salt tolerant genotypes and one single cluster containing all susceptible rice genotypes (Figure 4). The salt tolerant genotypes were more diverse than the salt susceptible panel in our study. Cluster I consisted of 15 tolerant genotypes containing 2 sub clusters (IA and IB). IA sub-cluster contained 6 genotypes, viz. two Indian- Kalo Nuniya, Pokkali and 4 exotic-Taangteikpan, Erati, Tarome and Talay, while IB sub-cluster contained 9 genotypes, viz. 4 exotic- Cypress, Dom Sofid, Hasawi, Som and 5 Indian- SR26B, CSR10, CSR30, CSR23 and Nona Bokra. Interestingly the smallest cluster (cluster II) contains exclusively two salt tolerant genotypes FL478 and Kala Rata. On the other hand, the largest cluster (cluster III) incorporated all salt susceptible genotypes in the study. Interestingly, all aromatic basmati rice genotypes (Pusa Basmati 1121, Pusa basmati I and Basmati 370) were grouped closely in the same sub cluster. Similarly, IR36, IR64 and IR50 were clustered together. Hence, it is distinct from the genetic diversity analysis using the 18 cgSSR markers that those markers are able to distinguish rice genotypes on the basis of salt sensitivity.

For overall representation of diversity, principal coordinates analysis (PCoA) which requires Euclidean distance

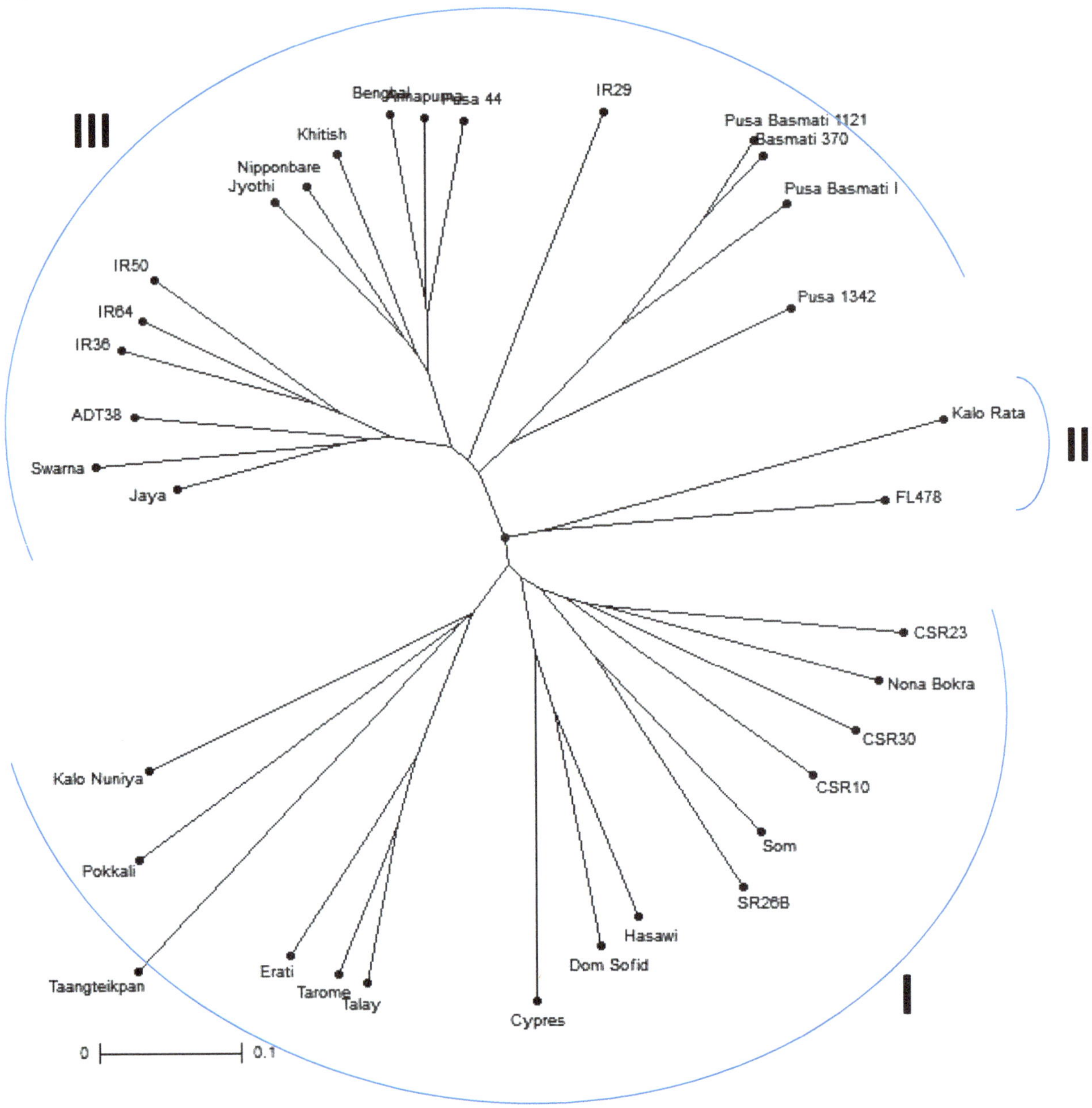

Figure 4 Dendrogram generated from an unweighted pair group method analysis (UPGMA) cluster analysis s based on salt responsive cgSSR markers. First two clusters showing all tolerant genotypes, whereas third cluster showing all susceptible genotypes.

between units has been performed. PCoA revealed distinct separation between each two rice genotypes (Figure 5). In accordance with the dendrogram, the PCoA also clearly divided the susceptible and tolerant panel without a single intermixing. Despite of being its one of the parent, susceptible IR29 was grouped separately from tolerant FL478. In a similar fashion, susceptible parent Jaya was in different group from the tolerant descendant CSR10. So it is clear that the developed cgSSRs from salt responsive genes distinguish the genotypes on account of their behavior in salt stress.

Discussion

Ubiquitously, no toxic substance restricts plant growth more than does salt [35]. Salt stress is an emerging threat not only to rice but also to all glycophytes. Salinity has a tremendous effect on plant growth and reproduction as it imposes two simultaneous stresses- one in the form of toxic salt ions and other in the form of water stress caused by a certain drop in water potential value of the soil solution. Although a majority of rice abiotic stress biologists focused on deciphering the mechanism, developing resistance and identifying candidate genes and QTL

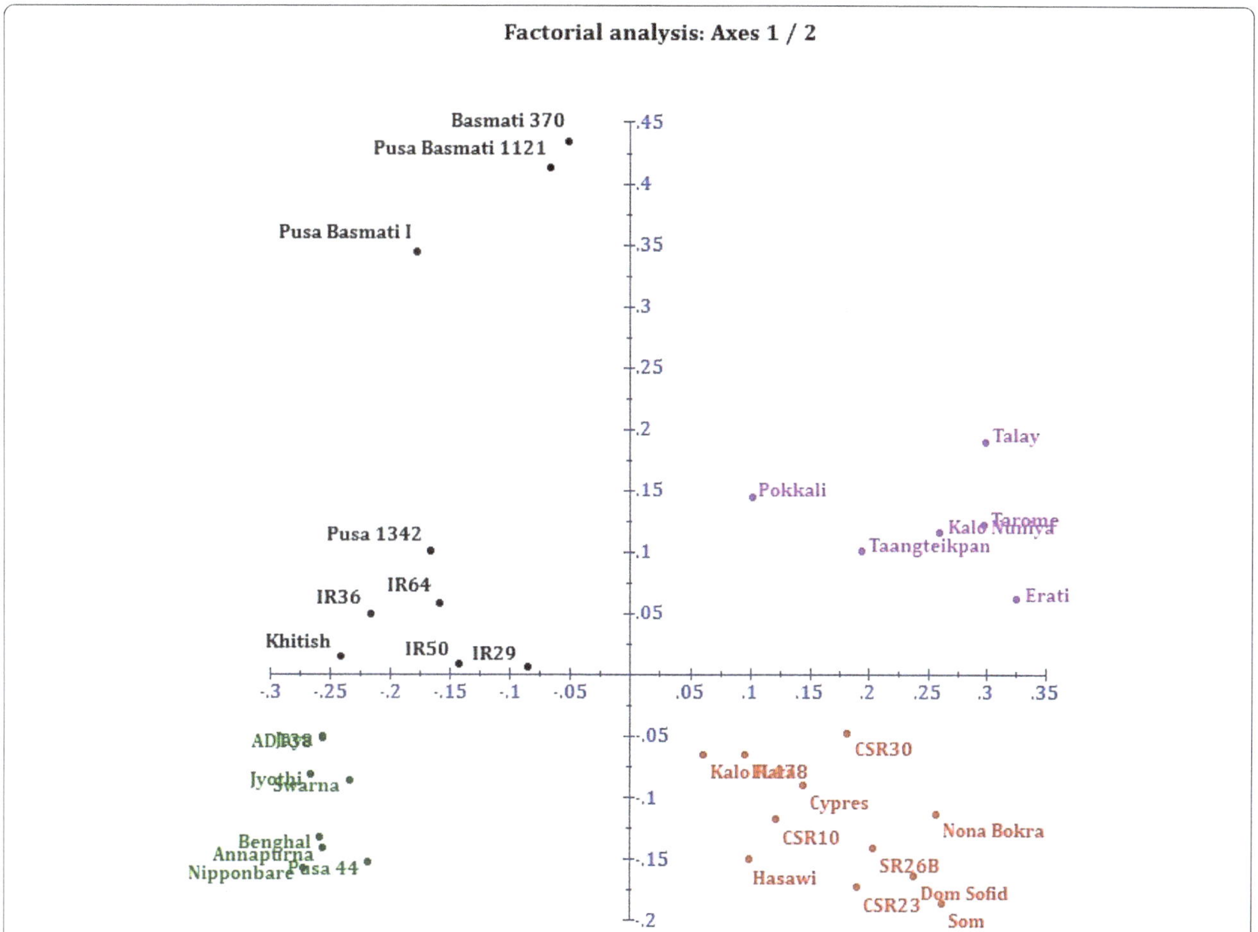

Figure 5 Two-dimension plot generated from principal coordinate analysis (PCoA) for all 34 rice genotypes. Red and violet colour was used for salt tolerant genotypes, while black and green colour was used for salt sensitive genotypes.

involved in salt stress, yet, very few salt tolerant commercially available varieties have been developed. In order to enrich the genomic resource for developing salinity tolerance in rice, here we report the development of salt responsive candidate gene based SSR markers (cgSSRs) in rice for the first time. However, in maize, SSR markers were identified from genes involved in zinc and iron transporter [26] and from candidate genes related to tryptophan and lysine metabolic pathways [27]. Unlike the previous reports, all types of characterized candidate genes including transporter, transcription factor, antioxidant, DNA/RNA modifying which showed differential regulation under salinity stress in rice were selected from published literature and their sequence were used to mine SSR loci. Result of our study showed that tri-nucleotide repeats (56.11%) are more abundant than di- (42.11%) and tetra- (2.8%) nucleotide repeats which is in accordance with previously published reports on rice SSR [17,36] and common bean genic SSR [22]. Similar kind of result was demonstrated in an *in silico* analysis of cereals (rice, wheat, maize, barley, oat and rye) EST derived SSRs showing tri-nucleotide were the most frequent (54–78%) followed by di- (17.1–40.4%) and tetra- (3–6%) nucleotide [37]. However, contrastingly, it has been reported that the number of tri-nucleotide repeats was lesser than the number of di-nucleotide repeats in rice genic non coding microsatellites (GNMS) [38]. Most of the tri-nucleotide motifs were found in CDS (59%) followed by in 5′UTR (21%), intron (11%) and 3′UTR (9%). In this respect, the result of our study is in agreement with an earlier report in wheat [39]. Other studies in rice also support our finding of highest frequency of occurrence of tri-nucleotide repeats in CDS region than any other region like 5′UTR, intron and 3′UTR [38,40]. The phenomenon of copiousness of tri-nucleotide repeats in CDS could be attributed to the selection pressure against frame shift mutation in coding regions resulting from length changes in nontriplet repeats [41]. A previous study of Fujimori et al. [40] proposed that there is a gradual reduction of microsatellite density along the direction of transcription in plant. However, in our study, except the highest frequency in CDS, microsatellite density declines along the direction of transcription (5′UTR—›Intron—›3′UTR) (Figure 1B). In our study, arginine coding (CGG) and alanine coding (GCC) tri-nucleotide repeat motifs were found as two most abundant classes which is in accordance of a previous study of unigene derived microsatellites in cereals [42].

Keeping in mind to validate those salt responsive cgSSRs, we analyzed their possible role to distinguish salt tolerant and susceptible rice genotypes. We speculated the repeat length variations in those cgSSR loci may play role in the manifestation of differential behavior of rice genotypes to salt stress. In order to demonstrate experimental evidence on the speculation, 19 selected cgSSRs were tested in two contrasting panels of rice genotypes which differ in salt sensitivity. The selection of those cgSSRs was based on the notion that SSR loci with more repeats tend to be more polymorphic [43]. SSR loci with 9 or more repeats have been selected to study polymorphism. With the exception of one which failed to amplify, remaining all 18 cgSSRs exhibited polymorphic banding pattern supporting our speculated hypothesis regarding the high level of diversity in salt responsive genes. Among the 18 cgSSRs, six was comprised of tri-nucleotide motif and twelve was with di-nucleotide motif (Table 1). As evidenced from PIC value, polymorphism level varies from primer to primer. Usually di-nucleotide repeats containing SSRs are more prone to mutation and as a result they show more polymorphism than tri-nucleotide repeats containing SSRs [43,44]. However, in our study, the mean difference of PIC value between di-nucleotide and tri-nucleotide containing cgSSRs was not quite statistically significant (*p* value 0.0627). The mean PIC value of all 18 cgSSR primers in the present study was 0.278 which is higher than the previous report describing salt responsive miRNA-SSR markers in rice [31]. Nevertheless, higher PIC values for SSR primers from genomic sequences of rice were reported in earlier studies [7,45]. This might be due to the fact that genic SSRs usually reveal less polymorphism in comparison with genomic SSRs as reviewed in a previous report [16]. The average number of alleles per locus was 21.8 in the present study. This average value is higher in comparison with the average value published in earlier reports [7,27,46,47]. However, producing more alleles than the presence of its repeats by SSR markers is also well documented in literature [48,49] which corroborate our present findings.

It is noteworthy of the present study that the 18 cgSSR markers were remarkably capable of indicating the variation or diversity among rice genotypes in relation to their salinity responsive characters. In the present study, dendrogram generated by UPGMA clearly established relationship between different rice genotype according to their salt sensitivity (Figure 4). Of the three clusters generated in the dendrogram, two contained all tolerant genotypes and another one was comprised of all salt susceptible genotypes exclusively. Our result is also in accordance with the result about the similar clustering pattern of Nona Bokra and Pokkali [50], CSR23, CSR10 and SR 26B [31], Hasawi and SR26B [51] and susceptible IR36 and IR64 [52]. Cluster II was consisted of two tolerant genotypes FL478 and Kala Rata. Similar grouping was observed in the report of salt responsive miRNA-SSR [31]. Remarkably, no single genotype from a particular panel (salinity tolerant or susceptible) is intermixed with another panel. However, out-grouping and intermixing of quite a few salt tolerant and susceptible genotypes were reported previously [31,53,54]. The similarity value between genotypes in the present study ranged from 19% to 89%. In this regard, our result is comparable to the reports published previously [55]. Of all 18, only one cgSSR, CML11 is located at a nearby position (17.58 Mb) of the well known major QTL Saltol (10.8-16.4 Mb) on chromosome 1 indicating it's possibility of being used in MAS programme [56]. Microsatellite within genes can play vital role in gene regulation for controlling a particular trait. SSR in CDS can lead to a gain or loss of gene function via frameshift mutation, SSR in 5′-UTRs can affect transcription and translation, SSR in 3′-UTRs can disrupt splicing and, possibly, disrupt other cellular function and intronic SSRs can affect gene transcription and mRNA splicing [23]. The diversity analysis in the study was based on cgSSR markers representing all classes (CDS, 5′UTR, 3′UTR and intronic) (see Table 1). As no null alleles were obtained, the possibility of the presence or absence of a particular allele in contrasting genotypes is discarded. Hence, our result clearly provoked a thought that the variation present in the salt responsive gene's microsatellite loci may be a key role player in the behavioral response of rice genotypes to salinity. The variation may also play important role in the differential molecular regulation of those genes in rice which differs in salt sensitivity.

Conclusion

To conclude, the present study represents an extensive identification of salt responsive candidate gene based SSR (cgSSR) and their validation as a remarkable tools to distinguish salt susceptible and salt tolerant rice genotypes. The cgSSRs developed here distinctly demarcated the distance between rice genetic resources which show different response to salinity. Identification of these types of allelic variations within salt responsive candidate genes from contrasting panel can provide unique genomic resources with delivering novel alleles to develop improved varieties for salt tolerance. Those developed cgSSRs markers have high potential of linkage and can be utilized for gene pyramiding in breeding programme for salt tolerance trait in rice. They may be proved as an aid in robust functional diversity analysis in the available array of rice germplasms and also in natural population. As there is a high chance of being conserved in nature, the cgSSRs markers are hypothesized to be highly transferable to other cereals which also face tremendous yield losses from salinity. To insight the exact molecular mechanism of that variation in the microsatellite loci governing different sensitivity to salinity, further intense investigation is required.

Availability of supporting data

Salt tolerance ad susceptible panel of rice germplasms used for validation of cgSSR markers are available in Additional file 1. Name of all 106 salt responsive genes, LOC number, number and position of SSR and their respective references are available in Additional file 2. All 19 cgSSR marker used in the present study to construct phylogenetic tree can be found as NCBI probe- Pr032302526- Pr032302544. The phylogenetic tree for the study have been submitted to DRYAD http://datadryad.org/review?doi=doi:10.5061/dryad.dj51c.

Additional files

Additional file 1: Rice genotypes used in diversity analysis using salt responsive cgSSR markers.

Additional file 2: Salt responsive genes with their LOC number, function and their number, locations.

Additional file 3: Bar diagram showing frequency of different types of cgSSR motifs found in salt responsive candidate genes of rice.

Competing interests

The authors declare that they have no competing interests.

Authors' contributions

KM is responsible for planning, performing the work, analysis of data, writing the manuscript, AD is responsible for performing the work, SG is responsible for planning as well as identification of salt responsive genes, TKM is responsible for conceiving, planning, analysis of data the manuscript. All authors read and approved the final manuscript.

Acknowledgement

The authors thank Dr. K.V. Bhat, Head, Division of Genomic Resources, NBPGR, New Delhi for his support and advice to carry out this work. Mr. Showkat Ahmad Ganie is grateful to the Department of Biotechnology, Government of India for the award of Senior Research Fellow.

References

1. Fresco L. "Rice is life". J Food Compos Anal. 2005;18(4):249–53.
2. Li J-Y, Wang J, Zeigler RS. The 3,000 rice genomes project: new opportunities and challenges for future rice research. Gigascience. 2014;3(1):1–3.
3. Zeng L, Shannon MC. Salinity effects on seedling growth and yield components of rice. 2000.
4. Redfern SK, Azzu N, Binamira JS. Rice in Southeast Asia: facing risks and vulnerabilities to respond to climate change. Build Resilience Adapt Climate Change Agri Sector. 2012;23:295.
5. IRGS. The map-based sequence of the rice genome. Nature. 2005;436(7052):793–800.
6. Powell W, Machray GC, Provan J. Polymorphism revealed by simple sequence repeats. Trends Plant Sci. 1996;1(7):215–22.
7. Das B, Sengupta S, Parida SK, Roy B, Ghosh M, Prasad M, et al. Genetic diversity and population structure of rice landraces from Eastern and North Eastern States of India. BMC Genet. 2013;14(1):71.
8. Choudhary G, Ranjitkumar N, Surapaneni M, Deborah DA, Vipparla A, Anuradha G, et al. Molecular genetic diversity of major Indian rice cultivars over decadal periods. PLoS One. 2013;8(6):e66197.
9. Babu BK, Meena V, Agarwal V, Agrawal PK. Population structure and genetic diversity analysis of Indian and exotic rice (Oryza sativa L.) accessions using SSR markers. Mol Biol Rep. 2014;41(7):4329–39.
10. Zhang P, Liu X, Tong H, Lu Y, Li J. Association Mapping for Important Agronomic Traits in Core Collection of Rice (Oryza sativa L.) with SSR Markers. PLoS One. 2014;9(10):e111508.
11. Agrama HA, Eizenga GC, Yan W. Association mapping of yield and its components in rice cultivars. Mol Breed. 2007;19(4):341–56.
12. Wang Z, Cheng J, Chen Z, Huang J, Bao Y, Wang J, et al. Identification of QTLs with main, epistatic and QTL× environment interaction effects for salt tolerance in rice seedlings under different salinity conditions. Theor Appl Genet. 2012;125(4):807–15.
13. Thomson MJ, de Ocampo M, Egdane J, Rahman MA, Sajise AG, Adorada DL, et al. Characterizing the Saltol quantitative trait locus for salinity tolerance in rice. Rice. 2010;3(2–3):148–60.
14. Bernier J, Kumar A, Ramaiah V, Spaner D, Atlin G. A Large-Effect QTL for Grain Yield under Reproductive-Stage Drought Stress in Upland Rice. Crop Sci. 2007;47(2):507–16.
15. Dutta S, Kumawat G, Singh BP, Gupta DK, Singh S, Dogra V, et al. Development of genic-SSR markers by deep transcriptome sequencing in pigeonpea [Cajanus cajan (L.) Millspaugh]. BMC Plant Biol. 2011;11(1):17.
16. Varshney RK, Graner A, Sorrells ME. Genic microsatellite markers in plants: features and applications. Trends Biotechnol. 2005;23(1):48–55.
17. Cho YG, Ishii T, Temnykh S, Chen X, Lipovich L, McCOUCH SR, et al. Diversity of microsatellites derived from genomic libraries and GenBank sequences in rice (Oryza sativa L.). Theor Appl Genet. 2000;100(5):713–22.
18. Yu J-K, La Rota M, Kantety R, Sorrells M. EST derived SSR markers for comparative mapping in wheat and rice. Mol Genet Genomics. 2004;271(6):742–51.
19. Gao L, Jing R, Huo N, Li Y, Li X, Zhou R, et al. One hundred and one new microsatellite loci derived from ESTs (EST-SSRs) in bread wheat. Theor Appl Genet. 2004;108(7):1392–400.
20. Castillo A, Budak H, Varshney RK, Dorado G, Graner A, Hernandez P. Transferability and polymorphism of barley EST-SSR markers used for phylogenetic analysis in Hordeum chilense. BMC Plant Biol. 2008;8(1):97.
21. Zhao Y, Williams R, Prakash C, He G. Identification and characterization of gene-based SSR markers in date palm (Phoenix dactylifera L.). BMC Plant Biol. 2012;12(1):237.
22. Blair MW, Hurtado N, Chavarro CM, Muñoz-Torres MC, Giraldo MC, Pedraza F, et al. Gene-based SSR markers for common bean (Phaseolus vulgaris L.) derived from root and leaf tissue ESTs: an integration of the BMc series. BMC Plant Biol. 2011;11(1):50.
23. Li Y-C, Korol AB, Fahima T, Nevo E. Microsatellites within genes: structure, function, and evolution. Mol Biol Evol. 2004;21(6):991–1007.
24. Sharopova N. Plant simple sequence repeats: distribution, variation, and effects on gene expression. Genome. 2008;51(2):79–90.
25. Tranbarger TJ, Kluabmongkol W, Sangsrakru D, Morcillo F, Tregear JW, Tragoonrung S, et al. SSR markers in transcripts of genes linked to post-transcriptional and transcriptional regulatory functions during vegetative and reproductive development of Elaeis guineensis. BMC Plant Biol. 2012;12(1):1.
26. Sharma A, Chauhan RS. Identification of candidate gene-based markers (SNPs and SSRs) in the zinc and iron transporter sequences of maize (Zea mays L.). Curr Sci. 2008;95:1051–9.
27. Babu BK, Agrawal PK, Gupta HS, Kumar A, Bhatt JC. Identification of candidate gene-based SSR markers for lysine and tryptophan metabolic pathways in maize (Zea mays). Plant Breed. 2012;131(1):20–7.
28. Sambrook J, Russell D. Molecular Cloning a laboratory manual. 3rd ed. New York, USA: CSHL Press; 2001.
29. Kawahara Y, de la Bastide M, Hamilton JP, Kanamori H, McCombie WR, Ouyang S, et al. Improvement of the Oryza sativa Nipponbare reference genome using next generation sequence and optical map data. Rice. 2013;6(1):4.
30. Temnykh S, DeClerck G, Lukashova A, Lipovich L, Cartinhour S, McCouch S. Computational and experimental analysis of microsatellites in rice (Oryza sativa L.): frequency, length variation, transposon associations, and genetic marker potential. Genome Res. 2001;11(8):1441–52.

31. Mondal T, Ganie S. Identification and characterization of salt responsive miRNA-SSR markers in rice (Oryza sativa). Gene. 2014;535:204–9.
32. Botstein D, White RL, Skolnick M, Davis RW. Construction of a genetic linkage map in man using restriction fragment length polymorphisms. Am J Hum Genet. 1980;32(3):314–31.
33. Perrier X, Flori A, Bonnot F. Data analysis methods. In: Hamon P, Seguin M, Perrier X, Glaszmann JC, editors. Genetic Diversity of Cultivated Tropical Plants. Montpellier: Enfield Science Publishers; 2003. p. 43–76.
34. Saitou N, Nei M. The neighbor-joining method: a new method for reconstructing phylogenetic trees. Mol Biol Evol. 1987;4(4):406–25.
35. Zhu J. Plant salt stress, Encyclopedia of life sciences. 2007.
36. Singh H, Deshmukh RK, Singh A, Singh AK, Gaikwad K, Sharma TR, et al. Highly variable SSR markers suitable for rice genotyping using agarose gels. Mol Breed. 2010;25(2):359–64.
37. Varshney RK, Thiel T, Stein N, Langridge P, Graner A. In silico analysis on frequency and distribution of microsatellites in ESTs of some cereal species. Cell Mol Biol Lett. 2002;7(2A):537–46.
38. Parida SK, Dalal V, Singh AK, Singh NK, Mohapatra T. Genic non-coding microsatellites in the rice genome: characterization, marker design and use in assessing genetic and evolutionary relationships among domesticated groups. BMC Genomics. 2009;10(1):140.
39. Yu J-K, Dake TM, Singh S, Benscher D, Li W, Gill B, et al. Development and mapping of EST-derived simple sequence repeat markers for hexaploid wheat. Genome. 2004;47(5):805–18.
40. Fujimori S, Washio T, Higo K, Ohtomo Y, Murakami K, Matsubara K, et al. A novel feature of microsatellites in plants: a distribution gradient along the direction of transcription. FEBS Lett. 2003;554(1):17–22.
41. Metzgar D, Bytof J, Wills C. Selection against frameshift mutations limits microsatellite expansion in coding DNA. Genome Res. 2000;10:72–80.
42. Parida SK, Kumar KAR, Dalal V, Singh NK, Mohapatra T. Unigene derived microsatellite markers for the cereal genomes. Theor Appl Genet. 2006;112(5):808–17.
43. Schug MD, Hutter CM, Wetterstrand KA, Gaudette MS, Mackay TF, Aquadro CF. The mutation rates of di-, tri- and tetranucleotide repeats in Drosophila melanogaster. Mol Biol Evol. 1998;15(12):1751–60.
44. Gadaleta A, Mangini G, Mulè G, Blanco A. Characterization of dinucleotide and trinucleotide EST-derived microsatellites in the wheat genome. Euphytica. 2007;153(1–2):73–85.
45. Jain N, Jain S, Saini N, Jain R. SSR analysis of chromosome 8 regions associated with aroma and cooked kernel elongation in Basmati rice. Euphytica. 2006;152(2):259–73.
46. Yang G, Maroof MS, Xu C, Zhang Q, Biyashev R. Comparative analysis of microsatellite DNA polymorphism in landraces and cultivars of rice. Mol Gen Genet. 1994;245(2):187–94.
47. Nagaraju J, Kathirvel M, Kumar RR, Siddiq E, Hasnain SE. Genetic analysis of traditional and evolved Basmati and non-Basmati rice varieties by using fluorescence-based ISSR-PCR and SSR markers. Proc Natl Acad Sci. 2002;99(9):5836–41.
48. Li FP, Lee YS, Kwon SW, Li G, Park YJ. Analysis of genetic diversity and trait correlations among Korean landrace rice (Oryza sativa L.). Genet Mol Res. 2014;13(3):6316–31.
49. Yasmin F, Islam MR, Rehana S, Mazumder RR, Anisuzzaman M, Khatun H, et al. Molecular characterization of inbred and hybrid rice genotypes of Bangladesh. SABRAO J Breed Genet. 2012;44(1):163–75.
50. Lisa LA, Seraj ZI, Elahi CF, Das KC, Biswas K, Islam MR, et al. Genetic variation in microsatellite DNA, physiology and morphology of coastal saline rice (Oryza sativa L.) landraces of Bangladesh. Plant and Soil. 2004;263(1):213–28.
51. Chattopadhyay K, Nath D, Mohanta RL, Bhattacharyya S, Marndi BC, Nayak AK, et al. Diversity and validation of microsatellite markers in'Saltol'QTL region in contrasting rice genotypes for salt tolerance at the early vegetative stage. Aust J Crop Sci. 2014;8(3):356–62.
52. Shanthi P, Jebaraj S, Geetha S, Aananthi N. DNA Finger Printing of Salt Tolerant and Susceptible Genotypes Using MicroSatellite Markers in Rice (Oryza sativa L.). Int J Plant Breed Genet. 2012;6:206–16.
53. Kanawapee N, Sanitchon J, Srihaban P, Theerakulpisut P. Genetic diversity analysis of rice cultivars (Oryza sativa L.) differing in salinity tolerance based on RAPD and SSR markers. Electron J Biotechnol. 2011;14(6):2–2.
54. Sudharani M, Reddy P, Reddy G. Identification of genetic diversity in rice (Oryza sativa L.) genotypes using microsatellite markers for salinity tolerance. Int J Sci Innov Discov. 2013;3:22–30.
55. Das B, Sengupta S, Prasad M, Ghose TK. Genetic diversity of the conserved motifs of six bacterial leaf blight resistance genes in a set of rice landraces. BMC Genet. 2014;15(1):82.
56. Bonilla PDJ, Mackill D, Deal K, Gregorio G. RLFP and SSLP mapping of salinity tolerance genes in chromosome 1 of rice (Oryza sativa L.) using recombinant inbred lines. Philipp Agric Sci. 2002;85:68–76.

Evidence of capsaicin synthase activity of the *Pun1*-encoded protein and its role as a determinant of capsaicinoid accumulation in pepper

Kana Ogawa[1], Katsunori Murota[1], Hanako Shimura[1], Misaki Furuya[1], Yasuko Togawa[1], Takeshi Matsumura[2] and Chikara Masuta[1*]

Abstract

Background: Capsaicinoids, including capsaicin and its analogs, are responsible for the pungency of pepper (*Capsicum* species) fruits. Even though capsaicin is familiar and used daily by humans, the genes involved in the capsaicin biosynthesis pathway have not been well characterized. The putative aminotransferase (pAMT) and Pungent gene 1 (Pun1) proteins are believed to catalyze the second to last and the last steps in the pathway, respectively, making the Pun1 protein the putative capsaicin synthase. However, there is no direct evidence that Pun1 has capsaicin synthase activity.

Results: To verify that the Pun1 protein actually plays a role in capsaicin production, we generated anti-Pun1 antibodies against an *Escherichia coli*-synthesized Pun1 protein and used them to antagonize endogenous Pun1 activity. To confirm the anti-Pun1 antibodies' specificity, we targeted *Pun1* mRNA using virus-induced gene silencing. In the *Pun1*-down-regulated placental tissues, the accumulated levels of the Pun1 protein, which was identified on a western blot using the anti-Pun1 antibodies, were reduced, and simultaneously, capsaicin accumulations were reduced in the same tissues. In the *de novo* capsaicin synthesis *in vitro* cell-free assay, which uses protoplasts isolated from placental tissues, capsaicin synthesis was inhibited by the addition of anti-Pun1 antibodies. We next analyzed the expression profiles of *pAMT* and *Pun1* in various pepper cultivars and found that high levels of capsaicin accumulation always accompanied high expression levels of both *pAMT* and *Pun1*, indicating that both genes are important for capsaicin synthesis. However, comparisons of the accumulated levels of vanillylamine (a precursor of capsaicin) and capsaicin between pungent and nonpungent cultivars revealed that vanillylamine levels in the pungent cultivars were very low, probably owing to its rapid conversion to capsaicin by Pun1 soon after synthesis, and that in nonpungent cultivars, vanillylamine accumulated to quite high levels owing to the lack of Pun1.

Conclusions: Using a newly developed protoplast-based assay for *de novo* capsaicin synthesis and the anti-Pun1 antibodies, we successfully demonstrated that the *Pun1* gene and its gene product are involved in capsaicin synthesis. The analysis of the vanillylamine accumulation relative to that of capsaicin indicated that Pun1 was the primary determinant of their accumulation levels.

Keywords: *Capsicum*, pAMT, Pun1, Protoplast assay, Pungency, Vanillylamine, Virus-induced gene silencing

* Correspondence: masuta@res.agr.hokudai.ac.jp
[1]Graduate School of Agriculture, Hokkaido University, Sapporo 060-8589, Japan
Full list of author information is available at the end of the article

Background

Capsaicin is the compound responsible for the pungency of peppers [1]. Capsaicin and its analogs are called capsaicinoids, and the fundamental structure of most capsaicinoids is a branched-chain fatty acid amide of vanillylamine. Capsaicinoids, produced only by species of the genus *Capsicum*, are specifically synthesized in placental tissues, mostly between 20 and 30 days after flowering (daf) in the pepper fruits of pungent cultivars [2]. Capsaicinoids are used in the human diet for their distinct pungent taste and bioactivities, such as thermogenesis [3,4] and the suppression of fat accumulation [5]. In addition, they are used for pharmaceutical purposes because they have potential bioactivities that are ascribed to antioxidants [6] and anti-cancer agents [7]. Although pungency in peppers is a desirable trait, nonpungent cultivars also have been developed as vegetables. Nonpungent cultivars have been shown either to accumulate very few capsaicinoids or to accumulate capsinoids, such as capsiate, which are capsaicinoid analogs (branched-chain fatty acid ester of vanillyl alcohol) that lack the stimuli of the capsaicinoids [8].

Two pathways are involved in capsaicin biosynthesis: (1) the phenylpropanoid pathway derived from phenylalanine leading to vanillylamine; and (2) the branched-chain fatty acid pathway derived from valine leading to 8-methyl-6-nonenoyl-CoA [9-11]. The condensation reaction of vanillylamine with 8-methyl-6-nonenoyl-CoA is thought to be catalyzed by a coenzyme A-dependent acyltransferase. Although the genes involved in capsaicinoid synthesis have been extensively studied, the gene responsible for the acylation of vanillylamine remained unknown until Kim *et al.* [12] reported the SB2-66 clone as a possible candidate. Stewart *et al.* [13] finally showed that SB2-66 was the putative acyltransferase involved in capsaicinoid production and encoded by *AT3* gene, namely *Pungent gene 1* (*Pun1*) as capsaicin synthase (CS). The expression profile of *Pun1* correlates with pepper pungency and the deletion or down-regulation of the *Pun1* gene results in a decreased accumulation of capsaicinoids. In addition to capsaicinoids, *Pun1* has been reported to control capsinoid synthesis in nonpungent pepper cultivars [14,15].

Another important step in capsaicin biosynthesis is the conversion of vanillin to vanillylamine, and a putative aminotransferase (pAMT) has been proposed to be the enzyme responsible for vanillin's transamination. The full-length cDNA clone of *pAMT* was identified from a cDNA library by differential display [16]. Abraham-Juárez *et al.* [17] showed reduced capsaicinoid levels using virus-induced gene silencing (VIGS) against *pAMT*, providing evidence that the putative *pAMT* gene was involved in capsaicinoid biosynthesis. In addition, Lang *et al.* [18] reported that the capsaicinoids were poorly synthesized in a spontaneous mutant, cultivar 'CH-19 Sweet', and that this phenotype was due to the loss-of-function of the *pAMT* gene in the particular mutant. Using this mutant, Tanaka *et al.* [19] analyzed the defective *pAMT* gene in detail and found that a single amino acid substitution was responsible for the capsaicin deficiency.

Very recently, pAMT was demonstrated, using an *Escherichia coli*-expressed recombinant enzyme, to act as a vanillin aminotransferase [20]. However, *Pun1* is considered a putative gene because its encoded enzyme has not been biochemically analyzed, even though functional genomics studies indicate that *Pun1* is responsible for the acylation of vanillylamine and vanillyl alcohol in *Capsicum sp.* [13-15]. In the present study, we used protoplasts and anti-Pun1 antibodies to verify that the *Pun1* gene is actually involved in the crucial step of capsaicin biosynthesis. In addition, we investigated the expression profiles for two important genes, *pAMT* and *Pun1*, in capsaicin biosynthesis and discussed the roles of pAMT and Pun1 in various pepper cultivars that differ in pungency.

Results

Preparation of anti-Pun1 antibodies against the Pun1 protein to antagonize endogenous Pun1 activity in an *in vitro* assay

To verify that the Pun1 protein (syn. AT3) is the actual acyltransferase for capsaicin production, we first developed an *in vitro* assay system for capsaicin synthesis using recombinant Pun1 proteins, which are produced from the cDNA clone of *Pun1* inserted into an *E. coli* expression vector. However, in preliminary experiments, we were not able to use recombinant Pun1 proteins in the enzymatic activity assay because they were mostly insoluble when expressed in *E. coli*, as reported previously [13]. Alternatively, we produced anti-Pun1 antibodies using the *E. coli*-synthesized Pun1 protein and tried to use them as antagonists of endogenous Pun1 activity in the *in vitro* assay system. To test the specificity of the created antibodies for the Pun1 protein, we first conducted a western blot analysis using total proteins from a nonpungent bell pepper, which is defective in the *Pun1* gene, as a true negative control. As shown in Figure 1, we detected a very strong band in proteins from a pungent cultivar ('Chosen') at the expected size of 52 kDa, whereas there was no 52 kDa-signal in the bell pepper. We tested whether the antibodies cross-react with another acyltransferase, hydroxycinnamoyl transferase (HCT), which is also listed as a candidate enzyme in the capsaicinoid synthesis pathway [21,22]. The cDNA of the pepper HCT (EU616565) with the FLAG tag sequence at the 3′ end was polymerase chain reaction (PCR)-amplified and then synthesized as the HCT-FLAG protein from *in vitro*-transcribed HCT-FLAG

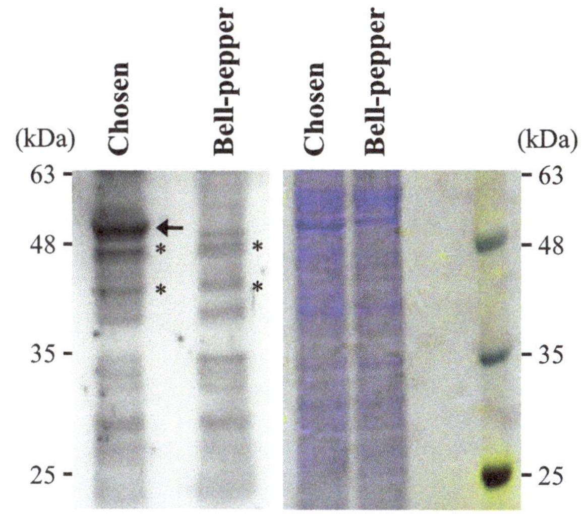

Figure 1 Western blot analysis of the Pun1 protein in the pungent variety 'Chosen' and the nonpungent bell pepper using anti-Pun1 antibodies. As indicated by the arrow, the Pun1 protein was detected in the total proteins isolated from placental tissues of 'Chosen' by polyclonal antibodies against the *E. coli*-synthesized Pun1 protein (left). The Coomassie brilliant blue-stained gel is shown as a loading control (right). Asterisks show nonspecific bands.

mRNA in a wheat germ system. The HCT-FLAG protein was clearly detected by an anti-FLAG antibody, but not by the anti-Pun1 antibodies, suggesting that the anti-Pun1 antibodies did not cross-react with the pepper HCT (Additional file 1: Figure S1). To further confirm the antibody specificity, we used VIGS to specifically reduce the expression level of *Pun1* mRNA and analyzed the capsaicinoid accumulation in the virus-infected pepper placenta. A 95-nt sequence of *Pun1* was inserted in the *Cucumber mosaic virus* (CMV) vector (CMV-Yd:CS95) to induce RNA silencing against *Pun1* mRNA. The pepper fruits infected with the virus vector containing the *Pun1* insert showed no apparent differences from those infected with the empty vector. Compared with healthy fruits, the virus-infected fruits were a little smaller but no great differences were observed. The quantitative reverse-transcription (RT)-PCR analysis confirmed that the *Pun1* mRNA levels were greatly reduced in the CMV-Yd:CS95-infected placental tissues (Figure 2A) in which the capsaicin accumulation was significantly reduced (Additional file 2: Figure S2). We next examined the Pun1 protein accumulation levels using a western blot analysis with the anti-Pun1 antibodies. As shown in Figure 2B, the Pun1 accumulation levels were indeed reduced in the CMV-Yd:CS95-infected placental tissues compared with those of the healthy and CMV-Yd-infected control tissues. These results suggest that the *Pun1* expression was specifically reduced by VIGS, and that the anti-Pun1 antibodies could detect the reduced levels of the Pun1 protein in a western blot. Consistent with those results, the capsaicinoid content in the same placental tissues infected with CMV-Yd:CS95 was also reduced to between ~25% and 33% of the control content (Figure 2C). The reduced level of the Pun1 protein detected by our antibodies reasonably reflected the reduced levels of capsaicinoid, and thus, we concluded that the specificity of the antibodies for the Pun1 protein was satisfactory.

Evidence that the *Pun1*-encoded enzyme can catalyze capsaicin synthesis

As described above, we produced anti-Pun1 antibodies using the *E. coli*-synthesized Pun1 protein and used the antibodies as antagonists of Pun1 activity. To develop an *in vitro* assay system for capsaicin synthesis, we first prepared cell-free crude enzyme extracts from the placental tissue of a pungent pepper and then added vanillylamine. However, we were not able to detect any *de novo*-synthesized capsaicinoid, probably owing to an unknown inhibitor(s). We therefore isolated protoplasts from placental tissues and used them for the assay (Figure 3A). In this assay, CS would be released from the protoplasts soon after the cells are broken, which occurs when vanillylamine is added. In the protoplast-based assay, we were able to detect *de novo*-synthesized capsaicin (Additional file 3: Figure S3). Capsaicinoid synthesis requires 8-methyl-6-nonenoyl-CoA, as well as vanillylamine, as substrates, but 8-methyl-6-nonenoyl-CoA is not commercially available. In our protoplast-based assay, 8-methyl-6-nonenoyl-CoA was provided from the placental protoplasts, leading to capsaicin synthesis. Using this assay system, we analyzed whether the addition of the anti-Pun1 antibodies to the reaction would prevent *de novo* capsaicinoid synthesis. As we expected, the addition of anti-Pun1 antibodies significantly reduced capsaicin synthesis to less than half of the control without anti-Pun1 antibodies (Figure 3B, Additional file 3: Figure S3), suggesting that the Pun1 protein plays an essential role in capsaicin synthesis.

Comparisons of expression levels of *pAMT* and *Pun1* between pungent and nonpungent cultivars

To understand the roles of pAMT and Pun1 in capsaicinoid accumulation, we analyzed the gene expression levels of *pAMT* and *Pun1* in various pepper cultivars that differ in pungency. We extracted RNA from the placental tissues 25 daf and analyzed the expression levels of *pAMT* using quantitative RT-PCR (Figure 4A). The expression levels of *pAMT* were significantly higher in the pungent varieties than in the mildly pungent and nonpungent varieties, while *pAMT* transcripts were barely detectable in the nonpungent varieties, suggesting that the more pungent the variety, the higher the *pAMT* expression. We next measured the *Pun1* transcript levels using the same RNA used for the analyses of *pAMT*. As

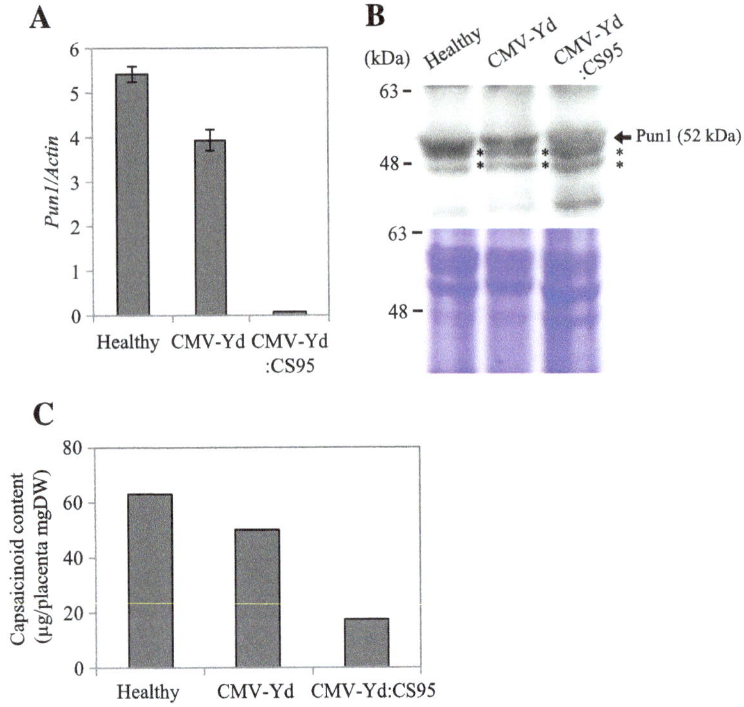

Figure 2 Effect of virus-induced gene silencing against the *Pun1* gene in pepper placental tissues. (A) *Pun1* mRNA accumulation levels in placental tissues infected with CMV-Yd:CS95. CMV-Yd and CMV-Yd:CS95 are the empty vector and the vector containing a 95-nt sequence of the *Pun1* gene, respectively. The *Pun1* mRNA levels are shown relative to the *Actin* mRNA level in placental tissues that were all harvested at 20 to 25 days after flowering. Values are means ± SD obtained from three replicates. **(B)** Western blot analysis of the Pun1 protein in CMV-Yd:CS95-infected placental tissues. The Coomassie brilliant blue -stained gel containing the same proteins as in the blot is shown as a loading control. Asterisks indicate nonspecific bands. **(C)** Capsaicinoid accumulation levels in CMV-Yd:CS95-infected placental tissues evaluated by HPLC. The same tissues were used for HPLC analysis as those used for western blotting.

shown in Figure 4B, the *Pun1* transcript levels were relatively high in the pungent cultivars, but very low in the nonpungent cultivars. Just like for *pAMT*, the more pungent the cultivar, the higher the *Pun1* expression. High-performance liquid chromatography (HPLC) analyses showed that the pungent cultivars produced higher levels of capsaicin (Figure 4C). Thus, the accumulation levels of capsaicinoids seem to correlate with the expression levels of *Pun1* and *pAMT*.

When we examined the expression levels of *pAMT* and *Pun1* over time using quantitative RT-PCR (Figure 5A), as expected, the expression levels of *pAMT* and *Pun1* were very high in the pungent cultivar 'Chosen' but barely present in the nonpungent cultivar 'Fushimi'. It is noteworthy that *Pun1* mRNA in 'Chosen' decreased after 25 daf when capsaicin reached its maximum level in the placenta (Figure 5B), whereas *pAMT* mRNA continued to increase even at 35 daf.

Comparison of vanillylamine contents between pungent and nonpungent varieties

To confirm that *Pun1*, rather than *pAMT*, is the primary determinant of capsaicinoid accumulation during capsaicin synthesis, we compared vanillylamine levels between pungent and nonpungent cultivars. The HPLC analyses showed significantly greater levels of vanillylamine in the placental tissues of nonpungent cultivars at 25 daf (Figure 6A). This result was unexpected because we had initially thought that the vanillylamine contents would be relatively low owing to the low *pAMT* expression in nonpungent cultivars. When we analyzed vanillylamine levels in immature placental tissues at 11 daf, vanillylamine was relatively abundant in the pungent cultivar compared with the nonpungent cultivar (Figure 6B). Thus, we assumed that most of the vanillylamine was efficiently converted to capsaicinoids by Pun1, which was present at high levels soon after vanillylamine synthesis in pungent cultivars at 25 daf. However, the very low levels of Pun1 present in nonpungent cultivars were probably not sufficient to produce capsaicin despite the high accumulation of vanillylamine. However, low levels of pAMT were still capable of converting vanillin to vanillylamine in nonpungent cultivars. Thus, we hypothesize that Pun1 is a more critical regulator of capsaicin synthesis than pAMT.

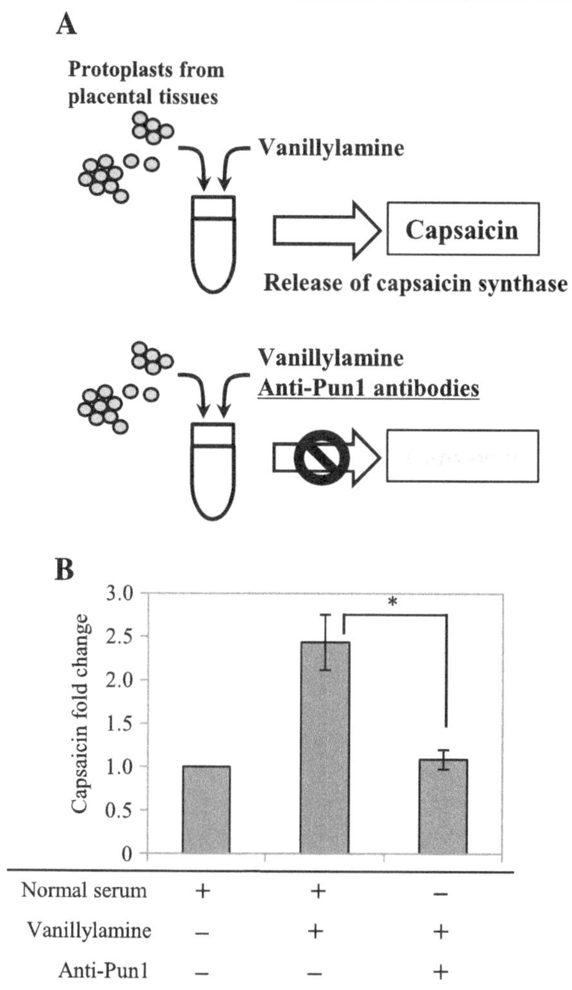

Figure 3 Quantification of *de novo*-synthesized capsaicin in protoplasts treated with or without anti-Pun1 antibodies. (A) Strategy for the cell-free assay of capsaicin synthesis. Protoplasts prepared from placental tissues of pungent peppers were opened in a hypotonic solution, including vanillylamine as a substrate, thus releasing enzymes immediately converted the exogenously added substrate to capsaicin using the endogenous 8-methyl-6-nonenoyl-CoA. Anti-Pun1 antibodies would inhibit this reaction by binding to the Pun1 protein. **(B)** Inhibition of capsaicin synthesis by anti-Pun1 antibodies. In the cell-free assay using protoplasts, *de novo*-synthesized capsaicin was analyzed in HPLC and quantified using a capsaicin standard calibration curve. The graph shows the data as fold change compared with the control [means ± SD relative to a control reaction treated with normal serum (set to 1.0) derived from four separate experiments]. The reduction in capsaicin synthesis in protoplasts with anti-Pun1 antibodies was statistically significant compared with control protoplasts using Student's t-test. Asterisks indicate a significant difference at 0.05. A representative HPLC chromatogram is shown in Additional file 3: Figure S3.

Discussion

Previous knockdown experiments of the *pAMT* and *Pun1* genes by VIGS demonstrated that the down-regulation of these genes resulted in reduced capsaicinoid contents in fruits [13,17], indicating that both genes are responsible for the last steps in capsaicinoid synthesis. In this study, we obtained a similar result with the VIGS targeting of the *Pun1* gene using the CMV vector. The whole-genome sequencing of pepper revealed that there were three copies of the *AT3* (*Pun1*) gene [23]. According to their genome structures and fruit-specific gene expression levels, AT3-D1 and AT3-D2 may be functional in capsaicinoid biosynthesis. Because they are very similar at the 95 nt long target sequence for VIGS, we assumed that both genes were simultaneously down-regulated by VIGS. We then compared the expression levels of the *pAMT* and *Pun1* genes and the capsaicin accumulation levels between pungent and nonpungent cultivars. The quantitative RT-PCR analyses indicated that the capsaicinoid levels correlated well with the expression levels of both *pAMT* and *Pun1*, with expression of the two genes in the pungent cultivars being 50- to 100-fold higher than in the nonpungent cultivars. These results strongly suggest that capsaicinoid levels (pepper pungency) are primarily determined by the expression levels of both *pAMT* and *Pun1*, which are exclusive to the pungent cultivars. Presently, we cannot completely eliminate the possibility that the enzymatic activities of pAMT and/or Pun1 differ between pungent and nonpungent cultivars. However, it is unlikely that the enzymatic activities of pAMT and/or Pun1 are the determinants of pepper pungency over their transcriptional levels because their mRNA levels in the nonpungent cultivars ranged from extremely low to undetectable in our experiments. Thus, the enzymes per se do not accumulate in the placenta, regardless of their activity levels. The accumulation of the AT3 (Pun1) protein is reported to be closely correlated with the level of *AT3* (*Pun1*) transcripts in the pungent cultivar 'Thai Hot' [13]. Additionally, *Pun1* is highly expressed in the pungent cultivar *C. chinense* 'Habanero' at the transcript and protein levels, but was hardly detectable in the nonpungent cultivars *C. annuum* 'Maor' and *C. chinense* NMCA30036 [24]. These reports indicate that capsaicinoid biosynthesis may be regulated primarily at the transcription level. Although the *pAMT* expression level in the nonpungent cultivar *C. annuum* 'CH-19 Sweet' was reported to be comparable with that in the pungent cultivars, this would be an exceptional case because 'CH-19 Sweet' accumulates capsinoids instead of capsaicinoids [18].

The transamination of vanillin to vanillylamine and the acylation of vanillylamine with a branched-chain fatty acid are critical steps in the biosynthesis of capsaicinoids [10,11]. The most plausible candidate genes for the acylation, *Pun1*, and the putative aminotransferase gene, *pAMT*, have been highly correlated with capsaicinoid accumulation in several genetic studies of mutant alleles that corresponded to either *pAMT* or *Pun1* [13,18,19]. However, because the enzymatic activities of pAMT and/or Pun1 may differ between the pungent

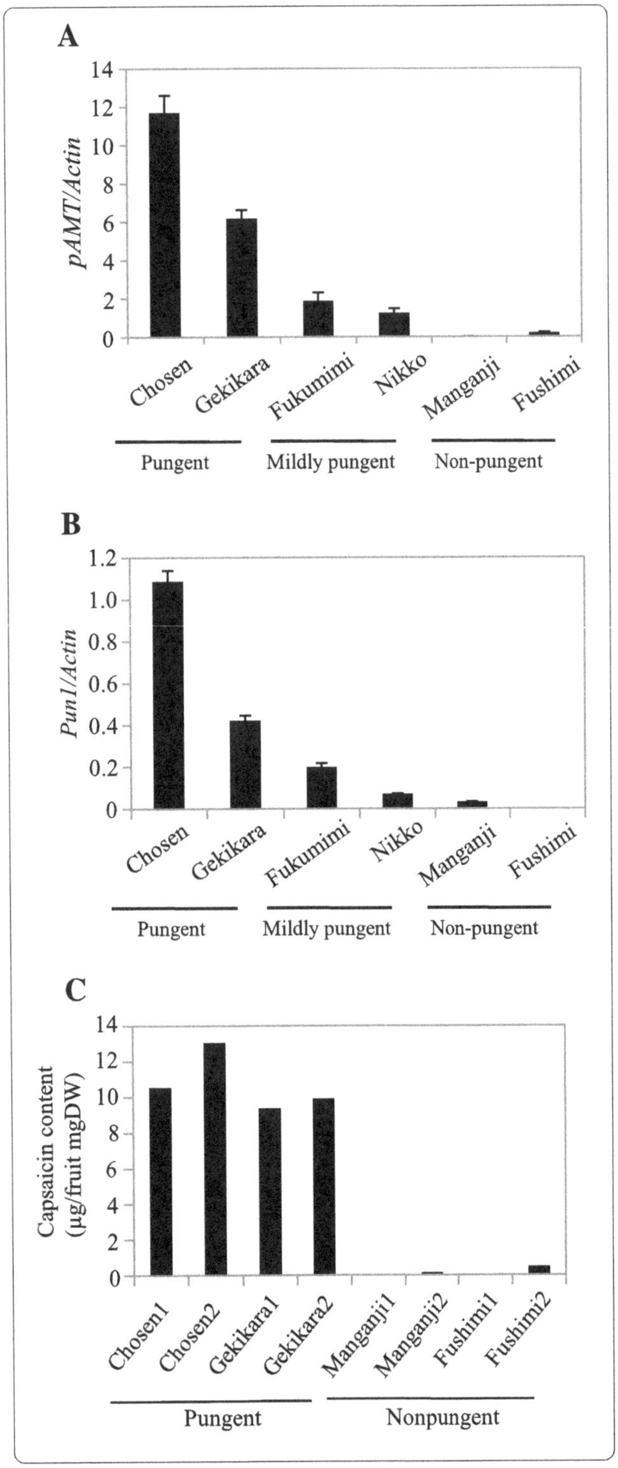

Figure 4 Relationships between the expression levels of *pAMT* and *Pun1* genes, and the accumulation of capsaicin in different pepper cultivars. (A) *pAMT* expression in different pepper cultivars. RNA transcript levels for the *pAMT* gene at 25 days after flowering in placental tissues were determined using quantitative RT-PCR (mean ± SE; $n = 3$). 'Chosen' and 'Gekikara': pungent pepper cultivars; 'Nikko' and 'Fukumimi': mildly pungent pepper cultivars; 'Fushimi' and 'Manganji': nonpungent pepper cultivars. **(B)** *Pun1* gene expression in different pepper cultivars. *Pun1* transcript levels were determined using quantitative RT-PCR (means ± SE; $n = 3$) with the same RNA as that used for the *pAMT* analyses. **(C)** Capsaicin levels in pungent and nonpungent pepper cultivars as determined by HPLC.

and nonpungent cultivars, we should be able to measure the enzymatic activities of the enzymes. Thus, biochemical verifications will still be necessary to determine whether these putative genes are the real players in the pathway. Recently, based on bioinformatics analyses, single nucleotide polymorphisms in the *Pun1* locus were found to distinguish pungent from nonpungent cultivars [25,26]. This suggests that *Pun1* is indeed the locus responsible for the qualitative effect on pungency in pepper cultivars and that the enzymatic activity of Pun1 may be categorized into two levels, which determine whether a cultivar will be pungent or not. Since Kim *et al.* [14] reported that cDNA clone SB2-66 is identical to the *Pun1* gene, Pun1 has been suspected to be the actual CS, but there has been no cell-free system to synthesize capsaicin. Purified proteins with CS activity have suffered from contamination with endogenous capsaicinoids; therefore, researchers must use a radioactive substrate, which is not commercially available. Here, we overcame these technical difficulties by using protoplasts isolated from placental tissues of a pungent cultivar. This protoplast-based assay is so simple and easy that comparative experiments assessing Pun1 activities among pepper cultivars are now possible. In addition, this system can be used for other purposes, such as measuring other enzymatic activities and screening for capsaicin synthesis inhibitors.

Even though the transcript levels of both *pAMT* and *Pun1* were very low in the nonpungent cultivars compared with the pungent cultivars, the vanillylamine levels were five times higher in the nonpungent cultivars than in the pungent cultivars at 25 daf (Figure 6). Although these results seem to contradict each other, one possible explanation is that the vanillylamine synthesized in the pungent cultivars was quickly converted to capsaicin by the high Pun1 enzyme activity (Additional file 4: Figure S4). According to this hypothesis, which is based on an expression time course of *pAMT* and *Pun1* relative to the vanillylamine accumulation levels, vanillylamine must be efficiently synthesized in nonpungent cultivars even when there is a very low *pAMT* expression level at 25 daf. From this point of view, the nonpungent feature is substantially determined by the loss of function

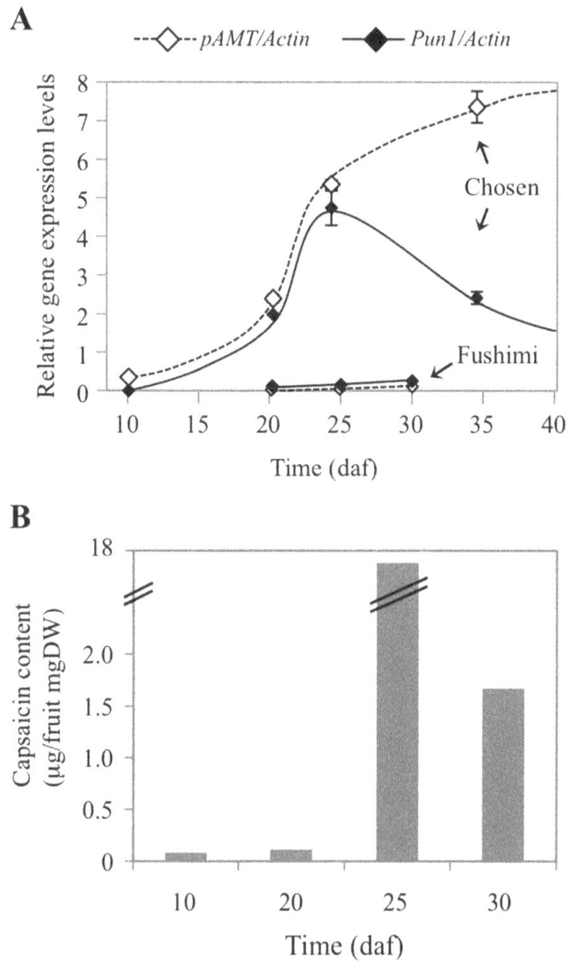

Figure 5 Gene expression levels of *pAMT* and *Pun1* and capsaicin levels during fruit development. (A) Quantitative RT-PCR to determine RNA levels of *pAMT* and *Pun1* in placental tissues (means ± SE; $n = 3$). Samples of 'Chosen' (pungent) were harvested at 10, 20, 25 and 35 days after flowering, and samples of 'Fushimi' (nonpungent) were harvested at 20, 25 and 30 days after flowering. **(B)** Capsaicin levels in the fruits of 'Chosen' (pungent) during fruit development as analyzed by HPLC.

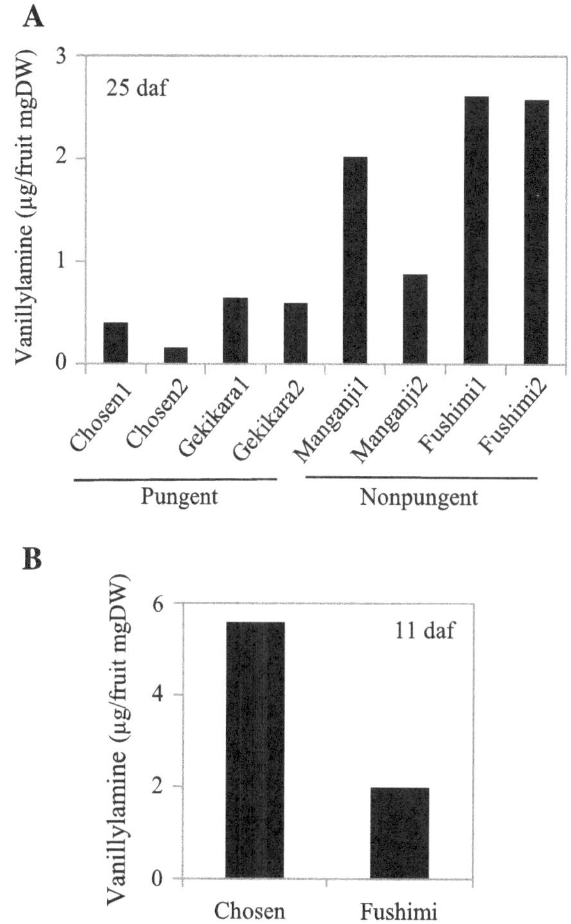

Figure 6 Vanillylamine levels in pungent and nonpungent pepper cultivars. (A) Vanillylamine levels at 25 days after flowering. **(B)** Vanillylamine levels at 11 days after flowering. Freeze-dried pepper fruits were ground, and vanillylamine was extracted with 0.1% acetic acid methanol and then analyzed by HPLC at 280 nm. The *x*-axis in **(A)** shows individual fruits. Three fruits were used for **(B)**.

of Pun1 (or very low *Pun1* expression levels), and thus the nonpungent phenotype, with relatively high vanillylamine levels, was determined essentially by low expression levels of *Pun1* but not of *pAMT*. Therefore, Pun1 would be the primary determinant of vanillylamine, as well as capsaicin, accumulation. A nonpungent cultivar *C. annuum* 'CH-19 Sweet' [18] was shown to accumulate less vanillylamine than pungent cultivars, but this case is exceptional because 'CH-19 Sweet' must have enough Pun1 activity to accumulate a high capsinoids (capsiate) content, unlike the nonpungent cultivars used in this study that showed very low *Pun1* expression levels.

Conclusions

To verify that the *Pun1* gene is actually involved in the final step of capsaicin biosynthesis, we raised antibodies against the *E. coli*-synthesized Pun1 protein and used them as antagonists of endogenous Pun1 activity. After confirmation of the antibodies' specificity, we developed an *in vitro* cell-free assay for *de novo* capsaicin synthesis using protoplasts from placental tissues. The addition of anti-Pun1 antibodies significantly reduced *de novo* capsaicin synthesis in the protoplast assay, demonstrating that the *Pun1*-encoded protein played an essential role in capsaicin synthesis. Analyses of the accumulation of vanillylamine, a precursor of capsaicin, revealed that nonpungent cultivars could accumulate much higher vanillylamine levels than pungent cultivars. This observation indicates that *Pun1* is the primary determinant of vanillylamine and capsaicin accumulation and is more important than *pAMT* in maintaining capsaicin-free fruit in most nonpungent cultivars.

Methods

Plant materials

Capsicum annuum (L.) cultivars 'Chosen,' 'Gekikara' and 'Kaien' were used as the pungent cultivars, 'Nikko' and 'Fukumimi' as the mildly pungent, and 'Fushimi' and 'Manganji' as the nonpungent cultivars. *C. annuum* var. *grossum* (bell pepper) was also used as a capsaicinoid-free control pepper. Plants were cultured in an incubation room under 16 h of fluorescent light at 24°C and 50% relative humidity. Green fruits at ~25 daf were used for the experiments unless otherwise indicated.

Cloning of the *pAMT* and *Pun1* genes

Total RNA was extracted from the placental tissues of pepper fruits using TRIZOL reagent (Ambion) according to the manufacturer's instructions. For cloning *pAMT* and *Pun1*, total RNA extracted from 'Chosen' was used as a template for reverse transcription with AMV Reverse Transcriptase (NIPPON GENE). The full-length cDNAs for *pAMT* and *Pun1* were amplified by PCR with the primer set 5′-ATGGCCAATATTACTAATG-3′ and 5′-TTAATGCTTCTGAGAC-3′ and the primer set 5′-ATGGCTTTTGCATTACCATC-3′ and 5′-TTAATTAGGCAATGAACTCAAGG-3′, respectively. The pGEM-T Easy Vector System I (Promega) was used for the subcloning and subsequent sequencing of the PCR products. Sequencing was performed with the BECKMAN COULTER CEQ8800 system. We designed all primers based on published sequences of the *pAMT* (AF085149) and *Pun1* (AY819029) genes.

Quantitative RT-PCR analysis

cDNA was synthesized from DNase-treated RNA and amplified using KOD SYBR qPCR Mix (Toyobo) according to the manufacturer's instructions. mRNA levels were quantified by quantitative RT-PCR using the StepOne Real-Time PCR System (Applied Biosystems). The *Actin* gene was used as an internal control. Primer sets for each gene amplification were as follows: *Actin*, 5′-GGTTAAGGCTGGATTTGCTG-3′ and 5′-ATGCATCCTTTTGACCCATC-3′; *pAMT*, 5′-GTAAGTTCCACTGGTGATCATG-3′ and 5′-TTACTGCTTCTGAGAC-3′; and *Pun1*, 5′-AGGCATCATCAATGCTAC-3′ and 5′-ATGTTAGTTGCTTCTATGGAG-3′.

Virus-induced gene silencing (VIGS) against *Pun1*

The *Cucumber mosaic virus* (CMV) vector (CMV-A1 vector) had been used previously for VIGS experiments in pepper plants [27]. Here, we modified the CMV-A1 vector by changing an R residue at position 46 into a C residue in the 2b protein (2b), which weakened the RNA silencing suppressor activity of 2b so that the viral symptoms became milder in the pepper. The modified CMV-A1 vector was designated CMV-Yd. We used a pungent pepper cultivar 'Kaien' for the VIGS experiments because CMV-Yd can systemically infect 'Kaien' and easily penetrate into placental tissues. A 95-bp fragment from the *Pun1* gene was amplified by the primer pair 5′-CGCACGCGTGAAGGAAGTTGAGGTGGCATA-3′ and 5′-CGAGGCCTGAGCAGTTTCCCTTCTCTCATTG-3′ and inserted between the *Sma*I and *Mlu*I sites in the CMV-Yd vector to create CMV-Yd:CS95.

Extraction of vanillylamine and capsaicin

Fruits were harvested, frozen and dried completely in a freeze-dryer (FDU-1200, EYELA) for 2 d. Dried fruits were then ground in a blender (Microsmash MS100, TOMY) at room temperature. For the vanillylamine extraction, 1 ml of 0.1% acetic acid methanol was added to 100 mg of dry fruit powder. For the capsaicin extraction, 0.1% acetic acid acetonitrile was added instead. After the samples were mixed, they were allowed to settle for 1 h at room temperature. The supernatant was passed through a 0.2 μm-pore membrane filter before being used in HPLC (JASCO LC-2000 system).

HPLC analysis

The samples were separated in a SHIM-PACK VP-ODS column (150 mm × 4.6 mm; Shimazu). For vanillylamine, the eluent was a mixture of methanol/10 mM phosphate buffer pH 2.6 (3:2 v/v), and the flow rate was 0.8 ml/min. For capsaicin, the eluent was a mixture of acetonitrile/1% acetic acid (2:1 v/v), and the flow rate was 1 ml/min. All eluates were monitored at 280 nm using a UV detector. External standards were prepared by dissolving commercial vanillylamine (Sigma-Aldrich) and capsaicin (Sigma-Aldrich) in methanol and acetonitrile, respectively.

In vitro cell-free assay for capsaicin synthesis

Protoplasts were isolated from fruits of the pungent cultivar 'Chosen' as previously described [28], except that cells were not shaken during the enzyme digestion and the reaction was stopped when the mass of cells (roughly 20–30 cells/mass) were released from the tissues. A 500-μl sample of the protoplasts was transferred to a 1.5-ml tube and centrifuged (3 min, 100 × *g*, 4°C) to remove the supernatant. Protoplasts were then resuspended in 200 μl solution containing 1 mM aqueous vanillylamine (Sigma-Aldrich). Protoplasts were ruptured to release cellular enzymes, including CS, by vortexing the mixture 4 times for 5 s each. After a 2-h incubation at room temperature, followed by centrifugation (2 min, 6,000 × *g*, 4°C) of the reaction mixture, the supernatant was collected and extracted with the ethyl acetate. The extract was dried overnight at room temperature in the dark, resuspended in 100 μl of 0.1% acetic acid acetonitrile and then passed through a 0.2-μm-pore membrane filter before HPLC. To inhibit CS activity, anti-Pun1 antibodies (described next) were added to the reaction mixture at a final dilution of 1:2000 before the *in vitro* capsaicin synthesis started.

Expression of Pun1 in *E. coli* and production of anti-Pun1 antibodies

The open-reading frame of the *Pun1* gene was amplified with primer set 5′-CGGAATTCATGGCTTTTGCATT ACCATC-3′ containing an *Eco*RI site and 5′-CGGGAT CCTAATTAGGCAATGAACTCAAGG-3′ containing a *Bam*HI site, and cloned between the *Eco*RI and *Bam*HI sites of pMAL-c2x (New England Biolabs). The N-terminal maltose binding protein (MBP)-fused Pun1 recombinant protein was then expressed in *E. coli* (BL21) and affinity-purified using the pMAL Protein Fusion and Purification System (New England BioLabs). Anti-Pun1 antibodies, prepared by Frontier Institute Co. (Sapporo, Japan), were raised by immunizing a rabbit with the purified recombinant protein.

Western blots

Total proteins from pepper placentas, extracted as described by Masuta *et al.* [29], were separated on 10% sodium dodecyl sulfate polyacrylamide gels (SDS-PAGE) and blotted on an Immobilon-P membrane (Millipore). The blots were then treated with primary antibodies raised against the *E. coli*-synthesized Pun1 protein, and subsequently with goat anti-rabbit IgG-alkaline phosphatase-conjugate (Takara Bio) as a secondary antibody. Signals were visualized using CDP-Star (Roche) as a substrate.

Additional files

Additional file 1: Figure S1. Western blot analysis of the *in vitro* synthesized pepper hydroxycinnamoyl transferase (HCT) protein using the anti-Pun1 antibodies. The cDNA clone of HCT was PCR-amplified and inserted in the pEU plasmid vector (CellFree Sciences, Japan). After *in vitro* transcription from the recombinant plasmid, HCT-fused to a C-terminal FLAG peptide (HCT-FLAG) was *in vitro* synthesized in the wheat germ cocktail (CellFree Sciences, Japan) according to the manufacturer's instructions. The HCT protein with a FLAG tag was then subjected to detection either by the anti-Pun1 antibodies (left) or by an anti-FLAG antibody (right). Note that anti-Pun1 antibodies did not react with the HCT protein. The arrow indicates the HCT protein.

Additional file 2: Figure S2. Accumulation levels of capsaicinoids in placental tissues infected with CMV-Yd:CS95. The graph shows the data as fold change compared with the healthy control (means ± SD from three separate experiments using 3 to 5 pepper fruits for each treatment). The reduction in capsaicin accumulation in CMV-Yd:CS95 was statistically significant compared with the empty vector control (CMV-Yd) using Student's t-test. Asterisks indicate significance at 0.01.

Additional file 3: Figure S3. HPLC chromatograms of *de novo* capsaicin synthesis products from the *in vitro* cell-free assay using protoplasts. The main peak (red) in each chart represents capsaicin. This experiment was repeated twice and produced the same results.

Additional file 4: Figure S4. Hypothesis to explain the accumulation of vanillylamine in nonpungent pepper cultivars. In pungent peppers, vanillylamine is quickly converted to capsaicin by highly active CS, while in nonpungent cultivars, vanillylamine is synthesized even with a very low pAMT activity level. Although vanillylamine is more abundant in nonpungent cultivars than in the pungent cultivars at 25 daf (dotted line), it will not be converted to capsaicin owing to the very low level of CS activity. This result is supported by the gene expression time-course of *pAMT* and *Pun1* (Figure 5).

Abbreviations

CS: Capsaicin synthase; daf: Days after flowering; pAMT: Putative aminotransferase; Pun1: Pungent gene 1; HCT: Hydroxycinnamoyl transferase; VIGS: Virus-induced gene silencing.

Competing interests

The authors declare that they have no competing interests.

Authors' contributions

CM and TM conceived and designed the research. KO, KM, MF and YT conducted the experiments. CM and KO analyzed the data. KO, HS and CM wrote the manuscript. All authors read and approved the manuscript. The first and second authors contributed equally.

Acknowledgments

We thank the Northern Advancement Center for Science and Technology (Noastech), Japan, for financial support. We also thank Dr. Kiyoshi Masuda for helping us operate the HPLC. We are grateful to Dr. Tatsuo Watanabe for providing us the capsaicin analog, octanoyl-CoA, which was used to practice the *in vitro* capsaicin synthesis.

Author details

[1]Graduate School of Agriculture, Hokkaido University, Sapporo 060-8589, Japan. [2]Plant Molecular Technology Research Group, Research Institute of Bioproduction, National Institute of Advanced Industrial Science and Technology, Sapporo 062-8517, Japan.

References

1. Thresh CF. Capsaicin, the active principle of capsicum fruits. Pharmaceut J. 1876;315:21.
2. Iwai K, Suzuki T, Fujiwake H. Formation and accumulation of pungent principle of hot pepper fruits, capsaicin and its analogues, in *Capsicum annuum* var. annuum cv. Karayatsubusa at different growth stage after flowering. Agric Biol Chem. 1986;43:2493–8.
3. Kawada T, Sakabe S, Aoki N, Watanabe T, Higata K, Iwai K, et al. Intake of sweeteners and pungent ingredients increases the thermogenin content in brown adipose tissue of rat. J Agric Food Chem. 1991;39:651–4.
4. Westerterp-Plantenga M, Diepvens K, Joosen AMCP, Bérubé-Parent S, Tremblay A. Metabolic effects of spices, teas, and caffeine. Physiol Behav. 2006;89:85–91.
5. Kawada T, Hagiwara K, Iwai K. Effects of capsaicin on lipid metabolism in rats fed a high fat diet. J Nutr. 1986;116:1272–8.
6. Srinivasan K. Antioxidant potential of spices and their active constituents. Cit Rev Food Sci Nutr. 2014;54:352–72.
7. Wutka A, Palagani V, Barat S, Chen X, Khatib ME, Gotze J, et al. Capsaicin treatment attenuates cholangiocarcinoma carcinogenesis. PLOS ONE. 2014;9:e95695.
8. Kobata K, Todo T, Yazawa S, Iwai K, Watanabe T. Novel capsaicinoid-like substances, capsiate and dihydrocapsiate, from the fruits of a nonpungent cultivar, CH-19 Sweet, of pepper (Capsicum annuum L.). J Agric Food Chem. 1998;46:1695–7.
9. Aza-Gonzalez C, Nunez-Palenuis HG, Ochoa-Alejo N. Molecular biology of capsaicinoids biosynthesis in chili pepper (Capsicum spp.). Plant Cell Rep. 2011;30:695–706.
10. Bennett DJ, Kirby GW. Constitution and biosynthesis of capsaicin. J Chem Soc. 1968;C:442–6.
11. Leete E, Louden M. Biosynthesis of capsaicin and dihydrocapsaicin in *Capsicum frutescens*. J Am Chem Soc. 1968;90:6837–41.
12. Kim M, Kim S, Kim S, Kim BD. Isolation of cDNA clones differentially accumulated in the placenta of pungent pepper by suppression subtractive hybridization. Mol Cells. 2001;11:213–9.

13. Stewart C, Kang BC, Liu K, Mazourek M, Moore SL, Yoo EY, et al. The Pun1 gene for pungency in pepper encodes a putative acyltransferase. Plant J. 2005;42:675–88.
14. Han K, Jeong HJ, Sung J, Keum YS, Cho MC, Kim JH, et al. Biosynthesis of capsinoid is controlled by the Pun1 locus in pepper. Mol Breeding. 2013;31:537–48.
15. Kobata K, Sugiwara M, Miura M, Yazawa S, Watanabe T. Potent production of capsaicinoids and capsinoids by Capsicum peppers. J Agri Food Chem. 2013;61:11127–32.
16. Curry J, Aluru M, Mendoza M, Nevarez J, Melendre M, O'Connell MA. Transcripts for possible capsaicinoid biosynthetic genes are differentially accumulated in pungent and non-pungent *Capsicum* spp. Plant Sci. 1999;148:47–57.
17. Abraham-Juarez MR, Rocha-Granados MC, Lopez MG, Rivera-Bustamante RF, Neftali OA. Virus-induced silencing of Comt, pAMT, Kas genes results in a reduction of capsaicinoid accumulation in chili pepper fruits. Planta. 2008;227:681–95.
18. Lang Y, Kisaka H, Sugiyama R, Nomura K, Morita A, Watanabe T, et al. Functional loss of pAMT results in biosynthesis of capsinoids, capsaicinoid analogs, in Capsicum annuum cv. CH-19 sweet. Plant J. 2009;59:953–61.
19. Tanaka Y, Hosokawa M, Miwa T, Watanabe T, Yazawa S. Newly mutated putative-aminotransferase in nonpungent pepper (*Capsicum annuum*) results in biosynthesis of capsinoids, capsaicinoid analogues. J Agri Food Chem. 2010;58:1761–7.
20. Weber N, Ismail A, Gorwa-Grauslund M, Carlquist M. Biocatalytic potential of vanillin aminotransferase from Capsicum chinense. BMC Biotechnol. 2014;14:25.
21. Hoffmann L, Besseau S, Geoffroy P, Ritzenthaler C, Meyer D, Lapierre C, et al. Silencing of hydroxycinnamoyl-coenzyme A shikimate/quinate hydroxycinnamoyltransferase affects phenylpropanoid biosynthesis. Plant Cell. 2004;16:1446–65.
22. Mazourek M, Pujar A, Borovsky Y, Paran I, Mueller L, Jahn MM. A dynamic interface for capsaicinoid systems biology. Plant Physiol. 2009;150:1806–21.
23. Qin C, Yu C, Shen Y, Fang X, Chen L, Min J, et al. Whole-genome sequencing of cultivated and wild peppers provides insights into Capsicum domestication and specialization. Proc Natl Acad Sci USA. 2014;111:5135–40.
24. Stewart C, Mazourek M, Stellari GM, O'Connell M, Jahn M. Genetic control of pungency in *C. chinense* via the Pun1 locus. J Exp Bot. 2007;58:979–91.
25. Reddy UK, Almeida A, Abburi V, Alaparthi SB, Unselt D, Hankins G, et al. Identification of gene-specific polymorphisms and association with capsaicin pathway metabolites in *Capsicum annuum* L. collections. PLoS ONE. 2014;9:e86393.
26. Hill TA, Ashrafi H, Reyes-Chin-Wo S, Yao J, Stoffel K, Truco MJ, et al. Characterization of *Capsicum annuum* genetic diversity and population structure based on parallel polymorphism discovery with a 30 K unigene pepper genechip. PLoS ONE. 2013;8:e56200.
27. Hong JS, Rhee SJ, Kim EJ, Kim TS, Ryu KH, Masuta C, et al. Application of a reassortant *Cucumber mosaic virus* vector for gene silencing in tomato and chili pepper plants. Plant Pathol J. 2012;28:81–6.
28. Pitzschke A, Persak H. Poinsettia protoplasts - a simple, robust and efficient system for transient gene expression studies. Plant Methods. 2012;8:14.
29. Masuta C, Tanaka H, Uehara K, Kuwata S, Koiwai A, Noma M. Broad resistance to plant viruses in transgenic plants conferred by antisense inhibition of a host gene essential in S-adenosylmethionine-dependent transmethylation reactions. Proc Natl Acad Sci USA. 1995;92:6117–21.

Permissions

All chapters in this book were first published in BMCPB, by BioMed Central; hereby published with permission under the Creative Commons Attribution License or equivalent. Every chapter published in this book has been scrutinized by our experts. Their significance has been extensively debated. The topics covered herein carry significant findings which will fuel the growth of the discipline. They may even be implemented as practical applications or may be referred to as a beginning point for another development.

The contributors of this book come from diverse backgrounds, making this book a truly international effort. This book will bring forth new frontiers with its revolutionizing research information and detailed analysis of the nascent developments around the world.

We would like to thank all the contributing authors for lending their expertise to make the book truly unique. They have played a crucial role in the development of this book. Without their invaluable contributions this book wouldn't have been possible. They have made vital efforts to compile up to date information on the varied aspects of this subject to make this book a valuable addition to the collection of many professionals and students.

This book was conceptualized with the vision of imparting up-to-date information and advanced data in this field. To ensure the same, a matchless editorial board was set up. Every individual on the board went through rigorous rounds of assessment to prove their worth. After which they invested a large part of their time researching and compiling the most relevant data for our readers.

The editorial board has been involved in producing this book since its inception. They have spent rigorous hours researching and exploring the diverse topics which have resulted in the successful publishing of this book. They have passed on their knowledge of decades through this book. To expedite this challenging task, the publisher supported the team at every step. A small team of assistant editors was also appointed to further simplify the editing procedure and attain best results for the readers.

Apart from the editorial board, the designing team has also invested a significant amount of their time in understanding the subject and creating the most relevant covers. They scrutinized every image to scout for the most suitable representation of the subject and create an appropriate cover for the book.

The publishing team has been an ardent support to the editorial, designing and production team. Their endless efforts to recruit the best for this project, has resulted in the accomplishment of this book. They are a veteran in the field of academics and their pool of knowledge is as vast as their experience in printing. Their expertise and guidance has proved useful at every step. Their uncompromising quality standards have made this book an exceptional effort. Their encouragement from time to time has been an inspiration for everyone.

The publisher and the editorial board hope that this book will prove to be a valuable piece of knowledge for researchers, students, practitioners and scholars across the globe.

List of Contributors

Eva Žižková
Institute of Experimental Botany, Academy of Sciences of the Czech Republic, Prague 165 02, Czech Republic

Petre I Dobrev
Institute of Experimental Botany, Academy of Sciences of the Czech Republic, Prague 165 02, Czech Republic

Yordan Muhovski
Département Sciences du vivant, Centre wallon de Recherches Agronomiques, Gembloux B-5030, Belgium

Petr Hošek
Institute of Experimental Botany, Academy of Sciences of the Czech Republic, Prague 165 02, Czech Republic
Department of Biomedical Informatics, Faculty of Biomedical Engineering, Czech Technical University in Prague, Kladno 272 01, Czech Republic

Klára Hoyerová
Institute of Experimental Botany, Academy of Sciences of the Czech Republic, Prague 165 02, Czech Republic

Daniel Haisel
Institute of Experimental Botany, Academy of Sciences of the Czech Republic, Prague 165 02, Czech Republic

Dagmar Procházková
Institute of Experimental Botany, Academy of Sciences of the Czech Republic, Prague 165 02, Czech Republic

Stanley Lutts
Groupe de Recherche en Physiologie Végétale (GRPV), Earth and Life Institute - Agronomy (ELI-A), Université catholique de Louvain (UCL), Louvain-la-Neuve 1348, Belgium

Václav Motyka
Institute of Experimental Botany, Academy of Sciences of the Czech Republic, Prague 165 02, Czech Republic

Imène Hichri
Groupe de Recherche en Physiologie Végétale (GRPV), Earth and Life Institute - Agronomy (ELI-A), Université catholique de Louvain (UCL), Louvain-la-Neuve 1348, Belgium
Institut National de la Recherche Agronomique, Institut Sophia Agrobiotech (ISA), UMR INRA 1355, CNRS 7254, Université de Nice-Sophia Antipolis, 400 route des Chappes, BP167, F-06903 Sophia-Antipolis Cedex, France

Jin Hee Shin
Center for Applied Genetic Technologies & Department of Crop and Soil Science, University of Georgia, Athens, GA 30602, USA

Justin N Vaughn
Center for Applied Genetic Technologies & Department of Crop and Soil Science, University of Georgia, Athens, GA 30602, USA

Hussein Abdel-Haleem
Center for Applied Genetic Technologies & Department of Crop and Soil Science, University of Georgia, Athens, GA 30602, USA

Carolina Chavarro
Center for Applied Genetic Technologies & Department of Crop and Soil Science, University of Georgia, Athens, GA 30602, USA

Brian Abernathy
Center for Applied Genetic Technologies & Department of Crop and Soil Science, University of Georgia, Athens, GA 30602, USA

Kyung Do Kim
Center for Applied Genetic Technologies & Department of Crop and Soil Science, University of Georgia, Athens, GA 30602, USA

Scott A Jackson
Center for Applied Genetic Technologies & Department of Crop and Soil Science, University of Georgia, Athens, GA 30602, USA

Zenglu Li
Center for Applied Genetic Technologies & Department of Crop and Soil Science, University of Georgia, Athens, GA 30602, USA

Shizuka Kimura
Division of Agronomy and Horticultural Science, Graduate School of Agriculture, Kyoto University, Uji, Kyoto 611-0011, Japan

Yuki Higashino
Division of Agronomy and Horticultural Science, Graduate School of Agriculture, Kyoto University, Uji, Kyoto 611-0011, Japan

Yuki Kitao
Division of Agronomy and Horticultural Science, Graduate School of Agriculture, Kyoto University, Uji, Kyoto 611-0011, Japan

Taro Masuda
Division of Agronomy and Horticultural Science, Graduate School of Agriculture, Kyoto University, Uji, Kyoto 611-0011, Japan

Reiko Urade
Division of Agronomy and Horticultural Science, Graduate School of Agriculture, Kyoto University, Uji, Kyoto 611-0011, Japan

Bing Bai
Ben-Gurion University of the Negev, Jacob Blaustein Institutes for Desert Research, French Associates Institute for Agriculture and Biotechnology of Drylands, Midreshet Ben-Gurion 84990, Israel
Current address: Department of Molecular Plant Physiology, Utrecht University, Utrecht 3584 CH, The Netherlands

David Toubiana
Ben-Gurion University of the Negev, Jacob Blaustein Institutes for Desert Research, French Associates Institute for Agriculture and Biotechnology of Drylands, Midreshet Ben-Gurion 84990, Israel

Tanya Gendler
Ben-Gurion University of the Negev, Jacob Blaustein Institutes for Desert Research, French Associates Institute for Agriculture and Biotechnology of Drylands, Midreshet Ben-Gurion 84990, Israel

Asfaw Degu
Ben-Gurion University of the Negev, Jacob Blaustein Institutes for Desert Research, French Associates Institute for Agriculture and Biotechnology of Drylands, Midreshet Ben-Gurion 84990, Israel

Yitzchak Gutterman
Ben-Gurion University of the Negev, Jacob Blaustein Institutes for Desert Research, French Associates Institute for Agriculture and Biotechnology of Drylands, Midreshet Ben-Gurion 84990, Israel

Aaron Fait
Ben-Gurion University of the Negev, Jacob Blaustein Institutes for Desert Research, French Associates Institute for Agriculture and Biotechnology of Drylands, Midreshet Ben-Gurion 84990, Israel

Marie JJ Huysman
Department of Plant Systems Biology, Flanders Institute for Biotechnology (VIB), 9052 Ghent, Belgium
Department of Plant Systems Biology, VIB, and Bioinformatics, Ghent University, 9052 Ghent, Belgium
Protistology and Aquatic Ecology, Department of Biology, Ghent University, 9000 Ghent, Belgium
Environmental and Evolutionary Genomics Section, Institut de Biologie de l'Ecole Normale Supérieure, Centre National de la Recherche Scientifique, Unité Mixte de Recherche 8186, Institut National de la Santé et de la Recherche Médicale U1024, Ecole Normale Supérieure, 75230 Paris, Cedex 05, France

Atsuko Tanaka
Environmental and Evolutionary Genomics Section, Institut de Biologie de l'Ecole Normale Supérieure, Centre National de la Recherche Scientifique, Unité Mixte de Recherche 8186, Institut National de la Santé et de la Recherche Médicale U1024, Ecole Normale Supérieure, 75230 Paris, Cedex 05, France
Current address: Muroran Marine Station, Field Science Center for Northern Biosphere, Hokkaido University, Muroran, Hokkaido 051-0013, Japan

Chris Bowler
Environmental and Evolutionary Genomics Section, Institut de Biologie de l'Ecole Normale Supérieure, Centre National de la Recherche Scientifique, Unité Mixte de Recherche 8186, Institut National de la Santé et de la Recherche Médicale U1024, Ecole Normale Supérieure, 75230 Paris, Cedex 05, France

Wim Vyverman
Protistology and Aquatic Ecology, Department of Biology, Ghent University, 9000 Ghent, Belgium

Lieven De Veylder
Department of Plant Systems Biology, Flanders Institute for Biotechnology (VIB), 9052 Ghent, Belgium
Department of Plant Systems Biology, VIB, and Bioinformatics, Ghent University, 9052 Ghent, Belgium

Jiří Bárta
Faculty of Science, University of South Bohemia, Branišovská 31, České Budějovice 37005, Czech Republic

James D Stone
Institute of Experimental Botany CAS, Rozvojová 263 6-Lysolaje, Praha 16502, Czech Republic
Institute of Botany CAS, Zámek 1, Průhonice 25243, Czech Republic

Jiří Pech
Faculty of Science, University of South Bohemia, Branišovská 31, České Budějovice 37005, Czech Republic

Dagmara Sirová
Faculty of Science, University of South Bohemia, Branišovská 31, České Budějovice 37005, Czech Republic

Lubomír Adamec
Institute of Botany CAS, Section of Plant Ecology, Dukelská 135, Treboň 37982, Czech Republic

Matthew A Campbell
Institute of Fundamental Sciences, Massey University, Private Bag 11222, Palmerston North 4442, New Zealand

Helena Štorchová
Institute of Experimental Botany CAS, Rozvojová 263 6-Lysolaje, Praha 16502, Czech Republic

Qibin Yu
University of Florida, Institute of Food and Agricultural Sciences, Citrus Research and Education Center, Lake Alfred, FL 33850, USA

Anne Plotto
USDA-ARS Horticultural Research Laboratory, Fort Pierce, FL 34945, USA

Elizabeth A Baldwin
USDA-ARS Horticultural Research Laboratory, Fort Pierce, FL 34945, USA

Jinhe Bai
USDA-ARS Horticultural Research Laboratory, Fort Pierce, FL 34945, USA

Ming Huang
University of Florida, Institute of Food and Agricultural Sciences, Citrus Research and Education Center, Lake Alfred, FL 33850, USA

Yuan Yu
University of Florida, Institute of Food and Agricultural Sciences, Citrus Research and Education Center, Lake Alfred, FL 33850, USA

Harvinder S Dhaliwal
College of Agriculture, Punjab Agricultural University, Ludhiana, Punjab 141004, India

Frederick G Gmitter Jr
University of Florida, Institute of Food and Agricultural Sciences, Citrus Research and Education Center, Lake Alfred, FL 33850, USA

Tomasz Bieluszewski
Department of Biotechnology, Institute of Molecular Biology and Biotechnology, Adam Mickiewicz University, Umultowska 89, 61-614 Poznań, Poland

Lukasz Galganski
Department of Biotechnology, Institute of Molecular Biology and Biotechnology, Adam Mickiewicz University, Umultowska 89, 61-614 Poznań, Poland

Weronika Sura
Department of Biotechnology, Institute of Molecular Biology and Biotechnology, Adam Mickiewicz University, Umultowska 89, 61-614 Poznań, Poland

Anna Bieluszewska
Department of Molecular Virology, Institute of Experimental Biology, Adam Mickiewicz University, Umultowska 89, 61-614 Poznań, Poland

Mateusz Abram
Department of Biotechnology, Institute of Molecular Biology and Biotechnology, Adam Mickiewicz University, Umultowska 89, 61-614 Poznań, Poland

Agnieszka Ludwikow
Department of Biotechnology, Institute of Molecular Biology and Biotechnology, Adam Mickiewicz University, Umultowska 89, 61-614 Poznań, Poland

Piotr Andrzej Ziolkowski
Department of Biotechnology, Institute of Molecular Biology and Biotechnology, Adam Mickiewicz University, Umultowska 89, 61-614 Poznań, Poland
Department of Plant Sciences, University of Cambridge, Downing Street, CB2 3EA, Cambridge, UK

Jan Sadowski
Department of Biotechnology, Institute of Molecular Biology and Biotechnology, Adam Mickiewicz University, Umultowska 89, 61-614 Poznań, Poland

Miao He
College of Life Science, Capital Normal University, Beijing 100048, China

Chong Zhu
College of Life Science, Capital Normal University, Beijing 100048, China

Kun Dong
College of Life Science, Capital Normal University, Beijing 100048, China

Ting Zhang
College of Life Science, Capital Normal University, Beijing 100048, China

Zhiwei Cheng
College of Life Science, Capital Normal University, Beijing 100048, China

Jiarui Li
Department of Plant Pathology, Kansas State University, Manhattan, KS 66506, USA

Yueming Yan
College of Life Science, Capital Normal University, Beijing 100048, China
Hubei Collaborative Innovation Center for Grain Industry, 434025 Jingzhou, China

Igor A Fesenko
Department of Proteomics, Shemyakin-Ovchinnikov Institute of Bioorganic Chemistry of the Russian Academy of Sciences, 16/10, Miklukho-Maklaya, GSP-7, Moscow 117997, Russian Federation

Georgij P Arapidi
Department of Proteomics, Shemyakin-Ovchinnikov Institute of Bioorganic Chemistry of the Russian Academy of Sciences, 16/10, Miklukho-Maklaya, GSP-7, Moscow 117997, Russian Federation
Moscow Institute of Physics and Technology, 9 Institutskiy per., Dolgoprudny, Moscow Region 141700, Russian Federation

Alexander Yu Skripnikov
Department of Proteomics, Shemyakin-Ovchinnikov Institute of Bioorganic Chemistry of the Russian Academy of Sciences, 16/10, Miklukho-Maklaya, GSP-7, Moscow 117997, Russian Federation
Biology Department, Lomonosov Moscow State University, Moscow 199234, Russian Federation

Dmitry G Alexeev
Research Institute of Physical-Chemical Medicine, Federal Medical & Biological Agency, 1a, Malaya Pirogovskaya, Moscow 119992, Russian Federation
Moscow Institute of Physics and Technology, 9 Institutskiy per., Dolgoprudny, Moscow Region 141700, Russian Federation

Elena S Kostryukova
Research Institute of Physical-Chemical Medicine, Federal Medical & Biological Agency, 1a, Malaya Pirogovskaya, Moscow 119992, Russian Federation

Alexander I Manolov
Research Institute of Physical-Chemical Medicine, Federal Medical & Biological Agency, 1a, Malaya Pirogovskaya, Moscow 119992, Russian Federation
Moscow Institute of Physics and Technology, 9 Institutskiy per., Dolgoprudny, Moscow Region 141700, Russian Federation

Ilya A Altukhov
Research Institute of Physical-Chemical Medicine, Federal Medical & Biological Agency, 1a, Malaya Pirogovskaya, Moscow 119992, Russian Federation
Moscow Institute of Physics and Technology, 9 Institutskiy per., Dolgoprudny, Moscow Region 141700, Russian Federation

Regina A Khazigaleeva
Department of Proteomics, Shemyakin-Ovchinnikov Institute of Bioorganic Chemistry of the Russian Academy of Sciences, 16/10, Miklukho-Maklaya, GSP-7, Moscow 117997, Russian Federation

Anna V Seredina
Department of Proteomics, Shemyakin-Ovchinnikov Institute of Bioorganic Chemistry of the Russian Academy of Sciences, 16/10, Miklukho-Maklaya, GSP-7, Moscow 117997, Russian Federation

Sergey I Kovalchuk
Department of Proteomics, Shemyakin-Ovchinnikov Institute of Bioorganic Chemistry of the Russian Academy of Sciences, 16/10, Miklukho-Maklaya, GSP-7, Moscow 117997, Russian Federation
Research Institute of Physical-Chemical Medicine, Federal Medical & Biological Agency, 1a, Malaya Pirogovskaya, Moscow 119992, Russian Federation

Rustam H Ziganshin
Department of Proteomics, Shemyakin-Ovchinnikov Institute of Bioorganic Chemistry of the Russian Academy of Sciences, 16/10, Miklukho-Maklaya, GSP-7, Moscow 117997, Russian Federation

Viktor G Zgoda
Institute of Biomedical Chemistry RAMS im. V.N. Orehovicha, 10, Pogodinskaya Street, Moscow 119121, Russian Federation

Svetlana E Novikova
Institute of Biomedical Chemistry RAMS im. V.N. Orehovicha, 10, Pogodinskaya Street, Moscow 119121, Russian Federation

Tatiana A Semashko
Research Institute of Physical-Chemical Medicine, Federal Medical & Biological Agency, 1a, Malaya Pirogovskaya, Moscow 119992, Russian Federation

Darya K Slizhikova
Department of Proteomics, Shemyakin-Ovchinnikov Institute of Bioorganic Chemistry of the Russian Academy of Sciences, 16/10, Miklukho-Maklaya, GSP-7, Moscow 117997, Russian Federation

Vasilij V Ptushenko
A. N. Belozersky Institute of Physico-Chemical Biology, M.V. Lomonosov Moscow State University, Leninskye Gory, House 1, Building 40, Moscow 119992, Russian Federation

Alexey Y Gorbachev
Research Institute of Physical-Chemical Medicine, Federal Medical & Biological Agency, 1a, Malaya Pirogovskaya, Moscow 119992, Russian Federation

Vadim M Govorun
Department of Proteomics, Shemyakin-Ovchinnikov Institute of Bioorganic Chemistry of the Russian Academy of Sciences, 16/10, Miklukho-Maklaya, GSP-7, Moscow 117997, Russian Federation
Research Institute of Physical-Chemical Medicine, Federal Medical & Biological Agency, 1a, Malaya Pirogovskaya, Moscow 119992, Russian Federation
Moscow Institute of Physics and Technology, 9 Institutskiy per., Dolgoprudny, Moscow Region 141700, Russian Federation

Vadim T Ivanov
Department of Proteomics, Shemyakin-Ovchinnikov Institute of Bioorganic Chemistry of the Russian Academy of Sciences, 16/10, Miklukho-Maklaya, GSP-7, Moscow 117997, Russian Federation

Feng-Quan Tan
Key Laboratory of Horticultural Plant Biology (Ministry of Education), Key Laboratory of Horticultural Crop Biology and Genetic Improvement (Central Region) (Ministry of Agriculture), College of Horticulture and Forestry Sciences, Huazhong Agricultural University, Wuhan 430070, China

Hong Tu
Key Laboratory of Horticultural Plant Biology (Ministry of Education), Key Laboratory of Horticultural Crop Biology and Genetic Improvement (Central Region) (Ministry of Agriculture), College of Horticulture and Forestry Sciences, Huazhong Agricultural University, Wuhan 430070, China

Wu-Jun Liang
Key Laboratory of Horticultural Plant Biology (Ministry of Education), Key Laboratory of Horticultural Crop Biology and Genetic Improvement (Central Region) (Ministry of Agriculture), College of Horticulture and Forestry Sciences, Huazhong Agricultural University, Wuhan 430070, China

Jian-Mei Long
Key Laboratory of Horticultural Plant Biology (Ministry of Education), Key Laboratory of Horticultural Crop Biology and Genetic Improvement (Central Region) (Ministry of Agriculture), College of Horticulture and Forestry Sciences, Huazhong Agricultural University, Wuhan 430070, China

Xiao-Meng Wu
Key Laboratory of Horticultural Plant Biology (Ministry of Education), Key Laboratory of Horticultural Crop Biology and Genetic Improvement (Central Region) (Ministry of Agriculture), College of Horticulture and Forestry Sciences, Huazhong Agricultural University, Wuhan 430070, China

Hong-Yan Zhang
Key Laboratory of Horticultural Plant Biology (Ministry of Education), Key Laboratory of Horticultural Crop Biology and Genetic Improvement (Central Region) (Ministry of Agriculture), College of Horticulture and Forestry Sciences, Huazhong Agricultural University, Wuhan 430070, China

Wen-Wu Guo
Key Laboratory of Horticultural Plant Biology (Ministry of Education), Key Laboratory of Horticultural Crop Biology and Genetic Improvement (Central Region) (Ministry of Agriculture), College of Horticulture and Forestry Sciences, Huazhong Agricultural University, Wuhan 430070, China

Eun-Young Jeong
Department of Chemistry, Seoul National University, Seoul 151-742, South Korea

Pil Joon Seo
Department of Chemistry, Seoul National University, Seoul 151-742, South Korea

Je Chang Woo
Department of Biological Science, Mokpo National University, Jeonnam 534-729, South Korea

Chung-Mo Park
Department of Chemistry, Seoul National University, Seoul 151-742, South Korea
Plant Genomics and Breeding Institute, Seoul National University, Seoul 151-742, South Korea

Rhonda C Foley
CSIRO Agriculture Flagship, Centre for Environment and Life Sciences, Floreat, WA, Australia

Jose C Jimenez-Lopez
The UWA Institute of Agriculture, University of Western Australia, Crawley, WA, Australia

Lars G Kamphuis
CSIRO Agriculture Flagship, Centre for Environment and Life Sciences, Floreat, WA, Australia
The UWA Institute of Agriculture, University of Western Australia, Crawley, WA, Australia

James K Hane
CSIRO Agriculture Flagship, Centre for Environment and Life Sciences, Floreat, WA, Australia
Current address: Curtin Institute for Computation and Centre for Crop & Disease Management, Department of Environment & Agriculture, Curtin University, Bentley, Western Australia, Australia

Su Melser
CSIRO Agriculture Flagship, Centre for Environment and Life Sciences, Floreat, WA, Australia

Karam B Singh
CSIRO Agriculture Flagship, Centre for Environment and Life Sciences, Floreat, WA, Australia
The UWA Institute of Agriculture, University of Western Australia, Crawley, WA, Australia

Jieyu Yue
Tianjin Key Laboratory of Animal and Plant Resistance, School of Life Sciences, Tianjin Normal University, Tianjin 300387, China

Hong Sun
Tianjin Key Laboratory of Animal and Plant Resistance, School of Life Sciences, Tianjin Normal University, Tianjin 300387, China

Wei Zhang
Tianjin Key Laboratory of Animal and Plant Resistance, School of Life Sciences, Tianjin Normal University, Tianjin 300387, China

Dan Pei
Tianjin Key Laboratory of Animal and Plant Resistance, School of Life Sciences, Tianjin Normal University, Tianjin 300387, China

Yang He
Tianjin Key Laboratory of Animal and Plant Resistance, School of Life Sciences, Tianjin Normal University, Tianjin 300387, China

Huazhong Wang
Tianjin Key Laboratory of Animal and Plant Resistance, School of Life Sciences, Tianjin Normal University, Tianjin 300387, China

Liyan Su
Université de Toulouse, INP-ENSA Toulouse, UMR990 Génomique et Biotechnologie des Fruits, Avenue de l'Agrobiopole, CS 32607, F-31326 Castanet-Tolosan, France
Actual address: Department of Life Sciences, Xi'an University of Arts and Science, Xi'an 710065, PR China

Gianfranco Diretto
Italian National Agency for New Technologies, Energy, and Sustainable Development, Casaccia Research Centre, 00123 Rome, Italy

Eduardo Purgatto
Department Food and Experimental Nutrition; NAPAN/FoRC - Food Research Center, Universidade de São Paulo, School of Pharmaceutical Sciences, Av. Prof. Lineu Prestes 580, Butantã, CEP 05508-000 São Paulo, SP, Brazil

Saïda Danoun
Université de Toulouse; UPS; UMR 5546; Laboratoire de Recherche en Sciences Végétales (LRSV), 24 Chemin de Borde Rouge, F-31326 Castanet-Tolosan, France

Mohamed Zouine
Université de Toulouse, INP-ENSA Toulouse, UMR990 Génomique et Biotechnologie des Fruits, Avenue de l'Agrobiopole, CS 32607, F-31326 Castanet-Tolosan, France
INRA, UMR990 Génomique et Biotechnologie des Fruits, 24 Chemin de Borde Rouge, CS 52627, F-31326 Castanet-Tolosan, France

Zhengguo Li
Genetic Engineering Research Centre, Bioengineering College, Chongqing University, Chongqing 400044, PR China

Jean-Paul Roustan
Université de Toulouse, INP-ENSA Toulouse, UMR990 Génomique et Biotechnologie des Fruits, Avenue de l'Agrobiopole, CS 32607, F-31326 Castanet-Tolosan, France
INRA, UMR990 Génomique et Biotechnologie des Fruits, 24 Chemin de Borde Rouge, CS 52627, F-31326 Castanet-Tolosan, France

Mondher Bouzayen
Université de Toulouse, INP-ENSA Toulouse, UMR990 Génomique et Biotechnologie des Fruits, Avenue de l'Agrobiopole, CS 32607, F-31326 Castanet-Tolosan, France
INRA, UMR990 Génomique et Biotechnologie des Fruits, 24 Chemin de Borde Rouge, CS 52627, F-31326 Castanet-Tolosan, France

Giovanni Giuliano
Italian National Agency for New Technologies, Energy, and Sustainable Development, Casaccia Research Centre, 00123 Rome, Italy

Christian Chervin
Université de Toulouse, INP-ENSA Toulouse, UMR990 Génomique et Biotechnologie des Fruits, Avenue de l'Agrobiopole, CS 32607, F-31326 Castanet-Tolosan, France
INRA, UMR990 Génomique et Biotechnologie des Fruits, 24 Chemin de Borde Rouge, CS 52627, F-31326 Castanet-Tolosan, France

Tahise M Oliveira
Universidade Estadual de Santa Cruz-UESC, Rodovia Ilhéus-Itabuna, Km 16, Salobrinho, Bahia, Brazil

Fernanda R da Silva
Centro de Biotecnologia, Departamento de Biologia Molecular e Biotecnologia, Universidade Federal do Rio Grande do Sul-UFGRS, Avenida Bento Goncalves, 9500 Porto Alegre, Rio Grande do Sul, Brazil

Diego Bonatto
Centro de Biotecnologia, Departamento de Biologia Molecular e Biotecnologia, Universidade Federal do Rio Grande do Sul-UFGRS, Avenida Bento Goncalves, 9500 Porto Alegre, Rio Grande do Sul, Brazil

Diana M Neves
Universidade Estadual de Santa Cruz-UESC, Rodovia Ilhéus-Itabuna, Km 16, Salobrinho, Bahia, Brazil

Raphael Morillon
IVIA; Centro de Genomica, Ctra. Moncada-Náquera Km 5, 46113 Moncada, Valencia, Spain
CIRAD, UMR AGAP, Avenue Agropolis - TA A-75/02 – 34398, Montpellier Cedex 5, France

Bianca E Maserti
Dipartimento di Scienze BioAgroAlimentari, CNR-IPSP, Istituto per la Protezione Sostenibile delle Piante, Area della Ricerca CNR, Via Madonna del Piano 10, Via Madonna del Piano n 10, 50019 Sesto Fiorentino, FI, Italy

Mauricio A Coelho Filho
Embrapa Mandioca e Fruticultura, Rua Embrapa, s/n, Cruz das Almas 44380-000, Bahia, Brazil

Marcio GC Costa
Universidade Estadual de Santa Cruz-UESC, Rodovia Ilhéus-Itabuna, Km 16, Salobrinho, Bahia, Brazil

Carlos P Pirovani
Universidade Estadual de Santa Cruz-UESC, Rodovia Ilhéus-Itabuna, Km 16, Salobrinho, Bahia, Brazil

Abelmon S Gesteira
Embrapa Mandioca e Fruticultura, Rua Embrapa, s/n, Cruz das Almas 44380-000, Bahia, Brazil

Jian Zeng
The Genetic Engineering International Cooperation Base of Chinese Ministry of Science and Technology, The Key Laboratory of Molecular Biophysics of Chinese Ministry of Education, College of Life Science and Technology, Huazhong University of Science and Technology, Wuhan, China

Cheng Wang
The Genetic Engineering International Cooperation Base of Chinese Ministry of Science and Technology, The Key Laboratory of Molecular Biophysics of Chinese Ministry of Education, College of Life Science and Technology, Huazhong University of Science and Technology, Wuhan, China

Xi Chen
The Genetic Engineering International Cooperation Base of Chinese Ministry of Science and Technology, The Key Laboratory of Molecular Biophysics of Chinese Ministry of Education, College of Life Science and Technology, Huazhong University of Science and Technology, Wuhan, China

Mingli Zang
The Genetic Engineering International Cooperation Base of Chinese Ministry of Science and Technology, The Key Laboratory of Molecular Biophysics of Chinese Ministry of Education, College of Life Science and Technology, Huazhong University of Science and Technology, Wuhan, China

Cuihong Yuan
The Genetic Engineering International Cooperation Base of Chinese Ministry of Science and Technology, The Key Laboratory of Molecular Biophysics of Chinese Ministry of Education, College of Life Science and Technology, Huazhong University of Science and Technology, Wuhan, China

Xiatian Wang
The Genetic Engineering International Cooperation Base of Chinese Ministry of Science and Technology, The Key Laboratory of Molecular Biophysics of Chinese Ministry of Education, College of Life Science and Technology, Huazhong University of Science and Technology, Wuhan, China

Qiong Wang
The Genetic Engineering International Cooperation Base of Chinese Ministry of Science and Technology, The Key Laboratory of Molecular Biophysics of Chinese Ministry of Education, College of Life Science and Technology, Huazhong University of Science and Technology, Wuhan, China

Miao Li
The Genetic Engineering International Cooperation Base of Chinese Ministry of Science and Technology, The Key Laboratory of Molecular Biophysics of Chinese Ministry of Education, College of Life Science and Technology, Huazhong University of Science and Technology, Wuhan, China

Xiaoyan Li
The Genetic Engineering International Cooperation Base of Chinese Ministry of Science and Technology, The Key Laboratory of Molecular Biophysics of Chinese Ministry of Education, College of Life Science and Technology, Huazhong University of Science and Technology, Wuhan, China

Ling Chen
The Genetic Engineering International Cooperation Base of Chinese Ministry of Science and Technology, The Key Laboratory of Molecular Biophysics of Chinese Ministry of Education, College of Life Science and Technology, Huazhong University of Science and Technology, Wuhan, China

Kexiu Li
The Genetic Engineering International Cooperation Base of Chinese Ministry of Science and Technology, The Key Laboratory of Molecular Biophysics of Chinese Ministry of Education, College of Life Science and Technology, Huazhong University of Science and Technology, Wuhan, China

Junli Chang
The Genetic Engineering International Cooperation Base of Chinese Ministry of Science and Technology, The Key Laboratory of Molecular Biophysics of Chinese Ministry of Education, College of Life Science and Technology, Huazhong University of Science and Technology, Wuhan, China

Yuesheng Wang
The Genetic Engineering International Cooperation Base of Chinese Ministry of Science and Technology, The Key Laboratory of Molecular Biophysics of Chinese Ministry of Education, College of Life Science and Technology, Huazhong University of Science and Technology, Wuhan, China

Guangxiao Yang
The Genetic Engineering International Cooperation Base of Chinese Ministry of Science and Technology, The Key Laboratory of Molecular Biophysics of Chinese Ministry of Education, College of Life Science and Technology, Huazhong University of Science and Technology, Wuhan, China

Guangyuan He
The Genetic Engineering International Cooperation Base of Chinese Ministry of Science and Technology, The Key Laboratory of Molecular Biophysics of Chinese Ministry of Education, College of Life Science and Technology, Huazhong University of Science and Technology, Wuhan, China

Kutubuddin Ali Molla
Division of Genomic Resources, National Bureau of Plant Genetic Resources, IARI Campus, Pusa, New Delhi 110012, India

Ananda Bhusan Debnath
Division of Genomic Resources, National Bureau of Plant Genetic Resources, IARI Campus, Pusa, New Delhi 110012, India

Showkat Ahmad Ganie
Division of Genomic Resources, National Bureau of Plant Genetic Resources, IARI Campus, Pusa, New Delhi 110012, India

Tapan Kumar Mondal
Division of Genomic Resources, National Bureau of Plant Genetic Resources, IARI Campus, Pusa, New Delhi 110012, India

Kana Ogawa
Graduate School of Agriculture, Hokkaido University, Sapporo 060-8589, Japan

Katsunori Murota
Graduate School of Agriculture, Hokkaido University, Sapporo 060-8589, Japan

Hanako Shimura
Graduate School of Agriculture, Hokkaido University, Sapporo 060-8589, Japan

Misaki Furuya
Graduate School of Agriculture, Hokkaido University, Sapporo 060-8589, Japan

Yasuko Togawa
Graduate School of Agriculture, Hokkaido University, Sapporo 060-8589, Japan

Takeshi Matsumura
Plant Molecular Technology Research Group, Research Institute of Bioproduction, National Institute of Advanced Industrial Science and Technology, Sapporo 062-8517, Japan

Chikara Masuta
Graduate School of Agriculture, Hokkaido University, Sapporo 060-8589, Japan